D0214677

The Hydrogeology of the Chalk of North-West Europe

Chalk of the Turonian, Coniacian, and Santonian stages in cliffs at Bantam Hole, near South Foreland, north of Dover (photograph by Dr E. R. Shephard-Thorn, reproduced by courtesy of the British Geological Survey).

The Hydrogeology of the Chalk of North-West Europe

Edited by

R. A. DOWNING,
M. PRICE,
and
G. P. JONES

CLARENDON PRESS • OXFORD
1993

Oxford University Press, Walton Street, Oxford OX2 6DP
Oxford New York Toronto
Delhi Bombay Calcutta Madras Karachi
Kuala Lumpur Singapore Hong Kong Tokyo
Nairobi Dar es Salaam Cape Town
Melbourne Auckland Madrid
and associated companies in
Berlin Ibadan

Oxford is a trade mark of Oxford University Press

Published in the United States
by Oxford University Press Inc., New York

A catalogue record for this book is available from the British Library

Library of Congress Cataloging-in-Publication Data
The Hydrogeology of the Chalk of North-west Europe / edited by R.A. Downing, M. Price, and G.P. Jones.
Includes bibliographical references and index.
1. Water, Underground — Europe, Northern. 2. Chalk — Europe, Northern. I. Downing, Richard A. (Richard Allen) II. Price, Michael III. Jones, G. P.
GB1081.H93 1993 551.49'094—dc20 92-44648
ISBN 0-19-854285-2

Typeset by Expo Holdings Malaysia
Printed in Hong Kong

Preface

The British Committee of the International Association of Hydrogeologists, in co-operation with colleagues in Belgium, Denmark, France, the Netherlands, and Sweden, has prepared this book reviewing the hydrogeology of the Chalk, and the associated carbonate facies, in the sedimentary basin around the North Sea. The book was conceived in response to a need for greater communication between hydrogeologists studying the hydrogeology of the Chalk in North-west Europe. After an initial series of chapters that discuss the aquifer as a whole, the hydrogeology of the Chalk in each country is reviewed. Final chapters describe the role of the Chalk as a hydrocarbon reservoir and reflect on the problems that are now affecting the aquifer as a source of water supply.

The Chalk's distinctive colour makes it recognizable in cliffs and quarries from Denmark to the Channel coast, and it enhances the scenery wherever it occurs. Perhaps more than any other rock, it symbolizes the unity of northern Europe, which will once again become a physical reality with the completion of the Channel Tunnel through the Chalk in 1994. However, few who recognize the Chalk in its outcrops across Europe appreciate its economic value, particularly as a major water reservoir but also, below the North Sea, as an oil reservoir. Regrettably, one of the recurrent themes in the book is the degradation of the once pristine quality of groundwater in the Chalk. The careless disposal of wastes on to and into the aquifer, as well as the use of nitrate fertilizers on its outcrop, together with modern agricultural practices, are steadily contaminating the vast volume of groundwater stored in the aquifer. Water that enters the Chalk as rain can remain in the aquifer for centuries, if not for millennia. While a pollution incident in a river is dispersed quite quickly, in an aquifer such as the Chalk the long residence time of the groundwater, coupled with the nature of the rock, means that restoration is difficult if not impossible.

The microscopic algae that make up virtually all the Chalk lived at a time when the temperature was some 10 °C higher than at present, when the carbon dioxide content of the atmosphere was also significantly higher, and when sea levels were much higher and northern Europe was covered by a sea in which only the highest ground remained as islands. It may be regarded as an extreme example of the consequences of natural global warming and serves to draw attention to the current concern about similar but less severe consequences that may occur through man's thoughtless activities. The risk of climatic change emphasizes the strategic value of the groundwater reservoir in the Chalk if the weather patterns do become more extreme, leading to longer, drier summers with more frequent droughts, as experienced in western Europe between 1988 and 1992. The advantage of maintaining the quality of the water in the Chalk will then be seen even more clearly. It is hoped that this book will encourage governments to pursue active groundwater protection policies aimed at preventing processes that have irreversible consequences and, thereby, avoiding further degradation of the quality of groundwater in the Chalk, preserving it as a source of water for the people of North-west Europe into the future. Perhaps the book will also encourage further co-operative studies of the hydrogeology of the Chalk on an international basis, drawing on the strengths of individual countries to contribute to a better understanding of the factors that control the flow of fluids in the aquifer.

Finally, we would like to acknowledge the co-operation of our co-authors and thank them for their contributions. On their behalf, we extend our appreciation to the many organizations and individuals who have provided assistance and offered advice during the preparation of the book. In particular, we would like to thank the British Geological Survey for providing administrative support. We also appreciate the invaluable advice and assistance of the staff of Oxford University Press in seeing the book through the press.

March 1993

R. A. D.
M. P.
G. P. J.

Contents

Contributors

G. Alcaydé
Muséum d'Histoire Naturelle, Paris

J. A. Barker
British Geological Survey

P. Bracq
Université des Sciences et Technologies de Lille

N. Crampon
Université des Sciences et Technologies de Lille

A. Dassargues
Université de Liège

F. Delay
Université Pierre et Marie Curie, Paris

R. A. Downing
British Geological Survey

W. M. Edmunds
British Geological Survey

S. S. D. Foster
British Geological Survey

O. Gustafsson
Geological Survey of Sweden

M. D'Heur
Petrofina, S. A., Brussels

J. M. Hancock
Imperial College of Science, Technology and Medicine, London

G. P. Jones
University College London

M. Lepiller
Université d'Orléans

J. W. Lloyd
University of Birmingham

G. Mary
Université du Maine, Le Mans

A. Monjoie
Université de Liège

R. N. Mortimore
University of Brighton

E. Nygaard
Geological Survey of Denmark

M. Price
University of Reading

L. Rasplus
Université Françoise Rabelais, Tours

P. van Rooijen
Geological Survey of the Netherlands

J. C. Roux
Bureau de Recherches Géologiques et Minières, Orléans

1. The making of an aquifer

R. A. Downing, M. Price, and G. P. Jones

Some 100 million years ago, in the period of geological time known as the Late Cretaceous, the area that now surrounds the North Sea lay about 10° further south in the latitude of the present north coast of the Mediterranean Sea. The average temperature was about 20°C and sea level was probably several hundred metres higher than today. Only the highest parts of the ancient massifs remained as land. These had low relief and, as the climate was arid, little erosion was taking place and the limited detritus was deposited close to the shoreline.

In the warm waters of this sea, over a period of about 35 million years, a soft white ooze formed from the accumulation of the skeletal plates of microscopic planktonic algae, and this ooze became the limestone known today as chalk, and which makes up the lithostratigraphic unit called the Chalk. Although it formed over much of what is today northern Europe, as well as over parts of North America and Asia, this book is concerned only with the Chalk deposited in and around the present North Sea Basin, during Upper Cretaceous and Lower Palaeocene (Danian) times.

The algae whose remains make up the Chalk secreted a form of calcite that had a low magnesium content, and which was therefore stable at low temperatures and pressures. Because of this, it has not recrystallized significantly since Cretaceous times, except at depths of more than 1000 m, so that much of the Chalk has remained soft and friable. Had it not retained these properties, the scenery around the North Sea would have been very different from that of today, and the hydrogeology of the Chalk would not have been so unusual.

Around the North Sea, the Chalk occurs in England, France, Belgium, the Netherlands, Germany, Denmark and Sweden (Fig. 1.1). This region forms the western part of the extensive low plain that extends across northern Europe. Almost the entire region underlain by the Chalk and its cover deposits is less than 300 m above sea level. The topography ranges from the rolling chalk hills of England and France to the coastal and deltaic plains of Belgium and the Netherlands. Glacial deposits influence the form of the topography in Denmark, Sweden, and Germany, as well as the Netherlands and England.

The region experiences an oceanic, temperate climate with mild winters and cool summers. Rainfall is fairly evenly distributed throughout the year and the annual total is generally less than 750 mm; evaporation is of the order of 400–550 mm a^{-1}. As evaporation is high in summer and exceeds the rainfall, most infiltration to the Chalk is in the winter months, from October to March.

General structure and stratigraphy

The Late Cretaceous and earliest Cenozoic evolution of North-west Europe was influenced by the gradual opening of the North Atlantic and the onset of the Alpine plate collision of Africa and Europe (Ziegler 1981). In the North Sea Basin an important structural and palaeogeographical feature is the north–south rift system. This initially developed in Late Triassic times, and rifting continued into the Cretaceous, creating the Viking and Central grabens which underlie the Late Cretaceous-Cenozoic Central Trough (Fig. 1.2). During Late Cretaceous and Early Palaeocene times North–west Europe was part of a passive continental margin, and subsidence occurred on a regional scale. Chalk formed throughout the region and gradually filled the Central Graben. Similarly, the Paris Basin and the Danish–Polish Trough (Fig. 1.2) continued to develop and fill. This was the general pattern throughout the Late Cretaceous. Although it was a period of relative stability, tectonic movements did occur on a local scale and at different times. Two particular phases have been recognized—the Sub-Hercynian tectonic phase, during the Senonian (which comprises the Coniacian, Santonian, and

Fig. 1.1. The distribution of the Upper Cretaceous and Danian Chalk in North-west Europe. A = Amsterdam; B = Brussels; C = Copenhagen; H = Hanover; L = London; O = Oslo.

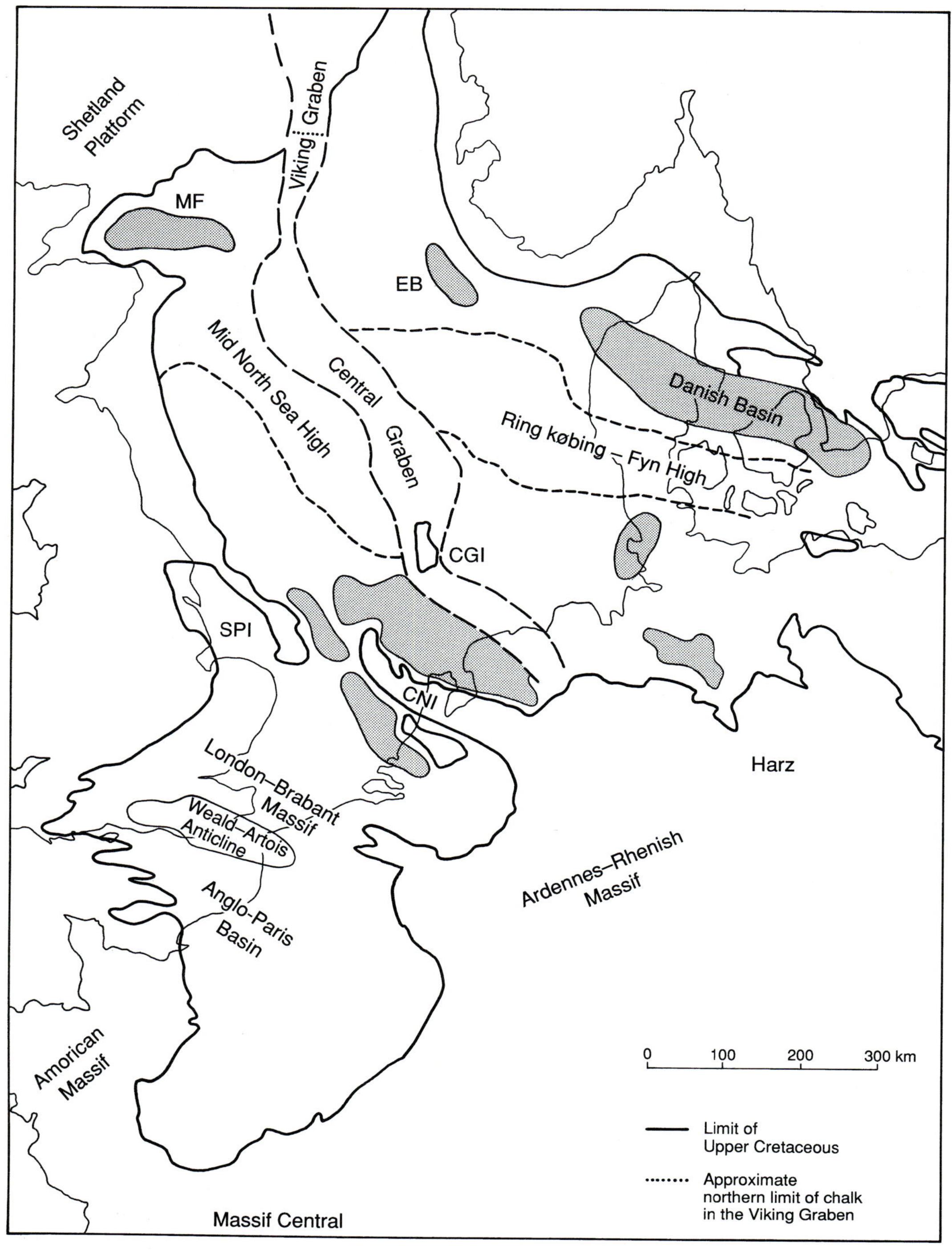

Fig. 1.2. Principal structural features affecting the Chalk in North-west Europe (after Ziegler 1990). Deep basins are stippled. CGI = Central Graben Inversion; CNI = Central Netherlands Inversion; EB = Egersund Basin; MF = Moray Firth Basin; SPI = Sole Pit Inversion.

Campanian stages), and the Laramide phase, in Late Palaeocene times (Ziegler 1990).

The Late Cretaceous saw the beginning of the tectonic uplift (or inversion) that continued in the Tertiary. Inversion movements of the Sub-Hercynian phase affected the southern part of the Central Graben and the depositional troughs at the margin of the basin, including the Sole Pit, the Central Netherlands and the Central English Channel basins, leading to erosion along the tectonic axes. During Late Senonian to Danian times, chalk was again deposited over these inverted eroded axes. Large-scale inversion of these basins occurred during Late Palaeocene times as a result of the Laramide event. These inversion movements coincided with the last phase of downfaulting in the Central Graben and were accompanied by uplift along the graben margins (Ziegler 1990).

The fall of the sea level during the Late Cretaceous and Early Palaeocene (Hallam 1984) brought an end to the deposition of chalk in the Viking Graben during the Danian and in the central and southern North Sea at the beginning of the Late Palaeocene. It also led to an important erosional episode. Large thicknesses of chalk were removed from the margins of the basin, an event represented by the major sub-Palaeocene unconformity. Chalk was also eroded from the Sole Pit axis, along a NW–SE axis through the Netherlands, and over an area in the southern North Sea (Fig. 1.2). Palaeotemperatures deduced from apatite fission track analysis (Green *et al.* 1993) suggest that uplift and erosion were not confined to the recognized inversion axes but extended over a wide area of North-west Europe. Drilling and seismic interpretation have revealed that the top of the Chalk has been subjected to karstification with collapse structures over a large area at the southerly extension of the Sole Pit inversion (Jenyon 1987).

Regional subsidence dominated the structural picture in the North Sea during Eocene to Recent times, leading to the formation of a symmetrical sedimentary basin with an axis coinciding with the Central Graben and containing up to 3.5 km of Tertiary sediments (Neilsen *et al.* 1986; Ziegler 1990). At the margin of this basin, deposition was concentrated in subsiding areas in the Low Countries, south-east England and the Paris Basin (Curry 1992). The Weald–Artois inversion was active in Miocene times but probably began in the Late Cretaceous. Inversion movements of the West Netherlands Basin and the inversion of the Central Channel Basin and Pays de Bray Anticline in the Paris Basin also occurred in the Tertiary (Ziegler 1990).

Tectonic movements continued into the Quaternary, mainly affecting the southern border of the basin, particularly the Rhine delta in the Netherlands. Quaternary sediments, predominantly glacial and fluvioglacial deposits, occur in eastern England and the North Sea lowlands, north of a line from south Essex through the Netherlands into Germany. They directly overlie the Chalk in Denmark, Sweden, north Germany and parts of England.

The inversion of the late Tertiary began to create the land surface as it is known today, continuing the erosion that began in the Late Cretaceous. Erosion was a dominant process in the marginal parts of the basin throughout the Cenozoic. All the Chalk formations vary in thickness and are absent locally because of intra- and post-Cretaceous unconformities; for example, unconformities exist between formations in the offshore Ekofisk Field. However, there is no evidence that the central offshore areas were ever exposed and subjected to the action of meteoric waters.

In the North Sea, the Chalk attains maximum thicknesses of 2000 m in the Viking Graben and Central Trough but 300–350 m is more usual on the flanks and in areas where it crops out (Fig. 1.3). This is a function of both the original mode of sedimentation and subsequent erosion. The considerable thicknesses of chalk in the Central Trough are due to redeposition following sliding, slumping, and the formation of debris flows and turbidites. Tectonic movements initiated repeated large-scale displacement of unstable chalk down the flanks of the trough to accumulate in the trough itself as complex allochthonous sequences. These allochthonous chalks were deposited very rapidly (in contrast to the natural sedimentation rates). As will be seen, this was a major factor in the development of oil reservoirs in the Chalk.

The distribution of carbonates in the Upper Cretaceous and Danian is illustrated in Fig 1.4. Changes in sea level and tectonic uplift have been invoked to explain the variation and distribution of the various lithologies (Hancock 1975a, b; Mortimore and Pomerol 1991). Although from a superficial examination the Chalk may appear to be

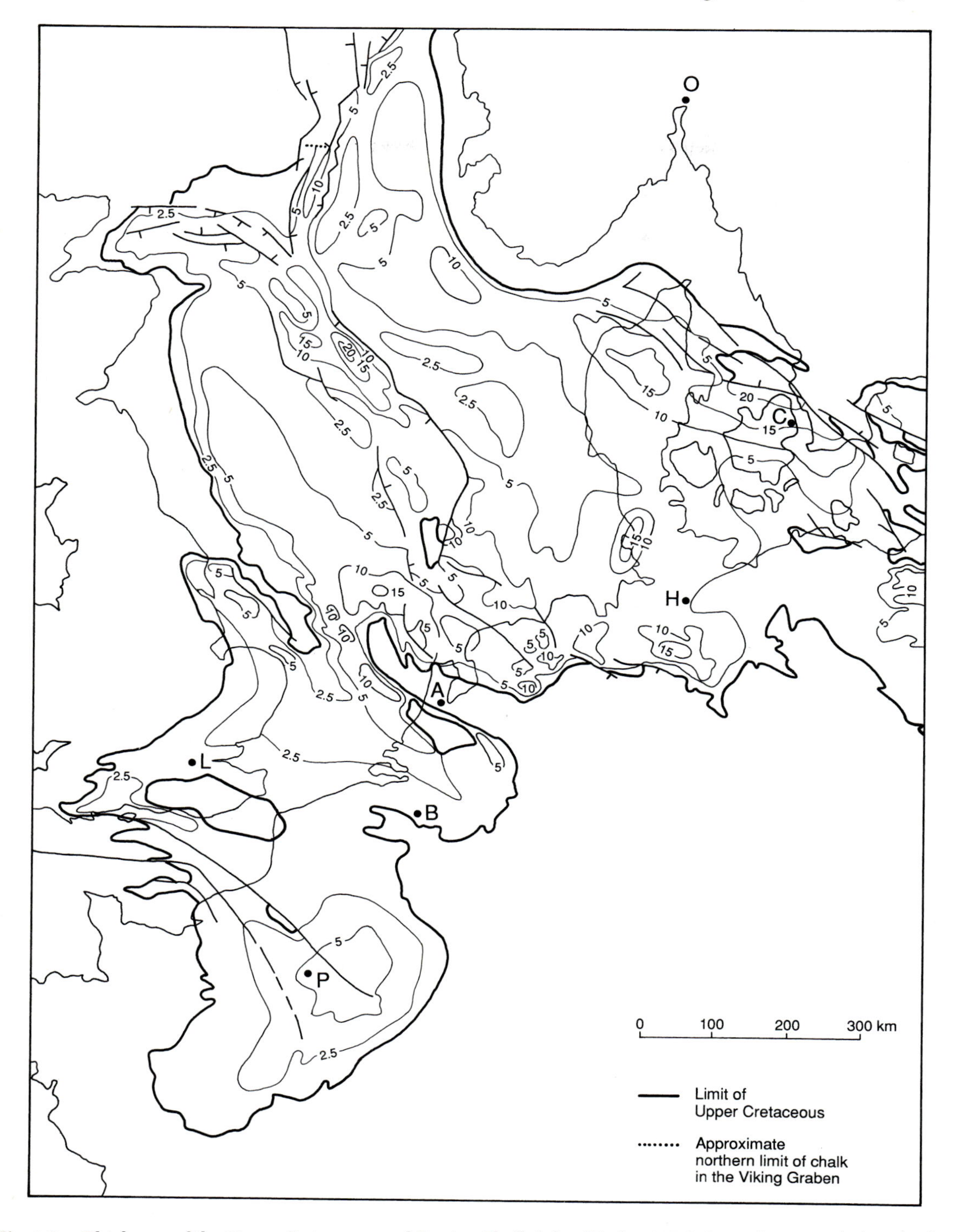

Fig. 1.3. Thickness of the Upper Cretaceous and Danian Chalk (after Ziegler 1990). Isopachytes are in hundreds of metres. A= Amsterdam; B; = Brussels; C = Copenhagen; H = Hanover; L = London; O = Oslo.

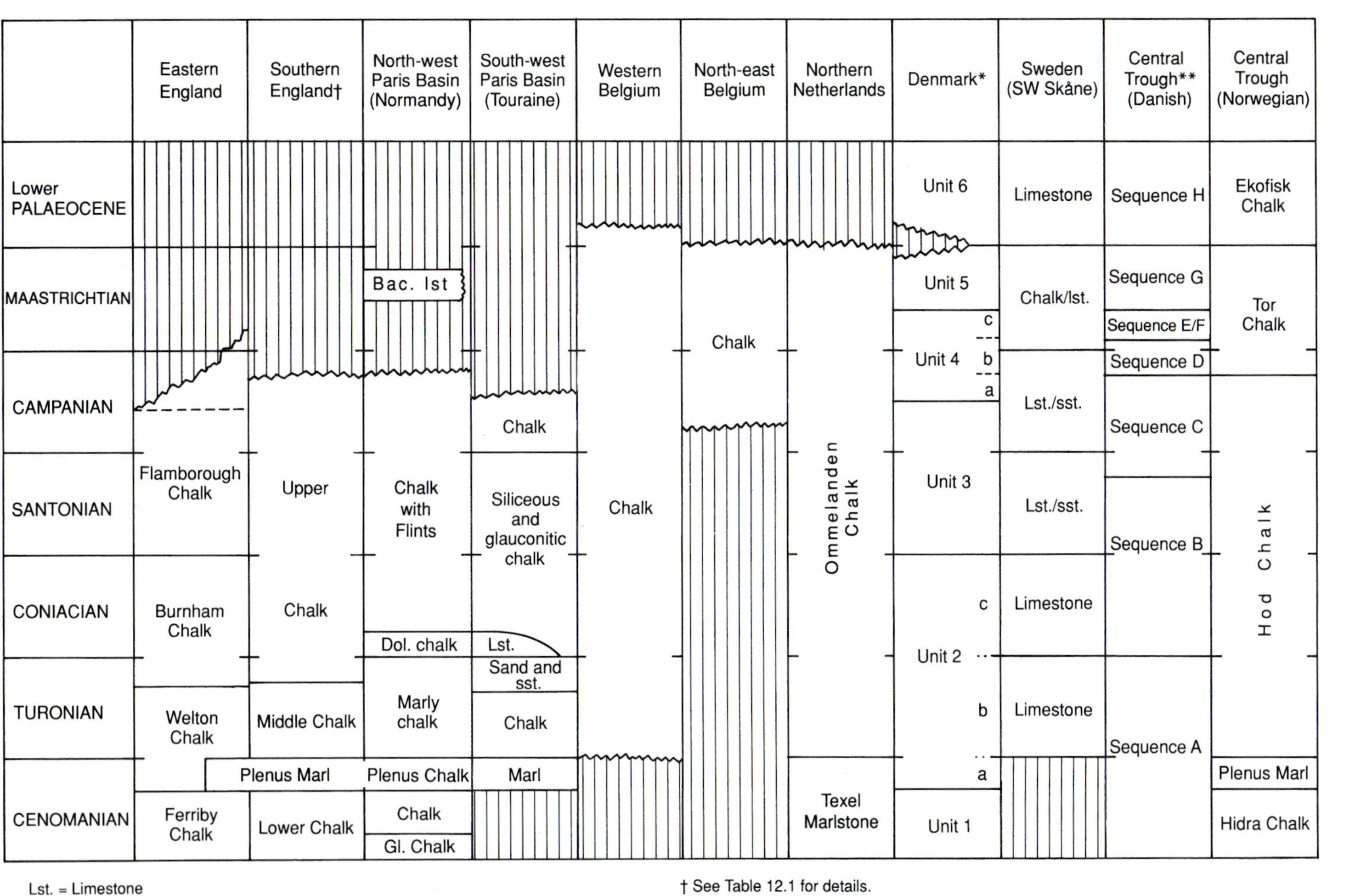

Lst. = Limestone
Sst.= Sandstone
Bac.lst.=Baculites limestone
Dol.=Dolomitic
Gl.=Glauconitic
The Coniacian, Santonian and Campanian are referred to as the Senonian

† See Table 12.1 for details.
* After Lieberkind (1982).
** After Nygaard *et al.* (1989).

Fig. 1.4. Representative stratigraphy of the Chalk and associated carbonates in North-west Europe.

a white homogeneous rock, in detail a wide range of lithologies are present in all units of the formation. The nature of the sediments reflects changes in thickness and the influence of local anomalies such as scour channels, pseudo-turbidites, phosphatic chalks and slump horizons. There is a close relationship between surface tectonic features, the sedimentary depositional centres, and condensed unusual lithologies, such as channels and hardgrounds. In the Anglo-Paris Basin changes in the thickness and lithology of the Chalk reflect the position of deep structures. Tectonic axes are the locus of unusual chalk sediments and are linked to intra-Cretaceous movements. There is probably an association between marine transgressions and chalk with marl seams, while regressions are more allied with rock bands, such as the Melbourn Rock. Thus chalk sedimentation was influenced by the broad tectonic setting (that is, whether on a basin margin or an axis), local tectonic setting (whether on the crest of a fold axis or in the cuvette formed by the accompanying syncline), and sea-level changes creating transgressions or regressions (Mortimore and Pomerol 1991). These authors point out that although the Upper Cretaceous is regarded as a quiescent period, the major tectonic axes continued to move and that structures previously interpreted as Tertiary are now seen to be Cretaceous. In the European context, two episodes stand out: a Late Turonian event and a Campanian event, both marked by widespread slumping and turbidity flows.

The outcome of the tectonic movements that affected the Upper Cretaceous and Danian sediments is that carbonates, predominantly chalks, outcrop or subcrop at relatively shallow depths around a deep basin centred in the North Sea. The Chalk at the onshore margins of this basin is used for water supply in England, France, Belgium, the Netherlands, Denmark, and Sweden (Fig 1.1). It also occurs below the North German Plain but there it is almost entirely overlain by some 1000–2000 m of Tertiary and Quaternary deposits and consequently is not used as an aquifer. However, halokinetic processes due to the movement of salt deposits have brought it to the surface above salt diapirs in Schleswig-Holstein, Hamburg, and Lower Saxony. Some 50 km north-west of Hamburg, a Middle Coniacian to Lower Maastrichtian chalk sequence, 400 m thick, is exposed in three very large quarries. The chloride content of the groundwater in this mass of chalk increases with depth. An interesting feature is that saline waters are ascending from the salt diapirs along faults and penetrating into the chalk matrix along microfractures or by diffusion, but only for a distance of 2 m from the fault planes (Schönfeld and Grube 1990).

Diagenetic modification

The initial porosity of the chalk oozes was as high as 80 per cent. A feature of the formation is that, even after diagenesis, values remain high, of the order of 25–40 per cent in onshore locations, and offshore some chalk sequences buried at depths of 2.5–3.5 km have porosities of up to 50 per cent.

Petrographic studies have recognized two phases of cementation—first, soon after the sediments were deposited on the sea floor; and second, during deep burial at a much later stage (Chapter 2; Scholle 1977; Taylor and Lapré 1987). Early diagenesis occurred within 10 m of the sea floor as a result of the circulation of sea water. Bioturbation tended to reduce the porosity by creating routes for the ingress of marine waters, thereby enhancing cementation. Horizons of sea-floor cementation are termed 'hardgrounds' (Bromley 1965). Porosity in them is reduced to between 10 and 20 per cent but the chalk, being of a brittle nature, can fracture cleanly giving zones of high permeability in onshore areas, although offshore the hardgrounds tend to be impermeable.

Physical compaction occurred between burial depths of 10 and 1000 m with a steady reduction of porosity. The carbonate oozes compacted until a rigid framework was created when porosity had been reduced to about 40 per cent at a burial depth of some 800–1000 m. Argillaceous layers underwent greater compaction. During this phase of physical compaction, shear fractures may have been established and subsequently infilled with a mineral assemblage. Some of the fluids associated with this precipitation were derived from the Zechstein and the Kimmeridge Clay, apparently entering the Chalk from below through well-developed fractures (Taylor and Lapré 1987).

The second phase of chemical compaction began at depths of 1000 m after the rigid framework had developed. The source of the carbonate cement was grain-to-grain pressure dissolution, with recycling of the carbonate through localized dissolution and precipitation. Chemical compaction was responsible for the reduction of porosity from 40 per cent to its present levels.

Despite chemical cementation, many chalks in the central North Sea have anomalously high porosities for their depths of burial. The allochthonous chalks deposited from slumps, slides, and debris flows were deposited very rapidly. Consequently, they had a high water content and the matrix was poorly packed. In thick sequences of redeposited chalk, the deeper material was not bioturbated; all these factors tended to preserve porosity.

Chemical cementation was also retarded by the occurrence of high fluid pressures and by the migration of hydrocarbons into the matrix. This began at burial depths of 600–1200 m, before chemical cementation had become a major factor. Hydrocarbons tended to move preferentially into more permeable parts of the reservoir and, therefore, accentuated these features which were originally caused by different modes of sedimentation. At depths of 3 km chalk that does not contain hydrocarbons (i.e is water-filled) has a porosity of 10–20 per cent. The burial and compaction caused by rapid sediment loading during the Tertiary produced the abnormal fluid pressures, of the order of 50 000 kPa at 3000 m. As will be seen, this was instrumental in maintaining open fractures and thereby creating a reservoir that could be exploited.

The influence of tectonics

At the present time in North-west Europe, the maximum horizontal stress has a NW–SE alignment because of the north-westerly pressure of the African Plate against Europe and the south-easterly pressure of the North Atlantic spreading ridge. The contrast between the maximum and minimum horizontal stresses encourages shear failure and causes strike-slip faulting on favourably orientated existing vertical faults (Evans 1988). This stress regime, initiated in Tertiary times, tends to produce NW–SE and NE–SW fracture and fault patterns and lines of incipient weakness in the Chalk. However, a dominant feature of southern England is the presence of a series of east–west flexures and fractures due to an Early Oligocene to Early Miocene episode of north–south compression. Towards the end of these movements, N–S and E–W fractures formed during a period of extension. But superimposed on these patterns is a vertical or steeply inclined NW–SE system of joints and faults associated with late Neogene movements that can also be recognized in northern France. There is also a series of NE–striking cross-joints resulting from a phase of stress relaxation after the formation of the NW system of fractures.

In northern France many lineaments exhibit a NW trend and coincide with major rectilinear drainage channels draining in a north-westerly direction into the English Channel. The valleys of northern France may be controlled by joint-swarms that possibly reflect buried faults in the Variscan Basement. On the other hand, the southern England tectonic pattern appears to be controlled by E–W faults in the basement (Bevan and Hancock 1986). In the Chalk of Humberside the dominant tectonic trends are both NW–SE and E–W (Patsoules and Cripps 1990). The depth to the Chalk around the Danish Basin is the key factor determining the use of the formation as an aquifer in Denmark.

The formation of joints has had an important bearing on the subsequent development of permeability in the Chalk. A genetic classification of joints in chalks of Israel (Bahat 1990) may have wider application and has indeed been applied to chalks in the Paris Basin. The joints are recognized as having formed during burial, subsequent uplift, and in response to major tectonic processes. During the initial burial stage, the sum cause was sedimentation, downwarping, and diagenesis, but probably most of the fracturing at this stage occurred through the release of fluid pressures following diagenesis. Spacing of these fractures is regular with maximum limits of several tens of centimetres, and fractures tend to be confined to one layer. Syntectonic fractures commonly reflect regional stress but they can correspond to local structures. Joints that form as a result of uplift tend to cut many layers, and the spacing is irregular and between 5 and 10 m.

In onshore areas, the various stress patterns have led to the formation of a basic orthogonal fracture system. As discussed below, this has been developed further by weathering, stress relief, and groundwater flow.

Evidence available from the offshore oilfields indicates that, although fractures exist, an extensive open fracture-joint system, such as that found onshore, does not occur. There are different views about the origin of the fractures in the offshore fields. According to Watts (1983a), there are two types of microfracture: steeply inclined shear fractures invariably infilled with calcite and/or clay; and tension fractures associated with stylolites and subsequently accentuated by the growth of salt domes. The shear fractures, being healed, do not contribute to the permeability, but the tension fractures are open. However, they are of very limited length and most likely only enhance the permeability laterally. The shear fractures are considered to be associated with Early Tertiary movements along the major NW–SE faults of the Central Graben system. Halokinesis occurred throughout the Cretaceous, continuing in a series of pulses until early in Miocene times, and movements of salt may have reactivated major faults. The tension fractures were possibly produced during Early Miocene times as a result of a combination of halokinesis and overpressures (Watts 1983a). There are possibly two orientations—one radial and the other concentric. Brewster *et al.* (1986) also recognized a series of small-scale, closely spaced, normal faults that provide paths for large-scale fluid movement.

A rather different emphasis has been given by Duval et al. (1990) based on studies of the Ekofisk Field. They maintain that the main fault system is NNE-SSW, with a secondary system trending NW-SE. The major faults are all open, at least in the crestal part of the field. The faulting is believed to be due to reactivation of basement faults by compressive movements in Upper Cretaceous and Lower Tertiary times. During Tertiary times, movement of salt diapirs may have played a part since most of the fields do appear to be located above diapirs. The planar tectonic fractures are probably mainly responsible for the enhancement of the permeability of the Chalk.

Whatever the origin of the fractures, knowledge of their patterns and form are critically important for the optimal management of the fields, including the design of water injection programmes for improving the secondary recovery of hydrocarbons.

Weathering and landscape

Following the regression of the sea and the inversions of the Late Cretaceous and Tertiary, the present-day land surface began to take form and has continued to evolve throughout Tertiary and Quaternary times. The Chalk landscape gives a very characteristic impression of rolling downland of convex slopes with steep valleys, often with relatively wide floors, particularly in the lower reaches. The rolling topography owes its origins to the erosional sculpture of a soft rock and the movement of soil water towards individual near-surface vertical fractures or fracture zones to give an apparently preferential orientation to the topography. Some of these features develop into the main valley systems but others are revealed only by hollows in the upland surface. Many of the valleys are now dry or only contain perennial flows of water in their lower reaches, with intermittent flows in the middle reaches. This is groundwater discharging from springs, for surface runoff in the form of overland flow does not occur. Major joint systems and karstic landforms, so characteristic of many limestone landscapes, are not a dominant feature, although they do occur, as for example in the Paris Basin. There is instead a network of very fine cracks that probably reflect incipient lines of weakness initiated by tectonic stresses. These have developed to a variable extent by subsequent contraction and solution processes and no doubt relaxation of stress through 'unloading' as erosion has proceeded. The joint and bedding plane discontinuities are very variable, for the lithology exerts a significant control over the degree of weathering and the subsequent fracturing. Tectonic stress has produced local zones of intense fracturing. Under extreme conditions mechanical and chemical weathering can lead to a breakdown of the rock into a structureless mass. Weathering and erosion during the Pleistocene had a marked effect on the structure of the near-surface chalk. A deeply weathered chalk profile, commonly overlain by periglacial deposits, exists below valley floors, including dry valleys (Mortimore 1990).

In firm, relatively massive chalk, the principal joint sets are vertical or steeply inclined with a spacing of 3–5 m, and bedding plane discontinuities between 0.5 and up to 10 m. However, discontinuities do develop with a variable spacing in response to the multiple stresses exerted on the rock and the extent of weathering. The variable nature of the chalk in the field results from the action of physical and chemical processes on a basically non-uniform, weak rock.

The aquifer properties

The aquifer properties of the Chalk have developed as a result of its origin, its subsequent diagenesis and modification by earth movements and weathering. Onshore, the result of these processes is the formation of an aquifer with a very fine-grained porous matrix intersected by a fracture network. This makes the Chalk a classic example of a dual porosity system, with one porosity due to the matrix pores and the other to the fractures. The porosity of the matrix is about 30–40 per cent, but that of the fractures is probably less than 1 per cent. The matrix is relatively impermeable but it represents a very large volume of storage. The fractures provide a distribution system for water; they gain water from the matrix when fluid pressures are lowered in them by, for example, pumping from wells. In saturated chalk, water moves preferentially through fractures, which are the paths of least resistance. The very high permeabilities in the Chalk (up to 10^{-3}m s^{-1}) are due to dissolution of calcium carbonate by meteoric water flowing along individual preferred fissures, often located at or near the water table. The apertures (or widths) of many of these fissures are only a few millimetres or less. Because of the fine-grained nature of the Chalk, the proportion of the total water in storage that can drain to wells is very small and amounts to only 1 or 2 per cent.

Offshore, the Chalk has not been widely subjected to weathering and, as already mentioned, the favourable reservoir (aquifer) properties are due to the mode of sedimentation, the formation of tension fractures by tectonic movements (including halokinesis), and the maintenance of open fractures by high fluid pressures, together with the retention of a high porosity of 20–50 per cent by the early migration of hydrocarbons. The matrix provides the storage for the oil, and the fractures provide the route to the wells. Without these fractures, commercial development of the oil would be impossible since matrix permeabilities are typically 1 millidarcy (mD). The fractures increase the effective permeability to over 100 mD.

Although the *modus operandi* of the Chalk for yielding both water and oil is similar, there are fundamental differences in the development of fissure permeability onshore and offshore. In the former dissolution and weathering have been paramount; in the latter the formation of salt domes and transpressional forces, together with high fluid pressures, influence the yields.

Development for water supply

The Chalk has always been recognized as a source of water supply, initially from scarp- and dip-slope springs or shallow wells. Many villages and towns owe their origin and development to the existence of a ready source of good quality water from the Chalk. With time, development schemes became more ambitious. In 1613 the New River, an artificial channel, began to bring water from the Chadwell and Amwell springs in Hertfordshire to Clerkenwell in London. By the middle of the seventeenth century the New River Company was distributing water through wooden pipes to a large part of the City of London. But it was the Industrial Revolution, in the second half of the eighteenth century, that created the need for a ready supply of water at many centres. The development of the steam engine, and hence steam-driven pumps, gave the stimulus to exploit the Chalk at greater depths and also below the Tertiary deposits. The cost of water from piped distribution systems was high, and as a consequence many private wells continued to be sunk. This trend continued until relatively recently, but now the tendency is to rely more and more on public water supply distribution systems, except for supplies for irrigation and industry.

The early deep wells had to be of large diameter to contain the pumping plant. The yield from the wells was often increased by driving adits (or tunnels) from wells at different levels and in different directions or by interlinking wells with adits (Fig. 1.5). Systems of adits, typically 1.2 m wide and 1.8 m high, are found, for example, in south-east England, Belgium, and the Netherlands. The water system providing supplies to Brighton includes 13.6 km of adits and in east London there are 18 km. Generally the adits were driven to intersect the principal fissure or fracture directions. With the introduction of improved methods of well drilling and the use of deep-well electrical submersible pumps, wells were drilled at smaller diameters, although high-yielding wells were still at least 450 mm in diameter and commonly 600 or 750 mm.

The increasing demand for groundwater from the Chalk led to undesirable consequences in the form

Fig. 1.5. Inflow of water from a fissure intersected in an adit driven from a well in the Lee Valley, in the London Basin.

of falling water levels, reduction of spring flows (and hence the flow of groundwater-dependent rivers on the Chalk), saline intrusion in coastal areas, and even slight ground subsidence in areas where the Chalk is overlain by Tertiary deposits, because of compaction of the Tertiary sediments as water drained from them into the Chalk.

The impact of these deleterious effects led, in the 1960s, to proposals in England for the management of the groundwater resources of the Chalk on a regional basis, taking into account not only the need to provide water supply but also the environmental and amenity benefits that were provided by the natural flow of water from the Chalk. Groundwater schemes have been promoted in both England and France to augment river flow by discharging groundwater from upland wells into the river system, generally with the water abstracted in the lower reaches for water supply. Along the South Downs of southern England, saline intrusion was reduced by abstracting from wells near the coast only when water levels were high (generally in the winter and spring) and switching to inland wells when water levels were low in the summer and autumn and there was a greater danger of saline intrusion.

Groundwater levels have fallen in many urban areas in North-west Europe as a consequence of heavy abstraction. This is commonly seen as an unwelcome outcome. However, groundwater levels must be lowered and artificially varied if the large groundwater resources stored in the Chalk are to be used. Falling water levels have to be seen in the context of the consequences of the decline on the yield of wells, on the quality of the water abstracted, on the effect on the environment, and on the feasibility of introducing regional management schemes. More recently, groundwater levels have risen in some areas as abstraction has been reduced, for example in the centre of the London Basin (Chapter 12) and in the Nord and Pas de Calais (Chapter 7).

Production of hydrocarbons

The first discovery of oil and gas in the North Sea was in 1966 but the first commercially viable find was made by Phillips in December 1969 when the Ekofisk Field was discovered, the first giant multi-billion barrel field in the North Sea. Since then a further 20 fields have been proved in the Chalk of the Central Trough, and almost all derive oil from allochthonous sequences. The oilfields in the Chalk in the Norwegian and Danish sectors contain about 20–25 $\times 10^9$ barrels of oil-in-place, of which about 20 per cent is expected to be recovered. Ninety-five per cent of the reserves are in the Danian and Maastrichtian, and some 75 per cent are in the Norwegian sector, with most of the remainder in the Danish sector; potential fields also occur in the UK sector.

The commercial development of oil in what is basically a very unfavourable reservoir rock is primarily due to the preservation of high porosities, in sediments that have undergone mass gravity movement, and the development of high fluid pressures that maintained an open and effective fracture pattern.

The source rocks are thick organic-rich shales in the Kimmeridge Clay of the Upper Jurassic. Maturation studies indicate that oil generation began in Early Tertiary times. In many of the fields long vertical oil columns occur (for example, more than 300 m in Ekofisk); yields from individual wells are up to 15 000 barrels per day.

The production of the oil and gas has caused large-scale compaction of the reservoirs in the Ekofisk Field, with subsidence of the sea floor of about 0.5 m a^{-1} (Jones 1990). The resulting geotechnical problems were imaginatively solved in 1987 by jacking up the platforms and the permanent extension of the legs of the Ekofisk complex by 6 m. Compaction of the reservoirs also led to deformation of wells both at the reservoir level and above the reservoirs.

The optimal management of a major oilfield in the Chalk is a particularly demanding task. A better understanding is still required of the movement of oil and gas between the matrix and fractures to facilitate this (Berget *et al.* 1990). Production enhancement by gas or sea-water injection is a major objective, as is the need to maintain reservoir pressure by reducing pressure depletion to a minimum. In this context, it is interesting that the oil displacement efficiency of water has been more effective than forecast by laboratory tests. The management programmes are constrained by time, for there is limited time to find the optimum methods of enhancing the recovery because the fields are in active production and different options have different optimal strategies; furthermore, the infrastructures in the fields themselves have a limited life (Berget *et al.* 1990).

The risk from pollution

During the mid-1970s it became apparent that groundwater in the Chalk was being contaminated by nitrate. Field experiments, beginning at that time throughout the Chalk area of North-west Europe, have confirmed that the cause is modern agricultural practice, including the use of nitrate fertilizers. Because of the aquifer's fine-grained matrix, the full impact of fertilizer applications is not felt immediately, for nitrate moves down through the unsaturated zone very slowly.

The role of the matrix in solute transport first became evident as a consequence of the release of tritium in the atmosphere by thermonuclear tests in the 1950s and early 1960s. The tritium in rainfall could be used as a tracer to study groundwater flow. At that time the Chalk was still regarded as an aquifer dominated by fissure flow, and the low concentration of tritium in springs discharging from the Chalk was a surprise. Dry coring of the Chalk revealed that most of the tritium was stored in the matrix in the upper part of the unsaturated zone and was moving slowly down through the zone at about 1 m a^{-1} by an apparent piston displacement (Smith *et al.* 1970). Only about 15 per cent was transported rapidly through fissures to the saturated zone. The importance of the matrix as a solute storage medium was clearly established and, not surprisingly, it was found that nitrate and other solutes behaved in a similar manner.

In many areas the concentration of nitrate in the saturated zone exceeds the European Community's maximum admissible concentration of 50 mg $NO_3\,l^{-1}$. Regrettably, the Chalk is also being contaminated in many areas by pesticides and organic compounds, particularly synthetic organic solvents. The latter can occur in both miscible and immiscible phases. The miscible phase can diffuse into the matrix from which it is subsequently released only very slowly. The immiscible phase, especially if denser and less viscous than water, can move rapidly in fissures to significant depths, where it can act as a source of the pollutant, possibly for decades.

It is the nature of the Chalk's matrix that makes it so difficult to predict the long-term consequences of pollution. A major factor in the deterioration in the quality of groundwater has been the difficulty of taking remedial action before irrefutable evidence is available. The avoidance of pollution is of paramount importance because of the long residence time of water in the aquifer, accentuated in the case of the Chalk by the diffusion of solutes between the fissure system and the storage represented by the matrix. Controlling and limiting pollution of the Chalk—one of the major sources of freshwater in Europe—is a cause that must be given maximum attention and widespread publicity.

A unique aquifer

The Chalk is a unique and fascinating rock, particularly in the context of aquifer hydrodynamics. In the onshore environment, it can be regarded as several interlinked aquifers. There is a very permeable zone that coincides with the alignment of the valley system; this is flanked by an aquifer of significantly lower permeability below the interfluves. The upper 50 m of the saturated zone generally has a much greater permeability than the deeper aquifer. But at depth, the more brittle chalk of the various rock bands, such as the Chalk Rock, Melbourn Rock and the Totternhoe Stone, represent discrete and important contributory horizons in their own right. These different sub-aquifers all depend on interconnected fissures for permeability, and leakage of water from storage in the matrix for long-term yields. Offshore, where weathering of the Chalk has not been a factor, sedimentation, tectonic movements and halokinesis have created a series of hydrocarbon reservoirs, again depending on fissures for economic flow to wells. In both domains transport of immiscible and miscible constituents takes place through layers with different properties. Dispersion and diffusion between the matrix and fractures control the distribution and rate of movement of these constituents. In the Chalk, preferred flow routes along discontinuities is the rule rather than the exception.

The yield of a water well may be as high as 150 litres per second ($l\,s^{-1}$) compared with maximum yields from oil wells of about $30\,l\,s^{-1}$. This reflects the different properties of the two fluids and the relative permeabilities of the Chalk in the onshore and offshore environments. Oil production is restricted to relatively small areas that are studied intensively. The development objective is to mine the resource at the most economic rate, which is generally as quickly as possible. The skill is to recover as much of the resource as possible using knowledge of the Chalk to manage efficiently the enhanced recovery methods that are available. In contrast, water development has a regional dimension. The objective is to conserve the resource and thereby ensure sustained development while protecting the environment. The relative values of oil and water have, of course, an overwhelming influence on the modes of operation in the two fields of endeavour.

Although the hydrodynamics of the Chalk is understood in a general sense and can be modelled with practical benefits, the movement of individual particles remains, and probably will always remain, uncertain. Precise knowledge of the aquifer must be elusive because information is available only for small isolated volumes such as around a well or between wells in a wellfield, so that data exist for only a small fraction of the total volume. Under such constraints, interpreting and managing the movement of fluids through the rock requires considerable ingenuity and not a little intuition based on experience; the answers cannot be guaranteed prior to implementation of an action. Consequences may be unpredictable because non-linear processes influence groundwater flow and solute movement, and small changes in the 'cause' can lead to large changes in the 'effect'. As knowledge of the complexity of flow processes in actual fissure systems materializes, where perhaps most of the flow is in only a limited 'channel' within the fissure, there is a realization that water may never take the same path twice—a characteristic of aquifers that is even more relevant to the Chalk. Flow predictions are valid in the regional context but not for individual particles. Increasingly, probability theory will be used to quantify this uncertainty, for the only statements that have any significance are statistical.

2. The formation and diagenesis of chalk

J. M. Hancock

Definition and meaning

Chalk is a sediment composed mainly of skeletal calcite of the phylum Haptophyta (also known as Prymenesiophyceae or Coccolithophoridae) with minor amounts of other biogenic fragments, the latter typically forming less than 10 per cent of the rock (Håkansson *et al.* 1974; Hancock 1975*b*; Aubry 1976; Scholle 1977; Scholle *et al.* 1983; Mortimore 1990).

In old English usage of the word 'chalk', one of its distinctive features was its softness. Some authors have tried to maintain an indication of the degree of hardness in the definition. Geologists working for the Deep Sea Drilling Project regard chalks as 'partly indurated oozes; they are friable limestones that are readily deformed under the fingernail or the edge of a spatula blade' (Winterer *et al.* 1973, pp. 9–10; Garrison 1981). If the sediment is harder than this, these geologists call it 'limestone'. This nomenclature allows for no distinction between lithified chalks and other limestones, which actually have very different textures, and hence markedly different porosities and permeabilities. The distinction between a hard chalk and ordinary hard limestone was already recognized by Whitehurst (1786).

More recently, Bromley and Gale (1982) have introduced the word 'chalkstone' to cover the 'massively lithified equivalent of chalk'. Although 'chalkstone' may often be a useful word, to exclude lithified chalks from chalk, the definition would mean that there is no chalk in Ireland or the North Sea; it would exclude the *Plänerkreide* of Germany and many of the chalks in the United States and much of the *Mel* in the former USSR (particularly in the Caucasus). This confusion over hardness is more difficult in English than in some other languages: the Germans use *Schreibkreide* for soft chalk, and the Russians use *pischi mel*, both meaning 'writing chalk'. Completely incoherent and water-rich chalk can be referred to as 'chalk-ooze'.

This is not to say that all Upper Cretaceous chalk is homogeneous. Although much of it is that very pure white micritic limestone that can be conveniently referred to as 'white-chalk', there are also hardground-chalks, phosphorite-chalks, chalk-marls and other sub-facies. Even white-chalks show variations, notably in the proportion of clay and the types of accessory biogenic components.

The Cretaceous system is named after the latin word *creta*, meaning chalk, but it is a mistake to believe that chalks are limited to the Cretaceous. The white limestone beds in the Kimmeridge Clay of Dorset are chalk. At the bottom of the oceans there are chalks of all periods from the Cretaceous to the present day.

In the discussion above, 'chalk', with an initial lower-case *c*, refers to a type of rock; 'Chalk', with a capital *C*, refers to a supergroup in lithostratigraphy. The confusion that can result in thinking that Chalk (or *La Craie* or *Die Kreide* or *Mel*) equates with an exact division of geological time is shown by the fact that in Ireland there is no pre-Santonian Chalk, in England none younger than early Maastrichtian, in the Paris Basin none above mid-Upper Campanian, whereas in Germany chalk extends into the Upper Maastrichtian, in Denmark and the Central Graben of the North Sea into the Palaeocene, and over the Russian platform the chalk facies only starts in the Turonian.

How much chalk?

Although chalk is not limited to the Upper Cretaceous, it certainly shows a great development of this facies that lasted some 33 million years. There is a much greater area of Chalk than is obvious from geological maps of onshore areas, for it is also one of the major offshore formations around the British Isles, and extends across the North German Plain under the cover of Cenozoic sediments (Fig. 1.3; see also maps in Cope *et al.* 1992 and enclosures 31 and 32 in Ziegler 1990). In spite of impressive cliff scenery in southern

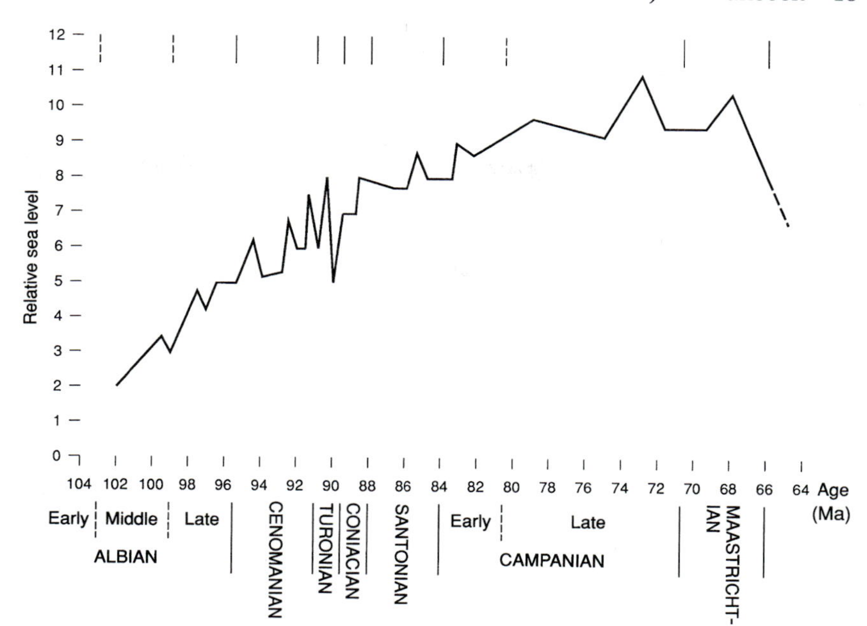

Fig. 2.1. Sea-level changes in northern Europe during Albian and Late Cretaceous times.

England and northern France, the surface exposures of onshore Chalk tend not to reveal the total thicknesses. The youngest Chalk exposed in Britain is mid-Lower Maastrichtian and in the Paris Basin mid-Upper Campanian; in Denmark (other than the island of Bornholm) only Upper Campanian and Maastrichtian chalks are exposed, and one would never guess from the little quarries in north Jutland that the Maastrichtian there totals some 700 m (Stenestad 1972). Although there is not a complete blanket of Chalk in the southern North Sea and across Denmark and north Germany, the thicknesses there are typically more than 400 m (Fig. 1.3; Hancock 1990, Fig. 9.8; Ziegler 1990, enclosure 41). The total thicknesses from surface outcrops in Britain and France are normally less: the Chalk just reaches 500 m in the Isle of Wight and 415 m in Humberside; more typically, the total does not exceed 300 m (Hancock 1975*b*). In the Yonne the Chalk possibly reaches more than 400 m (Hure 1933) but a more typical thickness in France is less than 250 m, as in Champagne (Lapparent 1943) and in the Mons Graben (Marlière 1967) (for some typical thicknesses in the Paris basin, see Mortimore and Pomerol 1987).

Why so much chalk?

There are two main reasons why the Upper Cretaceous in northern Europe is dominated by chalk, with so little in the way of clastic sediment. First, exceptionally high sea levels carried pelagic chalk, normally limited to the open ocean, on to the continental shelves and across regions that had not been under the sea since the Palaeozoic. Second, there was little erosion on those land areas that still existed.

High sea levels

Everyone agrees that sea levels were high, but just how much higher than the average of today (when the water-geoid itself varies through 180 m) is uncertain. Estimates range from 100–150 m (Schlanger *et al.* 1981) to 650 m (Hancock and Kauffman 1979). The latter assumes that sea level at the start of the Albian was about the same as today, and is a more practical figure for northern Europe (Fig. 2.1). Whatever the figure, most of the land that existed at the beginning of the Cretaceous was submerged by the Late Campanian when sea level was at its highest.

In northern Europe the only major areas that were land throughout the Cretaceous period were:

(1) the Northern and Grampian Highlands of Scotland;
(2) Norway and Sweden, except for Scania in southern Sweden;

(3) the Bohemian Massif;
(4) the Massif Central in France.

The actual limits of the seas are very uncertain because freshwater or shore-line sediments are wanting in most countries. After the Cenomanian even the deeper-water neritic facies are scarce. As a result there are substantial areas which may or may not have been land for all Late Cretaceous time: the Hatton and Rockall Banks; the Faeroe High; the Hebrides and Shetland Platforms; the higher parts of north Wales; the Armorican Massif; the Rhenish Massif. For examples of palaeogeographical reconstructions, see Pozaryski (1962), Colbeaux *et al.* (1977), Ziegler (1990, enclosures 31, 32 and references therein), Hancock (1990, Fig. 9.7), Cope *et al.* (1992).

During the latest Maastrichtian there was a large, rapid fall in sea level, so that in many areas chalk deposition ceased. However, in the deeper troughs, such as the Central Trough and the Danish Trough, chalk continued to accumulate during the Palaeocene.

Climate: non-seasonal and arid

There was some land available for erosion, particularly during the Cenomanian before sea level was really high, and again during the Late Turonian when the seas retreated. Two of the distinctive features of the marginal facies against these land areas were that there was very little detritus and almost no continental sediment. It was different in central Europe, where there are some 700 m of arenaceous facies in northern Czechoslovakia (Klein and Soukup 1966) and extensive freshwater basins, for example 340 m of the Klikov Formation in the South Bohemian Basin (Malecha 1966; Nĕmejc and Kvaček 1975). This scarcity of clastic sediment in northern Europe implies a non-seasonal climate (Wilson 1973). Of the three standard non-seasonal climates—glacial, desert and selva (tropical rainforest)—the most likely during the Late Cretaceous was an arid desert (Hancock 1975*a*).

There are four major basins in North-west Europe with great thicknesses of Upper Cretaceous siliciclastics:

(1) the Viking Graben, which contains up to 2000 m or more of the Shetland Group, largely a silty slipper-clay facies similar to the Gault of southern England.
(2) the Faeroe Basin, with up to 2000 m of a similar mudstone facies (Ridd 1981; Mudge and Rashid 1987).
(3) the West Shetland Basin, containing up to 1000 m of marine mudstones, mainly Campanian–Maastrichtian (Ridd 1981).
(4) the Rockall Trough (Megson 1987).

The onset of chalk deposition

Although it is clear that the two main controls that allowed a massive development of the Chalk were the rise of sea level and the lack of erosion on land areas, there must be at least one further control, yet to be detected, that is responsible for the onset of chalk deposition.

Large areas of the sea in North-west Europe during the Middle and Late Albian were deep enough, perhaps 125–200 m, to accumulate pelagic chalk, but there was a sufficient supply of silt and clay to deposit a slipper-clay or allied facies: the Gault of southern England, northern France, and the central part of the Niedersächsisches Becken in north Germany, where there is also the very high Albian silty Bemeroder Schichten facies (Kemper 1979). In these regions when there was insufficient detritus, another relatively deep water facies, the gaize (= *Flammenmergel*) was formed, not chalk (Fig. 2.2).

In eastern England and across a large part of the southern North Sea, the Middle and Upper Albian is represented by the Red Chalk (= Hunstanton Red Rock). This is a coccolith-rich limestone and therefore a true chalk, but it is remarkably thin, never more than a few metres thick; it has some of the features of the tethyan Ammonitico Rosso (Jeans 1973; Sellwood 1979). This appears to have formed in water deep enough to accumulate uncondensed chalk and sufficiently free from detritus, yet normal chalk did not form. On the Humberside coast there is a complete gradation upwards into a normal chalk in the Cenomanian—which is why Wood and Smith (1978) included both the Albian and Cenomanian sediments in a single lithic formation, the Ferriby Chalk.

Early in the Cenomanian some factor allowed a great increase in the production of coccoliths. The

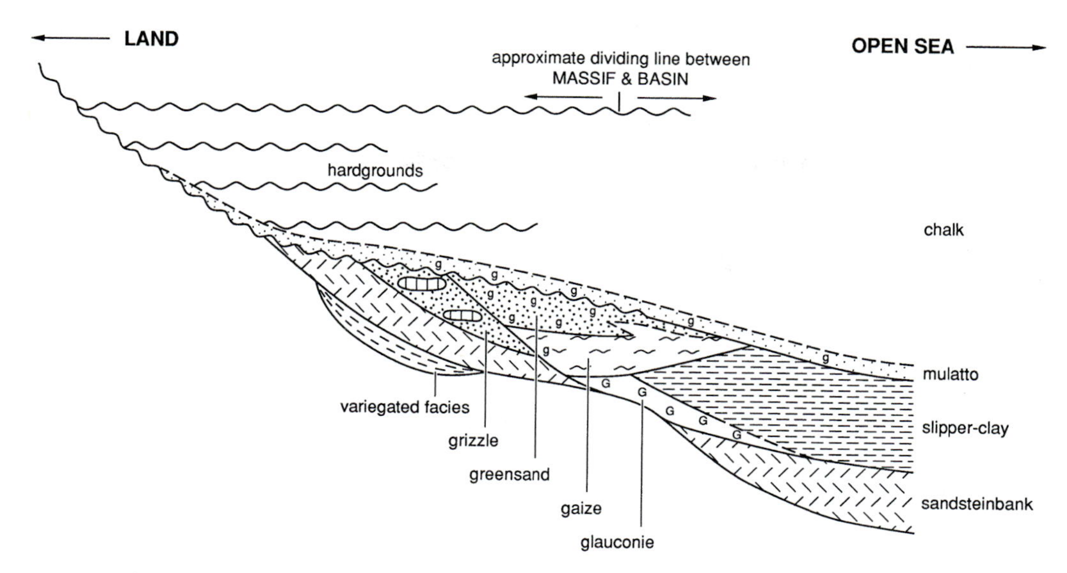

Fig. 2.2. Typical spatial relationship between facies in the Albian and Upper Cretaceous of northern Europe. The scale is approximate and the vertical is greatly exaggerated.
G = glauconite concentration
g = glauconite present in sufficient quantity to be obvious in the field
___ = sharp division between facies
– – – = transitional junction between facies

reality of this increase is evident from the absence of extensive development of chalk during Albian times, when suitable conditions also occurred.

Tectonic movements

The tectonic framework of northern Europe was established during the Triassic. Crustal movements during the Jurassic and Cretaceous merely accentuated aspects of this pattern. Major new developments—the continental rupture of the central Atlantic; the left-lateral transformation in north Africa and southern Spain; and the opening of the Alpine trough (all mid-Jurassic)—are outside our region (Dewey 1982). Ziegler's bird's-eye summaries emphasize this (see, for example, Ziegler 1990, enclosure 52): for northern Europe (Ziegler's North–west European Basin plus western shelves), he shows no folding phases until the very late Cretaceous movements in the Polish and 'marginal troughs'; even 'rifting and wrenching phases' are almost entirely pre-Albian. Only to the west of the British Isles were there significant stretching or rifting episodes. Thus the Late Cretaceous, or more strictly the Albian, was the beginning of a time of exceptional tectonic stability. Of course, such an assessment is relative, and not all the region was as stable as the Moscow Platform where most dips are still less than 0.5° (Gerasimov *et al.* 1962).

There are plenty of examples of small-scale fault movements within the largely stable massifs. In Antrim, in Northern Ireland, there is post-Cenomanian and pre-Campanian faulting that cuts out the Cenomanian Glauconite Sands and Yellow Sandstones west of a narrow coastal strip. Small Cenomanian movements are described by Smith (1957) in Devon, and discussed over a wider area of south–west England by Drummond (1970). In the Mons Basin in Belgium it was probably differential faulting on the north and south margins of the graben that produced different Chalk successions in the north and south of the basin. In southern Sweden there was faulting in the Fennoscandian Border Zone (Bergström 1985). Faulting on the northern border of the Harz Mountains was renewed during the Santonian, resulting in the formation of local conglomerates.

There were also some movements leading to doming or tilting. The Lippisch Swell, on the north–east side of the Münster Basin, was domed during the Coniacian; Upper Cretaceous sediments on its south–east flank slid into the basin to form the great slide-conglomerates now seen at Halle (Voigt 1962). The Bohemian Massif was tilted to the north or north–east so that thicker sediments developed in Saxony and northern Czechoslovakia than in eastern Bavaria.

All these movements, except the tilting of the Bohemian Massif, were on a small scale. None of the faulting seems to have involved vertical displacements of more than tens of metres.

Basinal movements

It has been recognized since the late nineteenth century that the Chalk in basinal regions is thicker than over stable platforms, many of which were land during part or all of the Early Cretaceous (Hume 1894). This implies that Chalk basins were subsiding and the palaeogeology cannot be interpreted solely in terms of sea levels. Relatively rapid lateral changes in thickness of individual units within single basins have also been known for some time; for example, Jefferies (1963) recorded that the Plenus Marls nearly double in thickness from Steyning to Beachy Head in Sussex, a distance of about 45 km. The extent of differential vertical movements within basins has become apparent only with the detailed measurements of Chalk lithostratigraphy by Mortimore and Pomerol (1987; 1991). Their observations extend over southern England and northern France, but they made detailed studies in East Sussex. With data from seismic sections they have been able to relate surface periclines, axes and chalk sedimentation to deeper underlying fault systems, involving horsts, grabens and half-grabens. The importance of this work for the interpretation and understanding of local lateral changes in Chalk stratigraphy cannot be overemphasized. At the same time the scale of the vertical movements needs to be kept in mind: it is difficult to disentangle the magnitude of Cretaceous movements from those of the Tertiary; certainly there are contemporaneous changes of thickness of 60 m between Portsdown and Culver, a distance of 22 km. This is less than the typical depth of the Chalk sea and less than the vertical eustatic movements of the sea. It is significant that none of these vertical movements has been demonstrated to bring the Chalk above sea level during the Cretaceous.

As our knowledge of local Chalk lithostratigraphy improves, these reactivations of older structures will probably be found in other regions. Although little described, it has been known for some time that such movements and structures occur in the Cenomanian of Normandy outside the area discussed by Mortimore and Pomerol.

Diapirism

A second important tectonic process is evaporite-diapirism. The classic examples of these plugs and domes are in north Germany, where diapirism has been the agent that has brought the Chalk to the surface through the 1–2 km cover of Cenozoic sediments. In the example at Lüneburg, 45 km south-east of Hamburg, which exposed a sequence from Campanian down to Albian resting on Keuper, there are steep dips and parts of the succession are inverted (Schmid 1963). The deep quarry at Hemmoor, little more than 1 km across, exposed over 140 m of Chalk, all Maastrichtian and all sensibly horizontal (Schmid 1982). The quarries around Lägerdorf and Kronsmoor, some 40–50 km north of Hamburg, are over a broad evaporite dome rather than a diapiric plug (Ehrmann 1986).

The clearest example in England is the inlier at Winterbourne Abbas, west of Dorchester (Robbie 1950). At this locality and at Misburg, near Hanover, the salt movement occurred during the Late Cretaceous.

The dome-like structures of the hydrocarbon plays in the Central Trough result from evaporite diapirism (Parsley 1990), but Duval *et al.* (1991) have emphasised inherited structures rather than evaporite-diapirism.

Retroversion

A third type of tectonic movement is that known as 'basin-inversion' or simply 'inversion', although better English would be 'retroversion' (John Ramsay, personal communication). This occurs in areas which have initially undergone major downwarping, allowing thick sedimentation; the downwarping has later been reversed and the area has been uplifted. Such uplifts are commonly elongated and are then called 'inversion axes'. A classic study is by Voigt (1963); for a recent general survey, see Cooper and Williams (1989).

Most of the inversion uplifts were Cretaceous-Tertiary events, but a few were Jurassic, for example Pompeckj's Swell in north Germany (Kemper 1979). The change from basin to inversion axis was usually episodic and there is no general pattern of timing (Heybroek 1974; Hancock and Scholle 1975; Glennie and Boegner 1981); for a different view see Ziegler (1990, pp. 134–42). On some, for example the southern Central Graben, the uplift was completed before the Late Palaeocene (Heybroek 1975), but in others the main uplift was mid-Tertiary, for example the Weald of south-east England. In the Sole Pit inversion the uplift shifted laterally south-westwards over a long time, with Cretaceous inversions very early in the period, then Turonian to Early Campanian; the Late Campanian transgressive peak submerged the area and it is difficult to decide how much of the Late Maastrichtian regression here is due to inversion and how much to the eustatic fall in sea level.

The principal inversions which were moving during the Late Cretaceous were: the Sole Pit inversion; the southern part of the Central Graben; the Broad Fourteens–West Netherlands Inversion (Bodenhausen and Ott 1981); the Lower Saxony Basin Inversion; the Harz Mountains; and the Polish Trough inversion (Pozaryski and Brochwicz-Lewiński, 1978). With the possible exception of the Harz Mountains, all these structures involve vertical movements in the range 1–3.5 km.

In general, evaporite-diapiric structures are more local and more sharply bounded than either inherited or inversion structures. Inversion structures are on a much broader and larger scale than inherited structures; for example, in the West Netherlands Basin the Chalk is absent over the inversion but up to 1500 m thick only a few kilometres away (Fig. 9.3). Nevertheless, there is not always a clear distinction between the three types of structure. Many relatively small uplifts that could well be inherited are ascribed to inversion, and in some areas there has been dispute between inversion and diapirism, for example in the Lower Saxony Basin (Boigk 1968).

Effects of uplift

Whatever type of uplift is involved, if it was moving during the Late Cretaceous it will obviously affect the accumulation and sedimentation of chalk. Over uplifts one can get erosional channels, hardgrounds, glauconitic or phosphatic sediment, even a complete restart of the facies succession, in addition to lateral changes in the thickness of units.

The pattern of facies

In spite of the movements described above, the broad tectonic stability in northern Europe means that the Upper Cretaceous sediments can be considered in only three structural settings: above and against massifs of Palaeozoic and older rocks; in basins between massifs; and in fault-bounded troughs within the basins.

The neighbourhood of massifs

The principal massifs in northern Europe are shown in Fig. 1.2. The pattern of Upper Cretaceous facies associated with massifs is similar across Europe from Ireland into eastern Europe (Fig. 2.2). The most complete succession of these standard facies is seen on parts of massifs that were submerged before the Campanian and with gently sloping flanks, for example the eastern flank of Cornubia or the north-eastern flank of the London–Brabant High. The actual age of any particular facies is, of course, dependent on the height and topography of the massif and may be diachronous. Thus sandsteinbank facies are Lower Albian in the Folkestone Formation of southern England but Upper Santonian in the Aachen Sands of north-east Belgium. For more detail, see Hancock (1975*a*).

A number of massifs have partly faulted borders across which rapid lateral changes of facies may occur, for example the south-eastern margin of the Palaeozoic massif south of the Highland Border Ridge in Antrim in Northern Ireland (Hancock 1961; Reid 1971); the western margin of the Ringkøbing–Fyn High (Michelsen and Anderson 1983) and its margins bordering the Horn Graben (Olsen 1983); and the northern margin of the Harz (Tröger 1975).

There are also various sorts of tuffeau facies which lie outside the common sequence and possibly formed in water only a few metres deep where there was almost no detritus draining off the land. Such a facies is found in the Upper Maastrichtian of Limburg in the south-east Netherlands and north-east Belgium (Albers and Felder 1979), and the Campanian–Maastrichtian of southern Sweden (Surlyk and Christensen 1974; Bruun-Peterson 1975). A different type of tuffeau, rich in cristobalite, occurs in the Middle Turonian of Touraine.

The transgressive surges during the Albian had submerged major areas of some massifs by the end of the Early Cretaceous: much of the Mid-North Sea High (but not the Dogger High); the eastern part of the Ringkøbing–Fyn High; the western half of the London–Brabant High; the eastern flanks of the Cornubian and Armorican massifs. Such massifs continued to exert an influence during the Late Cretaceous. Over them chalk sequences are thinner, often coarser, more liable to contain hardgrounds, and (very occasionally) show a reversion to shallower-water facies during a regressive trough, for example the Late Turonian.

Basins between massifs

Chalk was deposited in the basins between massifs (and over massifs already submerged by the beginning of the Late Cretaceous). Basins in this broad sense are obviously rather vague but the principal ones are: the Celtic Sea Basin; the Western Approaches Trough; the Anglo-Paris Basin (subdivided into Channel Basin, Wessex Basin, and Paris Basin); the southern North Sea and north German area with many subsidiary basins; the Horda–Egersund Basin; the Greater Moray Firth area (which is structually complicated) and the Dutch Bank Basin.

Typical complete thicknesses are 400–500 m, but post-Cretaceous pre-Eocene erosion has often reduced this to 250 m or less over large areas. The southern North Sea basin of about 180 000 km^2 is a similar area to the Danish–Polish Furrow and about twice the size of the Paris Basin. With a thickness of 400 m or more over about 100 000 km^2, it contains the greatest single basinal mass of chalk on the continental shelves of the world.

Some of these basinal regions between massifs continued to subside during the Late Cretaceous. Judged by the thicknesses of chalk, this basinal subsidence allowed up to four times the amount of accumulation over massifs, for example in the Wessex Basin during the Late Campanian compared with north Norfolk over the London–Brabant High. Actual subsidences may have been both more and less than this.

In the northern North Sea several basins show much less subsidence during the Late Cretaceous than during the Jurassic–Early Cretaceous, for example the Moray Firth Basin and Horda–Egersund Basins.

Fault-bounded troughs within the basins

Within the broad basinal regions there are a number of trenches with thicknesses of 1000 m or more: the Rockall Trough; the Porcupine Trough; the Celtic Sea Trough; the Viking Graben (with up to 2000 m of Shetland Group in the far north); the Central Trough (with more than 1200 m of Upper Cretaceous and 1,350 m of chalk including that of the Lower Palaeocene; up to 2000 m according to Ziegler's (1990) isopachyte map); the South Halibut Basin and Witch Ground Graben (each 1000 m) which were structurally not connected to the Central Graben; the Lower Saxony Basin; the Central Netherlands Basin, north of the Central Netherlands Inversion (more than 1200 m); the West Netherlands Basin, south of the Central Netherlands Inversion (more than 1600 m); the Sole Pit Basin, west of the Early Cretaceous Sole Pit Basin (1000 m); the Münster Basin (up to 2000 m).

The greatest trenches of all were: the Danish–Polish Trough with thicknesses of chalk up to nearly 2000 m; the Faeroe Trough that, according to Ridd (1981), contains up to 2000 m of mudstone of the Shetland Group, and up to 3000 m according to Ziegler (1990); and the West Shetland Basin with more than 5000 m (Ziegler 1990).

Some of these trenches, such as the Viking and Central grabens, are bounded by fault-zones on both margins; others, such as the Sole Pit Basin, have only a single flank marked by a major fault, often on the north-eastern flank. The extensional faulting that formed these grabens was largely pre-Late Cretaceous. Indeed, some of the basins and trenches that had accumulated a thick sequence of Jurassic and/or Lower Cretaceous sediments were inverted.

Table 2.1 Chemical composition of chalk: results of 13 analyses of English white-chalk

	Average (%)	Maximum (%)	Minimum (%)
SiO_2	0.63	2.12	0.15
Al_2O_3	0.33	1.01	0.08
Fe_2O_3	0.08	0.13	0.01
MnO	0.04	0.06	0.02
MgO	0.30	0.36	0.22
CaO	54.90	55.54	54.08
Na_2O	0.05	0.16	0.01
K_2O	0.03	0.05	0.01
P_2O_5	0.08	0.14	0.05
CO_2	43.54	43.85	42.86
	99.98		
($CaCO_3$)	98.17	99.09	96.77

The Chalk is primarily a pelagic sediment of accumulation, rather than a mechanical deposit. There is therefore no obvious reason why the troughs should accumulate greater thicknesses—and, as will be obvious from the figures listed above, much greater thicknesses—than the usual blanket cover of 400–500 m of the ordinary basinal regions. Research in the Central Trough has shown that these enormous thicknesses are the result of mechanical resedimentation of chalk from the flanks of the troughs.

Composition of chalk

The Chalk, being an extremely pure limestone, consists almost entirely of calcium carbonate (Tables 2.1 and 2.2), but the small proportion of other constituents can have a significant bearing on the composition of the groundwater. The occurrence of these constituents is discussed below.

Non-calcite, mainly inorganic mineralogy

Clay minerals

The smectite group usually dominates, and most of this is Ca-montmorillonite. The mica group is less important: muscovite is more common than illite and glauconite *sensu stricto*. Chlorite is rare, palygorskite very rare. In chalks with more than 2 per cent clay minerals these proportions may be different. Thus in much of the chalk-marl facies (Cenomanian in southern England) there is an illite–kaolinite–chlorite–vermiculite assemblage of which less than 20 per cent is smectite (Weir and Catt 1965; Jeans 1968; Morgan-Jones 1977; Pitman 1978b; Seibertz and Vortisch 1979; Pomerol 1984; Pacey 1984).

Whereas there is little doubt that the muscovite, illite and chlorite are all detrital, there has been much discussion and disagreement on the origin of the smectite group. Weir and Catt (1965) and Jeans (1968) both say that the present form of the crystals would probably not survive transport and concluded that this group was neoformed, precipitated from

Table 2.2 Typical mode of white-chalk

Biogenic components: 96–99% nearly all low-magnesian calcite	Haptophyta (coccolith-bearing 'algae')	80–92%
	Foraminifera (dominantly benthic in most stages)	5–10%
	Inoceramid bivalves—commonly 2%	0–5%
	Bryozoa	0–5%
Inorganic components:	Clay minerals (Ca-montmorillonite usually dominant, then clay-grade muscovite)	1–4%
	Silica minerals (+ flints) (chiefly clay grade quartz and cristobalite)	1–2%

solution within the pore-waters of the chalk-ooze. Evidence for a possible volcanic origin was found in the Maastrichtian of north Germany by Valeton as long ago as 1960, but the transfer of this idea to marl bands in the Central Trough and eastern England was largely guesswork until recently. In 1982 Jeans *et al.* showed that many grains of glauconite in the British Cretaceous are argillized particles of lava. Their only examples of Coniacian to Campanian age are all from Northern Ireland, and they left open the origin of the smectite in the main mass of the Chalk.

From his studies on the clays in the Chalk of Lincolnshire and Humberside, Pacey (1984) has produced evidence that the marl bands in the Turonian represent smectite formed *in situ* from volcanic ash: (1) good crystal form; (2) absence of mixed layering; (3) scarcity of accompanying detrital quartz and illite; (4) magnesium-rich montmorillonite which is characteristic of smectite grown in seawater; (5) presence of pyroclastic fragments from the marls, including glass spheres as are formed by Strombolian type volcanic activity. The possibility of a non-volcanic origin remains for the smectite in the main mass of the Chalk in the North Sea and eastern England. Similarly, one should remember that most of the smectite in southern England is a calcium-rich not a Mg-montmorillonite. Wray has shown that individual marl seams in the Turonian of south-east England and northern France each contain their own individual pattern of trace elements. He remarks: 'Whilst not proving a volcanic origin for the marls, it is difficult to envisage an alternative process which could produce a similar result,' (Leary and Wray 1989).

Hardman and Kennedy (1980) and Skovbro (1983) refer to authigenic clay in the Central Trough, but it is not clear how much is authigenic, nor if this is neoformation from solution or diagenetic clays. The possibility of an ordinary detrital origin for the smectite is seldom discussed (but see Parra *et al.* 1985).

Burki *et al.* (1982) and Glasser and Smith (1986) have discovered that there are coatings of smectite composition, only 4–5 nm thick, on the biogenic algal particles that make up the bulk of the Chalk. It is not yet known how widespread such coatings are, but they are an important protection against diagenesis since they shield the calcite from interstitial water. Variations in the amount of such coatings may be a major factor in variations in the hardness and porosity of chalks.

Distribution of clay

Up to 30 per cent clay is found in the marls of the chalk-marl facies in the Cenomanian of southern England and elsewhere, but in the white-chalk facies, which is what is usually thought of as chalk, the clay content is typically less than 2 per cent in both England and Denmark. Marly beds in the Turonian Middle Chalk of southern England contain up to about 12 per cent clay (Morgan-Jones 1977); in Lincolnshire 20–65 per cent (Pacey 1984). To some extent the clay content in many hardened chalks has been enhanced by solution diagenesis.

As one passes north from the Central Trough into the Viking Graben there is every gradation into the pure mudstone of the Shetland Group.

Zeolites

In the north-western part of the Paris Basin there are appreciable quantities of the zeolite mineral clinoptilolite in the insoluble fraction of the Cenomanian and Turonian chalks (Pomerol 1984). Trace amounts of the clinoptilolite-heulandite group have been found in the Cenomanian and Coniacian of the Chiltern Hills and the North Downs of the Weald in southern England (Morgan-Jones 1977). None has yet been recorded from North Sea Chalk. Shumenko (1980) has argued that the habit of clinoptilolite and the associated mineralogy points to a non-volcanic origin in chalks.

Silica minerals

Quartz occurs as clay and silt-grade particles and as pseudomorphs after foraminifera throughout the Chalk. Small amounts of neoformed cristobalite are almost universal but it is more obvious in chalk-marl facies, and occasionally reaches several percent in other chalks. Trydimite has been recorded from a few onshore chalks, but it is usually absent and always rare.

Rarely one can find unaltered siliceous fossils, such as sponge spicules or radiolaria, but most of the original biogenic silica has undergone a series of diagenetic changes to become concentrated into chert-nodules called 'flints'. The British Upper Chalk was at one time called the Chalk with Flints, but on a European scale the flints can be found at any level in the Chalk. The formation of flints is outside the scope of this chapter; for a modern discussion, see Clayton (1986).

Courses of flints always follow the original bedding where this has been tested, for example where chalk has slumped (Kennedy and Juignet 1974; Gale 1980). Individual tabular flints and courses of flints may have sufficient individual characters to be safely traced over tens of kilometres. With experience, and by association with other lithological features, it has been possible to trace some courses of flint right across the Anglo-Paris Basin (Mortimore and Pomerol 1987), but this is not a technique for the inexperienced.

Other inorganic components

Trace quantities of other minerals are found throughout the Chalk. Collophane, probably fluor-carbonate-apatite, occurs in two forms: (1) as concretions in mulatto facies at the base of many Chalk successions and occasionally on hardgrounds within the Chalk; (2) as coarse brown pellets, sometimes in sufficient concentration to form phosphorite-chalk. This second form is best known in the Santonian to Lower Campanian in the northern part of the Paris Basin (Jarvis 1980), but occurs sporadically elsewhere, for example at Taplow in Berkshire, in Strahan's Hardground at Lewes, and in the main hardground at Downend in Portsdown (Gale 1980).

Scattered crystals of dolomite are known from many places. There are thin beds of dolomitic chalk in the Central Trough and in Northern Ireland, but it is best and most clearly developed in coastal exposures of the Turonian in the Pays de Caux of northern France.

Feldspars are very rare but widespread accessories. Authigenic pyrite and pyrolusite are both common and widespread accessories, and barytes occurs in the Central Trough.

Biogenic calcite

Algae

Algae typically comprise 80–92 per cent of chalk (Table 2.2) and all consist of low magnesian calcite. 97 per cent of typical white-chalk—that is, the most calcite-rich variety—is biogenic: the hard parts of organisms. Commonly 80–90 per cent of the whole rock is the skeletal fragments of prymnesiophytes of the class Haptophyta, and their biology is important in the understanding of chalks as sediments (Parke 1961; Westbroek *et al.* 1984; Lowenstam 1986; Green 1986).

The Haptophyta are popularly known as the golden-brown algae but the word alga has lost its apparently precise meaning for a single group of single celled 'plants'. Algae include representatives of two kingdoms: the Monera (bacteria and blue-green algae which lack a nucleus); and the Protista (single-celled organisms with a nucleus, and including the Haptophyta) which include both 'plants' and 'animals' in older and still popular classifications. This new distinction is important because some Protista can obtain nourishment in ways that are associated with plants, such as photosynthesis, and in ways that are associated with animals, such as by "eating" other organisms—bacteria and plant cells up to 5 μm in diameter (phagotrophy). This is true of the Haptophyta. The geological importance of this is that one must not assume that the Haptophyta were limited to the shallower depths of the sea where light can penetrate in strengths adequate for photosynthesis.

The taxonomic classification of the Haptophyta is one of the most difficult of any group of organisms for three reasons:

1. Some secrete a skeleton, some do not.
2. Very different skeletal shapes can occur in one individual.
3. They show an alternation of generations; that is, a motile phase alternates with a sessile phase. The two generations may produce skeletons so different that they were formerly placed in separate families. Or one generation, normally a sessile one, may not produce a skeleton at all.

Classification is further complicated by the fact that the two largest families, the Syracosphaeraceae and Zygosphaeraceae, each with more than 50 living species, are almost unknown as fossils. In contrast, there are a number of families, important contributors to Cretaceous sediments, that became extinct at the end of the Cretaceous. Some of these groups have only recently been recognized to belong to the Haptophyta. Again, all this might seem irrelevant to the petrologist, but evolution in the Haptophyta is one of the controlling factors in the distribution of particle sizes in chalks.

Many of the Haptophyta secrete calcite in

distinctively shaped particles. The commonest arrangement, as seen in the order Coccosphaerales, is that each alga, from around 10 to perhaps 25 μm in diameter, is typically encased in the calcite sphere stuck to the surface. This sphere, called a 'coccosphere', is composed of some 7 to 20 rings called 'coccoliths', 1–20 μm in diameter. Each coccolith in turn is built of tablet-shaped crystals of calcite called 'rays', 'plates', or 'laths' that may be anything from 0.17 to 2.8 μm across, but are typically around 0.5 to 1 μm. Each coccolith is secreted deep inside the single cell and gradually works its way to the surface to become part of the coccosphere. Coccoliths are shed from time to time during the life of the alga, which usually breaks up into separate coccoliths on death, but both the separate coccoliths and complete coccospheres found in Cretaceous chalks probably originated from encysted algae. At the time they are shed from the alga each coccolith is coated with a film of polysaccharides and amino acids, but this is probably destroyed at an early stage during digestion by animal plankton.

Even in these Coccosphaerales the shapes of the coccoliths vary considerably: empty rings (for example *Loxolithus*), rings with a cross (*Staurolithites*), discs (*Pontolithina*) and mushrooms, often called 'rhabdoliths' (*Deflandrius*) (Black 1972). Many other calcitic shapes in the Chalk look so different from these that it was only gradually realized that they probably belong to other orders of Haptophyta, often now extinct.

Cylinders built of brick-shaped plates called *Nannoconus* range from the top of the Jurassic to Maastrichtian but are prominent in the Santonian. Stick-like objects, around 15–30 μm long, commonly called 'Cayeux's organism' and technically named *Microrhabdulus* and *Lithrophidites*, are specially abundant in the Upper Coniacian and Santonian, but are now extinct; Cayeux's organism seems to be associated with chalks that are finer grained and poor in foraminifera. Hollow globules, commonly 60–80 μm in diameter, are usually called 'calcispheres' but are also known as *Oligostegina*, *Palinosphaera*, and *Pithonella*; calcispheres, which are also now extinct, are sometimes abundant in hardground and nodular chalks where they take the place of the foraminifera.

It is not known why most Haptophyta secrete calcite, but the fact that the number of coccoliths is not a constant for any one species but is strongly dependent on the amount of light reaching the alga, suggests that the coccosphere acts as a control on the rate of photosynthesis. In the open sea they live chiefly in the top 100 m, with a maximum concentration around 50 m near the equator and 10–20 m in British waters. The quantity of calcite secreted varies little with temperature, but a high rate needs water low in nutrients, particularly phosphorus. They can release toxins, such as dimethyl sulphide, that inhibit the growth of other protists such as diatoms.

Complete coccospheres are preserved only rarely in Cretaceous chalks. Although complete coccoliths are more common, even the majority of these are broken up into their component laths. There is some indication that the degree of break-up increases with the depth of accumulation, and there are more complete coccoliths preserved in most American chalks than in Europe.

In the Maastrichtian and Lower Palaeocene chalks there are ball-shaped calcitic objects around 10 μm in diameter called Thoracospheres. These have also commonly been classed as Haptophyta, but are currently regarded as calcareous cysts of dinoflagellates of the order Peridionales.

Foraminifera

Foraminifera often represent 5–10 per cent of chalk, but sometimes less. The original mineralogy was nearly all calcite, but some were low-Mg calcite and some were high-Mg calcite. The foraminifera now found in Cretaceous chalks of Europe were dominantly benthic in the Early Cenomanian, Coniacian, Santonian and Maastrichtian; dominantly planktic in Early to Mid-Turonian; and approximately in equal proportions in Mid- to Late Cenomanian. But these proportions are not necessarily original; there has probably been a preferential loss by solution of planktic foraminifera in Coniacian to Campanian European chalks (see later discussion of diagenesis). There is a higher proportion of foraminifera in present-day chalks. Foraminiferal tests are commonly recrystallized, and are sometimes silicified or infilled with chert.

Bivalves

Bivalves comprise up to approximately 5 per cent, but commonly only up to 2 per cent of chalk. The original mineralogy of rock-forming species was low-Mg calcite + aragonite. The only rock-forming bivalves were the inoceramids, now extinct. The original inner aragonitic layer was lost soon after the death of the animal; the outer prismatic layer then usually broke up into its component calcitic fibres, which are now found dispersed through chalk, but sometimes have been concentrated by winnowing. In present-day chalks the bivalves are not a rock-forming component.

Bryozoa

Bryozoa comprise up to possibly 5 per cent of chalk. The original mineralogy was mainly low-Mg calcite. The bryozoa are very variable, both geographically and stratigraphically. Most of them were cyclostomes, but cheilostomes became important in the Coniacian, and even more so in the Maastrichtian onwards.

Other groups

The total range of fossils known from the Chalk is broad but only occasionally does another group become noticeable as a component. Håkansson *et al.* (1974) record an Upper Maastrichtian chalk with appreciable amounts of octocorals. Sponge spicules, usually calcitized, are widespread. The radiolaria of North Sea chalks are also usually calcitized. Echinoderm debris (originally high-Mg calcite) is often more conspicuous in the field than quantitatively important. Ostracods can quite often be found in thin section. Kennedy (1980) has even recorded vertebrate debris from a North Sea chalk.

Initial accumulation

Chalk ooze is so dominated by coccoliths and their component laths that discussion of initial accumulation can be concentrated on them.

Even complete coccoliths sink very slowly; in still water the maximum rate is 60 m per day, and most of them sink at less than 10 m per day. The individual laths that make up the bulk of European Cretaceous chalks would have settled very slowly indeed without some form of aggregation; spherical particles, 1 μm in diameter, have a theoretical settling rate of 77 mm per day, and tablet-shaped particles settle at about half this rate (Lerman *et al.* 1974). Therefore, without aggregation, the smaller laths would have taken more than 28 years to reach the bottom of a 200 m chalk sea free of all currents!

The clusters now visible in scanning electron micrographs of nearly all chalks would have speeded up the settling rate but little (Smayda 1971). However, there are abundant faecal pellets around 120 μm in length in some North American chalks (Hattin 1975*a*; 1981), but they are also known from chalk in the Central Trough (Kennedy 1980) and the Maastrichtian of Denmark (Bromley 1980). Hattin suggested, and Honjo (1975) confirmed, that the main aggregating agents were copepods, marine planktic crustacea up to a few millimetres in length that are so numerous in present-day seas that it has been suggested that there must be more copepods in the world than all other multi-cellular animals put together (Hardy 1956). Their faecal pellets are several hundred micrometres long and tens of micrometres wide. Cylindrical shapes sink faster than spheres, let alone tablets (Lal and Lerman 1975), and such pellets descend at rates of approximately 200 m per day, arriving at the sea floor in the Tongue of the Ocean in the Bahamas, for example, at approximately 250 per square metre per day. Each contains around 4×10^4 coccoliths (Honjo 1975). Some pelleting is also due to tunicates (salps), and the food of copepods is not limited to Haptophyta (Reinfelder and Fisher 1991).

This pelleting of coccolithic debris is important in the genesis of the Chalk because pellets took only about a day to sink from near-surface waters to the bottom of the sea. In such pellets no solution of the calcite can occur until they reach the bottom. Presumably some coccoliths are not eaten by copepods, and some of the pellets themselves start to break up during descent, shedding free coccoliths. Hence there is no simple progressive dissolution with depth (Honjo 1975). At various stations in the Pacific—where undersaturation with respect to low-Mg calcite begins at depths of 200–2000 m according to latitude (Takahashi 1975)—etched coccoliths appear at depths of 600–1000 m but the majority were complete and intact. Coccoliths of *Umbellosphaera* were easily disintegrated into their component laths at depths of more than 600 m, but intact specimens with no sign of damage were often found at 4000 m. All this variation in solution *during settling* seems to be unrelated to the actual size of the crystals in spite of the faster reactivity of very small crystals. Rather it is a combination of variations in two or possibly three factors: pelleted or unpelleted debris; the preservation of the organic skin around the coccoliths which would be destroyed by bacteria in the copepod gut but which would protect against solution before this; and a possible skin of smectitic

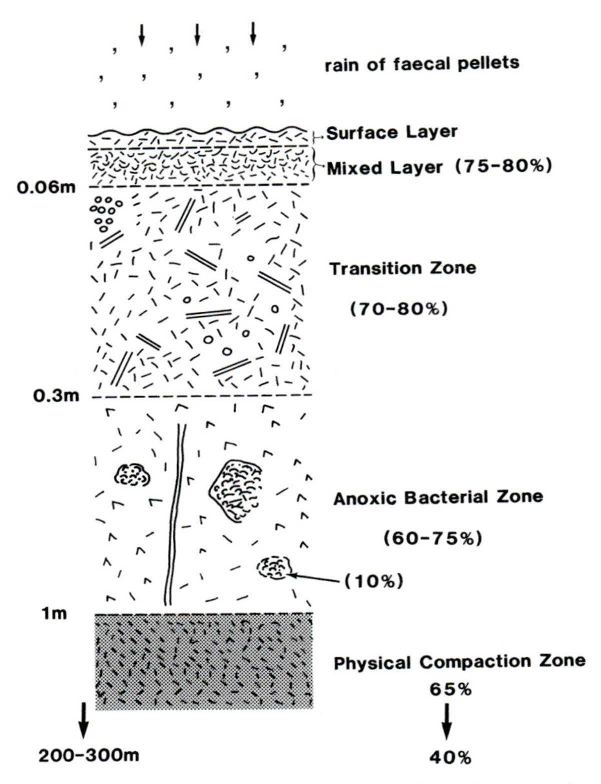

Fig. 2.3. Typical layers in white-chalks at the time of accumulation, with approximate values for porosity.

material—but this was probably not emplaced until incorporation in the sediment.

Foraminiferal tests sink even faster than pelleted coccoliths (Matter *et al.* 1975) so with these also there is negligible solution during settling (Adelseck and Berger 1975). Even the delicate spines of globigerinids, which often break away from the test, do not suffer corrosion (Black 1980).

Early changes in chalk oozes: intrinsic diagenesis

Physical conditions at the bottom of the Chalk sea

The initial chalk-ooze can be considered in a series of depth layers (Fig. 2.3):

1. *Top few millimetres* When the coccolith-loaded faecal pellets arrived at the bottom of the sea they did not always meet a simple liquid–solid interface. The top of the chalk-ooze was close to a suspension of faecal pellets in water with porosities of 80–90 per cent. This incoherent layer was only a few millimetres thick—sufficient to provide problems for sedentary animals that needed to keep themselves largely clear of ooze (see, for example, Carter 1972; summary in Hancock 1975*b*). It could not have been more than a few millimetres thick because burrowing echinoids, such as *Hagenowia*, possessed cutting and compacting spines on the rostrum which implies a cohesive sediment (Gale and Smith 1982); and because minute suspension-feeding brachiopods (Surlyk 1972) and bivalves (Heinberg 1979) are abundant. Material in these top few millimetres was easily resuspended by bottom currents, and could then have formed a nepheloid layer, and left a true liquid–solid interface. Therefore, this top layer was not always present and may have been absent for extended periods.

2. *0 to 0.05–0.08 m.* On and within this layer (porosity 75–80 per cent) lived the majority of animals represented by body fossils in the Chalk. Bioturbation was so intense that no individual structures are preserved and the sediment is virtually homogenous—this is the 'mixed layer' of Scholle *et al.* (1983). Any original surface trails have been destroyed in this interval.

3. *0.05–0.08 to 0.2–0.35 m.* This is the 'transition zone' (porosity 70–80 per cent) of Scholle *et al.* (1983) in which deeper burrowing animals lived and whose traces now dominate the ichnology of the Chalk. At this depth the interstitial water may already have been anoxic.

4. *0.2–0.35 to approx 1 m.* Few burrowers reached this depth (porosity approximately 60–75 per cent) of which the most important during the Cretaceous were the larger *Thalassinoides* and the pogonophores (= *Bathicnus*), but it was still a depth with considerable bacterial activity. The interstitial water would definitely have been anoxic.

5. *Approximately 1 m to 200–300 m.* This was a zone (porosity approximately 40–65 per cent) of physical compaction, but compaction without any solution rarely reduced the porosity to below 40 per cent.

6. *Approximately 300–1000 m.* This was a relatively stable zone in which diagenesis was at a

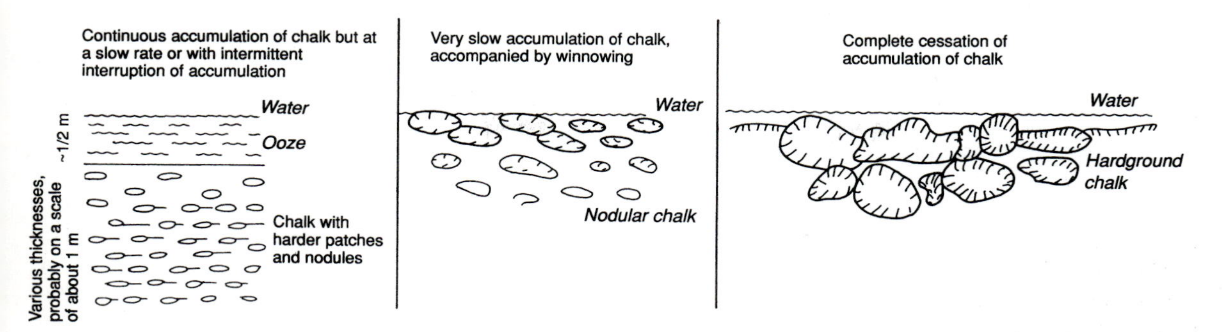

Fig. 2.4. Progressive development of more nodular chalk eventually leading to a hardground as the rate of accumulation of chalk decreased.

minimum, but the porosity was reduced to 30–40 per cent.

7. *Around 1000 m depth.* If the chalk was buried to these depths, various types of later diagenesis would have taken over (See the section below on later diagenesis).

What is meant by intrinsic diagenesis?

The diagenesis in the top five layers, listed above, describes the inevitable changes associated with the normal conditions of accumulation of chalk-ooze; these changes are intrinsic to chalk. The later diagenesis (as discussed below) did not occur unless special conditions were imposed on the chalk-ooze.

Bacterial activity

The faecal pellets formed the food for a variety of sediment feeders and in most European chalks the pellets themselves have disappeared. Their preservation in the Greenhorn Limestone, Fairport Chalk, and Smoky Hill Chalk in Kansas (Hattin 1975*b*; 1981; 1982) indicates that the Kansas chalks have suffered less diagenesis. The markedly higher proportion of complete coccoliths in these American chalks shows that coccolith break-up was not an automatic part of early diagenesis, but was associated with the intense biological activity in European chalks. In addition to macro- and micro-sediment feeders, there was moderately intense bacterial activity, breaking down organic material and disintegrating the pellets (Froelich *et al.* 1979). However, the constant churning of the top 5–8 cm meant that chemical equilibrium was maintained between interstitial water and seawater.

In the transition zone the destruction of crustacean carapaces was by chitinoblastic bacteria (Seki and Taka 1965). This destroyed most fish scales and fossil fish; the preservation of scales in *Terebella* burrows was because the burrowing animals carried the scales below the depths at which chitinoblastic bacteria lived. Probably at these depths also, and certainly below 0.35 m and down to a metre or more, there would be nitrate-reducing and especially sulphate-reducing bacteria (Lippmann 1973; Jeans 1980). The more important sulphate-reducing bacteria transformed sulphate ions to H_2S, which normally escaped into the overlying aerobic zone where it was reoxidized to sulphurous and sulphuric acids:

$$SO_4^{2-} + H_2O + 2[C] \rightarrow CO_3^{2-} + CO_2 + H_2S$$
$$SO_4^{2-} + CO_2 + 8[C] \rightarrow CO_3^{2-} + 3H_2O + H_2S$$

The new carbonate ions combined with available calcium and were precipitated as calcite. This early cementation at depths around 0.35–1 m beneath the sediment–water interface was probably the principal origin of nodular chalks. If the process lasted long enough a nearly continuous bed of lithified chalk formed, which, followed by sea-bed erosion, resulted in a rocky hardground or conglomerate on the sea floor (Fig. 2.4).

Why is all the Chalk not lithified in this way? Two major controls probably existed: first, normal chalk accumulation was too fast to provide sufficient organic food for the bacteria; and second, chalks that accumulated below perhaps a water depth of 100 m had a low bacterial content, the bacteria grew very slowly, and the high hydrostatic pressures inhibited bacterial activity (Jannasch and Wirsen 1977).

Since such slightly shallower regions were also more likely to be subjected to eroding currents, hardgrounds formed most commonly in areas over massifs or where there was contemporaneous uplift during chalk accumulation.

Movement of grains

Large numbers of animals (for example, echinoids, asteroids, bivalves, worms and fish) constantly churned and/or ingested the mixed layer at the top of the sediment column; crustacea and other burrowing animals disturbed the transition zone below. All this mechanical agitation helped to dewater the sediment. The porosity was probably not reduced by more than about 20 per cent—say, from 80 per cent to 60 per cent.

Burrows of *Chondrites*, *Planolites* and *Thalassinoides* are essentially universal. *Zoophycos* is known from a number of horizons, but is usually concentrated at particular levels, for example Cuilfail *Zoophycos* layers in the Upper Turonian of Sussex (Mortimore 1986), but it is widespread in the Chalk of the Central Trough. The presence of lamination, not destroyed by these burrowers, is itself evidence of resedimentation.

It was probably at this time of intense burrowing that the skins of smectite were emplaced on the coccoliths, but it is not known how.

Chemical changes

Many of the chalks that are now onshore, and certainly most of those in the North Sea, were deposited at depths of hundreds of metres. At these depths the water was no longer saturated with respect to $CaCO_3$, though it was above the lysocline—the depth where dissolution starts to be rapid. Whether the undersaturation was sufficient to trigger chemical diagenesis in the top few metres of ooze, independent of bacterial activity, is still controversial. Some studies of recent oozes, mainly in the Pacific, indicate little or no dissolution in the top 50 m (Matter *et al.* 1975; Garrison 1981), with calcium concentrations identical with the overlying seawater. The application of this evidence to Cretaceous chalks is dangerous because so little is known of depths at which undersaturation began during the Late Cretaceous, although some studies suggest that the calcite compensation depth itself may well have been 1–2 km shallower than today (Thierstein and Okada 1979; Roth and Bowdler 1981); this would imply quite shallow depths for undersaturation with respect to low-Mg calcite.

Several lines of evidence suggest much earlier dissolution than in present-day oozes of moderate depth:

1. Fossils with aragonite shells are normally preserved only in special sub-facies of the Chalk, such as nodular chalks and hardground-chalk. It is difficult to conceive of a complete absence of living gastropods and ammonites in the main body of the Chalk sea over shelf regions; in nearer-shore chalks, as in the Upper Campanian of Co. Derry in Ireland, gastropods, scaphopods and ammonites are common.

2. Marls in the Mid-Turonian and Lower Campanian of England have lost many of their species of planktic foraminifera. The marls themselves seem to be a concentration of clay through the early loss of $CaCO_3$ in the top few centimetres of accumulating chalk (Curry 1982).

3. There is some direct evidence of very early dissolution. Eller (1981) has recorded the truncation of clay micro-seams, formed as a result of dissolution by *Chondrites* burrows, but this seems to be rare. Both burrows and solution seams are commonly cut by small accommodation faults that seem to be the result of penecontemporaneous compaction (Hancock 1975*b*, Plate 2d). Similarly, Watts (1983*a*) demonstrated that steeply dipping shear-fractures, with accompanying solution effects, in the Albuskjell Field in the Norwegian North Sea, were developed early and pre-date significant pressure-solution. Ekdale and Bromley (1988) show that although their solution seams cut burrows, both the solution seams and accompanying conjugate sets of micro-fractures are developed in chalks which have never been under significant overburden (Jørgensen 1987).

In the assessment of the timing of solution effects it is necessary to keep in mind that there has also been later solution, sometimes associated with Tertiary tectonism, that may overprint early solution effects.

That near-surface solution is not limited to deep-sea carbonates below the lysocline is shown by the work of Alexandersson (1978): he demonstrated solution of calcite in water depths of 0–40 m in the eastern North Sea at the present day.

State of chalk after intrinsic diagenesis

A combination of the factors mentioned above, with an overburden up to a few hundred metres, leaves chalk in the sort of condition in which one now finds it over most of the onshore outcrop in southern England, northern France, Denmark and north Germany. Almost all of it has been intensively burrowed, though the sediment is often too homogeneous for the burrows to show in the field; sometimes the burrows can be shown up by painting oil on the rock (Bromley 1980; Ekdale and Bromley 1983). The porosities are in the range 35–47 per cent, and matrix permeabilities around 2–6 mD (Scholle 1974; Hancock 1975*b*; Price 1987). At some levels there are early lithified lumps in nodular chalks or continuous hard beds that formed hardgrounds, but these are scarce in open basins.

Early changes in chalk-oozes: non-intrinsic diagenesis

Nodular chalks and hardgrounds

Scattered through the Chalk of Northern Ireland, southern England, France and Belgium, there are nodular chalks and hardgrounds. The best developed, such as the Melbourn Rock and Chalk Rock of England, have long been used as markers to subdivide the lithostratigraphy of the Chalk.

There is every gradation from an omission surface, shown by no more than a line of *Thalassinoides* burrows, through nodular chalks, to coalescence of nodules into a hardground (Fig. 2.4). A slowing down of coccolith accumulation gives sufficient time for bacteria, in the layer 0.2–1 m beneath the sediment–water interface, to form nodules (as discussed earlier). If the supply of coccoliths stops completely, or slows down sufficiently for bottom currents to prevent them settling, a hardground will form which, if exposed long enough on the sea floor, could become glauconitized and/or phosphatized (Voigt 1959; Bromley 1967; 1975; Kennedy and Garrison 1975; Bromley and Gale 1982).

Lithification of the nodules in the Chalk was a very early process because they were avoided by burrowing animals descending from levels some fraction of a metre above. The absence of fossils attached on their surfaces shows that they had formed beneath the ooze–water interface. Unlike the isolated nodules, laterally continuous hardgrounds did form a hard floor to the sea bed once loose sediment had been swept away because fossil animals, such as oysters, that use a hard surface, can be found attached to them; hardgrounds were both contemporaneous and submarine. Scholle and Kennedy (1974) measured the $\delta^{18}O$ content as –1.5 to –2.8‰; these are original values that confirm the very early cementation; such values persist even when the surrounding chalk contains more negative $\delta^{18}O$ values from later diagenesis.

Casual examination of any particular succession can give an impression that the distribution of these harder chalk sub-facies is haphazard. In fact, nearly every example of such condensation or cessation of chalk accumulation can be ascribed to one of three causes: a widespread fall of sea level (Barrois 1877; Hume 1894; Hancock 1990); a local diapiric uplift (Gale 1980); or rejuvenation of uplift along an earlier tectonic axis (Mortimore and Pomerol 1991). Earlier proposals for organically bound carbonate banks in the Pays de Caux (Kennedy and Juignet 1974) fail to stand up to re-examination (Quine and Bosence 1992). Clearly, those that result from eustatic falls of sea level will be geographically widespread. The others may be very local, the evidence for them being limited in some cases to a single quarry. An important factor in hydrogeology is that hardgrounds may be more fractured than ordinary chalk, giving rise to high mass permeabilities (Price *et al.* 1977).

The stronger bottom currents needed to allow hardgrounds to form are probably limited to water depths of less than about 200 m. As a result hardgrounds are absent from chalks formed in deeper waters. They are scarce in the Upper

Campanian and Lower Maastrichtian. They are absent or rare in the Chalk of Humberside (where Early Campanian depths were possibly 600 m). In north Germany, the nearest to a hardground seems to be omission surfaces rich in burrows and sponges known as 'grabganglage'.

Early to late solution diagenesis

There can be some solution of the little grains of calcite shortly after deposition, concentrating the accessory clay. If this occurred without accompanying redeposition of calcite nearby, wispy 'horse-tails' formed. However, if the solution was accompanied by even a little early redeposition of calcite in the neighbouring chalk (and this was common), or if the solution increased the matrix permeability along its path, solution was concentrated along surfaces where solution had already started. The result is a distinctive pattern with low-angle intersection veins of marly clay. This macrotexture has been given a variety of names: 'flasers of marl' (Hancock 1975*b*, Plate 2a); 'solution seams' (Kennedy 1969; Scholle 1977, Fig. 7); 'flaser' (Garrison and Kennedy 1977).

There is a subtle gradation between, on the one hand, almost unaltered chalk in decimetre patches separated and isolated by solution seams, and, on the other, more or less hardened nodular chalk where the solution seams accentuate the nodularity. This more extreme diagenesis grades into forms comparable to the 'knollenkalke' of the Ammonitico Rosso facies in the Tethyan Jurassic (Jenkyns 1974; Scholle *et al.* 1983, Fig. 93) and the 'griotte-texture' in the Devonian of the Montagne Noir (Tucker 1974). However, there is no automatic increase in this sort of solution diagenesis with an increase in load imposed on the chalk. Early forms and patterns of solution can be retained while the chalk itself becomes hardened by burial diagenesis. Although these solution seams can pass into stylolites, the two are normally separate phenomena, even when they occur together.

All harder chalks, for instance in Humberside, the White Limestone of Northern Ireland, in the Central Trough of the North Sea, and the Plänerkalk of Germany, are affected by stylolites. Solution seams, and stylolites possibly to a lesser degree, act as barriers to the vertical movement of liquids. However, the pattern of oil impregnation in North Sea chalk reservoirs shows that the presence of solution seams and stylolites eases and enhances the horizontal flow of oil into the high-angle fractures that cut them.

Facies transitional from white-chalk

Tuffeau

In the Turonian of Touraine, the Maastrichtian of Limburg, and the Upper Campanian–Maastrichtian of southern Sweden, there are transitions from white-chalk to a tuffeau facies. Tuffeau is something of a sack-word since it embraces both the curious cristobalite-rich Tuffeau Blanc of Touraine and the simpler biocalcarenite facies of the Kristianstad region. The latter formed in near-shore current-swept regions where little or no detritus was being eroded from the land (Hancock 1975*a*).

In addition to these obvious examples, there are substantial variations in the proportion of comminuted macro-fossils and larger micro-fossils in what has simply been recorded as 'chalk'. Unfortunately, there is no systematic survey of occurrences of such coarser chalks which will have higher matrix permeabilities than pure white-chalk.

Chalk-marl

The most common transition from white-chalk to another facies, either laterally or vertically, is to an alternation (often rhythmic) of marly clay and chalk. This chalk-marl facies (also known as 'periodite' facies, see Kennedy 1987) was originally described in the Lower Cenomanian Chalk Marl of southern England, but the same facies forms the Fort Hayes Limestone (Coniacian) in Kansas (Hattin 1982), the middle Hod Formation (Turonian?) in the Norwegian sector of the Central Trough (Kennedy 1987); and the Teplice Formation (Turonian) of northern Czechoslovakia (Čech *et al.* 1980).

With increasing thickness and clay content of the marls between the beds of chalk, the chalk-marl facies passes into clays or silty clays (slipper-clay facies). Such transitions occur in the northern part of the Central Trough and the southern part of the Viking Graben as the Chalk Group passes into the Shetland Group.

Although chalk-marl is a transition to a clastic

facies, it implies no more than an increase in the supply of clay (and silt): it may still be a deep-water facies. True deep-sea rhythms of chalk and clay may actually reflect variations in the calcite compensation depth, but shelf-sea examples are likely to have had a climatic control which was related to orbital forcing (Robinson 1986; Gale 1990).

Clearly, any bed of clay will form a near-impermeable barrier to water, but so long as chalk preponderates over clay in the formation, joints and fractures should allow a relatively high mass permeability to be maintained. However, early, steeply inclined, shear-fractures are sealed by clay in the Albuskjell Field in the North Sea (Watts 1983*a*).

Resedimented chalk

It has been increasingly recognized in recent years that many chalks are not simply the result of a gentle rain of pellets to the sea floor. Chalk-ooze is incoherent. The upper few millimetres would be resuspended by the most gentle currents, and even to a depth of 0.08 m by moderate currents. Clouds of coccolithic debris would then form. There is no obvious way of knowing how often this has happened, for once resettled the ooze would be again subjected to the processes described earlier. But with a pre-existing gradient from local uplift or at the borders of a fault-bounded trough, resedimentation would take place on a larger scale, producing new facies of chalk. Most of the studies of these chalks have been in the Central Trough, (see, for example, Hardman 1982; Skovbro 1983; Nygaard *et al.* 1983; Kennedy 1987; Taylor and Lapré 1987), although various forms of resedimented chalks are also widely known onshore (Voigt 1962; 1977; Voigt and Häntzschel 1964; Juignet and Kennedy 1974; Hancock 1975b; Nygaard and Frykman 1981).

A progressive series of resedimented chalks has been drawn up by Kennedy (1987):

1. *Slide-sheets*. These are difficult to detect, apart from the basal shear-zone; they are known in thicknesses up to tens of metres. Tectonic mimics of slide-sheets can be developed where chalk has been subject to glacial deformation, e.g. on the north Norfolk coast, Møn Klint in Denmark, Rügen in Germany.

2. *Slump deposits*. In these there are slump-folds on a scale from a decimetre to tens of metres. Some slumps are horizontally isoclinal. Every gradation from plasticity to brecciation is known.

3. *Debris flows*. Mass flows of chalk differ from their terrigenous clastic equivalents (Nygaard *et al.* 1983): uncemented chalk clasts may be reshaped (phacoids of Voigt 1962) or broken up during transport, producing more homogenous chalks with a shredded fabric which itself may be laminated. Debris flows may contain exotic pebbles in addition to the chalk itself (for example, mélange of Hancock and Scholle 1975). A considerable range in thickness is possible. Some of the best onshore examples are actually to be seen in the semi-chalk facies of the Campanian on the coast between Hendaye and Biarritz in south-west France.

4. *Turbidites*. These have all the features familiar in their clastic equivalents. Those in the Münster basin described by Voigt and Häntzschel (1964) are only centimetres thick, but in the Tor Formation (Maastrichtian) they range up to several metres, and in some successions they can be stacked on top of one another.

5. *Laminated chalks*. These are the most widespread of resedimented chalks, but also the most contentious because some of them turn out to be diagenetic rather than depositional. Centimetre to millimetre laminations have been widely recorded by Kennedy (1987). The fact that they show clear alternations with bioturbated chalk is good evidence of their primary origin. Still finer laminations are sometimes actually cut by burrows, again proving them to be primary. Onshore examples are also widespread, for example, in the Turonian on the Humberside coast. Current-aligned inoceramid fragments are also common. However, there are also many occurrences of micro-laminations that cut across pre-existing burrows of *Planolites* and *Zoophycos*, and these laminations are micro-stylolites (Ekdale and Bromley 1988). In clean white-chalk it may be impossible to distinguish primary from diagenetic laminations.

Resedimented chalks have greater porosities and higher matrix permeabilities than autochthonous chalk (Chapter 13; D'Heur 1984; Taylor and Lapré 1987). All the published measurements have been made on chalks in the Central Trough, where the importance for oil reservoirs is obvious. Even after considerable burial diagenesis (see below), the primary fabric of a resedimented chalk maintains an enhanced porosity (for instance 30 per cent instead of 5 per cent) and matrix permeability (5 mD instead of 0.1 mD) compared with the autochthonous chalk churned by myriads of burrowing animals. Superficial examination of onshore laminated chalks shows that the higher initial matrix permeability has allowed greater interstratal solution.

Later diagenesis

General controls

Most limestones start as carbonate sediments composed of aragonite and high-Mg calcite, with no more than accessory low-Mg calcite. Within days of the breakdown of any protective organic sheaths, the carbonate minerals begin to invert to low-Mg calcite. But chalk-ooze starts as low-Mg calcite and is therefore stable at surface temperatures and pressures. Apart from intrinsic and other early diagenetic changes, described above, Cretaceous chalk may have changed little from the time of deposition to today. A useful measure of the degree of diagenesis is the ^{18}O:^{16}O ratio. As diagenesis increases $\delta^{18}O$ becomes increasingly negative (Scholle 1974; Hancock and Scholle 1975).

Later diagenesis can be caused by three factors: weight of overburden (burial diagenesis); influx of heat (thermal diagenesis); or tectonic stresses (tectonic diagenesis). The results are similar in all three and there are controlling factors in common:

1. *Composition of pore fluids.* Undersaturation with respect to calcium will increase the reactivity. Therefore, freshwater will be more reactive than seawater. Chalk in which the marine-derived pore-water has been flushed out by freshwater suffers a much more rapid loss of porosity with burial than chalks in which marine-derived water has been retained (Scholle *et al.*, 1983; Fig. 94; Mimran 1985).

Following the work of Lippman (1960), it has been recognized that magnesium in solution can inhibit the crystallization of calcite in favour of aragonite. It does not follow, as has generally been believed (see, for example, Neugebauer 1974), that magnesium can equally inhibit the diagenesis of low-Mg calcite.

In the absence of aragonite, strontium can substitute for calcium in low-Mg calcite more readily than is often supposed. In recent chalk-ooze the strontium content is normally less than 1000 ppm (Baker *et al.*, 1982), but some Humberside chalks contain 1520 ppm (Mimran 1978), and some Central Trough chalks more than 3000 ppm (Taylor and Lapré 1987). These increases are accompanied by marked reductions in porosity.

2. *Size of the crystals.* Smaller crystals have a larger surface in relation to their mass than larger crystals, and hence have enhanced surface energies, which means that they are chemically more reactive. Small coccolith laths are more likely to be dissolved or altered than larger ones. The effect is only great with crystals less than 3 μm in diameter. However, coccoliths seem to be less affected than foraminifera, especially planktic foraminifera (Schlanger and Douglas 1974; Curry 1982). This difference can be related to the very small size of the component crystals in foraminiferal tests (Neugebauer 1975).

3. *Pore fluid pressure.* Compaction leading to, or accompanying, mechanical damage and chemical solution is dependent on high pressures being developed at grain-to-grain contacts. If a fluid, oil or water, becomes trapped within the pore spaces, the fluid takes up the pressure and prevents the grains from pressing against one another. Overpressures are well known in the hydrocarbon geology of the central North Sea.

Diagenetic changes

The principal factors or features are:

1. *Spot welding.* This is only visible at magnifications greater than 30 000. It is probably an early stage of the following process (Mapstone 1975).

2. *Corrosion and growth of crystals.* Some coccolithic laths show embayments; others show overgrowths or become rhombohedra (organically grown laths do not form rhombohedra). There is often considerable variation in the degree of corrosion and overgrowth, even within one field of a scanning electron microscope (Hancock and Kennedy 1967; Scholle 1974; Aubry 1976; Scholle *et al.* 1983, Figs. 95, 96; Bonnemaison 1989).

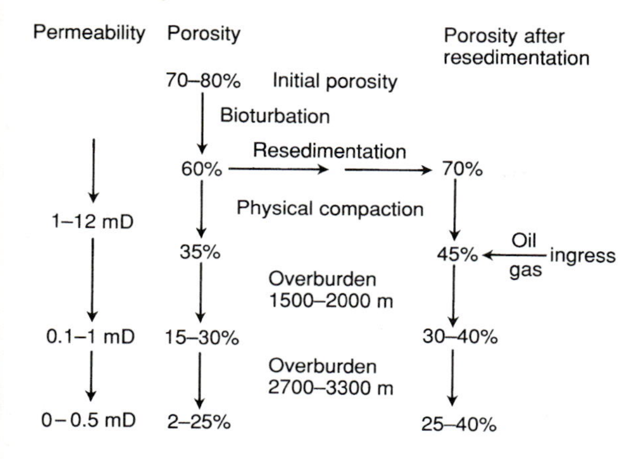

Fig. 2.5. Changes of porosity and permeability in chalks with increasing diagenesis and the advantage of resedimentation.

3. *Sparry infill.* In regions of only moderate diagenesis there may be sparry infill of the chambers of foraminifera or lining the external moulds of aragonitic fossils.

4. *Stylolites.* In this account the term 'stylolites' has been limited to macro-stylolites, visible to the naked eye, which are the 'sutured-seam solution' of Wanless (1979). 'Non-seam solution' (Wanless 1979) produces the micro-stylolites of early diagenesis, with or without horse-tails (Ekdale and Bromley 1988). Macro-stylolites develop only in previously lithified chalk.

The overall effect of these diagenetic changes is to harden the chalk, reducing the porosity and matrix permeability. Permeability is then effectively due to fractures. Since many of these fractures are microscopic, it can be difficult to distinguish matrix from mass permeability (Koestler and Ehrmann 1987).

Causes of later diagenesis

Burial diagenesis

There is still much to discover about the relationships between depth of burial and the degree of diagenesis in chalks. Much the best general account is by Scholle (1977), who makes the important distinction between straight mechanical compaction and solution transfer (= chemical compaction).

In present-day deep-sea oozes and chalks there is no systematic correlation of depth with lithification and decrease in porosity, and complete reversals in hardness can occur with depth (Garrison 1981). Nevertheless, in Cretaceous chalks there is a broad decrease in porosity and an increase of hardness with depth *in a particular region* (see, for example, Chapters 10 and 13); for relatively shallow depths, see Carter and Mallard (1974, Fig. 3) and for depths into thousands of metres, see Scholle (1977, Fig. 5).

It is to be expected that there will be an overlap between mechanical compaction and solution transfer. Any pressure on grain-to-grain contacts should ease solution of the calcite and this should start as soon as there are grain-to-grain contacts, but the amount of pressure solution compared with solution not dependent on grain-to-grain pressure will be slight until there is an appreciable overburden. Neugebauer (1974) argued that pressure solution would be inhibited by magnesium poisoning and that it could not effectively start until an overburden of 1000 m or more had been reached. No such relationship has been found in deep-sea cores, although Beall and Fischer (1969) did find that solution transfer only began below a depth of burial of 250 m in an area east of the Bahamas. The experimental work of Neilson (1990) suggests that magnesium is a weaker inhibitor of calcite nucleation than formerly supposed.

Typically, early diagenesis and an overburden of 250 m reduced the porosity to 35–50 per cent, with a matrix permeability around 1–12 mD, depending in part on how the measurements are made. An increase of overburden to 1000 m gives only a slight further reduction of porosity to around 30–40 per cent. With overburdens of 1500–2000 m the porosity falls to 15–30 per cent and the matrix permeability to 0.1–1 mD; overburdens of 2700–3000 m reduce the porosity to 2–25 per cent and the matrix permeability to 0–0.5 mD (Fig. 2.5).

Thermal diagenesis

The Chalk in Northern Ireland is very much harder than any other onshore chalk. The $\delta\,^{18}O$ of this White Limestone is in the range of –3.3 to –8.14‰, which corresponds to recrystallization attained after burial to depths of 1400–7300 m. Even the average $\delta^{18}O$ of –5.6‰ corresponds to more than 4000 m of overburden, whereas there is no evidence that the basalts overlying the Chalk were ever more than 1800 m thick. The obvious explanation is the anomalously high heat-flow that existed in the region in Tertiary times (Hancock 1975*b*).

Similarly high thermal diagenesis has occurred in the Albuskjell Field of the North Sea (Watts 1983a). The general hardness of the Humberside chalk is not so easily explained. The degree of diagenesis seen in scanning electron micrographs from samples on the coast corresponds to an overburden of only about 2000 m but Apatite Fission Track Analysis has shown that elevated palaeotemperatures occurred in Late Cretaceous to Tertiary times over a wide area including north-eastern England and the Southern North Sea. These high temperatures, due to the depths of burial, could be the cause of the hardness of the Humberside chalk (Green *et al.* 1993).

Tectonic diagenesis

In Dorset the Chalk seems to have been hardened by local tectonic stresses. As the dip increases from 0° to 90° the bulk density increases from 2.0 to 2.6 gcm^{-3} and calcispheres change from being circular to elliptical. Mimran (1975) estimated that 50 per cent of the chalk in this area has been lost by solution. However, mere tilting is not sufficient, for the effect is not seen to anything like the same degree in the Isle of Wight, where the chalk is vertical. It is now known that there is more Tertiary macro-faulting and shearing in south Dorset (Jones *et al.* 1984; Ameen 1990).

3. The Chalk as an aquifer

M. Price, R. A. Downing, and W. M. Edmunds

Introduction

Some aquifers and petroleum reservoirs consist of porous rock which is traversed by fissures. It is common for the rock mass to possess significant porosity but low to moderate permeability, and for the fissures to contribute little to the total porosity of the rock but to contribute most of the permeability.

The Chalk is an example of these aquifers or reservoirs, which are said to possess *dual porosity* or *double porosity*. The porosity and permeability components contributed by the fissures are referred to as the *fissure* (or *fracture*) porosity and permeability. The blocks bounded by the fissures are usually described as *matrix blocks*, and the non-fissured fraction of the porosity and permeability as the *matrix* (or *matric*) *components* (Fig. 3.1a).

Other aquifers and reservoirs exhibit heterogeneity, with permeable intervals alternating with layers of lower permeability. The permeable intervals may have higher primary permeability than the others, or they may be fissures or beds displaying extensive fissuring. These formations are said to exhibit *dual permeability* or *double permeability* (Fig. 3.1b).

The concept of dual-porosity reservoirs appears to have been first developed by Barenblatt *et al.* (1960) and Warren and Root (1963), but the dual-porosity nature of the Chalk aquifer had been recognized informally before this (see, for example, Ineson 1959). In a classic dual-porosity aquifer the matrix pores provide storage and the fissures provide the permeable pathways for flow. When a well is pumped or allowed to flow, fluid flows through the fissures to the well; the resulting reduction in head in the fissures causes fluid to drain into them from the adjacent matrix blocks. Conversely, in a dual-permeability reservoir, flow occurs towards the well in both the highly permeable and the less permeable layers.

The depositional, diagenetic, and tectonic history of the Chalk have created an unusual dual-porosity reservoir rock. The shape and composition of the coccolith particles that make up much of the matrix gave rise to high depositional porosities that, even after diagenesis, remain typically in the range 20–45 per cent (Fig. 3.2). The small size of the particles, however, means that the pores, and especially the interconnecting pore-throats, are also small—the effective pore-throat diameter being typically in the range 0.1–1 μm (Fig. 3.3). The pore-throat sizes are generally very uniform (Price *et al.* 1976; Ballif 1978). The matrix permeability is correspondingly low, typically in the range 0.1–10 mD; for relatively shallow groundwaters this corresponds to a hydraulic conductivity range of around 10^{-9}–10^{-7} m s^{-1} (10^{-4}–10^{-2} metres per day).

The Chalk functions as an aquifer and a hydrocarbon reservoir because it is fissured. The origin of the basic fissure system is tectonic, though in the North Sea some very fine fractures are apparently associated with stylolites. Most workers on the European Chalk usually refer to three sets of fissures, one set being parallel to the bedding, the other two more or less perpendicular to it and often more or less perpendicular to each other. In addition, high-angle joints may be present.

Where the Chalk is deeply buried, as in North Sea oilfields, fractures may increase the permeability of the rock mass to a value of the order of 100 mD (around 10^{-6} m s^{-1}). At lesser depths onshore, a combination of factors such as dissolution, stress release as overburden is removed, and weathering, can lead to significant enlargement of these initial fractures, either individually or in layers or zones. It is this enlargement that is responsible for the majority of the high transmissivities that characterize much of the Chalk aquifer at outcrop or beneath thin overburden in North-west Europe.

To distinguish between the relatively tight tectonic fractures and the enlarged passageways, some workers (for example, Reeves 1979) have referred to them as 'micro-fissures' and 'macro-fissures'. In

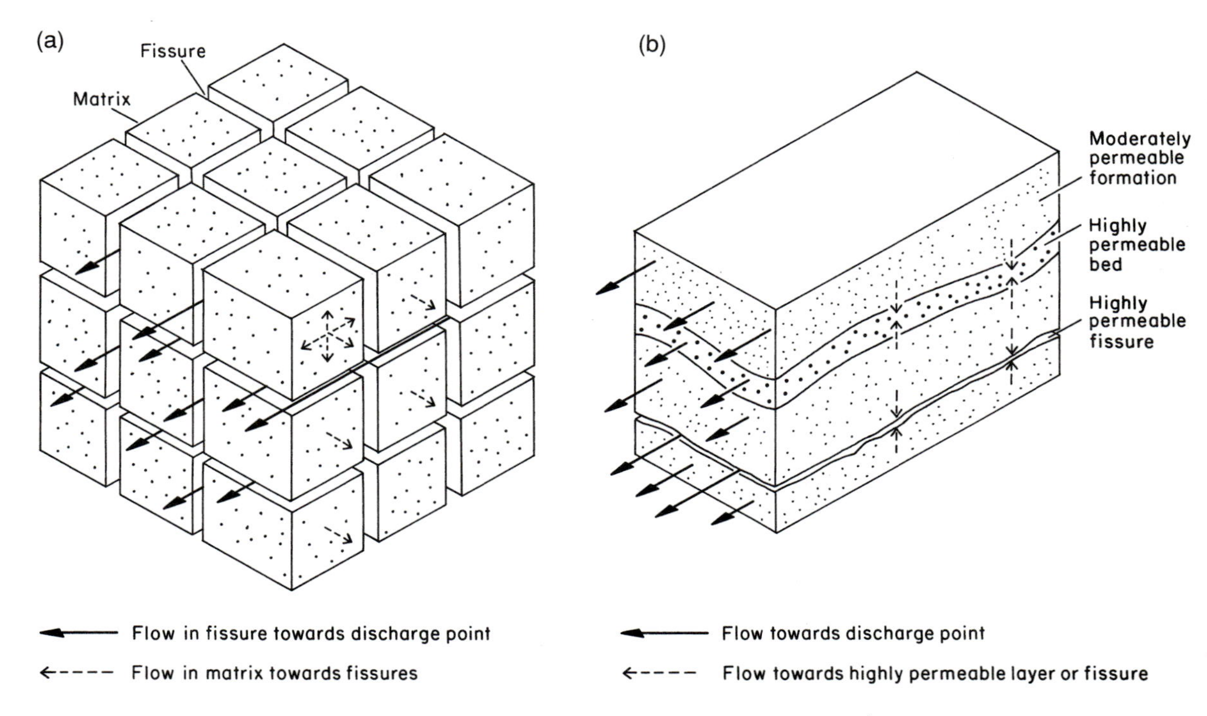

Fig. 3.1. (a) An idealized dual-porosity aquifer. (b) An idealized dual-permeability aquifer.

Fig. 3.2. Scanning electron photomicrographs: (a) Upper Chalk from Berkshire, England, Porosity 46 per cent, permeability 6.5 mD (hydraulic conductivity 5×10^{-8} m s^{-1}). The field of view is about 40 μm. Note the abundant well-preserved coccoliths, which are about 5 μm across, and the smaller fragments of broken coccoliths.
(b) Campanian Chalk from Humberside, England, Porosity 30 per cent, permeability 1 mD (hydraulic conductivity 7.5×10^{-9} m s^{-1}). The field of view is about 15 μm. This chalk appears to have undergone more alteration, and little original coccolith structure remains. (Photographs by courtesy of the British Geological Survey.)

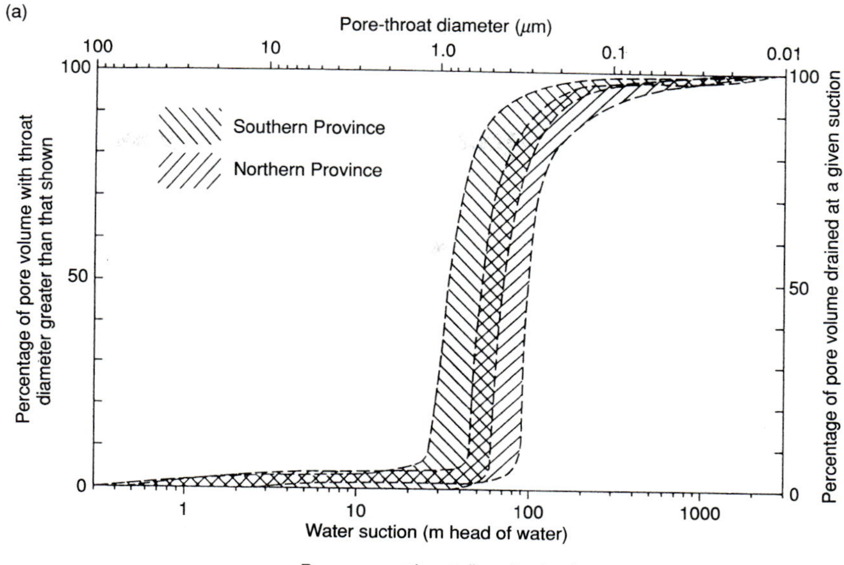

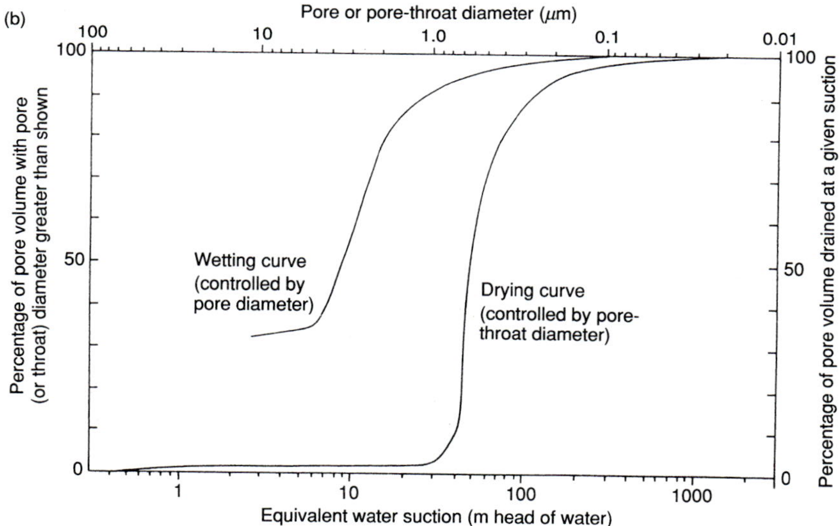

Fig. 3.3. (a)Mercury-intrusion capillary-pressure curves from the Upper Chalk of England. The envelope for the Southern Province is based on determinations on 21 samples from six localities and the envelope for the Northern Province on determinations on seven samples from seven localities (after Price *et al.*, 1976). (b) Mercury-intrusion and mercury-expulsion capillary-pressure curves for a sample of Middle Chalk from Fleam Dyke, Cambridgeshire, England, showing hysteresis. The sample has a porosity of 42 per cent.

an effort to distinguish more clearly their origin. Price (1987) referred to the two types as *primary fissures* and *secondary fissures*. The two classifications are not comparable; Reeves' microfractures actually refer to inferred, very fine, closely spaced cracks in the matrix.

Whichever nomenclature is adopted, it should be understood that in reality there is a more or less continuous range from tight fissures to greatly enlarged, even karstic openings. In at least some localities, however, there appears to be a preponderance of primary fissures; well-developed and highly permeable secondary fissures are concentrated at only a few discrete horizons. Figure 3.4 shows the relationship between fissure spacing, fissure aperture, porosity and hydraulic conductivity for a rock mass traversed by three sets of orthogonal discontinuities with smooth-sided plane parallel openings, assuming the rock to be saturated with pure water at 10°C. It must be understood that at any given locality there is likely to be a range of fissure spacings and fissure apertures, and that the apertures of individual fissures will vary from place to place. It must also be realized that the porosity

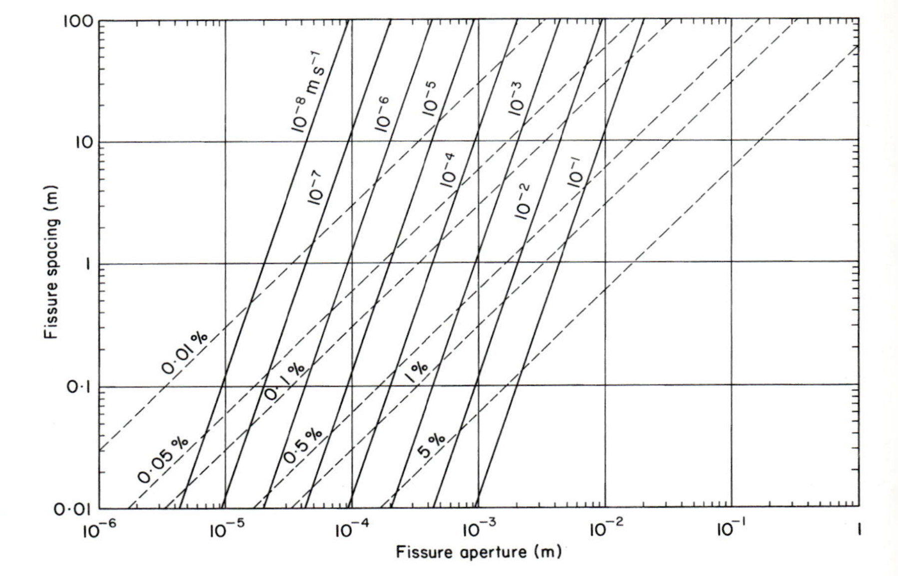

Fig. 3.4. The relationship between fissure spacing, fissure aperture, porosity and hydraulic conductivity (m s^{-1}) for a fissure system containing three plane, parallel, mutually perpendicular smooth-walled fissures filled with pure water at 10°C.

and hydraulic conductivity relationships will be significantly altered by wall roughness; in reality, fissure apertures, and therefore porosities, will be larger than those shown in Fig. 3.4 to compensate for the roughness and irregularity of fissure surfaces.

The presence of these two types of fissure gives the Chalk an additional dual-porosity behaviour. Rapid flow through the secondary fissures to wells can be followed by drainage of the primary fissures. In combination with the matrix porosity, the Chalk therefore frequently exhibits a double-dual-porosity behaviour, but variation in fissure sizes also means that it may exhibit dual-permeability behaviour.

The unsaturated zone

The Chalk is generally a permeable formation. Consequently, it displays low or moderate hydraulic gradients. In areas where the topography is undulating, this combination of a relatively flat water table and a varying ground surface leads to an unsaturated zone which varies in thickness from around zero in river valleys to more than a hundred metres in some tracts of hill country.

The unsaturated zone is defined as that part of the aquifer that lies above the deepest water table. It is thus a zone in which pore-water pressures are less than atmospheric pressure except locally where perched aquifers exist. It is important that this definition, based on pore-water pressures, is understood and accepted, because it will become clear from what follows that much of the 'unsaturated zone' of the Chalk aquifer is close to saturation for much of the time. It is also important to recognize that the unsaturated zone includes the soil.

The unsaturated zone is important for two principal reasons: first, the nature and state of the unsaturated zone are the major controls on whether or not infiltration will reach the water table and thus recharge the aquifer; second, the fate and behaviour of many potential groundwater pollutants are inextricably linked to the nature of and processes within the unsaturated zone. Moreover, monitoring of the unsaturated zone is an essential activity in identifying pollutants that have entered the ground and may be migrating to the saturated zone.

Unsaturated flow: General principles

In the unsaturated zone, flow is governed by Darcy's law, occurring in response to a hydraulic gradient, or change in hydraulic head, as it is in the saturated zone. Head can be defined as the mechanical energy of the fluid per unit weight, and, as in the saturated zone, can be thought of as having three

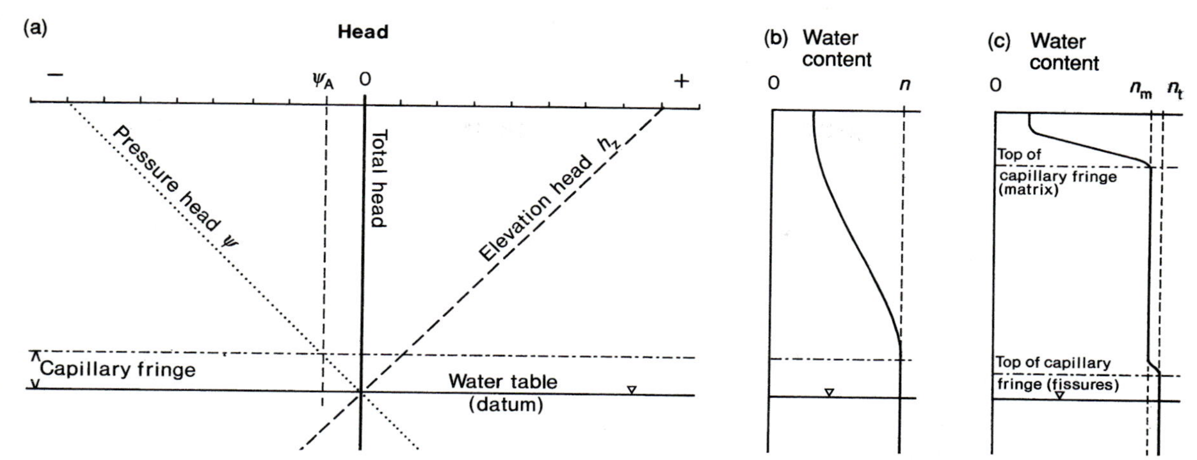

Fig. 3.5. Theoretical relationships between head, depth and water content in the unsaturated zone of an aquifer after drainage to equilibrium: (a) head relationships; (b) water content in a uniform granular aquifer; (c) water content in a dual-porosity aquifer.
n = porosity of granular aquifer
n_t = total porosity of dual-porosity aquifer
n_m = matrix porosity of dual-porosity aquifer

components, arising from elevation (elevation head, h_z), movement (velocity head, h_v), and pressure (pressure head, ψ). In the unsaturated zone, even more than below the water table, movement of subsurface water is generally so slow that the velocity head is negligible. Thus the total head, h, is in practice the sum of the elevation and pressure heads and can be written as

$$h = h_z + \psi \quad (3.1)$$

A good understanding of the behaviour of the unsaturated zone of the Chalk was provided by Wellings (1984), following the more general work by Wellings and Bell (1982). Much of the following discussion is derived from their work, but taking the water table as the datum rather than the ground surface. Use of the water table as the datum has the practical disadvantage that it is not fixed but rises and falls. However, it has the conceptual advantage that when the head in the unsaturated zone is positive, water will move down *towards* the water table; when it is negative, water will move up *from* the water table. The essential concepts are illustrated in Fig. 3.5.

Figure 3.5a shows a uniform *granular* aquifer which has undergone sufficient recharge to saturate the entire profile, and then been allowed to drain to equilibrium without further recharge. In this condition there is no vertical movement in the profile, so that the total head is constant throughout the profile and is everywhere the same as the head at the water table. Because the water table has been chosen as the datum, the total head is everywhere equal to zero. Below the water table, the pressure head, ψ, increases linearly with depth, and the elevation head decreases linearly, the sum of the two being constant and equal to zero. Above the water table h_z increases linearly and ψ decreases, that is, there is a negative pore pressure or a pore-water tension; the total head is again zero. At the water table, by definition, $\psi = 0$ (i.e. the pore pressure is equal to atmospheric pressure), $h_z = 0$, and the total head is zero.

The water content in the profile, expressed as a fraction of bulk volume, is shown in Fig. 3.5b. Below the water table the pore space is totally saturated, hence the water content, θ, is equal to the porosity, n. This situation persists also in the capillary fringe which is described as the 'tension-saturated zone' by some workers (such as Freeze and Cherry 1979), because the pore space is filled with water but the water is at less than atmospheric pressure. Above the top of the capillary fringe, the pore-water content reduces in accordance with the

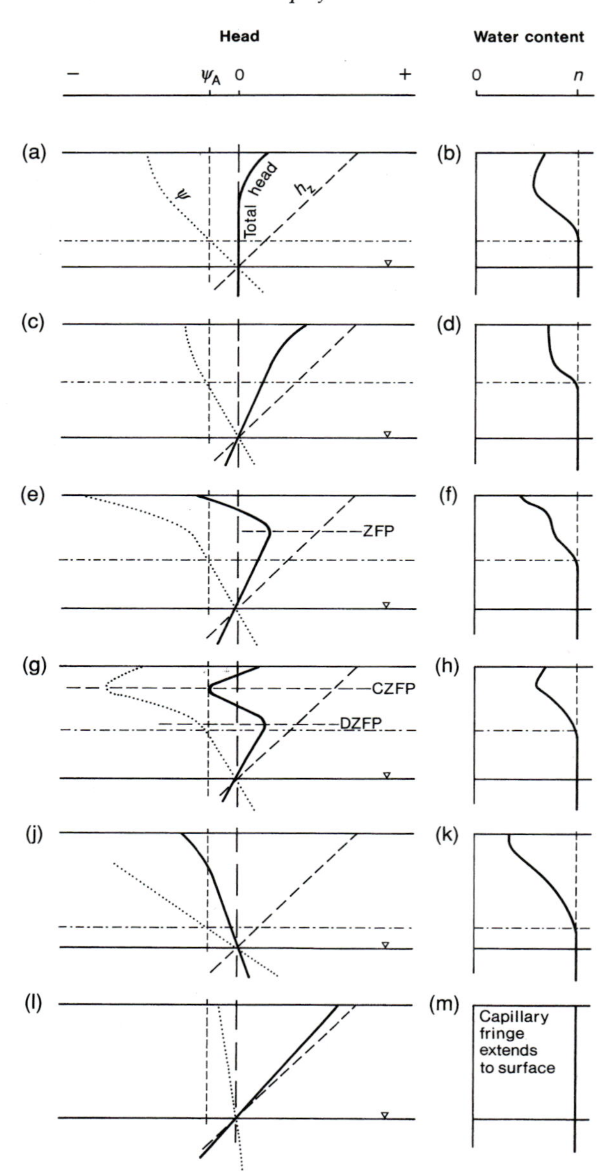

Fig. 3. 6. Theoretical relationships between head, depth and water content in the unsaturated zone of a granular aquifer for various conditions. For full explanation see text.

nature of the pore-pressure/moisture-content characteristic for the rock or soil.

The capillary fringe exists because the pore space does not begin to drain until the pore pressure falls below a limiting value. This pressure is known as the 'air-entry pressure' or the 'bubbling pressure', and the corresponding pressure head is ψ_A. The level in the aquifer where the pressure head falls below ψ_A defines the top of the capillary fringe. The air-entry pressure is lower (that is, more negative) for fine-grained materials than for coarse-grained materials, so that fine-grained materials remain fully saturated for a greater height above the water table, that is, they have a thicker capillary fringe.

This concept is of special interest when dealing with the Chalk aquifer, because of its dual-porosity nature. The small size of the pore throats that connect the pore spaces in the chalk matrix (Fig. 3.3) has been demonstrated by, for example, mercury porosimetry (Price *et al.* 1976; Price 1987), and it is the size of the throats, rather than of the pores, that determines the air-entry pressure. Pore-throat diameters of less than 1 μm correspond to values of ψ_A of more than 30 m, so that chalk aquifers have a very thick capillary fringe. The fissures in the unsaturated zone of the Chalk can be assumed to have apertures that typically are more than 50 μm. The value of ψ_A corresponding to a parallel-plate fissure with an opening of 50 μm is about 0.6 m. Therefore, although at equilibrium the pore spaces of the chalk matrix can be expected to be saturated to a height of about 30 m above the water table, the fissures will typically be saturated to a height of little more than 0.5 m above the water table. This leads to the water content in the profile looking something like that shown in Fig. 3.5c.

Such an equilibrium condition would never be attained in practice, because recharge and evapotranspiration would continually affect both the head profile and the water-content profile. In a homogeneous granular aquifer, the processes of recharge and evapotranspiration will cause the effects illustrated in Fig. 3.6.

The effect of recharge will be to create a downward gradient in the unsaturated zone (Fig. 3.6a), and to increase the water content (Fig. 3.6b). When recharge reaches the top of the capillary fringe (Fig. 3.6d), this downward gradient will persist through the saturated zone (Fig. 3.6c). When the rate of removal of water by evapotranspiration from the soil exceeds the rate of replenishment by rainfall, both the pressure head and the water content in the soil zone will be reduced and water will begin to move upwards (Fig. 3.6e, f).

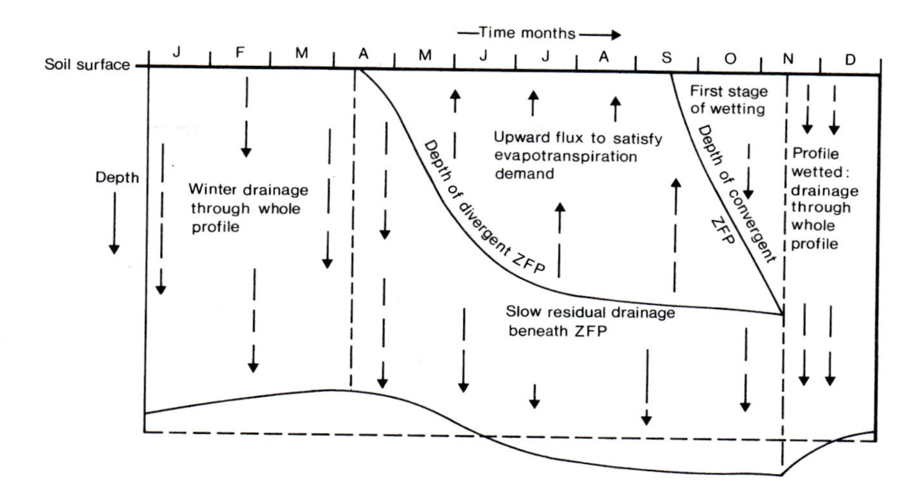

Fig. 3.7. The annual cycle of movement of the ZFP in a temperate climate (after Wellings and Bell 1982).

The surface labelled ZFP in Fig. 3.6e is known as the *zero-flux plane*. This is a maximum in the head profile; above it water is moving upwards, and below it water is moving downwards. As its name implies, no flow occurs across the ZFP itself. As flow is occurring away from the ZFP, it is a *divergent* zero-flux plane. A divergent ZFP forms at the soil surface early in the summer or growing season, and moves downwards (Wellings and Bell 1982) rapidly at first but then more slowly as evapotranspiration declines in response to the reduced upward flux of water as the moisture content, and hence the hydraulic conductivity, are lowered.

At the beginning of the next winter, or recharge season, water will again begin to move down through the soil and the unsaturated zone, eliminating the soil-moisture deficit. This will lead to the creation of a second, *convergent* zero-flux plane, above the divergent ZFP (Fig. 3.6g, h). This convergent ZFP will move down and overtake the divergent ZFP, so that by the end of the recharge season the situation will again be that of Fig. 3.6c. The entire process during the course of one year is illustrated in Fig. 3.7.

If the water table is shallow, or recharge is delayed, then the divergent ZFP may move down as far as the water table (Fig. 3.6j). In this situation water will be drawn upwards from the saturated zone, but because the water content in the unsaturated zone will be depleted (Fig. 3.6k) the hydraulic conductivity in the unsaturated zone will be reduced below the saturated value and the rates of upward flow are likely to be small.

Unsaturated flow: Chalk

There has been much discussion as to whether recharge in the Chalk takes place through the matrix or via the ubiquitous fissure system. The answer depends on the relative magnitude of the infiltration rate and the hydraulic conductivity of the material forming the unsaturated zone.

The average downward vertical hydraulic gradient through the unsaturated zone must be slightly less than unity. As Fig. 3.6l shows, for the gradient of total head in the unsaturated zone to be unity, total head would have to be equal to elevation head, so that pressure head would be zero (pore-water pressure would be equal to atmospheric pressure throughout the unsaturated zone). However, the water table is by definition the level at which pore-water pressure is equal to atmospheric pressure, so that if the pore-water pressure were to rise to zero throughout the unsaturated zone, the water table would rise to the ground surface (Fig. 3.6m).

Figure 3.8 illustrates the condition that will occur in a dual-porosity aquifer such as the Chalk as the downward gradient approaches unity. (In what follows, it is assumed that all the fissures have the same aperture.) The pore-water pressure is greater than the air-entry pressure for the chalk matrix, so

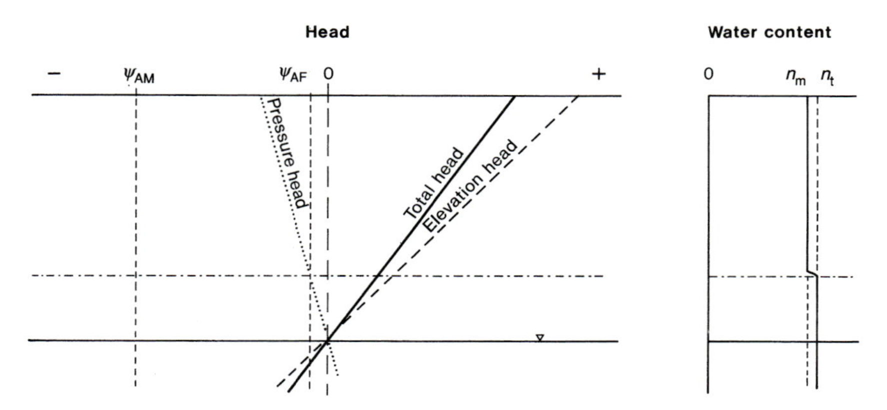

Fig. 3.8. Recharge in a dual-porosity aquifer. The maximum downward hydraulic gradient that can occur in an aquifer without surface ponding is unity. As the downward gradient approaches unity, the total head line approaches the elevation head line and the pore-water pressure becomes close to zero gauge pressure, that is, it approaches atmospheric pressure throughout the profile. Because the pore-water pressure is greater than the air-entry pressure for the matrix, the matrix pore space is water-filled to the ground surface. If the rate of water supply at the ground surface is greater than can be conducted by the fully saturated matrix, the pressure-head gradient will steepen further so that the pore-water pressure is greater than the air-entry pressure for the fissures throughout the profile; at this point the fissures will become saturated and will conduct water, and the rate of downward movement through the profile will be numerically equal to the total saturated hydraulic conductivity of the dual-porosity medium. If the rate of recharge is greater than this hydraulic conductivity, surface ponding will occur.

that the matrix pores are saturated throughout the profile (Fig. 3.8); as far as the matrix is concerned, therefore, the capillary fringe extends to ground level. However, throughout most of the profile the pore-water pressure is still less than the air-entry pressure for the fissures, so that the fissures are drained except just above the water table.

Therefore, the hydraulic conductivity of the unsaturated zone of the chalk in this example will be the same as the saturated hydraulic conductivity of the matrix, because the matrix pores are filled with water and the fissures (except those immediately above the water table) are empty. It follows that the maximum rate at which water can move down through the unsaturated zone of the Chalk, without fissure-flow being involved, is given by $K_m.I_v$ where K_m is the saturated hydraulic conductivity of the matrix, and I_v, the vertical hydraulic gradient, is slightly less than unity. Numerically, the maximum infiltration rate that can occur without fissure flow is therefore approximately equal to the matrix conductivity.

If the hydraulic conductivity of the matrix is less than the average rate of infiltration, then the hydraulic gradient will steepen. If the hydraulic gradient exceeds unity, the pore-water pressure throughout the profile will be greater than atmospheric pressure, the water table will rise to ground surface, and ponding and overland flow will occur. In a fissured aquifer like the Chalk, however, before this happens, the pore-water pressure will exceed the air-entry pressure for the fissures, which will therefore fill with water at less than atmospheric pressure. In effect, the capillary fringe in the fissures will rise to the ground surface. The fissures will, therefore, be available to conduct water so that the hydraulic conductivity of the unsaturated zone will increase markedly, and the unsaturated zone will conduct water downwards at a rate which is unlikely to be exceeded by the rate of infiltration. At pressure-head values lower than about −0.5 m of water, the fissure system will not conduct water (Cooper *et al.* 1990) and the conductivity will be that of the chalk matrix. At higher pressure heads, flow will occur through both the matrix and fissures. For much of the Chalk the saturated hydraulic conductivity of the matrix,

K_m is around 3–5 × 10^{-3} m d^{-1}, so when infiltration exceeds a threshold value of 3–5 mm d^{-1}, maintained over several days, the fissure system will fill and conduct water.

The preceding discussion assumes that all the fissures are of the same size, with the same air–entry pressure. In reality, there is a range of fissure apertures. This means that narrow fissures will become water-filled and water-conducting before wider ones. At times of exceptionally high recharge rates, therefore, the hydraulic conductivity will increase progressively from a value typical of the saturated matrix to a much higher value typical of the water-filled fissure system.

Implications of unsaturated zone behaviour

Water resources

As far as water resources are concerned, the actual mechanism—flow through fissures or flow through the matrix—by which water passes through the unsaturated zone and reaches the water table is somewhat academic. The behaviour of the unsaturated zone does, however, have implications for water resources. Firstly, the pore-size distribution of the chalk matrix is such that, although very little water will drain from it, a high proportion of the water content can be removed by the roots of plants. This means that large soil-moisture deficits can build up in chalk areas in dry summers and reduce much of the effectiveness of the following winter rains. Because most of the matrix pore space remains saturated over a wide range of suctions, the hydraulic conductivity of the matrix below the root zone is almost as high during periods when water moves upward as when it moves downward. This increases the degree to which the moisture content can be depleted during prolonged dry weather.

A second feature is that the fissures represent a one-way flow system; they can conduct water into the aquifer at times of high infiltration rates, but will always be empty, because of the high pore-water suctions (Fig. 3.6j) when water is being drawn up from deep in the unsaturated zone or from the water table. Were it not that this mechanism provides some limit to the degree to which the unsaturated zone can be depleted, even larger amounts of water might be lost to evapotranspiration.

Solute movement

Before about 1970, it was generally assumed that almost all infiltration to the Chalk took place through fissures. This implied rapid flow, with solutes travelling from the ground surface to the water table within days or weeks.

Two phenomena were seen as confirming this flow mechanism. One was the often-noted relatively rapid response of the water table to the onset of winter rain. It was assumed that the time between water entering the soil and the rise of the water table corresponded to the flow time. The second piece of evidence was that, after heavy rain, the water from some Chalk wells often became cloudy and contained soil bacteria or suspended solids, presumed to have been washed from the soil through fissures to the water table.

The availability of tritium analyses in the late 1960s began to cast doubt on the theory that all the infiltration was carried rapidly through the unsaturated zone. The tritium content of baseflow from the Chalk was generally not more than about 10 tritium units (TU), implying that only a small part of the rain that had fallen since the onset of thermonuclear weapon testing had reached the saturated zone. (A tritium unit represents a concentration of 1 tritium atom in 10^{18} hydrogen atoms.)

In 1970 Smith *et al.* produced a profile of the tritium content from the unsaturated zone of the Chalk. This showed a tritium peak at a depth that corresponded to a rate of downward movement of about 1 m a^{-1}. This is the rate that would occur if movement took place through the intergranular pore space, with water in the pore space being displaced downwards as more water was added from above. Such a 'piston-displacement' mechanism would explain the rapid rise in the water table after rainfall, but the water arriving at the water table would have been displaced from the bottom of the unsaturated zone instead of travelling quickly through it. From a study of the mass balance, Smith *et al.* (1970) concluded that some 85 per cent of the thermonuclear tritium that had entered the profile had either decayed or was still there; they concluded that about 15 per cent of recharge to the Chalk took place through fissures.

These results were met with a mixture of scepticism and consternation. Foster and Crease (1974) pointed out that if the bulk of thermonuclear tritium was still in the unsaturated zone, then other solutes and pollutants introduced at about the same time would also still be there. These included the bulk of the nitrate leached from arable farmland in the 1960s and early 1970s. Nitrate levels in chalk groundwaters were already rising in 1974; if this was in response to pre-1960 levels of fertilizer application, what would happen when the bulk of this nitrate from the unsaturated zone reached the water table?

Scepticism was based largely on what was seen as the failure of the 'piston flow' theory to explain observed phenomena such as the bacterial contamination after heavy rain. It was clear from this, and from field examination of fissures that showed the presence of clay presumably deposited by water flowing through them, that some recharge occurred through fissures; however, although it might be tempting to assume that almost all recharge travelled by that route, the evidence from the tritium studies was difficult to refute.

In 1975, Foster proposed a mechanism to reconcile fissure flow with the slow downward movement of the tritium peak through the unsaturated zone. He suggested that as recharge containing tritium moved down through fissures, a concentration gradient would exist between the fissure water and the relatively immobile water in the matrix pore space of the blocks adjacent to the fissures. Under the influence of this gradient, tritium would move down by continuous exchange between the matrix and the fissures through diffusion. This view was supported by Oakes (1977), who likened the unsaturated zone of the Chalk to a chromatographic column. Extensive field studies involving dry coring of chalk and the analysis of pore waters have indicated that, as with tritium, the movement of nitrate and other solutes is retarded; this has also been linked to diffusion into the matrix of the Chalk (Young *et al.* 1976a, b; Foster *et al.* 1986).

A sequence of chemical analyses of groundwater that included tritium, from several wells in the South Downs of England, provided evidence of rapid movement of water to the water table through fissures particularly after heavy rain (Downing *et al.* 1978). This study showed that after such infiltration events the tritium concentration in water reaching the saturated zone could be greater than in the rainfall, implying that water containing tritium, and presumably other solutes, was displaced from the matrix of the unsaturated zone into the fissures.

In the 1980s, evidence from physical measurements in the unsaturated zone increasingly indicated that at some sites only a small percentage of recharge did occur through the fissures (Wellings 1984). These sites were invariably on Chalk where the permeability of the matrix was relatively high. In these circumstances it seems that the hydraulic conductivity of the Chalk's matrix is able to cope with winter drainage rates. However, where the matrix permeability is low the evidence indicates that fissure flow is a common occurrence in winter, as demonstrated by Cooper *et al.* (1990) at a site in Cambridgeshire in eastern England. Cooper *et al.* (1990) also suggested that the thickness of soil, drift and weathered chalk overlying the undisturbed chalk is an important control on the likelihood of initiation of fissure flow. The much greater spread of pore size in these materials provides a buffer for storage of rainwater, allowing it to be released into the undisturbed chalk more slowly and hence reducing the frequency of occurrence of fissure flow. It is worth noting that the average rainfall intensity over most of the chalk areas of north-west Europe during winter is in the region of 2–3 mm d^{-1}; this value is slightly less than the hydraulic conductivity of the matrix of much of the Chalk. Flow in fissures probably occurs when the infiltration, averaged over a period of several days, exceeds 4 mm d^{-1} (Downing *et al.* 1978).

At the Cambridgeshire site, the stable isotopes ^{2}H and ^{18}O were also measured in pore waters from cores drilled in the unsaturated zone and compared with the isotopic composition of drainage from a lysimeter (Darling and Bath 1988). The isotopic composition of drainage from the lysimeter was not identical to that of pore waters from the matrix at the same depth. The pore waters from the matrix are likely to be representative of micropores in a continuum of matrix pore sizes (and drainage routes) while the lysimeter-sampled water is moving in larger pores and small fissures. The analyses also indicated that, at least at this site, isotopic equilibrium is not attained between the faster-moving water, represented by the lysimeter sample, and water in the small pores of the matrix.

The saturated zone

A characteristic feature of the saturated zone of the Chalk is the association of high permeability with

Table 3.1 General aquifer properties of the Chalk

Porosity of the matrix	0.3–0.4
Porosity of the fractures	10^{-4}
Permeability of the matrix (m s^{-1})	10^{-9}–10^{-8}
Permeability of fissured chalk (m s^{-1})	10^{-5}–10^{-3}

valleys, where the transmissivity can be of the order of $3 \times 10^{-2} m^2 s^{-1}$. Values reduce markedly away from the valleys to about $2 \times 10^{-4} m^2 s^{-1}$ beneath the higher parts of the interfluves. This relationship is very evident where the aquifer is unconfined but it also tends to be true for some major valleys where it is confined (Ineson 1962).

A second important feature is that it is generally the upper 50–60 m of the saturated Chalk that is the principal, or effective, aquifer. This has become evident from statistical studies of the variation of specific capacity with depth and from flow logging in boreholes (for example, Water Resources Board 1972; Owen and Robinson 1978). The permeability decreases with depth as the fissure density and fissure aperture reduce.

In the saturated zone the hydraulic conductivity of the matrix makes a negligible contribution to the transmissivity of the aquifer. The permeability of the unfractured matrix is only some $10^{-8} m s^{-1}$ (Table 3.1). Furthermore, the primary fissure component, although important, cannot account for the high transmissivity values that make the Chalk such a productive aquifer (Price 1987). Such high transmissivity is due to enlargement, by solution, of the primary fissure component (Foster and Milton 1974; Price *et al.* 1977; 1982). There is also evidence that the solution (or secondary) fissures are concentrated at the water table or within the zone of its seasonal fluctuation, as it exists now or existed in the past. The thickness of the 'effective aquifer' is related to the base level of erosion and, particularly near coastlines and along the lower reaches of major rivers, to Pleistocene base levels. For this reason the effective thickness near the coast can be of the order of 100 m (Monkhouse and Fleet 1975).

The development of secondary fissures is due to solution by water containing carbon dioxide. Although favourable conditions for carbonate solution exist especially at the top of the unsaturated zone, water containing high pCO_2 may reach the saturated zone, particularly where, and at times of the year when, rapid infiltration occurs through the unsaturated zone. Furthermore, groundwaters saturated with calcite but having different concentrations of HCO_3 do mix, and carbon dioxide may be released which is then available to dissolve more carbonate (Bogli 1964). Different solutions following different paths are constantly mixing in the Chalk's fissure systems but particularly near outlets from the aquifer along valleys. Solution of carbonates as a result of mixing of different solutions is believed to be important in the formation of karst.

Several factors are probably responsible for the high permeability along valleys. First, many valleys follow lines of structural weakness where primary fissures tend to be well developed. Second, erosion along valleys leads to a reduction in effective stress and consequently favours the opening up of near-horizontal incipient discontinuities. However, groundwater flow towards valleys is probably the most significant factor in the development of the permeability pattern. Price (1987) has applied to the Chalk the classic explanation by Rhoades and Sinacori (1941) of the mechanism whereby flow through an initially homogeneous isotropic carbonate leads to the enhancement of permeability near the area of groundwater discharge (Fig. 3.9). Flow-lines become more closely spaced at shallow depths and towards outlets from the aquifer, with the same volume rate of flow occurring through successively smaller areas of rock—producing a corresponding increase in velocity—as the discharge area is approached. The concentration of flow towards the river outlets leads to more rapid solution in the valleys and development of a single major fissure or fissure zone. Once this has formed, flow tends to be concentrated along this horizon leading to further growth away from the valley (Fig. 3.9).

When a secondary fissure or zone has developed, the water table tends to stay within or near it. There is evidence that such a zone controls the water table in the Candover Valley in Hampshire, England (Headworth *et al.* 1982). In many areas it is appropriate to accept that the water table lies within a zone of enhanced permeability rather than that a zone of enhanced permeability develops in the zone of water–table fluctuation (Morel 1980; Price 1987).

This explanation for the preferred alignment of high permeability with valleys can also be applied

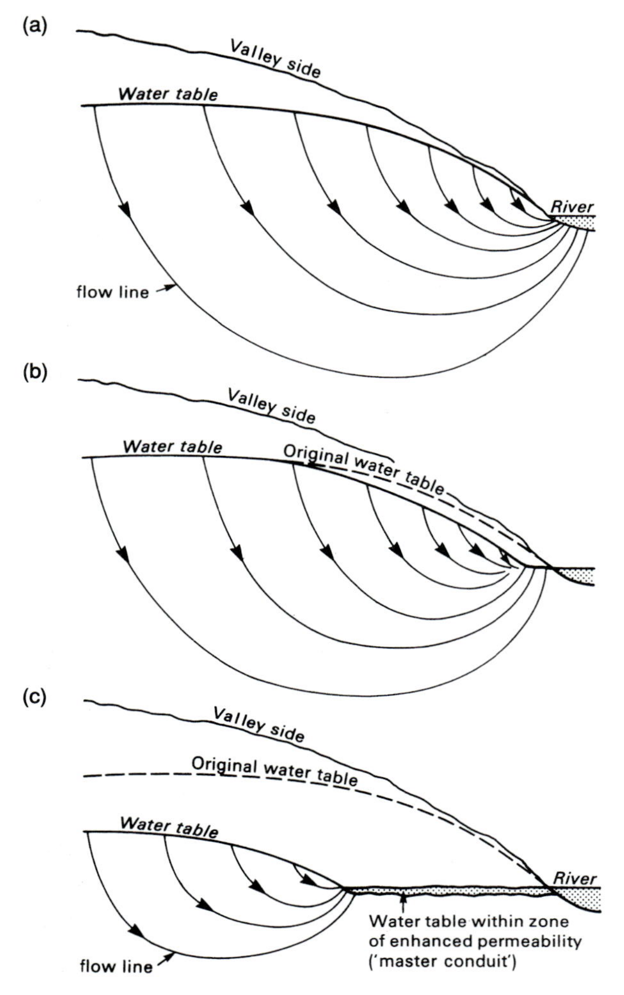

Fig. 3.9. Topographic control of the development of enhanced permeability in a carbonate aquifer; (a) flow pattern to a river in a homogeneous isotropic aquifer; (b) concentration of flow near the river leads to preferential solution at shallow depths along the valley, which enhances permeability and leads to further concentrations of flow; (c) eventually a highly permeable zone (or a single enlarged fissure) develops, and the water table is constrained to stay within this zone (after Rhoades and Sinacori 1941; and Price 1987).

to areas where the Chalk is overlain by less permeable materials. Valleys in these areas may follow lines of structural weakness, and fracture development may be enhanced by stress relief caused by erosion, leading to some preferential flow and solution. More important may be the fact that valleys incised in overlying materials will offer the easiest pathway for discharge from the confined Chalk because the thickness, and therefore the hydraulic resistance of the confining bed, will probably be at its least, and the vertical hydraulic gradient will be at a maximum. Groundwater flow and solution will, therefore, tend to be concentrated beneath valleys even in confined areas (Water Resources Board 1972).

In reality, the Chalk would not have been perfectly homogeneous and isotropic to begin with. Minor variations in lithology and hardness would have led to similar variations in the frequency and nature of jointing and fracturing, so that some of the primary fissures would be more open and more permeable than others. Thus, preferential pathways, and hence solution and enlargement, would have been selected on the basis of lithology or stratigraphy as well as topography.

There is clear evidence of higher permeability in units such as the Chalk Rock and Melbourn Rock, which probably fracture more cleanly because of their greater hardness. There is also some evidence of structural and lithological variations, in particular marl layers, influencing the distribution of enhanced permeability, with karstic features perhaps developing laterally along the marl layers where these have restricted water movement vertically; these factors are discussed more fully in Chapter 5.

At the edge of the Tertiary cover the concentration of acidic runoff from the generally less permeable Tertiary deposits frequently leads to the development of enhanced permeability. This has been a significant factor in the creation of rapid flow systems in the Chalk, as at Bedhampton in Hampshire (Atkinson and Smith 1974) and between Potters Bar and the Lee Valley in Hertfordshire (Water Resources Board 1972). In these and probably other areas, the secondary fissure systems may develop in a directional fashion with serpentine conduits (like flattened tubes) concentrated at particular fissured horizons. These tubes probably branch and rejoin. Such a system would explain the apparent directional nature of the rapid flow systems just referred to. In France solution of chalk in fissure systems has progressed to the stage where true karstic systems exist (Chapter 7). In Britain, solution hollows are present, and large openings have been revealed by excavation and as a result of ground collapses, but they are often filled or partially filled by sediment.

Solution of the Chalk may have been more active in the colder waters of the Pleistocene when frost-shattering encouraged the disintegration of the Chalk. Younger (1989) has suggested that in the Thames Valley considerable solution occurred during the late glacial (Devensian) stage because of concentrated flow of low temperature water below the main valley at a time when permafrost restricted flow below the interfluves. However, where the river was in a braided form, small channels probably froze in the winter and fused with the underlying permafrost. The alternate freezing and thawing produced putty chalk at the gravel–chalk interface. Below these areas the Chalk has a low permeability even in the main valley.

Although the upper 50–60 m of the saturated Chalk is the main aquifer, it must not be overlooked that significant yields can be obtained from greater depths, particularly from the principal hardgrounds, such as the Melbourn Rock and Totternhoe Stone, which, being hard and brittle, tend to fracture more readily than the softer horizons.

Because of the fine-grained nature of the Chalk matrix, little if any drainage occurs from it. Therefore the specific yield is low, probably being contributed largely by fissures, and is typically only about 0.01 to 0.02. Consequently, the seasonal water-level fluctuations in the unconfined aquifer, particularly in areas remote from discharge zones, are usually large, sometimes exceeding 30 m (see, for example, Giles and Lowings, 1990). This has a significant bearing on available water resources, given the relatively small effective thickness of the aquifer. In the Upper Thames Valley, the volume of available water remaining in storage in the Chalk under drought conditions only equals the average annual recharge (Downing *et al.* 1981). It is worth noting, however, that because the unsaturated zone of the Chalk is usually very thick, drainage of water from/ only a small fraction of the pore space in the matrix of the Chalk could represent in total a significant contribution to water resources. For example, drainage from pores representing 0.4% of bulk volume in an unsaturated zone 25 m thick would represent 0.1 m^3 per square metre of catchment, which is equivalent to 100 mm depth of water. The slow release of this water—equivalent to about half the annual recharge for much of the Chalk outcrop—could have a bearing on the reduced fluctuations of the water table noted in some areas (for example, Giles and Lowings 1990).

The rate of decline of the water level during extended dry periods in a well in the South Downs in England was interpreted to be reflecting drainage from large, medium-sized and small fissures (Thomson 1938). In the winter of 1989–90, when the depletion of groundwater levels in the well due to the drought of 1988–92 was interrupted by a period of heavy rainfall, the water level rose rapidly, but declined equally rapidly (Fig. 14.1; Institute of Hydrology/British Geological Survey, 1991). It appeared that only the larger fissures were replenished in the short time covered by the infiltration event. The same effect probably explains how saline intrusion into wells near the south coast of England can recover very quickly—only the major fissures are affected in the time available.

The seasonal variation in the quality of water from the Chalk of some wells in the South Downs suggests that the winter recharge tends to remain stratified at the top of the saturated zone and is discharged from the aquifer relatively quickly during the winter period in areas where the permeability is high. The 'base flow' through the aquifer is relatively constant and it is this deeper-seated water flowing below the zone that is seasonally replenished (Downing *et al.* 1978). Headworth *et al.* (1980) came to a similar conclusion in their study of the contamination of the aquifer in east Kent by saline mine-drainage water.

The thin, very permeable zone at the top of the saturated aquifer tends to have a higher specific yield; in the Candover Valley it is about 0.03–0.05, while values for the underlying aquifer are typically about 0.01. A similar pattern is found in the Upper Thames Valley, where values of 0.02 at shallow depths below valleys decline with increasing depth to less than 0.01, while laterally, below the interfluves, they reduce to 0.001 (Owen and Robinson 1978). A similar difference between values in valleys and below interfluves is found in the Paris Basin (Chapter 7).

It may be, however, that the main control on the magnitude of groundwater-level fluctuations is not the specific yield, but the transmissivity. Highly-transmissive zones discharge water rapidly, particularly near valleys, preventing large rises in the water table. Giles and Lowings (1990) discussed this and apparent exceptions. It is possible that a high specific yield is frequently assigned erroneously to the zone of water-table fluctuation, particularly during modelling studies, when what is really happening is that recharge is being discharged rapidly from the aquifer because of the high permeability of this zone.

The flow of fluids in fissures and the matrix has different characteristics. The speed of propagation of a head change in an aquifer is proportional to the hydraulic diffusivity, κ, given by:

$$\kappa = \frac{K}{S_S}$$

where K is the hydraulic conductivity and S_s is the specific storage, which is about $5 \times 10^{-6}\,m^{-1}$ for the confined Chalk. The diffusivity for fissures is much greater than that for the matrix. The rapid response of fissures to head redistribution due, for example, to a pumping well, contrasts with the delayed response in the matrix. This leads to a head differential between the matrix and the fissures which drives fluid from the matrix into the fissures. The quantity of water driven from the matrix to the fissures in the unconfined case as fissures drain is very small. However, Price (1987) has pointed out that most of the water released from elastic storage in the Chalk is actually derived from the compressibility of the fissures, with expansion of water in the matrix also making a significant contribution; compaction of the matrix framework is of lesser importance.

Studies of solute transport through the Chalk require information about the size of the chalk blocks that are separated by the primary and secondary fissures, but little information is available. Commonly, wells drilled to depths of some 50 m at valley sites obtain most of the total flow from one or two secondary fissures, generally near the water table. The source of water to these fissures is a ramifying system of primary fissures.

Details of fissures intersected in the extensive adit systems near Brighton have been given by Mustchin (1974). In some adits water-yielding fissures were encountered every 7–12 m but in others the interval was much larger, often 150–200 m, and in some sections over 500 m. Individual fissures yielded very large flows of the order of 5000 to 15 000 $m^3\,d^{-1}$. This indicates that dimensions of the blocks of chalk can be large even in the highly permeable zone near the top of the saturated zone. Similar evidence was found by Trotter *et al.* (1985) during excavation of caverns for storage of liquid gas on Humberside. They reported that many of the minor fissures had been sealed by deposition of calcite.

Where the Chalk is confined below Tertiary strata, the distance between major fissures is probably considerable, and even between active primary fissures it may be of the order of tens of metres or more.

Processes controlling the groundwater chemistry

Most of the Chalk is an extremely pure carbonate and this purity is the principal factor controlling the chemical composition of groundwater in the aquifer, particularly at outcrop. However, the nature of the trace minerals is also important, especially as the length of flow paths increases and the rate of flow decreases in the confined aquifer, thereby increasing the contact time between water and rock. Small amounts of clay minerals, particularly smectites, occur throughout the Chalk (see Chapter 2; Morgan-Jones 1977) and their presence is very evident at some horizons, for example in the Lower Chalk of England where it gives rise to a distinct colour change to blue-grey in the lower part (Young 1965; Weir and Catt 1965; Jeans 1968). Thin, extensive smectite marl bands occur throughout the sequence in eastern and probably also southern England (Pacey 1984). The presence of clay minerals, and also colloids, accentuates the very large surface area that is a characteristic of the chalk matrix. Typical values of $6.0 \times 10^4\,cm^2\,cm^{-3}$ (Edmunds *et al.* 1973) provide a large area for water–rock contact and reaction. The clays also have a considerable bearing on the regional variations in the chemistry of groundwater in the Chalk.

The composition of the Chalk has determined its hydrogeological character; in particular, it is the unique nature of the aquifer—a fine-grained, low-magnesian carbonate consisting largely of coccoliths or coccolith fragments—that has determined its response to diagenetic changes, to tectonism and to weathering, and led to the creation of the dual-porosity system represented by the matrix and fractures. This in turn controls the flow patterns and

Table 3.2 Comparison of the chemical composition of rainfall and groundwater in the Chalk, as mg l^{-1} (after Water Resources Board 1972).

	Rainfall	Composition of the infiltrate*	Composition of groundwater at outcrop
Calcium	1.4	3.9	100
Magnesium	0.3	0.8	2
Sodium	1.8	5.0	10
Potassium	0.3	0.8	1
Bicarbonate	–	–	280
Sulphate	6.3	17.6	15
Chloride	3.1	8.7	15
Nitrate (as N)	0.4	11.2	20
Ammonia (as N)	0.5	–	–

* Following concentration by evaporation and transpiration

hence the evolutionary changes that can be recognized in the chemistry of the groundwaters.

Initially, at outcrop, congruent solution reactions predominate and the very low Mg/Ca ratios in the groundwater closely reflect those of the parent rock. Downgradient, below the confining Tertiary deposits, incongruent solution and precipitation of carbonate, ion-exchange, and oxidation-reduction reactions predominate. In addition, the dissolved constituents are exchanged by dispersion and diffusion into and from the matrix. A sequence of hydrochemical changes along the flow lines can be recognized which relates to the residence time of waters in the formation.

Solution reactions

The soil zone exerts a major influence on the chemical composition of the groundwater. Carbon dioxide is produced by biological activity in the soil giving rise to concentrations often 10 to 50 times above atmospheric. This is dissolved to produce carbonic acid (H_2CO_3), which reacts rapidly with carbonate minerals. The pH of the soil zone is typically 7.5–8.3. The production of organic matter in the soil and nitrification (the conversion of organic nitrogen to nitrate) are also important processes.

The solution of calcium carbonate by carbonic acid and organic acids predominates in the soil and upper 2 m of the unsaturated zone and, in contrast, little reaction occurs beneath this depth in the unsaturated zone. Significant amounts of Na, Mg, K, SO_4, Cl and NO_3 in the infiltrating water may be derived from rainfall (Table 3.2); at the present day an appreciable proportion of the NO_3, SO_4 and K is derived from fertilizers.

The uptake of moisture and solutes by trees can modify the composition of the pore-waters considerably (Edmunds *et al.* 1987). Chloride concentrations approaching 1000 mg l^{-1} have been found in pore-waters below beech trees. Low nitrate concentrations reflect the biological uptake of NO_3, but Cl is concentrated as a result of the use of relatively pure water by the trees.

The carbonate chemistry of soil water in Lower and Middle Chalk has been studied in Humberside, England (Pitman 1978*a*). The measured soil log pCO_2 (atm) was found to vary between a minimum of –2.60 in winter and early spring and a maximum of –1.46 in late summer. A significant feature was that solutions collected in lysimeters were markedly undersaturated in calcite during the period from November to April. When soils were at field capacity in the winter months, rain passed through the soil, on average, in 15 hours, leading to undersaturation of the soil leachate. Conversely, during the summer, when the moisture content of the soil was low, soil leachate was super-saturated with calcite because of the high pCO_2, and the longer residence time. Thus, it is possible for seasonal changes in calcite saturation to occur in the upper

2–3 m of the unsaturated zone, where open system conditions occur, though studies in France and England indicate that calcite saturation is usually attained within the upper part of the unsaturated zone. Changes in the calcite saturation within the unsaturated zone may also occur, however, due to precipitation of secondary calcite.

The implication of these studies is that little free CO_2 is available to dissolve calcium carbonate in deep groundwaters, and that significant solution of carbonate can occur only at shallow depths. However, although a considerable proportion of the carbonate solution does occur in the upper few metres of the unsaturated zone, water within the zone can be unsaturated with respect to calcite (Edmunds *et al.* 1993). Studies in France using stable carbon isotopes and radiocarbon have shown that, following the attainment of saturation or supersaturation, there is a tendency for calcite to be precipitated, leading to a degree of undersaturation. This in turn allows the possibility of some solution of calcite at the water table, particularly if waters with different saturation states meet. Periodic rapid fluxes through the unsaturated zone of water containing carbon dioxide are probably an important means of carbonate solution in well-fissured zones in the Chalk. After periods of heavy rain, 'air' displaced from the unsaturated zone into the saturated zone by such infiltration fluxes can give groundwater a milky appearance in wells and lead to an increase in the Ca and HCO_3 concentrations (Downing *et al.* 1978).

A molar ratio in Chalk groundwater of Ca^{2+} to HCO_3^- of approximately 1:2 demonstrates that the source of both solutes is predominantly from the solution of calcite:

$$CaC0_3(s) + H_2CO_3 \rightleftharpoons Ca^{2+} + 2HCO_3$$

The amounts of impurities (Mg, Fe, and so on) taken into solution during the initial congruent reaction are similar to their concentrations in the Chalk itself. Sulphate concentrations in groundwater at outcrop can be derived from several sources but primarily from rainfall and traces of gypsum in the Chalk, accentuated in recent years by the use of chemical fertilizers. These various solution reactions lead to the Chalk at outcrop containing a calcium bicarbonate water with a varying proportion of sulphate.

Surface runoff from the overlying Tertiary deposits locally recharges the Chalk at the margin of the Tertiary outcrop. The acidic water introduced tends to have a high sulphate content and markedly increases not only sulphate, but also bicarbonate, calcium, and magnesium, and hence the total ionic content of the groundwater in the Chalk near and immediately downgradient of the Tertiary margin.

As mentioned previously, there is some evidence from the South Downs, near Brighton, England, that the seasonal winter recharge remains stratified at the top of the saturated zone in the highly permeable layer near the water table, and is discharged relatively quickly from the aquifer via shallow flow systems. The deeper groundwater tends to have a more constant and lower ionic composition (Downing *et al.* 1978).

Changes with depth

Analyses of pore-waters from the saturated Chalk, sampled in a cored borehole at outcrop in Berkshire (some 70 km west of London), revealed an increase in salinity with depth, up to a maximum sodium concentration of 300 mg l^{-1}. The water in the matrix was more mineralized than that in the fissures by as much as a factor of 10 (Edmunds *et al.* 1973). The ionic composition of water in the matrix also differed from that in the fissures, and the evidence points to a residual formation water as a source of the salinity.

At a second site in Berkshire, but where the Chalk is overlain by Tertiary deposits, the increase in salinity of pore-water with depth was much more obvious, with maximum values for total dissolved solids of some 2000 mg l^{-1} at a depth of about 300 m. The much higher values in the confined aquifer provide evidence for more restricted circulation below the Tertiary cover (Edmunds *et al.* 1987).

Pore-waters from the Chalk were sampled in a borehole in Norfolk, England, where the fully saturated aquifer is overlain by 42 m of Pleistocene deposits. They revealed that the salinity increased with depth, yielding what is possibly a true 'connate water' with a Cl content of 19 700 mg l^{-1}, very similar to that of seawater, at a depth of about 400 m (Bath and Edmunds 1981). There is a gradual enrichment of the stable isotopes ^{18}O and ^{2}H with increasing salinity. The stable isotope ratios of pore-waters in the depth profile showed progressive dilution of water with a marine composition. The data imply that the increasing salinity with depth is the result of mixing between fresh meteoric water and water trapped in the pore spaces of the sediment, possibly since deposition during the Cretaceous. Departures from the anticipated mixing sequence of meteoric water with seawater (for example, enrichment of Sr/Cl ratio) are believed to be caused by diagenetic changes through water–rock reactions. The salinity gradient in the profile could develop by diffusion over a period of no more than 1.5 million years.

The chemical evidence from these three sites implies that relict marine-derived waters remain in the chalk matrix and are slowly being diluted by exchange through diffusion with meteoric water circulating in the fissure system. The extent to which the exchange has been completed depends upon the scale (that is, the size and distribution) of the fissure systems and the rate of flow in them. This process also depends upon changes in hydraulic gradient with time and the position of the aquifer relative to sea level both now and in the past. Even at outcrop in the Berkshire Downs the process of flushing fluids derived from seawater from the matrix is not yet complete.

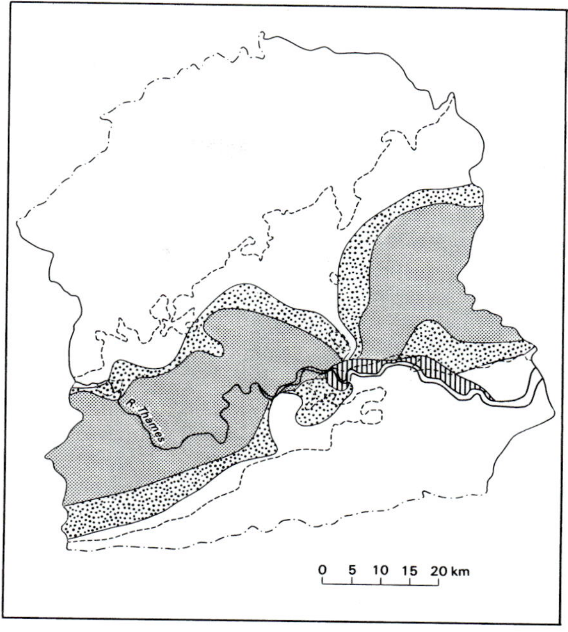

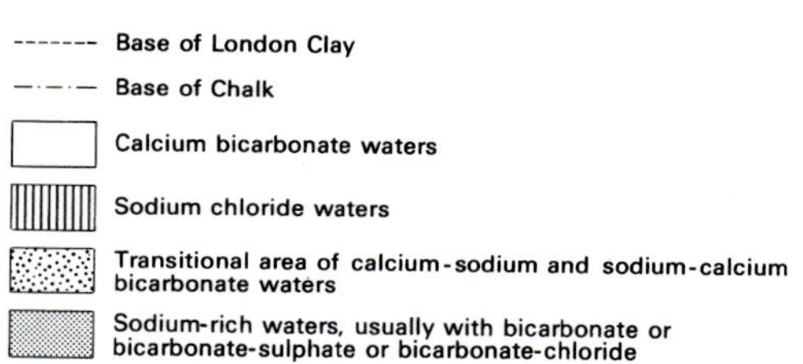

Fig. 3.10. Chemical types of groundwater in the Chalk of the central part of the London Basin (after Water Resources Board 1972).

Downgradient changes

The permeability of the Chalk reduces considerably where it is overlain by Tertiary deposits. As groundwater in the aquifer flows downgradient below the Tertiary cover, and the rate of flow decreases, the composition of the water changes as a result of interaction with the aquifer matrix (see, for example, Kimpe 1960; Ineson and Downing 1963). The dominant change is the conversion of the hard Ca-HCO_3 or Ca-Mg-HCO_3 water at outcrop to a soft Na-HCO_3 type in the confined aquifer as a consequence of cation exchange and other processes (Figs. 3.10; 3.11; Table 3.3). At increasing distances downgradient, mixing gradually occurs with a Na-Cl water that has a marine origin.

Groundwaters in the unconfined aquifer have a low molar Mg/Ca ratio but along flow paths this ratio increases to about 1 (Fig. 3.12). The Sr/Ca ratio increases similarly, but at a rate three times that recorded for Mg/Ca. These changes are brought about by a gradual recrystallization of calcite involving incongruent solution of carbonate, caused by an adaptation of a marine rock to a freshwater environment (Edmunds *et al.* 1987). In this process, the impurities in the Chalk are progressively released while a fractionally purer low-magnesian calcite is precipitated. Thus the Mg/Cl and Sr/Cl ratios may be used as qualitative indicators of residence time.

The conversion downgradient of the Ca-Mg-HCO_3 water (found in and near the outcrop of the aquifer) to a Na-HCO_3 water in the confined aquifer is due to the replacement by ion exchange of Ca ions in the water by Na in the matrix. The ion-exchange medium is believed to be small amounts of disseminated clay minerals in the Chalk matrix but clays lining the walls of fissures in the Chalk

Table 3.3 Chemical analyses of groundwaters from the Chalk illustrating downgradient changes below Tertiary cover (units are mg l^{-1} except the North Sea water which are mg kg^{-1}).

Location	1. Brightwalton	2. Bricket Wood	3. Shalford Farm	4. Mortimer	5. Brentwood	6. Dedham	7. North Sea
Water type	Ca-HCO_3	Ca-HCO_3	Ca-HCO_3	Na-Ca-HCO_3	Na-HCO_3	Na-Cl-HCO_3	Na-Cl
Calcium	124	126	52	52	6	35	16460
Magnesium	1.6	4	15	10	3	17	1340
Sodium	5.6	17	13	80	247	294	35080
Potassium	0.6	2	3.4	5	–	–	430
Bicarbonate	348	308	323	268	352	282	150
Sulphate	2.5	56	13	34	92	192	20
Chloride	11	27	16	72	123	256	92620
Nitrate	22	24	<0.1	<2	2	–	–

1. Chalk outcrop in Berkshire; from Edmunds *et al.* (1987)
2. Chalk outcrop in Hertfordshire; from Water Resources Board (1972)
3, 4. Below Tertiary cover in Berkshire; from Edmunds *et al.* (1987)
5, 6. Below Tertiary cover in central part of London Basin; from Water Resources Board (1972)
7. North Sea oilfield—from Chalk at a depth of 4403 m; from Egeberg and Aagaard (1989)

have been reported (Heathcote 1981), and where they occur they are probably also involved in the process (Howard and Lloyd 1983). The softening of water by ion exchange was first proposed by Thresh (Whittaker and Thresh 1916), although he believed the exchange medium was in the Thanet Formation of the Tertiary rather than the Chalk. The extent of cation exchange can be measured by the Na/Cl ratio. For example, the ratio increases in groundwater in the Chalk of Berkshire by a factor of 2, representing an increase of up to 30 mg l^{-1} of Na^+, which can be attributed to exchange reactions (Edmunds *et al.* 1987); in the centre of the London Basin the increase exceeds 100 mg l^{-1} Na^+ (Water Resources Board 1972).

High SO_4 contents in the Chalk water at depth may be partly caused by a reaction with gypsum cements that occur in hardgrounds (Morgan-Jones 1977) and by the oxidation of disseminated pyrite. However, where hydraulic gradients have been significantly modified by abstraction of groundwater, as in the central part of the London Basin, the natural changes just described are modified by the leakage of water into the aquifer from the overlying Tertiary cover. This introduces water into the aquifer with a high SO_4 content associated with Ca and Mg ions. These hard waters are softened by ion exchange not only in the Chalk but also as they migrate down through the sands that commonly occur at the base of the Tertiary sequence.

Trace-element concentrations provide additional evidence for the geochemical processes taking place in the groundwater (Fig. 3.13). Strontium is an important indicator of the extent of reactions between water and the carbonate of the aquifer, and similar effects are seen in the trends for F, I, Mn, and Ba in a downgradient direction. Potassium, NH_4 and probably Li indicate the extent of exchange with clay minerals in line with the increase in the Na/Cl ratio. Boron and bromine are useful indices of the increase in the component of saline water in the mixtures.

Groundwater in the unconfined aquifer usually contains several milligrams of oxygen per litre. The rate of reaction of this oxygen in the saturated zone has an important bearing on the evolution of the hydrochemistry (Edmunds *et al.* 1984). Removal of oxygen probably occurs through inorganic reduction by ferrous iron derived from impurities in the calcite.

An important hydrogeochemical feature is the presence of a redox boundary in the confined zone marking the point where oxygen is reduced to zero. This boundary has been described in the western

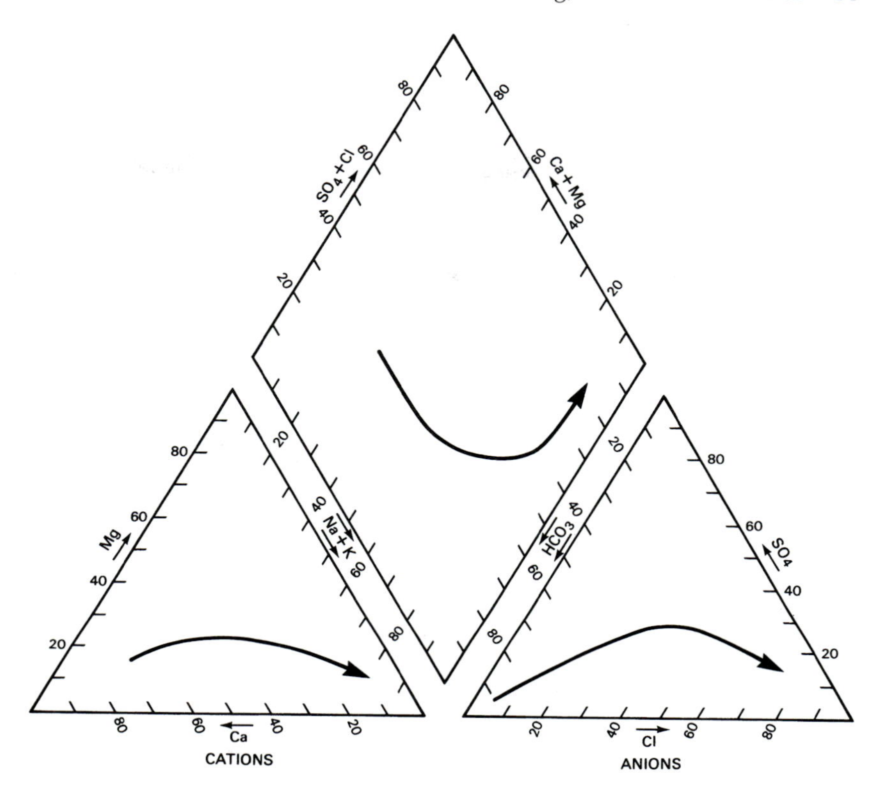

Fig. 3.11. Sequence of chemical changes of groundwater in a downgradient direction below increasing thickness of Tertiary deposits.

part of the London Basin (Edmunds *et al.* 1987) and no doubt it is a universal feature of flow systems in the Chalk (see, for example, Mariotti *et al.* 1988). In the outcrop zone, oxygenated groundwater occurs with Eh values of +330 to +420 mV but they decline downgradient in the confined zone to about +160 mV or less. Concurrently, dissolved oxygen declines from some 10 mg l^{-1} to below the level of detection as a result of the reactions described above. Dissolved iron increases significantly in the aquifer following the onset of reducing conditions. The complete elimination of oxygen leads to the relatively rapid reduction of NO_3. Strontium, Mn and Li, as well as other ions, all increase markedly at the redox boundary for a variety of reasons (Fig. 3.13; Edmunds *et al.* 1987). Uranium is found in the oxidizing zone of the aquifer but as the redox potential decreases there is a decrease in concentration since the reduced form is only slightly soluble; the $^{234}U/^{238}U$ activity ratio increases in the deeper groundwaters. Radium is also leached from the matrix of the Chalk and adsorbed and concentrated by clays lining the walls of fissures. This process is responsible for a relatively high release of radon (Cuttell *et al.* 1986).

The occurrence of nitrate-reducing and sulphate-reducing bacteria in the confined zone of the Chalk has been reported over the years (Ineson and Downing 1963; Smith *et al.* 1976). The presence of tritium in groundwater that has a low nitrate content, implying recharge since 1964 when high nitrate values would have been expected, has been taken as evidence that denitrification of nitrate is indeed occurring in the confined zone (Edmunds *et al.* 1987; Clark *et al.* 1991). The denitrification process in the aquifer has also been confirmed by an increase in the N_2/Ar ratio in the groundwater downgradient of the redox boundary, above the ratio that would be expected from the atmosphere (Towler 1982). At Emmerin in northern France, an exponential increase of ^{15}N, as NO_3 decreases, must be the result of denitrification (Mariotti *et al.* 1988). The common presence of hydrogen sulphide at well heads in the reducing zone also implies that sulphate reduction is occurring, although probably not to a significant extent. Small amounts of methane

Fig. 3.12. Interstitial (pore) waters and groundwater from the Chalk in Berkshire, England (after Edmunds *et al.* 1987).

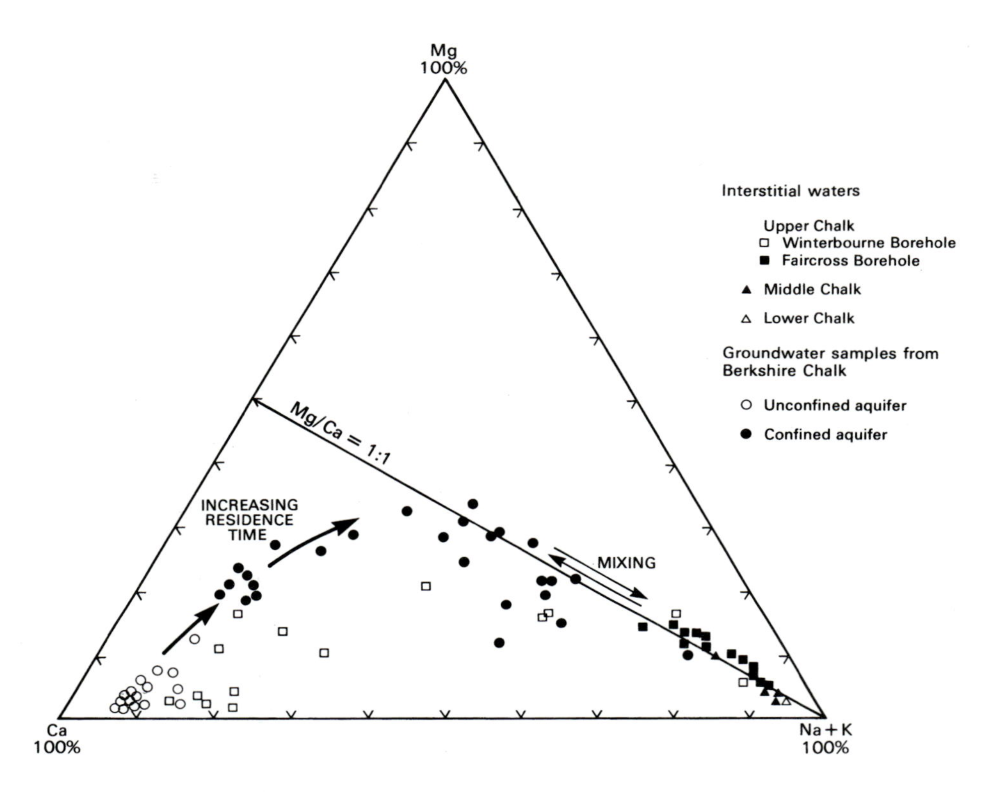

have been detected in deep Chalk groundwaters in East Anglia and the London Basin, implying at least limited methanogenesis (Bath *et al.* 1985; Darling 1985). Anaerobic bacterial processes are likely to be of greater importance in the deeper parts of the aquifer. The relative sizes of bacteria and the pore throats of the Chalk's matrix (Edmunds 1986) indicate that bacterial processes are likely to be restricted to fissure walls or the pore spaces immediately adjacent to fissures.

In detail, the chemical composition of groundwater in the Chalk below the Tertiary deposits depends to a significant extent upon the local geological conditions. This is well illustrated in the Lee Valley, in north-east London, where the influence of the overlying Tertiary sediments is very evident in the form of high sulphate concentrations caused by the oxidation of sulphide minerals that occur in the deposits (Flavin and Joseph 1983).

Downgradient, below confining beds, pore-waters in the Chalk become increasingly saline, eventually attaining and exceeding the concentration of seawater. The brines associated with the Chalk in the North Sea oilfields have a salinity of between about 50 and 100 $g\,l^{-1}$. They are Na-Cl waters with a significant calcium component (Table 3.3). The source of the salinity is believed to be a brine that formed in Permian times from the evaporation of seawater beyond the point of halite precipitation. This connate brine is considered to have migrated by advection and diffusion into the Chalk, where it has mixed with a Palaeogene meteoric water (Egeberg and Aagaard 1989). The formation waters in the Chalk may indeed be largely the result of a mix of such a two-component system but they are also likely to include components derived from Tertiary seawater and original Jurassic and Lower Cretaceous formation waters that have been modified and

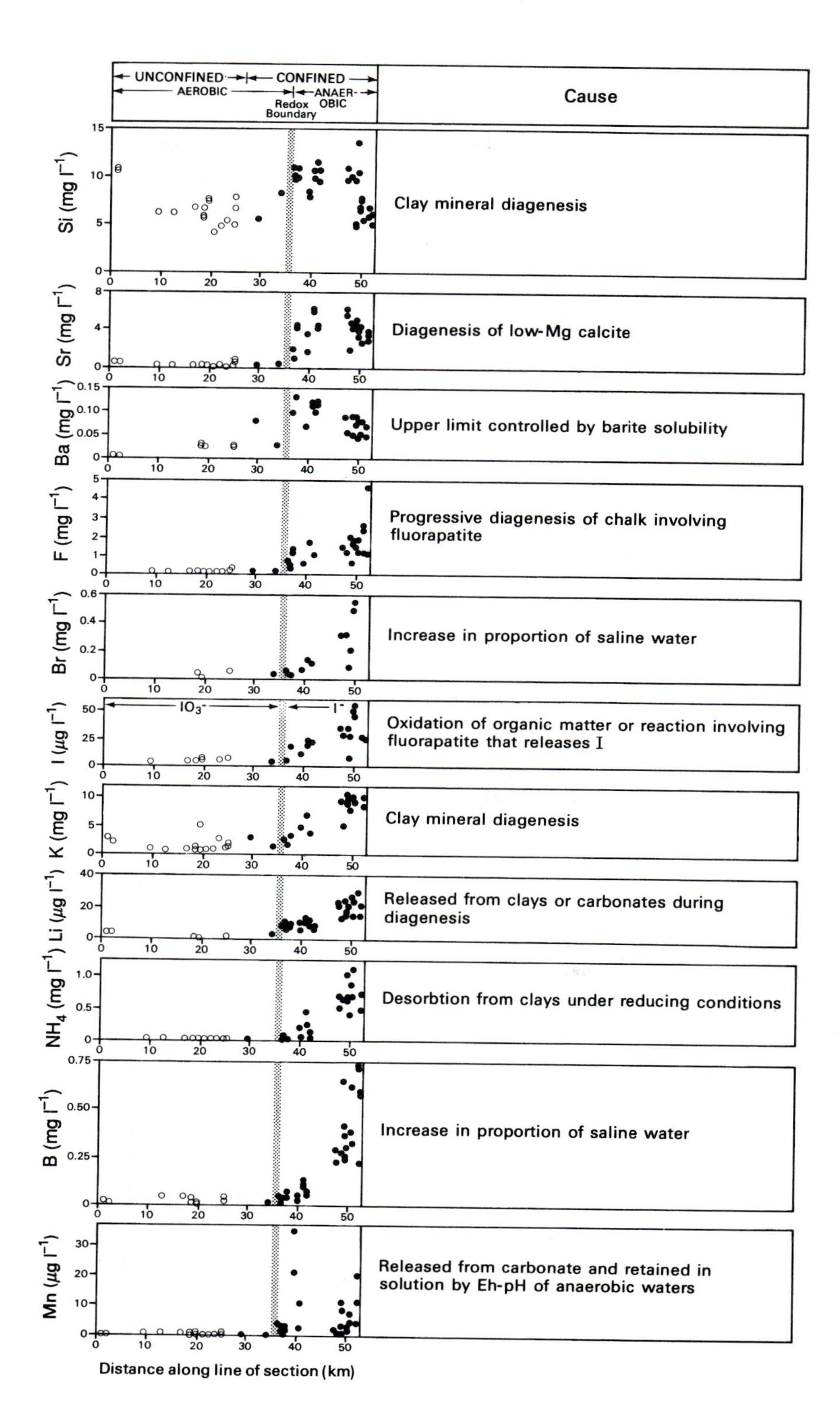

Fig. 3.13. Variation of minor elements along a downgradient flow path below Tertiary cover (after Edmunds *et al.* 1987). Open symbols denote unconfined and solid symbols confined groundwaters.

displaced upwards into the Chalk by compaction. The present composition of the waters in the Chalk is also the result of complex water–rock interactions (see Chapter 13).

At some locations around the present-day shoreline, groundwaters in the Chalk are also saline but this is due either to the intrusion of modern seawater or intrusion that occurred during high sea-level stands in the Pleistocene and Tertiary. Relict Tertiary waters occur in the Chalk below Tertiary deposits in East Anglia, and Pleistocene waters are found in north–east Lincolnshire and the Netherlands (Chapters 9 and 12).

Solute exchange processes and groundwater residence times

Most of the flow in the Chalk occurs in the upper 50 m of the saturated zone, where it is controlled by the widely differing porosities and permeabilities of the matrix and the fractures. The finer fractures and the matrix provide the storage, and the secondary fissures the water-distribution system. The interrelationship between the two has a paramount influence on the composition of the groundwater.

Water flowing in the secondary fissures disperses and diffuses into the primary fissures, and from these it diffuses into the intergranular pore spaces. There is a continuous exchange, in both directions, because of concentration gradients, between the water in the fissures and water in the matrix, as well as between water in dead-end (or semi-stagnant) zones in the main fissure distribution systems. For example, groundwater in the outcrop zone of the London Basin contains tritium but this disappears from the water in the fractures in a downgradient direction, apparently because it diffuses into the matrix.

In the long term, under natural conditions, an apparent state of equilibrium is attained and sequential chemical changes on a regional scale can be recognized as already described. However, because of the continuous interaction between water and rock, geochemical zones move very slowly downgradient in the confined aquifer. This is accentuated in areas where high abstraction has modified the flow pattern, as, for example, in the Lee Valley, to the north-east of London. There the Na-HCO_3 zone in the London Basin has been dissected by the penetration of Ca-HCO_3 water under the influence of a highly permeable zone along the valley and of the large abstraction for water supply (Fig. 3.10).

Preferred flow paths exist in the Chalk at the level of individual primary fissures but also at the regional level where a zone of preferred flow develops, for example, along a valley. If the rate of flow in fissures is more rapid than the rate of chemical reactions then a disequilibrium occurs between the composition of the matrix and fissure waters. Matrix diffusion is not a dominant process in fast-flowing channels where there is insufficient time for species to diffuse into the matrix.

The $\delta^{13}C$ value of the dissolved carbonate species changes in a downgradient direction from about –13‰ in and near the outcrop to a value less negative than –1‰, approaching that of the aquifer matrix in the confined zone. This implies that as the water moves through the aquifer there is a continuous precipitation and solution of carbonate minerals (Smith *et al.* 1976; Edmunds *et al.* 1987). This incongruent solution also affects the ^{14}C content of the dissolved carbonate so that the age of the water, derived using the conventional ^{14}C method, is too old and more refined models are needed. The increase in the concentration of the heavy isotope (^{13}C), as well as the Mg/Cl and Sr/Cl ratios, provide qualitative indices of groundwater residence times in the aquifer.

The ratios of the stable-isotope concentrations of $^{18}O/^{16}O$ and D/H in the groundwater in the deeper parts of the confined aquifer of the London Basin tend to be very slightly more negative than those at and near the outcrop (Smith *et al.* 1976, Figs. 6 and 7). The difference in the values of the $^{18}O/^{16}O$ ratio is between 0.6 and 0.9‰ and implies that the groundwater is probably mainly recent water transported in fissures, but includes a component of glacially derived water from the matrix (Edmunds *et al.* 1987).

The complex carbonate exchange between the groundwater and the matrix, as well as the mixing of fissure and matrix waters with different residence times as the water flows downgradient, prevents a conclusive interpretation of ^{14}C values with respect to the age of the water. Carbon-14 levels in samples from the centre of the London Basin are very near to the limit of detection, but it would appear that the mean age of the water in this central zone is greater than 10 000 years, and includes a component that pre-dates the last major glacial advance of the

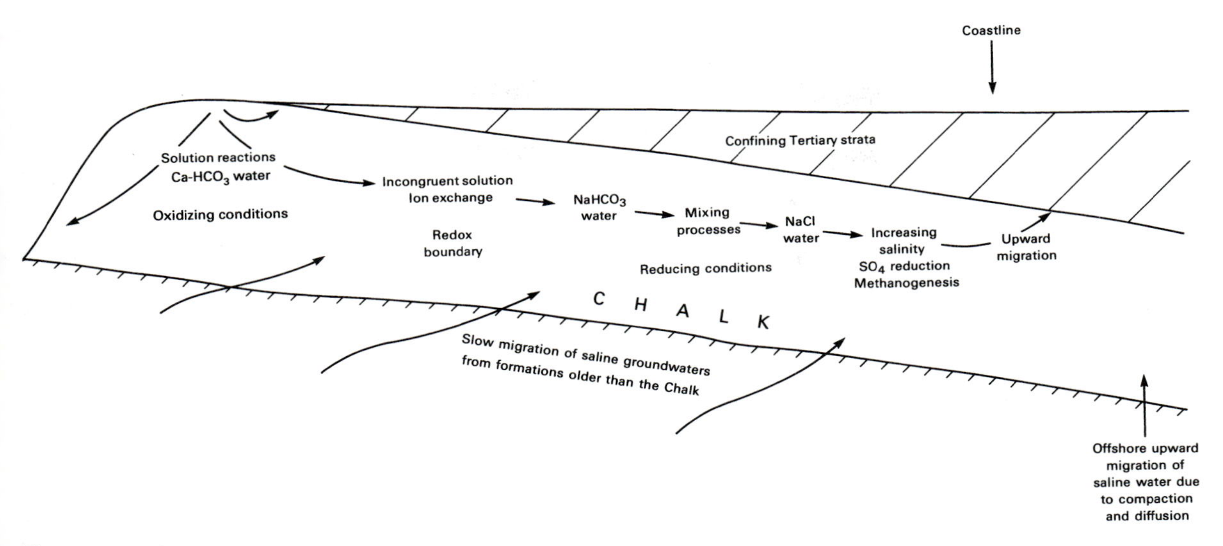

Fig. 3.14. Schematic diagram of downgradient chemical changes in groundwater in the Chalk.

Devensian (Downing *et al.* 1979; Edmunds *et al.* 1987; Elliot 1990). Although the stable-isotope data imply that the temperature at the time the water was recharged was lower than current temperatures, the strongest evidence for a Pleistocene component in the waters in the centre of the basin is afforded by the noble gas content, which indicates recharge temperatures as low as 5–6°C. The rather higher temperatures suggested by the stable isotopes are possibly due to the fact that values have been modified by admixture with an older water, or even a connate water, in the chalk's matrix which has less depleted values (Elliot 1990).

On balance, considering all the data from carbon-14, stable isotopes and noble gas analyses, some of the water in the central part of the basin was probably recharged during an interstadial in the Devensian glaciation, about 45 000 years BP.

The residence times derived from helium-4 concentrations are considerably in excess of the radiocarbon dates. There is a linear correlation of helium-4 and increasing chloride towards the centre of the basin and this suggests that part of the source of the helium-4 is connate water in the matrix. However, there is also probably an external source, and this is believed to be diffusion and/or advection of water from sediments below the Chalk (Edmunds *et al.* 1987; Elliot 1990; Downing and Penn 1992). Upward migration of saline water from the Carboniferous into the Chalk is also believed to occur in the Netherlands and Belgium (Chapters 8 and 9), and without doubt formation waters in the Chalk below the North Sea have diffused and migrated upwards because of the compaction of sediments, and hence include a component derived from strata older than the Chalk (Chapter 13).

In summary, the Chalk groundwater displays well-defined regional variations in chemical composition (Fig. 3.14). Dominating these variations is the sequence of changes that occur as meteoric water moves downgradient and interacts with water in the matrix. The resulting water ultimately mixes with a marine-derived water that has itself passed through a series of diagenetic changes. The chemical zonation of water in a downgradient direction reflects the existence in the matrix of waters of different ages with different compositions at different stages in their evolution. Waters moving rapidly in the secondary fissure system can be younger than those in the matrix at any particular location, but owe their chemical character to diffusion from the matrix. Thus, the water in the aquifer at any point is a complex mix reflecting its position on a flow-line and its depth in the aquifer.

Conclusions

Over the last 20 years, the nature of the flow processes in the Chalk, in both the unsaturated and saturated zones, has become clearer, but as a consequence its complexity, previously only suspected, has been revealed (Downing and Headworth 1990).

The evidence currently available about the nature of the infiltration process seems to imply that before a major infiltration event, when the unsaturated zone is at field capacity, the fissures have drained but the matrix remains virtually saturated. With average or below average infiltration rates, water moves from the soil zone into the matrix by piston displacement. However, once infiltration exceeds a threshold value, which is a function of the permeability, flow occurs in the fissures. If flow is activated in the major fissures, then velocities of the order of $50\,m\,d^{-1}$ can occur through the unsaturated zone. Once fissure flow is activated, exchange of solutes by diffusion probably occurs between fissures, particularly small fissures, and the small pores of the matrix. As fissure flow declines, the effect of the infiltration event continues to be seen in the form of water displaced from the base of the unsaturated zone by 'piston displacement'. Clearly the proportion of groundwater recharge carried rapidly through the fissures in the unsaturated zone varies from area to area but the average is probably about 10–15%. Where the chalk matrix is very permeable it can accept all the infiltration and flow in fissures may be exceptional.

In detail, the infiltration process is complex, with the form it takes depending upon factors such as the rate of infiltration (and hence the extent to which the fractures are filled with water), and, of course, the nature of the chalk. It must be accepted that at individual sites the process may not always take the same form because the underlying controlling factors are so diverse.

In the saturated zone, water moves very slowly downgradient through the matrix but more rapidly in the fissure system which acts as the main distribution network and imparts the high permeability. In addition to the physical displacement of water downgradient, there is a continuous exchange of water and solutes between the fissures and the matrix. Thus, although the bulk of the water in the matrix is moving only very slowly, it is involved in the transport process because it is moving by continuous exchange through diffusion. The actual rate of exchange, between fissures and matrix, depends upon the fissure aperture and spacing, and the water velocity in the fissures. Generally, advection and dispersion of solutes occurs in the fractures while diffusion is the dominant process in the matrix.

In the unsaturated zone, it is the matrix-fissure dual-porosity nature of the chalk that appears to dominate both hydraulic and hydrochemical behaviour. In the saturated zone, the hydraulic behaviour seems to be dominated by the dual-porosity interaction between primary fissures and secondary fissures, although dual-permeability effects may also be important; the hydrochemical behaviour seems again to be controlled principally by the matrix-fissure dual-porosity system. These aspects are discussed more fully in Chapter 4.

The very saline formation waters in the Chalk of the North Sea have developed from a connate Permian brine that has been modified by complex water–rock reactions and probably diluted with a Tertiary meteoric water (see Chapter 13).

4. Modelling groundwater flow and transport in the Chalk

J. A. Barker

At this stage it is convenient to review the transport mechanisms in the Chalk and the factors that have to be considered when creating models of the Chalk's flow systems. Modelling is a discipline that requires a clear understanding of the processes occurring in the Chalk. The increasing ability to construct sophisticated models has focused attention on these processes and led to more rigorous and realistic conceptual models of the aquifer for examining both flow and pollution problems.

Transport mechanisms

Flow of water

Intergranular flow in saturated chalk is invariably described by Darcy's law, which is also valid for flow in fissures provided the Reynolds number is less than about 2300. If this condition is satisfied, then each fissure can be ascribed a transmissivity, T_f. For a perfect planar fissure of uniform aperture, a, and with smooth surfaces, the transmissivity is given by the so-called *cubic law*:

$$T_f = \frac{ga^3}{12v}$$

where g is the acceleration due to gravity and v is the kinematic viscosity of the fluid. The equation implies that the transmissivity is very sensitive to the size of the aperture; for example, it gives a transmissivity of about $10^{-3}\,m^2\,s^{-1}$ for a fissure with an aperture of only 1 mm, and an enormous $1\,m^2\,s^{-1}$ for a 1 cm aperture. The equation is particularly useful for estimating fissure apertures from hydraulic conductivities (assumed to be imparted by the fissure system) when the fissure density can be estimated.

From tracer tests it is known that rates of movement of groundwater through the Chalk's *secondary* fissure system can often be in excess of $100\,m\,d^{-1}$. In the primary fissure system, however, a hydraulic conductivity, K, of the order of 10^{-7} to $10^{-5}\,m\,s^{-1}$, a porosity of 1 per cent and a gradient of 10^{-3} give rates of movement of the order of a few millimetres per day. For the same gradient, the rate of movement in the matrix (assuming K is $10^{-8}\,m\,s^{-1}$ and the porosity 35 per cent) will only be of the order of a millimetre per year. Consequently models of transport in chalk often consider the matrix water to be immobile.

Matrix diffusion

If the water in the chalk matrix is (effectively) immobile, then any matrix transport of solutes or heat can still take place by molecular or thermal diffusion, respectively. Under transient flow conditions, when the head in the fissures surrounding a matrix block decreases, water flows within the block quasi–radially towards the surrounding fissures, the water being released from elastic storage. (The water will actually be taken only from the near-fissure pores, normally only from within a small fraction of a millimetre of the block surface.) Since flow can be described by Darcy's law, which is a diffusive law, water transport in chalk matrix is by hydraulic diffusion.

So water, solutes and heat can all be diffusively transported in chalk matrix, and Table 4.1 summarizes the factors relating to their diffusion. For any diffusive process, a characteristic time for diffusion over distance x can be defined as x^2/D, where D is the diffusivity (the conductivity parameter divided by the storage parameter). Taking x as 1 m gives a characteristic time equal to the reciprocal of the diffusivity: typical values of that time are given at the bottom of Table 4.1.

For diffusion over distances other than 1 m it is only necessary to multiply the time values given in Table 4.1 by the square of the distance in metres. For example, the characteristic time for molecular diffusion over a 10 cm matrix block would be

Table 4.1 Diffusive transport processes and parameters

	Entity being transported		
	Water	Heat	Solute
Transport potential	Potentiometric head	Temperature	Concentration
Conductivity parameter	Hydraulic conductivity	Thermal conductivity	Diffusion coefficient
Storage parameter	Specific storage	Specific heat × density	Porosity
Matrix flux law	Darcy	Fourier	Fick
Characteristic time	5–500 s	10–100 d	50–500 a

between 6 months and 5 years. Diffusion over a distance of 100 m would take of the order of a half to a few million years. Since molecular diffusion is always active, such times can be used to put upper limits on the periods required for concentration changes to propagate through matrix pore-waters.

Hill (1984) reported measurements of the molecular diffusion coefficients of nitrate, chloride, sulphate and water (labelled using tritium) in samples of both fissured and unfissured chalk. The smallest values were for sulphate (0.28×10^{-10}–$1.47 \times 10^{-10}\,m^2\,s^{-1}$); similar values were found for nitrate (0.53×10^{-10}–$3.2 \times 10^{-10}\,m^2\,s^{-1}$) and chloride ($0.52 \times 10^{-10}$–$3.23 \times 10^{-10}\,m^2\,s^{-1}$), while tritiated water gave slightly larger values (0.6×10^{-10}–$3.51 \times 10^{-10}\,m^2\,s^{-1}$). These values represent the mass flux through the saturated matrix per unit concentration gradient in the water. In a critical discussion of Hill's results, Muller (1987) deduced that the ratio of the diffusion coefficients to the free-water diffusion coefficients is about 0.25. So, the diffusion coefficient in chalk can be estimated as one quarter of the free-water value, if this is known for a solute. Hill's values are appropriate for use in Fick's first law of diffusion and should be divided by the porosity to obtain a diffusivity for use in Fick's second law.

Adsorption

Most of the surface area of chalk on to which adsorption can take place is within the rock matrix. The area can be estimated by considering the matrix to be made up of spheres with radii of the order of the throat radius: for a radius of a few micrometres the surface area per unit volume is found to be of the order of $1\,m^2\,cm^{-3}$. Taking that area density and a typical molecular diffusion coefficient of $10^{-10}\,m^2\,s^{-1}$, the area accessible to a diffusing solute after 1 day is equal to the surface area of the blocks multiplied by a factor of about 10–100.

Mineral or organic deposits are quite often observed on the surfaces of fissures. When present these must have a significant impact on the amount of adsorption taking place, both by providing adsorption sites and by acting as a barrier for diffusion into the matrix. Some models include a *fracture skin* to take some account of this phenomenon.

Dispersion

The term *dispersion* refers to the process of spreading during transportation: solutes, particles and heat are all dispersed in groundwater. The processes of matrix diffusion and adsorption both have dispersive effects, but the term *hydrodynamic dispersion* refers more specifically to spreading caused by variations in the fluid flow velocity.

Hydrodynamic dispersion presents two major problems: first, there is no generally accepted physical and mathematical description for heterogeneous media (of which fissured rock is a fairly extreme case); and second, it is very difficult to perform experiments (for example, tracer tests) to quantify the dispersive characteristics of rocks.

Table 4.2 Dispersivity required to give the same order of magnitude of dilution as diffusion (matrix porosity = 35%, diffusion coefficient = 10^{-10} m^2 s^{-1})

Fissure aperture (mm)	Velocity in fissure (m d^{-1})	Dilution factor (diffusion)	Equivalent dispersivity (m)
0.1	1	10^8	10^{14}
0.1	10	10^7	10^{12}
0.1	100	10^6	10^{10}
1	1	10^6	10^{10}
1	10	10^5	10^8
1	100	10^4	10^6
10	1	10^4	10^6
10	10	10^4	10^4
10	100	10^3	10^2

The normal approach to modelling transport in a dispersive medium is via the so-called convection–dispersion equation, which contains a characteristic dispersion coefficient, D_L. This coefficient is normally considered to increase in proportion to the absolute value of the velocity, υ, so $D_L = \alpha\,|\upsilon|$ where α is known as the *dispersivity*, and has units of length. Experimental determinations of dispersivity in chalk were reviewed by Black and Kipp (1983), who also carried out experiments of their own. Values vary from a few centimetres, for laboratory-scale measurements, to a little over 100 m, at the field scale.

An indication of the relative importance of molecular diffusion into the rock matrix and hydrodynamic dispersion in chalk fissures can be provided by considering the following idealized situation. Water flows steadily through a uniform fissure into which a pollutant is instantaneously injected, with uniform concentration, over a 10 metre length interval. The concentration of that pollutant is monitored 1 km downstream of the centre of the injection interval and the maximum concentration attained is noted. A measure of the amount of dispersion that has occurred is the dilution: the initial concentration divided by the maximum concentration at the observation point. That dilution has been calculated for two distinct cases: when the only dispersive mechanism is molecular diffusion into the matrix pore-water (assuming a uniform concentration across the fissure); and when the only mechanism is hydrodynamic dispersion, which is characterized by a dispersivity value. In order to make a direct comparison, dilutions have been computed for the diffusive case and then the dispersivity values that would give the same dilutions have been found.

The results are given in Table 4.2 for various flow velocities and fissure apertures. These were computed from a standard analytical solution of the convection–dispersion equation, and, for diffusion, from a solution (Barker 1982) based on an assumption of a constant concentration of solute for a period equal to the fissure velocity divided by the polluted interval length (10 m). Because the latter solution is only approximate, values in Table 4.2 have been rounded to the nearest order of magnitude.

It is evident from Table 4.2 that diffusion can give rise to very large dilutions (even though a relatively small diffusion coefficient has been adopted). The dispersivity values required to give the same dilutions are in most cases unrealistically large since they exceed the transport distance of 1 km. So dispersion, although not negligible, can normally be ignored, in relation to matrix diffusion, when modelling solute transport in chalk.

These results also indicate that in any attempt to determine dispersivity from a field experiment, great care must be taken to account for diffusion.

Types of model

Regimes of behaviour in fissured systems

In order to understand why various types of model are used to simulate fissured systems such as the Chalk, it is necessary to consider how the behaviour of a fissured system is related to the time-scales of the transport processes.

It is particularly important to establish whether advection or diffusion (molecular or thermal) is the dominant transport mechanism in the matrix. As stated earlier, the time taken for diffusion over a distance x is of the order of x^2/D, where D is the diffusion coefficient. The time for advection through a matrix block of dimension x in the direction of flow is simply xn/Ki, where n is the porosity, K is the hydraulic conductivity, and i is the hydraulic gradient. The ratio of the diffusion to

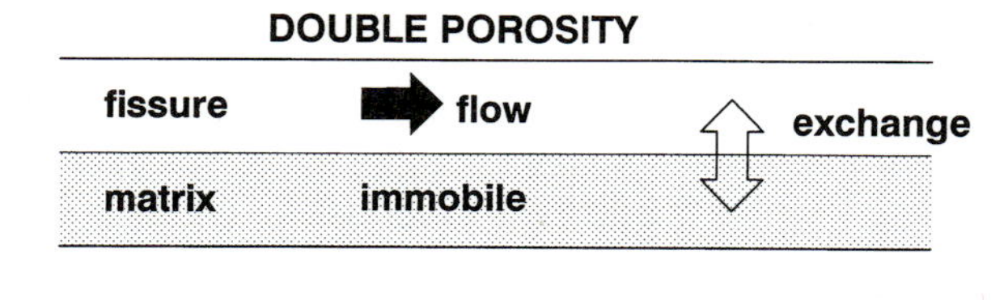

Fig. 4.1. Schematic representation of the difference between double-porosity and double-permeability models.

advection times, often referred to as the Peclet number, is therefore xKi/nD. Diffusion is dominant when the Peclet number has a value much less than unity. Even taking relatively large values for chalk matrix of $K = 10^{-7}\,\mathrm{m\,s^{-1}}$ and $i = 0.01$, the Peclet number for solute transport is less than unity for block sizes, x, less than about 10 cm. For thermal diffusion, the Peclet number is about three orders of magnitude smaller, so thermal transport is always predominantly by thermal diffusion.

Whenever the Peclet number is much less than unity, the assumption of immobile matrix water is appropriate and the resulting model is normally referred to as a *double–porosity* model. If on the other hand the rate of advection in the matrix is not negligible, a *double–permeability* model is appropriate (Fig. 4.1). These and other forms of model are discussed later.

This discussion does not apply to the transport of water in the matrix which, when described by Darcy's law, is always diffusive. Consequently the terms 'double porosity' and 'double permeability' have been used almost interchangeably when only flow is being considered (such as in pumping-test analysis). However, a distinction can be made between double-porosity models, in which matrix flow is directly towards or away from the local fissure system (and is represented by a source term in the fissure-flow equation), and double-permeability models, in which the matrix flow is in the same direction as flow in the fissures.

All double-porosity systems, with advective transport in the fissures and diffusive transport in the matrix, exhibit different forms of behaviour depending on the time-scale of the process under consideration in relation to the characteristic times for diffusion across a fissure or a matrix block. The time-scale referred to here might, for example, be a period of oscillation (as in an aquifer influenced by tidal movements) or a half-life (as for transport of a radioactive isotope). Many systems will exhibit a range of time-scales, and for those the shortest time scale will normally be the one of interest.

Four regimes of double-porosity behaviour can be identified and models suitable for these regimes are indicated (the models are described later):

1. If the time-scale is small with respect to the characteristic time for diffusion across the fissure width, the effects of the porous matrix can be ignored (because of both the restricted diffusion out of the fissure and the small volume of matrix accessed, in relation to the fissure volume). Under such conditions (which rarely exist outside the laboratory) chalk can be modelled by an *equivalent porous medium* (EPM) model with a porosity equal to the fissure porosity.

2. If the time-scale is a small fraction of the time for diffusion across a matrix block, then only the matrix/fissure surface area per unit volume of the rock is important, not the block size or shape. *Diffusive-type double-porosity* models, simplified by assuming an infinite matrix, are appropriate.

3. If the time-scale is similar to the time for diffusion across a matrix block, then the sizes and shapes of the matrix blocks become important. Then a general *diffusive-type* double-porosity model should be used, although a *quasi-steady-state* (QSS) double-porosity model may be adequate over some periods.

4. If the time taken for diffusive equilibrium between fissures and matrix is small in relation to the time for any significant change in the fissure system, then the chalk will behave as a (locally) homogeneous medium characterized by the total porosity. A QSS double-porosity model should prove to be more than adequate and an EPM model might also be suitable.

For some processes there is a very wide distribution of characteristic time-scales, and only the diffusive double-porosity models can then be used with confidence.

Equivalent porous medium (EPM) models

An equivalent porous medium model is simply a homogeneous model with parameters chosen to be characteristic of the fissured rock. Such models can be employed provided that either the fissures act independently of the matrix or the two act in unison, the latter condition being more likely than the former. For example, regional water resources models (Nutbrown *et al.* 1975; Morel 1980) can replace the fissured system by a homogeneous one with storage and permeability values characteristic of the matrix and fissures combined.

Double porosity (DP) models

Barenblatt *et al.* (1960), in modelling pumping tests in fractured porous rock, envisaged two overlapping continuous media (the matrix and fracture phases corresponding to the *primary and secondary porosity*, respectively) with an exchange mechanism. Such *double-porosity* models can be divided into two categories depending on the physical (and mathematical) description of that mechanism:

Diffusive type

Diffusive DP models are those for which the transport in the blocks of rock matrix can be described by one of the flux laws given in Table 4.1. The potential (head, concentration, or temperature) within a matrix block is controlled by the variations within the fissure system, and the two potentials are normally assumed equal at the surfaces of the blocks. The shapes of the matrix blocks affect the behaviour (Barker 1985a; 1985b).

Fissured rocks have often been represented by sets of identical, equally-spaced parallel conduits (see, for example, Grisak and Pickens 1980; Barker 1982). These models are mathematically equivalent to and correctly described as double-porosity models. Bibby (1981) used such a (numerical) model to investigate pollution of the Chalk in east Kent, and Muller (1987) applied a very similar model to pollution in the Chalk in Cambridgeshire. Barker and Foster (1981) applied a numerical double-porosity model to transport in the unsaturated zone of the Chalk.

Given the small characteristic times for hydraulic diffusion in the chalk matrix (Table 4.1), only pumping-test data for tests recording very rapidly changing conditions will be significantly influenced by matrix diffusion. Using a double-porosity flow model, Barker and Black (1983) showed that a standard slug-test analysis in a double-porosity fractured rock will always tend to overestimate the transmissivity.

Quasi-steady-state (QSS) type

In QSS models, the matrix is characterized by a single potential (for example, concentration) and the diffusive flux between the matrix and fissures is taken to be proportional to the difference between their (local) potentials. When viewed as an approximation to a diffusive model, the QSS model is valid only when changes within the fissures are slow in relation to the time for diffusive equilibrium (Table 4.1) across a matrix block or other immobile fluid zone: the diffusive and QSS models are incompatible for relatively fast changes within the fissures (Barker 1985a; 1985b).

The Water Research Centre's model of nitrate transport (Oakes 1982; 1990) includes the effect of matrix diffusion by an exchange term of the QSS type. Black and Kipp (1983) described a tracer experiment on an irrigated plot on the Lower Chalk and a QSS DP model fitted the data well.

Given the fast hydraulic reaction time in chalk matrix (Table 4.1) it is reasonable to adopt a QSS model (eg Barenblatt et al. 1960) for all but the most rapid transient pumping tests.

Double-permeability models

In double-(or *dual*-) permeability models the advective velocity in the matrix, although much less than that in the fissures, is not regarded as negligible. Such models will occasionally be useful for solute transport in very porous chalk with large blocks (and hence large travel times). Double-permeability concepts are more applicable to the interaction between the primary and secondary fissure systems in chalk than to that between the matrix and fissures, for which the permeability contrast is much larger.

Figure 4.1 schematically contrasts double-porosity and double-permeability models, as de scribed here.

Hard-rock models

A number of modelling concepts have been specifically developed for the study of groundwater in hard rocks. Much of this work has been in response to recent interest in radioactive waste disposal. Such systems differ from chalk mainly in that fractures are generally less frequent and hard rocks have little intrinsic porosity, although they do often contain a dense network of micro-fissures concentrated near the fractures which can impart a double-porosity character. Such models are valuable when considering rapid flow in the Chalk under karstic conditions (such as occur in France) when matrix porosity may not be significant.

It is often appropriate to use discrete rather than continuum concepts when modelling fracture systems with virtually no porosity; in particular, much effort has gone into the study of transport in computer-generated network models. A major difficulty with these models is that of calibration (see, for example, Long and Billaux, 1987). These discrete concepts might be applied to the chalk macro-fissures, which are typically of three types: pipe or channel fissures; vertical or high-angle cracks or joints; and bedding-plane fissures.

Channel models are somewhat less general and less complex than network models. Such models were introduced as a result of various observations that indicate that flow in fractured rock is often restricted to relatively small channels within the rock. These models are very sensitive to aperture variations along their length (Tsang and Tsang 1987). Channelling is important because it reduces the dynamic porosity and hence increases velocities; also, it limits the contact area between mobile and immobile water, which controls the amount of sorption and diffusion into the matrix. The possibility of channelled flow in chalk should be taken into account when estimating travel times for pollutants.

Multiphase flow

Many contaminants, such as organic solvents and oil, enter groundwater systems in the immiscible phase. The physical parameters affecting such flows are: density, viscosity, interfacial tension, wetting properties, and fissure widths, wall roughnesses, and the dip of the strata. There are as yet very inadequate data on the corresponding characteristic parameters for systems such as the Chalk (see, for example, Chilton *et al.* 1990).

Multiphase flow modelling of double-porosity formations is still in its infancy in hydrogeology. However, almost identical problems have been studied by petroleum reservoir engineers concerned with the processes of gas/oil drainage and water/oil imbibition (which are analogous to the hydrogeological processes of unsaturated flow and pollution by immiscible fluids, respectively). There are, however, continuing difficulties; in particular, there is disagreement over the degree of capillary continuity between neighbouring matrix blocks.

The complex qualitative behaviour of dense immiscible liquids in fissures has been investigated by Schwille (1988) using laboratory models. Further experiments need to be carried out specifically for the Chalk.

Dense solvents are among the most frequently occurring organic contaminants of groundwater and are therefore of particular interest. Because of their high density and the fact that they do not wet saturated chalk, these liquids will penetrate deep into a chalk aquifer. They will spread over a wide area and adopt a complex geometry within the fissures. They will then slowly dissolve into the groundwater and thereby act as a secondary source. The rate of dissolution into the passing water will depend on the area of contact and the flow velocity. However, diffusion through the chalk matrix is also likely to be important because of the relatively large surface areas over which it can occur. Given the limited knowledge of the structure of chalk fissure systems, it is not yet possible to model solvent transport with any confidence, especially in the region of the source. However, attempts at modelling will have to be made because of the need to estimate the lifetimes of pollution problems and to aid the design of effective clean-up strategies.

Unsaturated zone

The behaviour of groundwater in the unsaturated zone is significantly different from that in the saturated zone, mostly as a result of capillary forces. The flow patterns can be exceedingly complex even in idealized laboratory systems or in computer simulations. Exotic theories involving such concepts as percolation, fractal geometry, and chaos have all been brought to bear on these problems. In the face of such difficulties, practical models have had to be based on fairly crude physical concepts.

Oakes *et al.* (1981) described a model for the leaching of nitrate in which 5–10 per cent of the leachate is passed directly to the water table (through larger fissures) and the rest is passed downwards in a piston-like manner. Barker and Foster (1981) used what was essentially a saturated double-porosity model in a quantitative investigation of the diffusion-exchange model proposed by Foster (1975) to resolve the tritium anomaly. Oakes (1977) had previously considered a simpler form of the same model which assumed instantaneous diffusive equilibrium between the fissures and the matrix, and was therefore essentially an equivalent porous medium model.

Isotope dating

The dating of groundwater, on the basis of exponential decay of isotopes with time, is in principle very straightforward: the *apparent* isotope age is given by $t_{\frac{1}{2}} \ln(c_i/c)/\ln 2$ where $t_{\frac{1}{2}}$ is the half life, c_i is the input concentration (assumed constant), and c is the measured concentration. However, there are four important difficulties: misunderstandings about the nature of the transport processes; ambiguities in the meaning of terms such as 'age of the water'; problems due to a lack of local concentration equilibrium; and chemical changes. The first three difficulties will be addressed here.

In any groundwater system, and especially in the Chalk, the water molecules assume a wide distribution of velocities and a correspondingly wide distribution of travel times. For steady-state flow, the mean travel time is given by the mean velocity of all the molecules that are free to pass through the system. Since the water in the chalk matrix is relatively immobile, this mean is approximately given by the (mean) fissure velocity multiplied by the ratio of the fissure porosity to total porosity. Although the *mean* travel time depends only on the relative porosities and flow velocities, the *distribution* of travel times is also affected by the fissure and matrix geometries and the value of the diffusion coefficient of the water molecules in the matrix water. This mean travel time seems deserving of the title 'age of the water' when it refers to a system extending from the recharge area to the point of interest. However, in the Chalk another time of interest is the mean travel time in the fissures, which represents the transit of particles which can move through the fissure system but which are too large to enter the matrix. The fissure-system travel time will be smaller than the mean transit time (of diffusing water molecules) by a factor equal to the ratio of the total and fissure porosities—several orders of magnitude smaller.

An experimental determination of water age may give the travel time averaged across the whole system or just across the fissures, or any time in between those two extremes. In past studies it has not always been made clear which age is intended. An important early discussion of the problem was given by Neretnieks (1981)—note that the term *residence time of the water* used in that paper refers to the mean travel time in the fissures.

The time taken to establish (molecular) diffusive equilibrium across a block of chalk with a thickness of a few metres will be no more than a few thousand years; so, under steady-state conditions, the fissure and matrix concentrations of an isotope will be locally similar if the half-life is a thousand years or greater. Therefore, the isotope age for carbon-14 (with a half-life of 5730 a) represents the average transit time, from the source (often the outcrop) to the sampling point, taken over both the fissures and the matrix, provided the fissure separation (block size) is no more than a few metres.

If the half-life is comparable to or less than the time for diffusion across a matrix block, the isotopic concentration will diminish away from the fissures due to decay. This will result in greater apparent ages for matrix water than for fissure water. This will usually be the case for tritium (with a half-life of 12.3 a), but for carbon-14 it will only apply when fissure separations exceed several metres. The situation with large fissure separations needs careful analysis since advection may have a significant effect relative to diffusion (the Peclet number may exceed unity). Under these conditions either a double-porosity or double-permeability model will be needed to interpret the apparent isotopic age. The data requirements of such models include information on the fissure sizes and separations as well as the matrix block shapes; for the Chalk, such parameters are little known and spatially variable. A thorough review of relevant double-porosity modelling was given by Maloszewski and Zuber (1985).

For the isotopic age to be characteristic of the fissure porosity alone, a simple diffusion model can be used to show that the half-life would have to be much less than one week when fissure apertures are less than 5 mm: this is unlikely to be realized in practice.

Adsorption of an isotope on to mineral surfaces magnifies the effective matrix porosity and hence tends to increase the isotopic age relative to the mean travel time of water.

Tritium is a rather different tracer from carbon-14. The input concentration has varied considerably over the time period extending back several half-lives, so it is not possible to derive a ratio of measured concentration with a unique input concentration to determine an age. The use of tritium in age dating is most commonly based on identifying the (spatial) peak position corresponding to the maximum input concentration during the period of intense thermonuclear testing in the atmosphere in 1963–4. In some studies the first rise in tritium concentration above a few tritium units has been associated with the start of thermonuclear testing in 1953. However, these markers will advance with a velocity less than that of the water in the fissures but greater than the mean velocity of the water in both the fissures and matrix.

Carbon-14 dating of groundwater in the Chalk and Pleistocene sands of Suffolk was considered as part of a study by Bath *et al.* (1985). They came to the conclusion that the apparent age probably reflects the average residence time in the fissures and matrix.

5. Chalk water and engineering geology

R. N. Mortimore

Introduction

To assess the potential engineering hazard created by water in the Chalk it is first necessary to identify the range of engineering operations. It is also necessary to establish the engineering characteristics of chalk that generate particular hazards in combination with pore-water and/or fissure water.

Engineering operations that are relevant cover underground excavations, surface earthworks, slope stabilization, preparation of foundations, waste disposal, water disposal in soakaways, and the abstraction of both water and hydrocarbons. Each of these operations is dependent on a knowledge of pore and fissure structure, porosity and permeability, and the degree of saturation. A knowledge of water chemistry and the interaction between the 'dual porosity' of pores and fissures is also often needed. This is in addition to understanding the gross lithology and structure of chalk and chalk terrain evolution.

Some of the questions that arise in engineering operations in the Chalk in relation to hydrogeology include:

1. Is artesian pressure present?
2. What buoyancy uplift or hydrostatic pressures will occur on civil engineering structures such as tunnels, foundations, and walls, for example with fluctuating water levels in intertidal zones or seasonally fluctuating water tables?
3. Will hydrostatic pressures on civil engineering structures change as water is released from elastic storage and/or will settlements of surface structures occur?
4. How much inflow might be expected in tunnelling operations and how will this affect sidewall stability and overbreak?
5. How effective is a grout curtain around a potential excavation and ahead of a tunnelling face?
6. How can excavations be dewatered?
7. In contrast, are there chalks that will not drain freely and thus cause problems, for example, in foundations?
8. Where should highway soakaways be located, what volume of water can be discharged into the Chalk and what pollution problems or other hazards in fissured rocks might this create?
9. What hazard will a landfill create at a particular site and are there measures that might protect the chalk aquifer for all time? What are safe procedures and what real data exist for setting criteria in landfill operations?
10. During ground investigations what drilling fluids can be used so as not to contaminate groundwaters? Similarly, what hydraulic fluids should be used in any equipment such as loading jacks that will be buried in boreholes?
11. Will major road constructions or tunnels which transgress catchments influence groundwater regimes in the Chalk?
12. How will engineering operations interact with the surrounding chalk and other, possibly discrete and/or interconnected, aquifers?
13. What is the influence of pore-waters in the matrix of the Chalk on physical properties and engineering behaviour in, for example, earthworks?

These questions can be grouped into two broad areas. The first deals with fissure porosity and permeability and includes water pressures on civil engineering structures or constructions; water (or fluid) discharges to the Chalk aquifer; and the wider impact of large-scale engineering structures on the aquifer. The second area involves the influence of pore-water on the behaviour of chalk as an engineering material, for example in earthworks.

This chapter first identifies the range of engineering characteristics of the stratigraphical units of the Chalk and terrains in or on which water will act or has acted. Case histories then illustrate the influence of both matrix and fissure water on engineering operations. Not all the above questions can be answered in the same detail, but reference is made to key papers dealing with the topics.

Chalk lithostratigraphic units, fracture characteristics and hydrogeological properties

Fracturing and lithology

It has often been suggested that different lithological units in the Chalk have quite different aquifer properties (see, for example, Price 1987; Mortimore 1979; 1990; Mortimore *et al.* 1990*a*, p. 50). This results from a lithological control on fracturing which is illustrated in Figs 5.1, 5.2 and 5.3 for chalks in the Anglo-Paris Basin. Not only does the fissure volume vary but the hydraulic continuity varies depending on the extent and tightness of fractures or the sealing effects of chalk-putty produced by the softer varieties of chalk. The chalk-putty may result from tectonic softening or weathering along fractures.

The Lower and Middle Cenomanian Chalk Marl within the main axis of the Anglo-Paris Basin comprises alternating layers of clay-rich marls and hard to very hard, 300 mm thick, limestone bands containing fossil sponges and abundant calcispheres. Subhorizontal joints are common in the marly layers and vertical joint sets are more common in the limestone bands. It has been noted in some tunnelling operations that particular, plastic, clay-rich layers in the lower Chalk Marl, traceable over many hundreds of square kilometres, give rise to perched water horizons. Water presumably collects at the interface between the marl and the overlying limestone as a result of both more intense fracturing and horizontal hydraulic continuity of fracture networks in the overlying limestone. These confined bedding fractures link to more profound water pathways along widely spaced master fractures and faults.

Another example of perched water in the Lower Chalk is the 'Cast Bed' near the top of the Chalk Marl which forms a spring line (for example, at Lydden Spout, Shakespeare Cliff, Dover; Birch

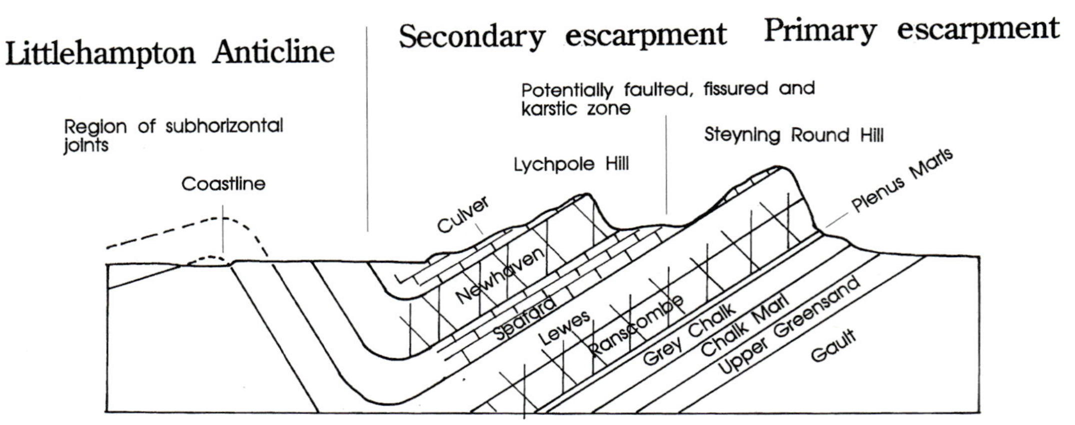

Fig. 5.1. A profile across the South Downs between Steyning in the north and the coast at Worthing in the south, showing the relationship between fracturing and lithology and the development of the primary and secondary escarpments.

1990). These examples of water in beds often considered to be almost impermeable could also mean that leachates escaping from a landfill site could follow the bedding more easily than might be expected (see later section on Landfill).

This relationship between more and less competent chalk layers and fracturing is even more conspicuous in East Sussex between the thick Plenus Marls and the overlying very hard Melbourn Rock and Holywell Beds (Fig. 5.3). The very hard

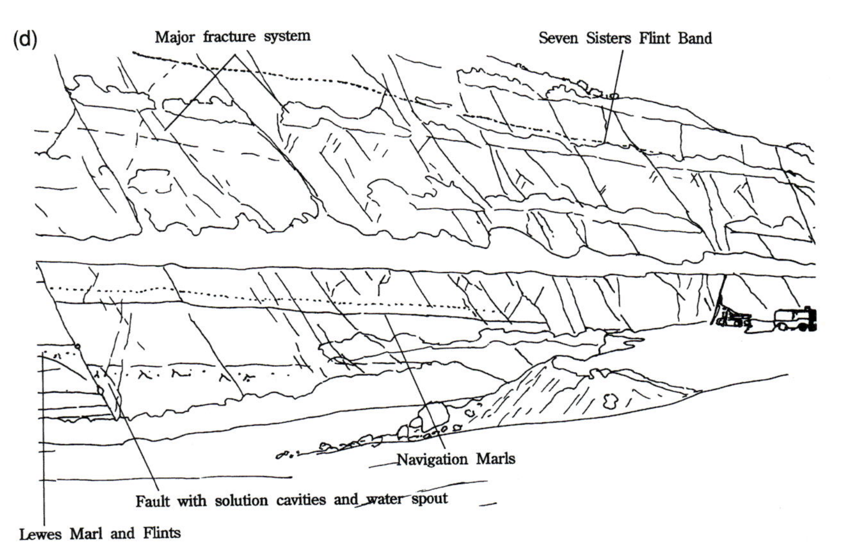

Fig. 5.2. Dimensions of different fracture systems in the main Chalk lithologies. (a) Blocky chalk with two regular, close- to medium-spaced joint sets at a depth of 12 m below the ground surface at Charmandean Chalk Pit, Worthing, West Sussex in the Culver Chalk Member. The joints become very closely spaced in the uppermost 3 m. This blocky chalk is typical of the relatively homogeneous pure white-chalks of the Seaford and Culver Chalk members. (b) Inclined conjugate joint sets and horizontal fractures along marl seams in the Newhaven Chalk Member at Newhaven, East Sussex. Such joint patterns are typical of the White Chalk members with marl seams (Ranscombe Chalk, lower Lewes Chalk and the Newhaven Chalk). Note the cavity developing at the intersection point of the joints. (c) Inclined conjugate joint sets in the Welton Chalk Formation of northern England, (Melton Ross Chalk Pit, South Humberside). This demonstrates the widespread occurrence of this joint pattern in chalks with marl seams. The Welton Chalk Formation is roughly equivalent to the Turonian Ranscombe Chalk of southern England and the Barton Marls equate with the Glynde Marls (Mortimore and Wood, 1986). (d) A major inclined set of joints in the Lewes Chalk and basal Seaford Chalk in Shoreham Cement Works Quarry, West Sussex. These joints form the dominant partner in a conjugate set. This also illustrates the depth to which joint systems are well developed and that major water courses operate along karstic openings along the major joint system to this depth.

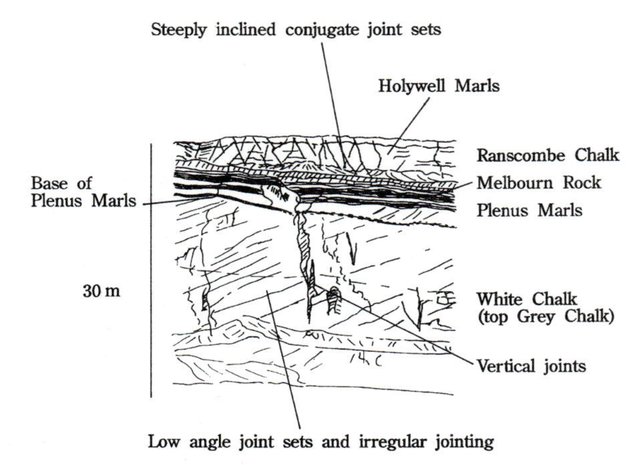

Fig. 5.3. Contrasting fracture styles in the Upper Cenomanian White Bed, Plenus Marls and Melbourn Rock and the overlying Turonian Ranscombe Chalk in Asham Chalk Pit 3, Beddingham near Lewes, East Sussex, England. The inclined conjugate joint sets in the Ranscombe Chalk feed water down to the Melbourn Rock on top of relatively impermeable Plenus Marls. Note that neither the Plenus Marls nor the White Bed and Grey Chalk are completely impermeable and are crossed by joints and faults.

rock layers are extensively fractured by steeply inclined conjugate sets of joints. These are not developed in the thick clay-rich Plenus Marls which have dissipated stresses sub-horizontally, causing the overlying fractures in the Melbourn Rock to open as a result of tension. This situation is exploited by constructing water-collecting adits along the Plenus Marls–Melbourn Rock interface at Holywell, Eastbourne. The same style and intensity of fracturing is present in the Melbourn Rock of Kent (Mortimore 1990, Fig. 8, p. 36).

Similar marl and fracture relationships are present in the Middle Turonian New Pit Beds where intense conjugate fracturing dissipates along the marl seams and water follows the fractures until it meets a marl seam (Fig. 5.4a, b; see also Isle of Wight at the same stratigraphical level (Barton 1990)). Such marl seams have solution-widened cavities along them. This pattern of fracturing and palaeokarst along marly surfaces is also present in the Late Santonian to Early Campanian Newhaven Chalk of southern England (Fig. 5.2b; 5.4c, d). Seepages along these marl seams can be seen in winter on aerial photographs taken along the dip slope of the secondary escarpment west of the Adur around Lancing Hill, West Sussex.

In contrast, the more homogeneous, pure white, soft to medium hard Seaford and Culver Chalk Members of the main axis of the Anglo-Paris Basin, without marl seams, contain regular, often orthogonal, joint sets (Figs. 5.2a; 5.5a, d; 5.6b). This generates a potentially greater fissure pore volume for water storage and possibly a slower rate of percolation (see Price 1987, p. 145, for calculation of hydraulic conductivity in this setting; see Giles and Lowings 1990, for transmissivity and storativity differences between catchments which may be partly lithostratigraphically controlled).

Very soft chalk units such as the Late Cenomanian White Bed of the North Downs, England (Mortimore *et al.* 1990a) and the northern part of the Paris Basin, and the very soft Margate Chalk (Kent–Essex) are irregularly fractured and the fractures are often sealed by a skin of remoulded chalk-putty. Thus, these very soft chalks can act as aquitards or even aquicludes. Putty-chalk along faults in this soft material can also act as a barrier to hydraulic continuity between major aquifer blocks in the North Downs of southern England, in northern France and in the Yonne (Fig. 5.5a, b). This could be important if, for example, the potential movement of pollution plumes needs to be calculated.

The chalks of the Anglo-Paris Basin with regular orthogonal joints are generally the soft to medium hard, pure varieties and these appear to have better-developed karstic systems than the softer chalks. A cave system has been mapped in the Seaford Chalk at Beachy Head (Reeve *et al.* 1980). The same chalk unit has well-developed karstic features along faults and master joints in Shoreham Cement Works, West Sussex (Figs. 5.2d; 5.4e). This relationship may be illusory as active sinkholes are known, for example, in the very soft chalks of the Ruislip Reservoir area (Hester 1980) and the Mimms Valley (Wooldridge and Kirkaldy 1937; Kirkaldy 1950). However, these are in river valleys, and it has been shown that away from river valleys transmissivity can be very low, as in the London Basin (Water Resources Board 1972) and southern East Anglia (Lloyd *et al.* 1981). Thus the harder chalks of, for example, the South Downs, the Aube and Marne may be very different, hydrogeologically, in contrast to the softer chalks at the

Fig. 5.4. (a) Middle Turonian New Pit Beds showing typical steeply inclined, conjugate fractures above the New Pit Marl 2 with water emanating from the solution-widened fractures and along the upper surface of the marl, Senneville, near Fécamp, Normandy. (b) Middle Turonian New Pit Beds at the north Portal of the Lewes Tunnel, Lewes, East Sussex showing intense steeply inclined, conjugate fracturing producing slope instability. (Compare with similar stratigraphic horizon in northern England, Fig. 5.2c and Fig. 5.6a). (c) Karstic surfaces along the upper surface of a marl seam in the Early Campanian Newhaven Chalk at Newhaven, East Sussex, England. (d) Early Campanian Newhaven Chalk showing typical steeply inclined, conjugate fractures. Sub-horizontal discontinuities, some of which have been widened by solution, are along marl seams . (e) Solution widening and calcite recementation along a small fault in the Seaford and Lewes Chalks of Shoreham Cement Works, West Sussex. The point where a water spout used to develop each winter is evident by the cavity towards the base of the picture and at the junction of fractures and the Bridgewick Marl 2.

Fig. 5.5.
(a) Extremely soft chalk at St Riquer near Abbeville, Somme, France, showing reqular joint sets which are generally closed and prone to putty fills.
(b) Very soft, homogeneous Early Campanian chalk near Pont sur Yonne and Sens, Yonne, France. Any fractures are usually tight and sealed.
(c) Hydrofracturing in Late Turonian chalk at Fontaine Luyères, Aube, eastern Paris Basin. The distinctive vertical fracture sets have been slightly bent downhill by surface creep. Note the Lewes Marl below in which a few tubular Lewes Flints also occur.
(d) Regular orthogonal joint sets producing blocky chalk in the late Coniacian soft, pure white-chalk at Poix, Picardy, France.
(e) Hydrofracturing in the medium hard Seaford Chalk in a road cutting at Patcham, Brighton, East Sussex, showing the same distinctive style of fracturing seen at Fontaine Luyères. This hydrofracture zone is accompanied by caverns filled with sediment and colour banded staining in the chalk.
(f) The hydrofractured zone at Patcham showing the sediment-filled cavern and colour-banded chalks.

same stratigraphical levels on the London–Brabant Platform and the Somme region.

Support for the observation that lithology has a strong control on fracture frequency and/or style and thus on permeability comes from permeability measurements made in boreholes in the Candover Valley, Hampshire (Price *et al.* 1977; 1982) where some of the highest permeabilities occur in the Chalk Rock (Price 1987).

In future it is hoped that, by fixing the altitude and the exact stratigraphy of the saturated rock column, the relationship between lithology and the key water-bearing horizons can be established.

Fracturing and tectonic structure

In addition to the broad lithological divisions of the Chalk, local zones of more or less intense fracturing, with a particular style of fracture, are related to the tectonic regime (Fig. 5.7). For example, intense hydraulic fracturing is present along the Metz Fault Zone in the Aube and the Marne of the northeastern Paris Basin (Fig. 5.5c, Coulon and Frizon de Lamotte 1988). Similar intense fracturing, probably initiated by high water pressure along and adjacent to a fault, was encountered at Patcham during construction of the Brighton Bypass (Fig. 5.5e, f). This zone of fracturing has subsequently developed dissolution cavities filled with residual Tertiary and/or Quaternary sediments. The surrounding chalk is heavily Liesegang banded. Many other examples of secondary karstification of particular fissures have been recorded (Headworth 1978; Headworth *et al.* 1982; Price 1987). Hydrofracturing, of the type identified by Coulon and Frizon de Lamotte (1988), is probably not reduced at depth and thus provides highly permeable zones right through the Chalk, unlike fractures related to lithology or geomorphology.

Fig. 5.6. (a) Steeply inclined conjugate fractures in the Middle Turonian Welton Chalk Formation, Melton Ross Quarry, South Humberside. The Barton Marls form the conspicuous layers at the top of the quarry face. Filled solution cavities are present along a bedding plane created by a marl seam towards the base of the quarry face. (b) Regular orthogonal joint sets in the Late Coniacian medium hard, homogeneous, pure white Seaford Chalk of Shoreham Cement Works, West Sussex. Some inclined conjugate joints are also present. The Seven Sisters Flint forms the continuous seam across the face behind Bernard Pomerol. (The exposure is about 20–25 m below the ground surface). (c) Weathered profile in the Early Campanian Newhaven Chalk adjacent to Brighton Marina, Sussex. The dark flinty loess passes down into pale-coloured Coombe Deposits with involutions. Beneath the involutions the very closely spaced fractures are typical of Mundford Grade IV Chalk. The marl seams form a boundary to Grade III more blocky chalk. Note the presence of inclined conjugate fractures which penetrate to the base of the involutions; one is filled with sheet flint.

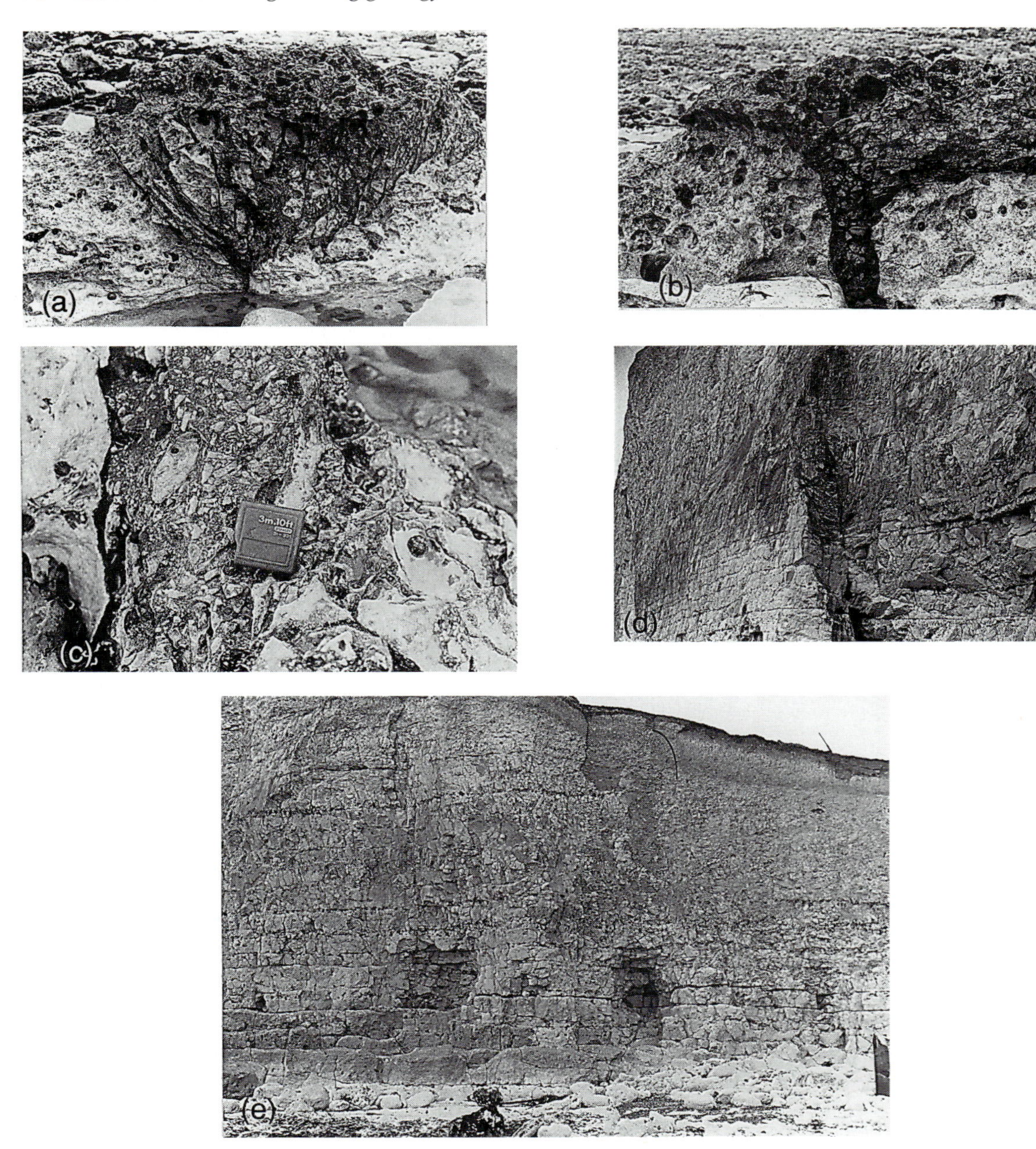

Fig. 5.7. (a), (b) Different styles of hydrofracturing and brecciation along a fault exposed in the wave-cut platform east of Birling Gap, Beachy Head East Sussex. (c) Breccia in the hydrofractured zone at Birling Gap. The breccia is in Late Coniacian chalk (*Volviceramus involutus* beds) and is similar to the breccia at Omey near Vitry-le-Francois on the Marne, northeast Paris Basin, the site of hydrofracturing described by Coulon and Frizon de Lamotte (1988). (d) Oval-shaped hydrofracturing along the fault zone at Birling Gap, Beachy Head, East Sussex, in the Late Coniacian Seaford Chalk. Note the regular jointing producing blocky chalk between the flint seams on the right-hand side of the picture. (e) A high-permeability zone along the west margin of Hope Gap Valley at Hope Gap, Seaford Head, East Sussex. This illustrates the intensity of open fractures and increased permeability typical of valley margins in the North and South Downs, England. The Chalk is progressively destroyed by weathering/alteration towards the axis of the valley. Could the putty-chalk forming the matrix in the valley floor have acted as an aquitard and confined converging flows to deeper levels, thus increasing the depth of weathering in the valley and causing the heave structures common in these situations?

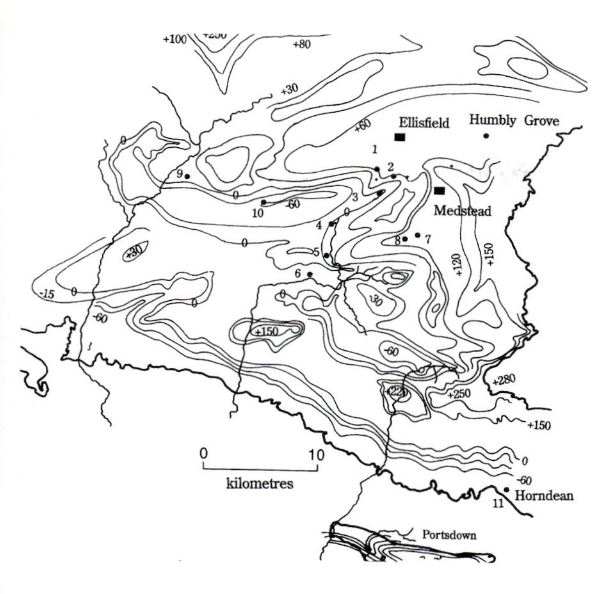

Fig. 5.8. Structure contour map of the Chalk in central Hampshire showing 'domes' rather than linear axes of folding. These domes broadly match the groundwater 'mounds' discussed by Giles and Lowings (1990). The datum is the Late Turonian Bridgewick Marl; contours are in metres. Boreholes: 1 Axford, 2 Bradley, 3 Wield, 4 Totford, 5 Abbotstone, 6 Itchen Abbas, 7 Frias F2, 8 Frias F4, 9 Barton Stacey, 10 Micheldever, 11 Horndean.

Headworth (1972, Fig. 1) and Giles and Lowings (1990) suggest that the coincidence of groundwater peaks with anticlines in the Itchen Catchment of Hampshire indicates that the Chalk's permeability is less on anticlines than in synclines. In the Itchen Catchment, topography appears to be concordant with structure (that is, low-lying areas, the river valleys, coincide with synclines). Thus, as indicated above for Ruislip and the River Mole, the valleys are more permeable than the surrounding hinterland. In other areas, however, this is not the case and topography can be inverted in relation to structure (for example, Mt Caburn at Lewes, East Sussex, is high ground in a syncline and the Kingston Anticline has been eroded along its axis).

The observations of Giles and Lowings (1990) are further supported for the Itchen area by constructing a more detailed structure contour map (Fig. 5.8). Dolfuss (1890; 1910), Wooldridge and Linton (1955), and Jones (1981) have all recognized that the folds in the Chalk form *en échelon* periclines rather than having the simple linear axes shown by Giles and Lowings (1990, Fig. 1, p. 620). These structures have a long history of growth through the Cretaceous and Tertiary (Mortimore and Pomerol 1991) influencing sedimentation and subsequent landscape evolution. The Itchen Catchment structure contour map (Fig. 5.8) identifies anticlinal 'domes' rather than simple linear fold axes and it is these domes that coincide with the groundwater 'mounds' of Giles and Lowings (1990, Fig. 2, p. 621). The Lasham and Preshaw groundwater mounds match the Lasham and Warnford anticlinal domes. The Medstead groundwater mound occurs towards the northern end of the Medstead anticlinal dome. This Medstead structure is complex and is probably a composite of two directions of gentle folding, north–south and east–west. This may partly explain the southerly and westerly flows indicated by the groundwater contours that puzzled Giles and Lowings (1990, p. 622).

The Preshaw and Froxfield groundwater mounds are on, or adjacent to, much stronger anticlinal domes (in amplitude) at Warnford and along the western end of the Fernhurst Anticline. Fracturing may be more extensive here than on the other weaker structures to the north. This may account for the difference in groundwater fluctuations noted by Giles and Lowings (1990). These fluctuations may also be partly due to lithology as the Medstead, Lasham, and Ellisfield mounds are in the homogeneous Seaford Chalk with regular blocky joints. The Preshaw and Froxfield mounds are in the upper half of the nodular Lewes Chalk with marl seams.

The tectonic folds are related to underlying much deeper faults (Chadwick 1986; Mortimore and Pomerol 1991). More work is required to establish the detailed relationship of the pattern of fracturing to particular structures in the Chalk but major faults penetrating the Chalk cannot be ruled out (for example, Gaster's Coldean Fault; Mortimore and Pomerol 1991). These must have an influence on groundwater behaviour. In Hampshire, sharp geniculations and re-entrants in the structure contours (Fig. 5.8) could be caused by faults. There are several potential fault lines.

Regional trends in fractures in the Chalk of southern England have been described by Wooldridge and Linton (1955), Middlemiss (1967; 1983) Fookes and Denness (1969), Fookes and Parrish (1969) and Ameen and Cosgrove (1990).

There are similar studies on the Humberside Chalk (Patsoules and Cripps 1990) and the Isle of Wight–Dorset Chalk (Mimran 1975). Middlemiss (1983) related many of the joint sets to regional folding. There is also evidence that particular fracture styles and sets characterize particular regions (Mortimore 1979; Bahat 1990) and these may reflect deeper-seated tectonic lineaments. Bahat (1990) has identified a way of determining the origin of fractures in terms of burial, uplift and tectonic episodes. Such studies are now being applied specifically to identify particular fracture zones in the Chalk which will have application to hydrogeological studies.

Studies of regional patterns of fracturing are essential in engineering operations if, for example, it is necessary to predict the occurrence of fracture types and zones which might cross the route of a tunnel. Such fractures would in turn influence hydrostatic pressures, inflow and stability of underground excavations.

Water-well geophysical logs and chalk structure

Gray (1958; 1965) recognized the stratigraphical significance of electrical resistivity marker bands in the Late Cenomanian and Turonian chalk of Hertfordshire and Leatherhead, Surrey. The relationship of the geophysical log signatures to both the broad and detailed lithostratigraphy of the Chalk was subsequently established in Sussex and Hampshire (Mortimore 1979; 1986), between the London Basin and Humberside (Murray 1986; Mortimore and Wood 1986), and in Lincolnshire (Barker *et al.* 1984). Further long-distance correlation was established across southern England and to the Paris Basin (Mortimore and Pomerol 1987). This work continues with the linking of water-well and oil-well logs for the entire Paris Basin.

The logs for Fetcham Mill, Leatherhead (Gray 1965; Murray 1986; Mortimore and Pomerol 1987) and those for Victoria Gardens and St Peters Church, Brighton (Southern Water Authority 1979*a*; Mortimore 1979; 1986) can be considered standards for southern England. These are linked to the Nibas 101 borehole log in Normandy (Fig. 5.9). Borehole logs from South Pickenham (Norfolk) are used as a standard for East Anglia. Similar studies linking the lithostratigraphy of the Chalk to bore-

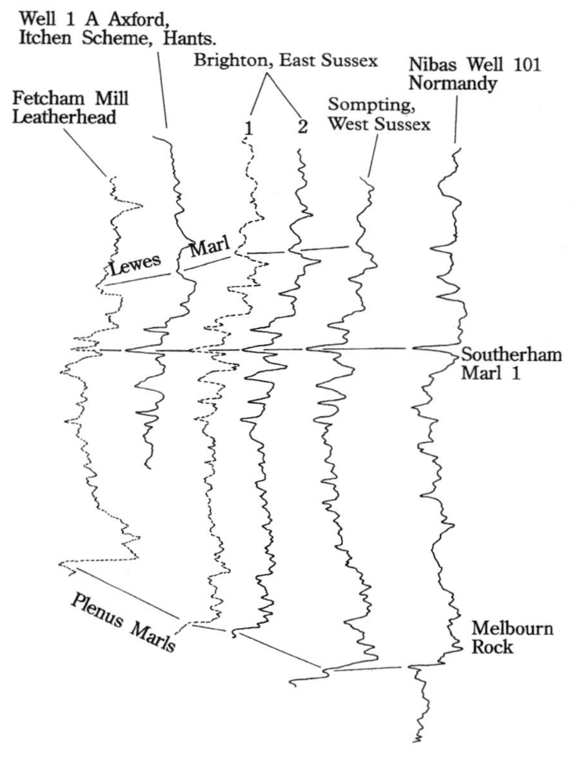

Fig 5.9. Electrical resistivity borehole logs showing key marker bands in the late Cenomanian, Turonian and Coniacian chalks of the Anglo-Paris Basin and also showing long-distance correlation. The peaks of key marker bands relate to marl seams and hard, nodular chalk bands.

hole geophysical logs have been published for the North Sea and Danish Trough (Nygaard *et al.* 1990).

The value of the electrical resistivity borehole logs as indicators of stratigraphical markers and larger-scale lithological changes is clear. Microresistivity logs have proved as good as, and in some cases better than, the original 16 inch/64 inch standard logs. These techniques can only be applied below the water table, but sonic logs have proved to be just as useful both above and below the water table—for example, in the Winterborne Kingston Borehole, Dorset (Rhys *et al.* 1982; Mortimore 1987; Mortimore and Pomerol 1987) and in the Marchwood Borehole, Hampshire (Mortimore & Pomerol 1987). Natural gamma logs have also been

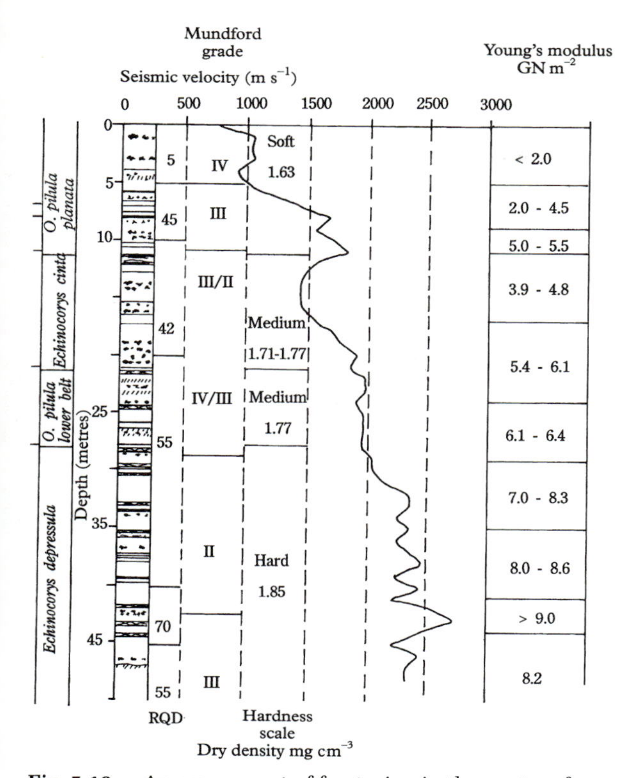

Fig. 5.10. An assessment of fracturing in the centre of an interfluve in the medium hard Early Campanian Newhaven Chalk with marl seams near Shoreham, West Sussex, England. This geomorphological setting should provide tighter rock than valley margins and valley floors. The lithological log is obtained from borehole cores which are also used to assess RQD (a measure of fracturing, 0–10 per cent highly fractured; 80-100 per cent widely spaced fractures) and Mundford Grades (refer to Table 5.1). The P-wave seismic, velocity profile is shown, and velocity increases with depth as the joints become generally tighter. Marl seams, possibly with karstic cavities and subhorizontal fractures, reduce velocities giving the profile its 'spiked' appearance. The seismic velocities, in conjunction with bulk density values (not the dry densities shown), are used to calculate Young's modulus. One particularly hard, tight bed is present at a depth of 43 m. This profile is taken in a particular geomorphological setting and in a particular part of the White Chalk. It cannot be extrapolated to every interfluve where the type of chalk may be different. Compare this profile with Fig. 5.12. (Data from L. G. Mouchel and Partners).

used and these sometimes work well but the gamma contrasts can be very low, making definitive identification of markers very difficult. This may often result from the wrong scale being set on the natural gamma record when many logs are run together in the same borehole, to get all logs on the same sheet of paper. Modern digital recording techniques should overcome this problem.

The relationship between neutron density/porosity logs and porosity measured on core samples of chalk has yet to be fully established, but this method works well in other formations (see, for example, Price 1985, p. 131). Neutron logs are now used more frequently in ground investigations to compare and contrast with other geophysical logs and physical property measurements (see Fig. 5.10). Such logs are, of course, routine in oil-well logging of chalk sequences in the North Sea (see, for example, D'Heur 1990).

As well as providing invaluable stratigraphical information, the combination of many borehole logs with core and outcrop information forms a basis for drawing structure contours of various horizons in the Chalk. Establishing the stratigraphical and structural frameworks is the first step in a hydrogeological, petroleum and engineering analysis of an area (see Fig. 5.10). Geophysical borehole logs from the Itchen Catchment in Hampshire can be used not only to test the accuracy of the structure contours but also to indicate which lithologies are present where unusual groundwater regimes occur.

Geophysical borehole logs are also used to monitor changes in salinity or chemistry, temperature, and flow rates of the groundwater, and to identify directions of flow and the presence of pollutants.

Engineering description of chalk ground conditions and relationship to hydrogeology

Features to be described and descriptive terminology

Chalk was, for many years, treated as a homogeneous material for engineering purposes, but even in the Symposium on Chalk in Earthworks and Foundations, organized by the Institution of Civil Engineers in 1966, Higginbottom (1966) recognized the importance of fracturing, weathering, particularly frost shattering, frost heave, solifluction deposits, and putty-chalk. Swallow-holes and pipes and their relationship to Tertiary and Quaternary

Table 5.1 The Mundford Grades (after Ward *et al.* 1968)

Grade V	Structureless melange. Unweathered and partially weathered angular chalk blocks and fragments set in a matrix of deeply weathered remoulded chalk. Bedding and jointing absent.
Grade IV	Friable to rubbly chalk. Unweathered or partially weathered chalk with bedding and jointing present. Joints and small fractures closely spaced, ranging from 1 to 6 cm apart. Joints commonly open up to 2 cm and infilled with weathered debris.
Grade III	Rubbly to blocky chalk. Unweathered medium to hard chalk with joints 60–200 mm apart. Joints open up to 10–30 mm, sometimes with secondary staining and fragmentary fillings.
Grade II	Medium hard chalk with widely spaced closed joints. Joints more than 200 mm apart. When dug out for examination purposes this material does not pull away along the joint faces but fractures irregularly.
Grade I	Hard, brittle chalk with widely spaced, closed joints. Details as for Grade II but the chalk is harder. Rock beds such as the Chalk Rock, Top Rock and other major hardgrounds all fall within this grade.

deposits capping chalk terrains were also described. Active swallow-holes have been known for a long time in the Mole Valley where, at times when the water table is low, the water from the River Mole partly flows in solution cavities (Dines and Edmunds 1933; Wooldridge and Kirkaldy 1937; Kirkaldy 1950; Higginbottom 1966). Ephemeral streams such as the Lavant near Chichester and the Winterborne at Lewes illustrate similar fluctuations in groundwater levels that have to be considered in engineering projects.

At Mundford, in south Norfolk, Ward *et al.* (1968) introduced a method of describing the ground profile in chalk terrains (Table 5.1). Their grade descriptions were based on a progressive reduction in fracture frequency with depth and an increase in fracture tightness and hardness of the chalk material with depth. These grades have been related in various ways to engineering hydrogeology, firstly by crudely determining the permeability of each grade, and then by using the knowledge of permeability and structure to plan dewatering strategies (Roberts and Preene 1990; Table 5.2) and to design highway soakaways. These grades have also been used to identify changes in Poisson's ratio and Young's modulus with depth, important properties in relation to the elastic behaviour of the Chalk aquifer (Price 1987) as well as for calculating potential foundation settlements.

Allen and Price (1990) describe the importance of investigating hydraulic conductivity within the weathered layers (Mundford Grades III–V) so that infiltration and migration of pollutants can be better understood. Clearly, accurate description of the ground is essential if more reliable predictions of water movement are to be achieved. Research since 1968 has shown that fracture frequency, tightness and chalk hardness do not vary linearly with depth (Mortimore *et al.* 1990b). On the contrary, there are distinct regions with particular hardnesses and fracturing which vary according to geomorphological setting, lithology, and tectonic structure (Fig. 5.11; Mortimore *et al.* 1990a).

Tables 5.3–5.5 provide recommended terminology for describing the material and mass characteristics of chalk on borehole logs and field sections. If Poisson's ratio and Young's modulus do vary

Table 5.2 Permeability of the Mundford Grades (m s^{-1}) (after Roberts and Preene 1990)

Grade V	Structureless melange	10^{-7} to 10^{-9}
Grade IV	Friable to rubbly chalk	10^{-5} to 10^{-3}
Grade III	Rubbly to blocky chalk	
Grade II	Medium hard chalk with widely spaced closed joints	erratic due to fissures
Grade I	Hard, brittle chalk with widely spaced, closed joints	

Table 5.3 Hardness and intact dry density classification of chalk (after Mortimore *et al.*, 1990*b*). The intact dry density values in the left-hand column are in megagrams per cubic metre.

Extremely soft <1.55	Crushed to putty when struck with the hammer head; steel pin pushed in by hand; geological pick penetrates with very light blows; crushed easily by hand and remoulds to putty easily.
Very soft 1.55–1.60	Crushed to putty when struck with hammer; steel pin hammered in by hand; putty splashes when struck with geological pick; broken with ease by hand.
Soft 1.60–1.70	Part fractured, part crushed, sometimes with a flinty ring when hammered. Steel pin and geological pick hammered in fairly easily; broken by hand with slight difficulty.
Medium hard 1.70–1.80	Fractured with a flinty ring when struck with hammer (clean, conchoidal fractures in purer chalks); some resistance to hammering in steel pin or pick end; broken with difficulty by hand.
Hard 1.80–1.95	Fractures when struck with hammer, chips and fragments fly off after heavy blows; great difficulty hammering in steel pin or pick; broken with great difficulty by hand.
Very hard 1.95–2.40	Slight ring when struck with hammer, and chips fly off when struck hard; steel pin and pick make only a slight dent; impossible to break with hand pressure.
Extremely hard > 2.40	Rings when struck with hammer, splinters after many heavy blows; impossible to hammer in steel pin, penetrate with geological pick, or break by hand.

significantly depending on fractured state and weathered conditions (Fig. 5.10), then accurate ground descriptions will be essential. This is in addition to the impact of fracturing and weathering on infiltration rates and hydraulic conductivity.

As well as field descriptions of the ground, geophysical techniques have been used to characterize the top 40 m of weathered chalks. In a region to the east of Reims, in the north-east of the Paris Basin, seismic refraction was used to identify superficial deposits and weathered zones on and in the Chalk (Coulon 1986). An interpretation of the seismic velocities in relation to possible weathering grades is shown in Fig. 5.12. This interpretation has to consider the complexity of the soils resting on the Chalk in this Champagne region, a complexity demonstrated in relation to wine growing (Pomerol 1989, pp. 58–60).

Cross-hole seismic techniques, although primarily used to obtain bulk modulus values at different depths in the Chalk, can provide an indication of fracture volume and/or fracture aperture widths. Such results are normally compared with fracture logs of cored boreholes, borehole caliper logs and any local fracture logs of field sections; together, these data will give a crude estimate of Poisson's ratio and Young's modulus (Fig. 5.10).

Table 5.4 Classification of fracture spacing (after British standard 5930:1981)

Structureless or poorly structured	
Extremely closed spaced	<20 mm
Very closely spaced	20–60 mm
Closely spaced	60–200 mm
Medium spaced	200–600 mm
Widely spaced	600 mm–2 m
Very widely spaced	>2 m

Solution softened or hardened chalk

Altered chalks are more difficult to define (Fig. 5.6c). Monciardini (1989) identifies areas of decalcification, silicification and dolomitization of chalk at points along the ECORS profile across the Paris Basin west of Paris. Some of these 'altera-

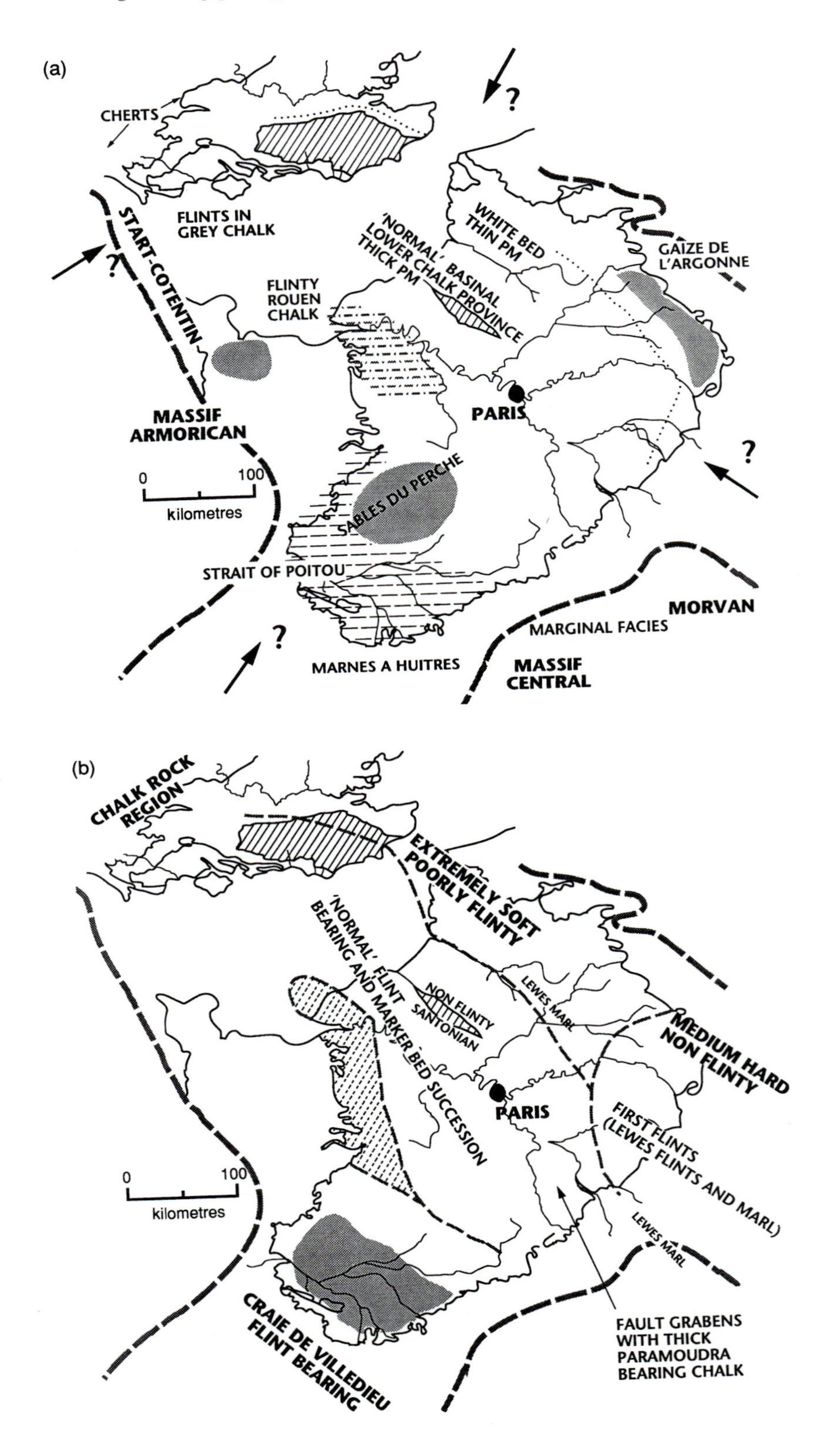
(a)
CHERTS
FLINTS IN GREY CHALK
START-COTENTIN
FLINTY ROUEN CHALK
'NORMAL' BASINAL LOWER CHALK PROVINCE THICK PM
WHITE BED THIN PM
GAIZE DE L'ARGONNE
MASSIF ARMORICAN
PARIS
0
100
kilometres
SABLES DU PERCHE
STRAIT OF POITOU
MARNES A HUITRES
MARGINAL FACIES
MORVAN
MASSIF CENTRAL
(b)
CHALK ROCK REGION
EXTREMELY SOFT POORLY FLINTY
'NORMAL' FLINT BEARING AND MARKER BED SUCCESSION
NON FLINTY SANTONIAN
LEWES MARL
MEDIUM HARD NON FLINTY
PARIS
FIRST FLINTS (LEWES FLINTS AND MARL)
0
100
kilometres
LEWES MARL
CRAIE DE VILLEDIEU FLINT BEARING
FAULT GRABENS WITH THICK PARAMOUDRA BEARING CHALK

Fig. 5.11. Maps showing distribution of the main hardness and fracture regions in the Chalk of the Anglo-Paris Basin.
(a) Simplified lithology of the Cenomanian.
(b) Simplified lithology of the Turonian–Campanian.
(c) Basic Structure: A = region of shallow basement, periclines on cross-faults, pro-glacial outwash and glacial to sub-glacial terrain features; B = region of deep basement, fault/thrust inversions with periclines and river drainage and periglacial geomorphology; C = region of the Metz Fault with hydrofractured hard, pure white flintless chalk; D = region of fault blocks, sometimes poorly fractured chalk but with areas of well-developed deep karst providing major sources of water.

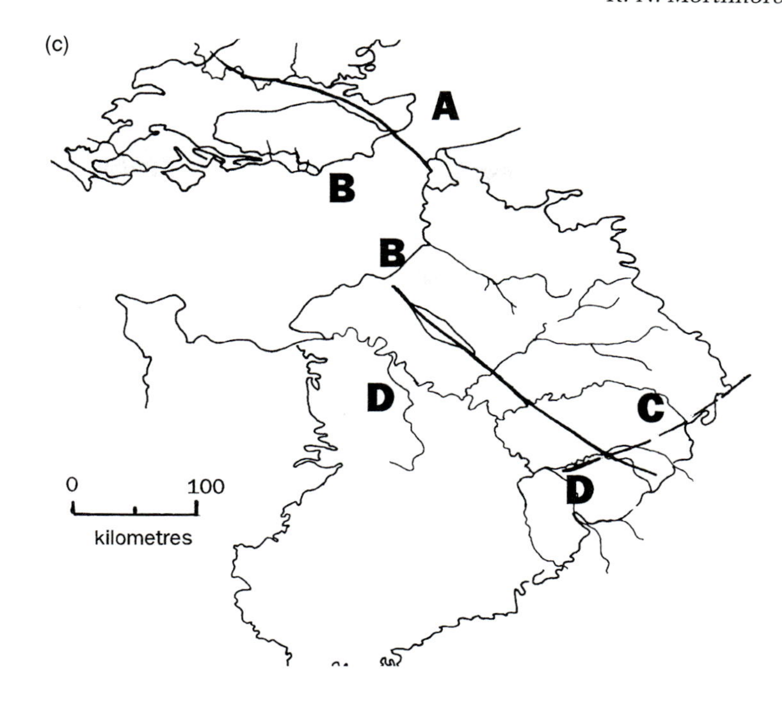

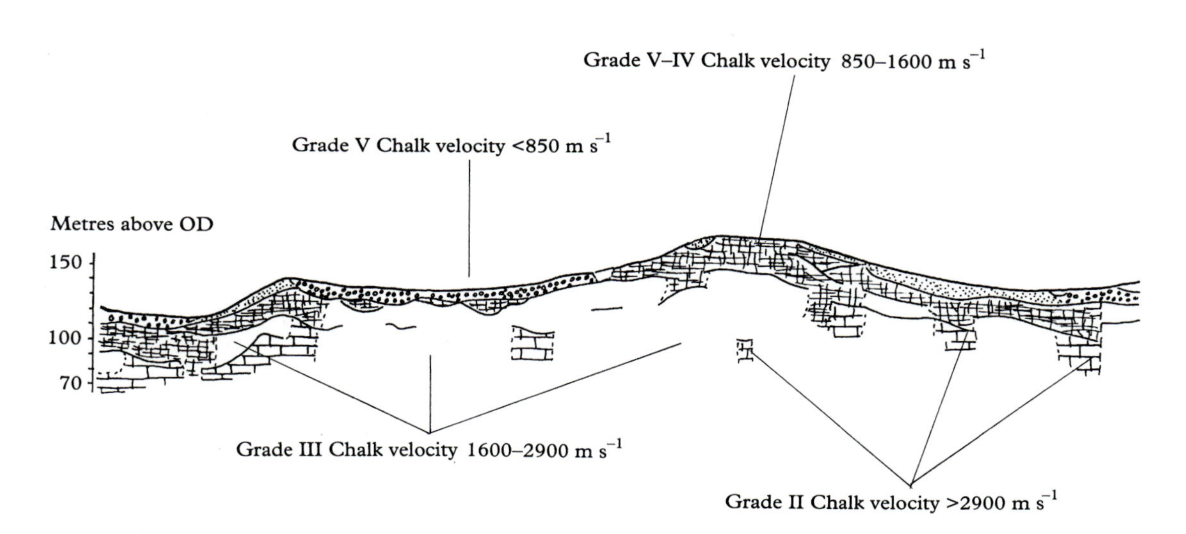

Fig. 5.12. Ground profiles in medium hard, pure white-chalk determined from seismic refraction and mapping in the region north of Montagne de Reims, near Reims, northeast Paris Basin (from Coulon, 1986). Coulon's divisions are reinterpreted in terms of Mundford Grades (see Table 5.1). Compare these seismic velocities with the borehole P wave profile in chalk of similar hardness but with marl seams, (Fig. 5.10).

Table 5.5 Classification of fracture tightness (after Mortimore *et al.* 1990b)

Open fractures	Degree of openness should be indicated in millimetres. Lightly to heavily filled with reconstituted chalk powder or fragments and/or putty-chalk. In other cases residual clays, sands and gravels fill fractures. The intact chalk can be heavily stained throughout and/or on fracture surfaces.
Closed fractures	Fracture surfaces in contact but either not held or only lightly held together so that core removed from liner falls apart along fractures. Chalk can be stained throughout.
Tight fractures	Chalk cannot be broken easily by hand along fractures which may be slickensided, stylolitic and/or stained.
Very tight fractures	Requires a machine to pull fractures apart and the chalk block may break internally rather than fail along the fracture.
Annealed fractures	Completely recemented fractures, calcite cement or 'chalky' veins along the fractures.

tions' may relate to synsedimentary processes while others relate to post-sedimentary weathering.

Areas containing extensive solution pipes often also contain chalks weathered by solution 'tubules' which may reduce the volume of chalk material by 30–40 per cent (also increasing permeability). Chalk surrounding pipes in this situation can be extensively softened and this may provide the clue to the chemical decomposition of chalks beneath dry valleys (Fig. 5.7e).

Areas of hardened chalks also occur where overlying beds have been decalcified and calcite has been reprecipitated (as calcretes), as, for example, beneath loess at Hope Gap, Seaford Head and beneath the Slindon Sands at Eartham, Chichester, West Sussex. The resulting extremely hard layers also have extremely low permeabilities as nearly all fractures and pores are filled with calcite cement.

Engineering/hydrogeological domains in chalk

Understanding landscape evolution provides a further tool for predicting ground conditions (Mortimore 1990, p. 37). Within chalk landscapes several different engineering domains are recognized and have been shown to influence physical properties (Mortimore and Jones 1989).

Wet zones in the Chalk cause considerable problems for earthworks and these are often related to materials capping the Chalk or to regions with solution pipes. Equally, the presence of an aquiclude beneath the Chalk, such as the Gault in the North Downs, causes slope-stability problems close to spring lines. Such situations have had to be thoroughly investigated in the construction of the Channel Tunnel and its associated facilities near Folkestone (see, for example, Hutchinson 1969; 1979; Jones *et al.* 1988; Birch 1990). Areas of the Chalk underlain by well-developed Upper Greensand, for example the Malmstone of West Sussex and Hampshire, have very different geomorphological profiles primarily related to the interaction of water with lithology and associated rock-mass structure. The secondary escarpment of the South Downs, as another example, probably relates closely to the control of water movement in the Newhaven Chalk with marl seams in contrast to the non-marly and well-jointed Seaford and Culver Chalks (Figs. 5.1 and 5.2).

In north-east England, Foster and Milton (1976) described the buried coastline at the eastern foot of the Humberside and Lincolnshire Wolds. This buried cliff and wave-cut platform of the Holderness Plain, with beach deposits resting on it, is considered to be a Quaternary interglacial feature (Catt and Penny 1966). Deep fissuring and the sandy, windblown beach deposits may result in exceptionally high permeability. The setting might generate engineering problems such as quicksands or artesian pressure.

The West Sussex Coastal Plain from Brighton to Bognor Regis to some extent mirrors the Holderness Plain in having an ancient Quaternary interglacial cliff-line and wave-cut platform with raised beach deposits resting on it. The Sussex situation is different because the geological structure is more complex. Synclines filled with Tertiary sediments, as well as chalk, also underlie raised beach deposits, and in addition more than one Quaternary cliff-line and wave-cut platform are present. Nevertheless,the engineering implications are similar to Holderness, and quicksands and artesian conditions may occur. Henry (1986, p. 19) reported quicksands (the Slindon Sands) on the Tertiary clays along the A27 road west of Worthing which had to be drained by ditching, causing significant delay to the engineering works, but which could have been drained by drilling through to the Chalk.

Artesian pressure from the Chalk caused construction problems for the Shoreham Harbour Lock Gates (Ridehalgh 1958; Henry 1986, p. 222). This artesian head occurred at the base of the Chalk along the South Downs on the edges of a syncline and from beneath Tertiary sediments which acted as an aquiclude. Increased abstraction and greater groundwater control have lowered the potentiometric surface since the 1950s, and such artesian conditions may be more limited in future. Other artesian situations were encountered during construction of Tilbury Docks and the Thames Barrier.

Answering the questions

Artesian ground conditions

Examples of artesian ground conditions in construction have been given above. Headworth (1978) describes two possible circumstances leading to artesian conditions in the Chalk: one is the presence of an overlying confining bed which, in some cases, is possibly a layer of weathered chalk; the other is upward groundwater flow beneath discharge areas, such as river valleys (perhaps in the manner illustrated by Price 1987, Figs. 5 and 6, p. 150). Of particular interest is Headworth's (1978) identification of upward leakage through less permeable chalk.

Converging flow towards valleys including vertical flow, both in confined and unconfined conditions, could have been partly or wholly responsible for the *in situ* chemical softening, disintegration and 'heave' structures common in cross-sections of valleys truncated by modern sea-cliffs (Fig. 5.7e). Subsequent exposure of valley floors to periglacial conditions and later weathering has probably modified the top layers by cryoturbation and decalcification.

Coastal engineering on chalk platforms

Coastal chalk platforms have been sites for marinas, harbour arms, pumping stations, sewage outfalls, nuclear power stations (in France), as well as coastal protection works. In this environment sealing the ground against water ingress and calculating permeability, abstraction rates, and the effects of cyclical hydrostatic pressures, are typical problems. As elsewhere, these problems are influenced by the different types of chalk and the variable ground profiles, particularly between valley floors and interfluves. Springs are also known offshore in the English Channel and along the coastal cliffs.

Redevelopment of the Portobello Outfall Pumping Station, Brighton, in the early 1980s (Fig. 5.13), required investigation of several aspects of the groundwater regime, including establishing the hydrostatic pressures that might be exerted on the floor and walls of the structure and whether sufficient screen-cleaning water could be abstracted by boreholes located in the wave-cut platform and within the security zone of the site.

The site was at the foot of a truncated dry valley as this gave easier access to the wave-cut platform than the high cliff interfluves. This meant that relatively deep weathering had taken place and the upper layers of the chalk contained many solution pipes. Red and brown clayey silts had been washed down into fractures and into joints widened by solution, nearly to beach level.

The pumping station was constructed in the solid chalk of the wave-cut platform at the base of the cliff at the western end of Telscombe Cliffs. On the basis of fracture frequency, the chalk at formation level (the level on which the foundation slab was constructed) could be described as Mundford Grade III to II. Steeply inclined sets of conjugate fractures,

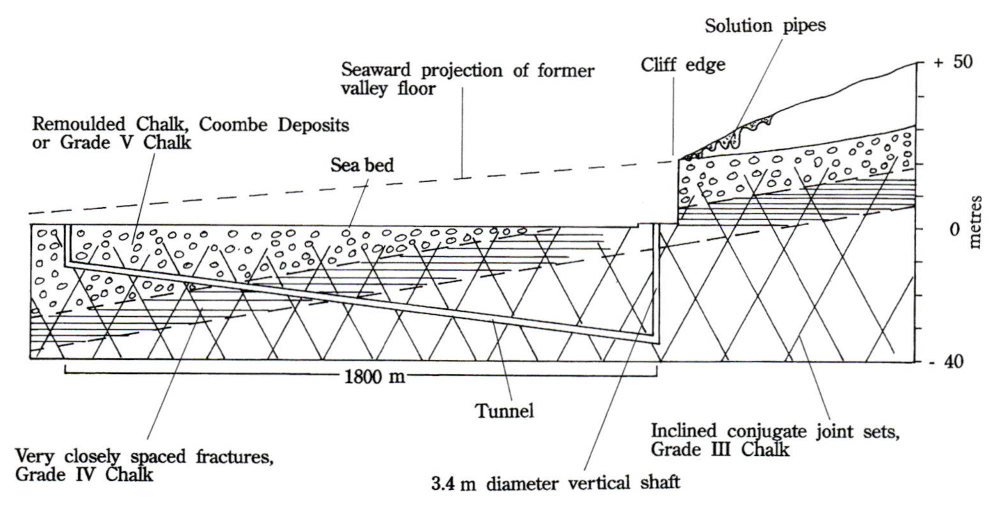

Fig. 5.13. The ground profile at Portobello, Telscombe Cliffs, near Brighton, East Sussex. This illustrates the effect of driving a tunnel towards the seabed at a point where a former valley floor with deeply weathered chalks is present. The permeable chalk was difficult to grout because of weathered putty-chalk around blocks and along joints. Marl seams in the Newhaven Chalk dip gently seawards and also contain karstic surfaces and subhorizontal fractures. Inclined conjugate joint sets are characteristic of the less weathered, deeper levels.

some filled with sheet-flint, were typical around formation level, as this was in the Newhaven Chalk Member (Mortimore 1986) containing regular marl seams. Fracture openings varied irregularly up to a few millimetres, but some were closed. Where unweathered, hard bands were layered in this chalk member, ranging from very soft to medium hard. Fractures were free of putty-chalk along surfaces in the harder varieties of chalk.

A very large excavation into the wave-cut platform some 60 m by 30 m and 6–8 m deep was required for the original pumping station and this had to be kept as dry as possible. A grout curtain had been injected in 1971 into the surrounding chalk to fill fissures and this could clearly influence subsequent abstraction from boreholes and permeability test results. Boreholes, for abstraction and permeability analyses, were sunk both inside and outside the grout screen. The results indicated that the area within the grout curtain acted as a separate reservoir from the general groundwater system (Harman and Vadgama, 1980) although some leakage through the grout was detected. Tidal cyclicity of water level was recorded in all boreholes but with a time-lag of between one and two hours.

Observations of importance at this site were that zones or fractures filled with putty-chalk were not readily grouted and water still seeped along the putty zones. Thus, full hydrostatic pressures could be expected on the walls of the pumping station. The undersea tunnel forming the long sea outfall to this pumping station (Haswell 1975; Tinkler 1976; Lake 1990) gradually rose offshore from a depth of 33 m in the Chalk onshore to only 12 m below the sea bed offshore. As it did so it encountered the offshore continuation of the dry valley (Fig. 5.13) with deeply weathered chalk containing zones of putty-chalk. Grouting was not completely successful in preventing water entering the tunnel workings and putty-chalk apparently 'leached' from sidewalls of probe holes and fissures. Thus the putty-chalk was not an effective seal to water inflow in an unconfined situation beneath the sea.

The collapse of a building at Percival Terrace, Brighton, in 1988 provides a second example of engineering problems related to wave-cut platforms in chalk and the deposits resting on them. Reasons for the collapse of part of a terraced house were not clear. One possibility was that sediments beneath the foundations were removed by water. The building was 18 m above a fractured, permeable chalk platform overlain by permeable raised beach gravels (interglacial) in turn overlain by periglacial solifluction Coombe Deposits. It is thought that tidal cycles created a hydraulic gradient within the raised beach and Coombe Deposits. It is suggested that this hydraulic gradient effectively 'sucked' down the relatively poorly consolidated Coombe Deposits, creating the cavity into which part of the building collapsed.

Underground excavations

It is often difficult, if not impossible, to predict precisely the degree of fracturing, number of fracture zones, and resulting permeability and water inflow into tunnels. It may, therefore, be difficult to predict groundwater yield, excavation stability, and overbreak. Predictions were fulfilled, for example, at South Killingholme, Humberside (Lake 1990; Varley 1990). Here the Welton Chalk Formation, at a depth of 200 m from ground surface, contains very widely spaced (> 5 m) and tight or very tight fractures. Thus water inflow was merely in the form of slow weeping along the inclined joints. The chalk was hard to very hard (that is, of low porosity) and, therefore, the pore-water contributed very little to any water generated by the excavation. At South Killingholme the hydrostatic pressure in the Chalk was used to contain the pressure of liquid petroleum gas stored in the caverns, which were left unlined for this purpose.

During construction of the Thames Cable Tunnel between Gravesend and Tilbury (Haswell 1969; 1975; Lake 1990) major fissure zones with inflows of groundwater were encountered while tunnelling in the saturated zone. These required the establishment of temporary bulkheads.

Thus identification of the fractured and weathered state of the chalk by precise descriptive terminology and measurement is of crucial importance.

The Channel Tunnel and the influence of water in the Chalk

Planned for more than 150 years, the Channel Tunnel will be in operation in 1994. It will comprise 160 km of completed tunnels. Although referred to as the 'Channel Tunnel', the project consists of three main tunnels, a service tunnel, and two running tunnels (Running Tunnel North and Running Tunnel South) for trains. In addition, access adits, shafts, pressure relief ducts, and cross-passages between tunnels were constructed. Two of the largest sub-sea-bed excavations, the 'crossover chambers', were also constructed. Most of the 160 km of tunnel was constructed in the Chalk Marl, the remainder being in Gault or higher Chalk formations (Fig. 5.14).

Because of its length (49.26 km, with 37 km under the sea), the Channel Tunnel crossed several engineering geology domains, both on land and

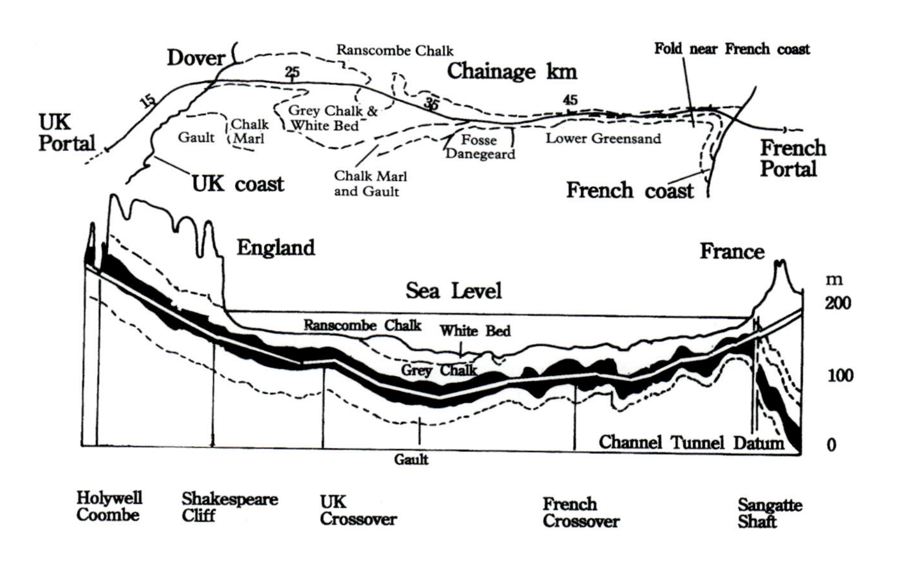

Fig. 5.14. Geological plan and section of the Channel Tunnel (modified from Mott, Hay and Anderson 1987, and Crawley and Pollard 1992). Chalk Marl is in black.

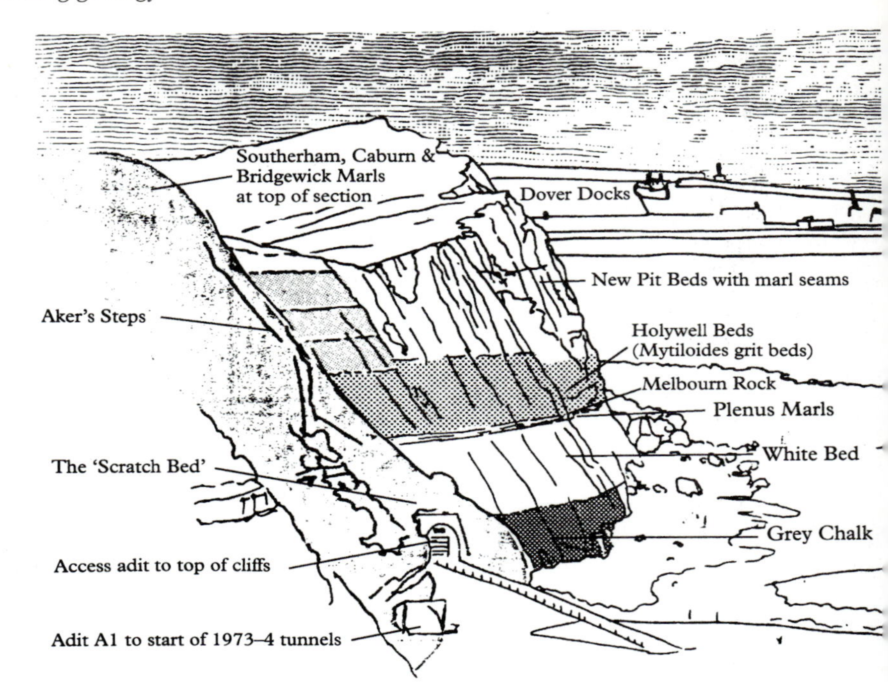

Fig. 5.15. Shakespeare Cliff, Dover, showing the broad Chalk stratigraphy exposed in the cliff above the Channel Tunnel workings. The geological section was formerly accessible to study via Aker's Steps. For Channel Tunnel construction, access to the base of the cliff for both the 1973–4 and 1987–93 operations was via the inclined adit indicated. Access to the underground works was via adit A1. A second adit, A2, was excavated for the 1987 start of tunnelling.

under the sea, in which water played an important part. These domains include the following:

1. A stress release domain along the Chalk escarpment near Folkestone, particularly through Castle Hill and Sugar Loaf Hill. This was associated with deep weathering, a combination which produced more permeable, fractured rock, even in the more marly Chalk Marl and Gault. This was also an area with several well-known ancient landslips which could have been reactivated by the construction of the tunnels.

2. A domain of thicker cover and potentially less permeable rock conditions beneath the North Downs between Folkestone and Shakespeare Cliff, Dover (Fig. 5.15). This length of tunnel passed beneath the Chalk aquifer in the Lydden Spout and Dour Valley catchments. Whether fracturing in the Chalk Marl also created permeable conditions with links to these overlying wet zones was uncertain.

3. A submarine domain in which permeability varied as a result of cover thickness, faults or fault zones, and more deeply weathered zones associated with ancient valleys cut in the floor of the Channel. These latter zones included the extension offshore from Dover of the Dour Valley and the deep 'fosse' in the central Channel known as Fosse Danegeard (Fig. 5.14).

4. A tectonic domain produced by an anticline close to the French coast which created different permeabilities in several ways. First, there was an increase in limestone content laterally towards the anticline. Such lateral variations across tectonic structures are common in the Chalk (see, for example, Mortimore 1986; Mortimore and Pomerol 1987; 1991), as early movements on folds produced low-amplitude features on the Chalk's sea bed. Marly layers in particular were attenuated or occluded against these features. Second, the more competent limestones fractured more readily than less competent marls, thus increasing permeability. Third, greater fracturing, as expected, occurred across the crest of the anticline, again enhancing permeability. Fourth, the reduction in the thickness of the Gault towards France reduced the 'smothering' effect of the formation on deeper-seated tectonic faults which may in turn have enhanced fracturing and permeability of the Chalk in the French sector.

During construction 'wet zones' were encountered which resulted in a debate about whether these were or could have been predicted. There was also uncertainty about the effect of freshwater and/or saltwater on the swelling properties of the more clayey layers in the Chalk Marl. Did the relatively dry early tunnels beneath Shakespeare cliff result from natural sealing of fissures by swelling clays? Because of the high costs of offshore marine drilling, the submarine section of the Channel Tunnel had investigation boreholes spaced at intervals of about 1 km with a similar spacing for seismic profiles. This spacing could only give a crude indication of potential ground conditions ahead of the working face.

Borehole-permeability tests were an important part of all the many ground investigations for the tunnels. It was always recognized that 'fissure water' rather than pore-water would be a key factor in the tunnel (Muir Wood and Casté 1970; Destombes and Shephard-Thorn 1971). Of the 144 'pumping-in' tests performed in 54 marine boreholes and the 66 pumping-in tests in 14 land boreholes during the 1964–65 investigations, 73 per cent gave a satisfactory value for the total permeability of the ground (Muir Wood and Casté 1970). Muir Wood and Casté were cautious about results in highly fissured zones and considered the 'pumping-out' test results more appropriate for predicting the inflow of water into the tunnel. Pumping-out tests, however, could only be performed on land-based boreholes. The pumping-out test results also suggested that 'local unsealing of fissures may cause the permeability to increase by a factor of 3 or 4' (Muir Wood and Casté 1970, p. 113).

Destombes and Shephard-Thorn (1971, Table 2) summarized the results from the 1964–5 *in situ* tests using broad stratigraphical units based on long test sections in boreholes. They also indicated the localized areas of comparatively high permeability and pointed out that the fissure pattern reflected the tectonic pattern and, as a result, these patterns should occur offshore. At Dover in the offshore extension of the Dour Valley, 'Sparker' surveys indicated north–south lineaments of weathered chalk (Destombes and Shephard-Thorn, 1971, Plate 1) which could have reflected an underlying joint pattern.

Subsequent ground investigations recognized the potential inadequacies of the long test sections, and the relationship of stratigraphy to permeability was reinterpreted and reinvestigated (Mott Hay and Anderson 1987). Permeability and other properties were then related to the micro-palaeontological subdivision of the Cenomanian (broadly the Lower Chalk), based on a series of foraminiferal zones (Carter and Hart, 1977). Lithological changes are not always related to the micro-palaeontological zones and there is evidence that fracturing and thus permeability is lithologically controlled, even in the Lower Chalk. Zone 8, the basal zone of the Chalk Marl, usually has very low permeability (5×10^{-8} m s^{-1}). This unit is characterized by the thickest 'marl' layers. Lake (1990) gave a range of permeabilities for the Chalk Marl of $1–5 \times 10^{-8}$ m s^{-1} over most of the Channel, increasing to $1–2 \times 10^{-7}$ m s^{-1} close to the French coast. Lake also pointed out the poor correlation between borehole records of fracture frequency and permeability.

It was also recognized that, although generally not part of the main Chalk aquifer, the Chalk Marl is permeable where fractures are opened by release of stress. This is the case along the escarpment near Folkestone where, for example at Cherry Garden reservoirs, the Chalk Marl is the source of water. Thus for the land-driven tunnels between Shakespeare Cliff and Folkestone, it was realized that a combination of stress release and deep weathering was likely to produce more permeable conditions, particularly in portal areas. It was the generally low permeability of the Chalk Marl, however, which attracted engineers to this horizon.

Other areas where permeability was expected to be higher were the deeply eroded and weathered chalk beneath the Fosse Danegeard, along fault zones, and across the anticline close to the French coast. The aborted 1973–4 tunnelling operations supported a generally held view that the UK side would be drier than the French side. All the early tunnelling beneath Shakespeare Cliff and the immediate offshore had been essentially dry. In contrast, the sinking of the shaft at Sangatte, and the initial tunnelling there, had proved wet ground. On these results different designs for tunnelling machines were prepared, the French machines being dual-mode, capable of operating in either sealed mode for wet conditions or open mode for dry conditions.

During construction, wet ground conditions were first encountered in the UK Marine Service Tunnel drive at about kilometre 20–21 (Fig. 5.16) some

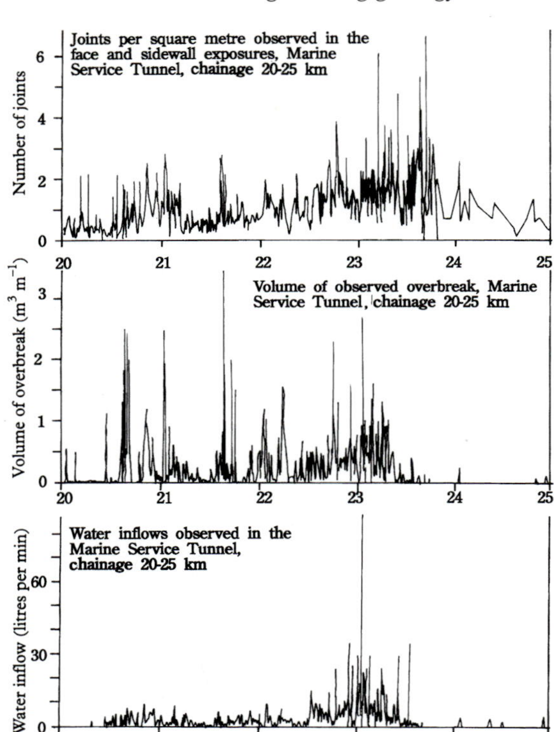

Fig. 5.16 Comparison of joint frequency, volume of overbreak and water inflow in the first 'wet zone' encountered on the Marine Service Drive of the Channel Tunnel 5 km off Dover (data from Crawley and Pollard, 1992).

5 km offshore from Shakespeare Cliff, Dover. Water was not expected to be a problem but progressive deterioration led to washed-out fractures, soaked miners and impeded grouting operations, which in turn led to loss of ground from the roof and sidewalls. Material fell on to the lining after the erection of support but before grouting was complete (see *New Civil Engineer*, 1 December 1988, p. 6). Crawley and Pollard (1992) have assessed the relationship between joint frequency, volume of overbreak and water inflows in this first 'wet' section of the tunnel (Fig. 5.16). The increase in the volume of joints (or possibly the presence of a fracture zone) over this section seems to correspond with an increase in both overbreak and water inflow.

To prevent loss of time in the construction of the two running tunnels through this and other wet zones, tube à manchette grouting was used for the ground treatment (Crawley and Pollard 1992). A row of three holes was drilled in a fan shape from the service tunnel, through the lining, at 3 m intervals, over the 700 m wet zone through which the running tunnels would have to be constructed. The holes were 17.8–20.5 m long and a special low-viscosity grout (Silacsol T) was used to give the required penetration of chalk joints; it also has a longer life than traditional silica grouts.

The design of the expanded, segmented concrete lining, the main support for the tunnels, ideally required a full-circle fit of the Chalk Marl on to the lining, without cavities. Thus ground-loss, partly caused by water inflow along fractures, delayed ring-building, made cavity grouting difficult, and required changes in the specifications for the grout mixture. The tunnel-boring machines also had to be modified with more robust pumps, sealed electrics and steel fingers to support the crown in the ring-erection area.

Forward-probing boreholes had been an essential requirement in all Channel Tunnel plans since the 1960s. Because of the wide spacing of the ground investigation boreholes, it was the forward probing boreholes that gave the first indications of wet conditions ahead in the Marine Service Tunnel. Initially this was based on probing ahead 120 m and driving 100 m before probing ahead again. Subsequently, the distance ahead was increased allowing more rapid progress.

During construction of the UK undersea crossover cavern (160 m long, 15.4 m high, 21.2 m span and 35 m below the sea bed) 7 km out from Shakespeare Cliff, an impermeable marl was unexpectedly discovered 5 m above the crown (Purrer *et al.* 1990). Water pressure on this layer caused increased settlement and local cracking of the shotcrete lining. Relief holes were drilled and further support given by additional rock dowels and shotcrete. This caused a temporary halt to cavern construction and illustrates the influence on water behaviour of layering in the Chalk Marl.

Groundwater was no less important in the land-driven tunnels. Very careful monitoring was required where the tunnels were to go through or close to known landslips such as at Castle Hill, Folkestone. The influence of water on swelling of the more clayey layers of the Chalk Marl was uncertain but Zone 6A at the top of the Gault showed some swelling characteristics when immersed in water.

An additional concern during the construction was the stability of the cliffs adjacent to Shakespeare Cliff (Fig. 5.15). A vital railway link ran through these cliffs, bringing segmented concrete rings to the site from the Isle of Grain. Cliff failures, which had occurred in the past, could have seriously affected the construction programme. Birch (1990) identified the trigger to the more important failures as high transient water pressures in the more permeable, fissured chalk above the Chalk Marl. An important spring horizon in relation to some of these cliff falls may be the 'Cast Bed' near the top of the Chalk Marl (see, for example, Birch 1990, Fig. 5, p. 548).

Thus study of the design and construction of the Channel Tunnel reinforces earlier observations that recognition and correct logging of lithological layers, fractures and the extent of weathering are essential to any engineering project and to the understanding of permeability. This needs to be combined with a knowledge of the processes that cause changes to these parameters if ground conditions are to be predicted with any certainty. The value of a pilot tunnel (in this case the Service Tunnel) in identifying locally difficult ground conditions is also emphasized.

Dewatering

To excavate soil and rock it is frequently necessary to lower the standing, or rest, groundwater level temporarily. Roberts and Preene (1990) described several case histories of well dewatering projects in the Chalk of southern England. Their terminology for describing the Chalk relied on Mundford Grades (Ward *et al.*, 1968; Wakeling 1970) and they related permeability to these grades. They also emphasized that it is often difficult to obtain reliable *in situ* permeability data and, therefore, the grade descriptions may provide the most accurate basis for assessing permeability.

Having identified the weaknesses in the original Mundford Grading scheme (see above) the improved engineering logging terminology suggested here (Tables 5.3, 5.4 and 5.5) should also improve estimates of permeability with depth in the Chalk. It was also suggested by Roberts and Preene (1990) that the highly weathered zones with putty-chalk were virtually impermeable. Ground heave might occur where a putty-chalk layer acts as a confining aquitard.

One of the cases quoted by Roberts and Preene (1990, p. 571) is the Port Solent Marina at Portsmouth, part of which required a 60 m by 26 m lock, 9 m deep. To prevent ingress of water and instability in the chalk within the excavation, ten wells were drilled to a depth of 17 m and an electric submersible pump installed in each. Three of the ten wells produced most of the water that was abstracted. Such variation in well production was repeated at a site in Littlehampton, Hampshire, and is typical of the Chalk. The wells at the Port Solent Marina achieved a drawdown of 9 m (the site conditions required a drawdown of 6.5 m) with a steady-state flow of 65 $l\,s^{-1}$. Such a drawdown and steady-state flow would have catered for the Peckham and New Cross Tunnels, described below, which had inflows of 45 $l\,s^{-1}$.

For dewatering of the Chalk to be successful, preconstruction trials are essential and a typical test arrangement suggested by Roberts and Preene is as follows:

Test well	300 mm diameter or greater
Test-well depth	1.5–2.5 times proposed excavation depth
Liner diameter	200 mm
Screen	1.5 mm slots, pea gravel surround
Discharge flow	10–30 $l\,s^{-1}$
Test duration	3–7 days
Piezometers	10 standpipes installed at two levels in five boreholes
Monitoring	24-hour monitoring of flow, drawdown in well and drawdown in piezometers at intervals of 1–2 hours

This test specification is recommended for structured chalk (Mundford Grades I to IV).

It is not discussed by Roberts and Preene, but perhaps, where practical in relation to an excavation, inclined boreholes might more successfully intercept water-bearing fractures. A nest of radiating inclined boreholes would be appropriate where the dominant fracture systems are vertical, as in the Seaford and Culver Chalks. The Port Solent Marina example was founded in Seaford Chalk. Vertical boreholes would probably be more successful in the Ranscombe and Newhaven Chalks which contain predominantly steeply inclined joint sets. Thus knowing the type and orientation of fractures might reduce costs and increase effectiveness of dewatering exercises.

Boxer (see Tinkler 1976, p. 63) described the problems of not carrying out a dewatering contract on the Peckham and New Cross tunnel prior to tunnel construction. Compressed air would have been used to prevent water ingress but gas was encountered, making compressed-air methods dangerous. As a result dewatering was necessary and it took an inordinate time for the wells to take command. At New Cross a shaft was sunk successfully by 'centre point sump pumping' (that is, pumping from a well in the centre of the shaft that is being excavated), but a second shaft to the north, sunk one year later, had its base blown out by artesian pressure. The northern section of the tunnel appeared to encounter much softer chalk and water ingress was at one point $45\,l\,s^{-1}$. Forward grouting and grouting from the surface proved unsuccessful and expensive. Compressed air proved to be successful in the end.

Chalks that do not drain freely

Some chalk terrains have very low permeabilities. Such areas include localized zones of putty-chalk, discussed above, but also include more widespread 'tight zones' and chalks which are very poorly fractured. The example of a very hard chalk with very widely spaced and tight fractures, the Welton Chalk Formation in the South Killingholme Caverns, has been given above. Similar widely spaced, steeply inclined, tight fractures occur in the Paris Basin around Sens but this time in very soft chalk. Some other tight regions such as interfluves and groundwater mounds have been referred to above.

Tight zones, particularly in the softer varieties of chalk, may have created problems for some forms of piling, particularly driven piles. It is possible that obstructions to driving piles, for example at Thurrock (Jones and Morrison 1990), described as flint bands and hardgrounds, may well be water trapped beneath the pile foot and the surrounding chalk. Water from pores in the intact chalk is released as the material is broken down and may not be easily expelled if convenient fracture(s) do not exist.

Highway soakaways

There could be conflict between the desire to protect the Chalk as an aquifer and the need to dispose of highway surface water containing tars, oils and other pollutants. Some soakaways are constructed as lagoons allowing direct access into the unsaturated zone of the Chalk and others are trenches filled with a filter medium or porous pipe. All require assessment of permeability, and this is usually related to the Mundford Grades. General permeability values for each Mundford Grade are given by Roberts and Preene (1990), and Price (1987).

Non-trunk road soakaways can often be located higher in the terrain in Grade IV and Grade III chalk. Major trunk-road soakaways are usually at low-altitude points and, therefore, are often in valleys. Although valleys are commonly floored with Coombe Deposits and Grade V chalk, satisfactory permeabilities have been obtained in soakaway lagoons and trenches. As indicated above, the valleys tend to be the most permeable regions in chalk terrains. In the Moulsecoomb Valley, Brighton, 800 mm diameter soakaway boreholes were excavated to a depth of 9.5 m on the margins of the valley but into Coombe Deposits or Grade V chalk. These were designed to accept $15\,l\,s^{-1}$.

It is the hazard created by direct drainage to the Chalk that is most uncertain. If soakaways are immediately up-valley of a housing estate, will there be the possibility of solution collapse down-valley over an extended period (as possibly in the Coombe Deposits of Percival Terrace described above)? In valleys prone to flooding would the increased discharge exacerbate that situation? Will plumes of contaminants develop with time from major soakaways and possibly link to one of the major fissure systems in the Chalk (see, for example, Price *et al.* (1992)). Each case needs to be considered individually in the light of these potential environmental impacts. Similarly, potential contamination problems have to be taken into account when considering proposals to store highway deicing salt and coated chippings in chalk pits.

Landfill

The pressure on local planning authorities to allow old chalk pits to be filled with domestic and other refuse is increasing. Consultation and planning involves advice from 'watchdog' bodies such as the National Rivers Authority in England. Specific data relating to the permeability of each site and potential wider interaction with the aquifer are frequently not available but should be obtained before decisions are made. The success of 'puddling' the chalk and/or coating the quarry with puddled clay or an artificial liner is not always certain. Even so-called impermeable beds such as the Gault and Chalk Marl have proved to be permeable in other engineering settings (see above).

Beard and Giles (1990) have shown that some direct discharge of sewage effluent to the chalk is successful but is dependent on the quality of the initial effluent (see also Chapter 6). In landfill operations the mixture of waste products generates a leachate which is chemically complex and commonly toxic. There is always a serious doubt whether such leachates should be allowed any possible access into the Chalk. Furthermore, many chalk pits are along the margins of valleys, very close to highly permeable zones and fault zones, thereby increasing the risk of contaminating groundwater.

Site investigation drilling and equipment fluids

Many site investigations take place in areas where chalk groundwater is used for water supply. Use of water as a flushing medium is sometimes prohibited and air-flush is used instead. Some ground testing involves burying loading jacks in boreholes and biodegradable oils are used to prevent pollution.

Intact chalk pore-water and chalk behaviour

Perhaps the most spectacular engineering behaviour of chalk has been in earthworks. Chalk excavated from cuttings is transported to embankment fill areas and subjected to rolling designed to compact it into dense 200–400 mm thick layers. Some soft, highly porous chalks have turned to a slurry during these operations both in France and England (Anon 1973; Clarke 1977; Clayton 1977). It is generally agreed that release of pore-water is responsible for the slurry; the higher the porosity the greater the amount of pore-water released. Measurements generally show that natural chalk is fully saturated for the complete range of porosities and/or intact dry densities in England and France (Mortimore 1979; Price 1987; Mortimore and Fielding 1990).

Engineering behaviour seems to be sensitive to relatively small changes in porosity/intact dry density. In hydrogeology it is the pore-throat diameters (Price 1987, Figs. 2–3) that are important in assessing the possible contribution of pore-water to aquifer behaviour. Earthworks behaviour is more dependent on porosity and intergrain bonding.

Discussion

A common description of the Chalk for engineering and hydrogeological purposes is emerging. The importance for ground engineering of the weathered zones in chalk terrains has been recognized since Mundford (Ward *et al.* 1968; Mortimore *et al.* 1990*b*) and is now being emphasized in hydrogeology (Allen and Price 1990; Lloyd and Hiscock 1990; Beard and Giles 1990; Giles and Lowings 1990). A greater appreciation of the variation in chalk terrains has led to a separation in descriptive logging of hardness, and fracture style, frequency and tightness (Mortimore *et al.* 1990*b*).

Geophysics is increasingly being employed to characterize lithology (in borehole logs, Figs. 5.9 and 5.10), weathered profiles (Coulon 1986), fracture frequency and thus Poisson's ratio and Young's modulus. These last two properties affect not only the calculation of potential stresses on civil engineering structures but also affect the elastic behaviour of the aquifer (Price 1987).

Giles and Lowings (1990, p. 622) have stressed the need for carefully prepared and detailed groundwater level maps. If these maps were to be supported by detailed lithological and fracture maps as, for example, used in the reservoir assessments of the Chalk in the North Sea (Watts 1983a; 1983b), and in geomorphological maps, then anomalies in groundwater and engineering behaviour could be more easily explained and targets for future water resources more accurately established.

A strong coincidence between tectonic structure and groundwater mounds has been identified but it is also the strength and true extent of the structure that probably influences storativity and transmissivity. The strength and extent of tectonic folds in the *en échelon* belt of the Anglo-Paris Basin can only be fully assessed by drawing structure contours. Combined with detailed maps of lithology and fractures, such contours provide an essential first step in an assessment of the interaction between engineering operations and the aquifer. This is particularly true of landfill, soakaway construction, and long tunnelling operations. It is also true of major highway constructions which might transgress catchments, groundwater mounds or aquifer boundaries.

The Chalk is still commonly treated as three formations—Lower, Middle and Upper—in the United Kingdom and as biostratigraphical units (Cenomanian to Danian) on the continent of Europe. It has been shown that these divisions are inadequate for engineering purposes in terms of material hardness/strength/porosity, and mass fracture characteristics (Mortimore *et al.* 1990*a*). The same inadequacies apply to hydrogeology. Clearly units with marl seams (Ranscombe, Lewes, and Newhaven Chalk members) react quite differently to stress and water compared with homogeneous, non-marly units (Seaford and Culver Chalk members). Detailed lithological mapping should, therefore, recognize the broader lithological units as well as specific marker marl seams which might be the cause of local perched water tables. Similar differences can be seen in the Chalk of Humberside between the relatively non-marly Burnham Chalk Formation and the Flamborough Formation with numerous marl seams. Tunnelling has emphasized the influence of marl seams as overbreak horizons and levels with perched water.

Conclusions

A combination of lithology, lithologically controlled fracturing, tectonic structure and geomorphological setting influences both the engineering behaviour and groundwater regime in chalk terrains. There is a need for a common approach to both ground descriptions and groundwater assessments. This can be achieved by application of the more detailed stratigraphy now available for much of the Chalk of Europe (see details, for example, in Mortimore 1990); application of techniques long used in the oil industry combining structure and sedimentation with stratigraphy (Krumbein and Sloss, 1956; Watts 1983*a*; 1983*b*; D'Heur 1984; 1990); and application of the same descriptive terminology for field and core-logging (Mortimore *et al.* 1990*b*).

Engineering operations are increasingly concerned with environmental impact, and the impact on groundwater is paramount. This impact, both short- and long-term, can only be assessed with more precise knowledge of the geology of the Chalk. The examples from construction projects given above also emphasize that, whether it is underground excavations, earthworks, foundations, or landfill, the way in which water influences construction is dependent on hydrostatic pressures, fracture intensity and tightness, state of weathering, and overall geological setting. These factors are also common to hydrogeology.

Water has had a major influence on the extent of fracturing and weathering of chalk. It is suggested above that the great extent of chalk decomposed *in situ* along valley floors and sides is related to the convergence of flow lines towards valleys. If this could be proved it might throw light on processes that form chalk landscapes.

Acknowledgements

In the early years of this research the late Stanley Holmes and the late Sydney Hester discussed, on many occasions, their wide experience of the effects of groundwater in the Chalk on engineering projects. The Hydrogeology Group of the British Geological Survey, particularly Michael Price, Michael Bird, Keith Murray, and also Chris Wood, gave invaluable support by providing details of cored boreholes and geophysical logs, and discussing problems. Similarly, the former Southern Water Authority, through Howard Headworth, Geoff Fox, and John Ellis, also gave borehole cores and geophysical logs to support the Chalk research programme. My understanding of the construction problems encountered in the Channel Tunnel has been enhanced by many discussions with Ron Williams (Mott Macdonald), Helen Nattrass, Paul Varley and Stewart Warren (TML) and Colin Warren (Eurotunnel). My knowledge of the Chalk in the Paris Basin comes from many years of joint fieldwork with Bernard Pomerol, to whom I owe a particular debt of thanks. I also thank John Lamont-Black and Stewart Ullyott for checking drafts of the manuscript and the late Dr Philip Drummond for much inspiration.

6. The Chalk aquifer—its vulnerability to pollution

S.S.D. Foster

Introduction

The very property of the Chalk that makes it an excellent aquifer—its well-developed fractures and fissures—also renders it especially vulnerable to pollution. The unusual characteristic of having superimposed and contrasting porosity/permeability elements results in distinctive behaviour in terms of pollutant transport, and greatly prolongs the persistence of pollution incidents by non-degradable contaminants.

This chapter identifies those properties of the Chalk, in its various common lithological and structural variants, which determine its capacity to transport, store, and attenuate water contaminants, and reviews the impact of different sources of pollution on the aquifer. Saline intrusion in coastal areas is not considered, although the controlling factors will be similar to those affecting the transport of mobile contaminants in the saturated zone, which are dealt with in this chapter.

Finally, the development of an aquifer protection strategy is briefly discussed, with particular reference to the problem of defining special protection areas around municipal wells and wellfields.

Origin of contaminant load

Diffuse sources

In North-west Europe, the outcrop of the Chalk forms extensive tracts of agricultural land. Groundwater recharge largely originates as rainfall infiltrating this land, and its quality is thus profoundly affected by agricultural land-use practices (Table 6.1). These have undergone, and continue to undergo, major changes and modification. For the most part the soils developed on the Chalk outcrop are thin and permeable, and there is a high probability of substantial leaching of naturally generated nutrients and soluble agrochemicals in the root zone, especially during the months of November to April. (Foster 1976; Person 1978; Lawrence and Foster 1987; Downing and Headworth 1990; Académie des Sciences 1991; Roux 1991).

The wet and dry fallout on the Chalk outcrop, like most of that in North-west Europe, is markedly acidic as a result of air pollution. This acidity is rapidly neutralized when rain infiltrates the Chalk subsoil, leading to the precipitation of any metals in solution. A considerable load of sulphur is also deposited. This may lead to substantial increases in the sulphate concentration of infiltration from some soils (Table 6.1), although this can be difficult to separate from the equally large amounts that were applied in past decades as inorganic fertilizer (Foster *et al.* 1985*a*).

Point sources

Much of the outcrop of the Chalk in North-west Europe has a relatively high population density, with a fairly frequent occurrence of medium-sized towns, and a wider distribution of major urban conurbations. The area is served by a dense network of highways, and as a result there are a large number of petrol-filling stations and many industrial premises located throughout the area. There are also a significant number of both civilian airports and military airfields.

Potential point sources of groundwater pollution (Table 6.1) are caused by the past disposal of both liquid effluents and solid wastes from these urban and industrial activities (Aspinwall 1974*a*; Roux 1977; Caous *et al.* 1978a; Department of the Environment 1978; Edworthy *et al.* 1978; Headworth et al. 1980; Caudron and Cremille 1985; Lawrence and Foster, 1987; Académie des Sciences 1991; Roux 1991), by leakages and spillages of liquid fuels and chemicals during transport, storage and handling, and by the disposal of polluted drainage waters from artificial impermeable surfaces (Aspinwall 1974b; Lawrence and Foster 1987; Académie des Sciences 1991; Roux 1991; Price *et*

Table 6.1 Summary of main activities that can generate a contaminated load on the Chalk

Potentially polluting process			
Sector	Activity	Type of pollution	Principal contaminants
Agricultural	Agrochemical applications	Diffuse	Nitrate, pesticides
	Sludge/slurry application	Diffuse	Pathogens, trace organics
	Livestock units	Point	Pathogens
	Farmyard effluent storage	Point	Nitrate, pathogens, organics
Urban and industrial	Drainage percolation	Multi-point	Hydrocarbons, salinity, trace organics
	Effluent discharge	Point/multi-point	Chlorinated organics, nitrate, salinity, pathogens
	Solid waste landfill	Point	Ammonium, chlorinated organics, salinity
	Atmospheric emission	Diffuse	Sulphate
	Chemical and oil storage	Point/multi-point	Hydrocarbons, chlorinated organics

al. 1992). The large number of farms throughout the area also create another series of potential sources of point pollution (Table 6.1).

In the older urban areas on the Chalk outcrop and along certain Chalk valleys which have long served as important lines of transport and development, the distribution and frequency of past point-pollution incidents may be such as to generate apparently diffuse contamination of the aquifer.

Vulnerability

The most consistent use of the term 'aquifer pollution vulnerability' is to reflect the susceptibility of groundwater (that is, water in the saturated zone) to deterioration of quality as a result of the imposition of a contaminant load at or near the land surface (Foster 1987; Foster and Hirata 1989; Adams and Foster 1992). It is, in effect, the inverse of the 'pollutant assimilation capacity of a receiving water-body' in the jargon of surface-water quality management.

On this basis, vulnerability is a function of a set of intrinsic characteristics of the strata separating the saturated aquifer from the land surface, reflecting the relative accessibility of the saturated zone, in a hydraulic sense, to the penetration of pollutants, and the attenuation capacity of these strata as a result of physicochemical retention and/or reaction with pollutants.

The vulnerability concept has significant limitations in rigorous scientific terms because 'general vulnerability to a universal contaminant in a typical pollution scenario' has little meaning. In practice, the aquifer system will respond differently to individual contaminants and pollution scenarios (Andersen and Gosk 1987; Foster and Hirata 1989; Adams and Foster 1992).

Groundwater pollution risks arise from the interaction between aquifer vulnerability and the contaminant load that is, will be, or might be, applied on the surface or in the subsurface as a result of human activity (Foster 1987; Foster and Hirata 1989). The vulnerability is intrinsic and therefore fixed, but the load can be modified by pollution control measures.

In this chapter the primary concern is the unconfined aquifer, and the focus is upon the transport and attenuation of various types of contaminant, associated with different pollution sources and scenarios, in the unsaturated zone. Suffice to say that in those areas where the Chalk is covered by Quaternary clays and exhibits semi-confined (or semi-unconfined) conditions, pollution vulnerability is greatly reduced and will depend on the characteristics of the clay cover, but this is not discussed further here. The lateral transport of contaminants in the saturated zone, and the likelihood of pollution affecting production boreholes, is a factor included in the looser definition of aquifer vulnerability used by some authors, and is thus also considered.

Aquifer characteristics

Water-bearing properties

The physical properties of the Chalk that control fluid flow in both the unsaturated and saturated zones are reviewed in Chapter 3. The intention here is to refer only briefly to the most important features relating to contaminant transport. The Chalk has two distinct porosity/permeability components (Chapter 3 and Fig. 6.1). In the context of vulnerability to pollution, it can be treated as a dual-porosity (but not dual-permeability) formation, in the sense that the matrix water is for the most part essentially immobile.

At shallow depths (down to about 10 m or so), and normally within the unsaturated zone, the rock mass is affected by recent weathering and in many areas Quaternary glacial and periglacial processes (Ward *et al.* 1968; Price 1987). Most commonly this results in increased hydraulic conductivity of the primary fracture system, but in some cases has led to the partial or total breakdown of the rock matrix into a melange or putty of reduced hydraulic conductivity, or to the formation of coombe and head deposits with distinctive physical properties.

In the saturated zone of the aquifer, and especially within 10–25 m of the water table in areas where the Chalk crops out, some fractures are normally enlarged by solution, and these transmit much of the groundwater movement in highly transmissive parts of the aquifer (Foster and Milton 1974; Owen and Robinson 1978; Headworth *et al.* 1982).

In areas where the local topography results in surface watercourses with acidic runoff draining

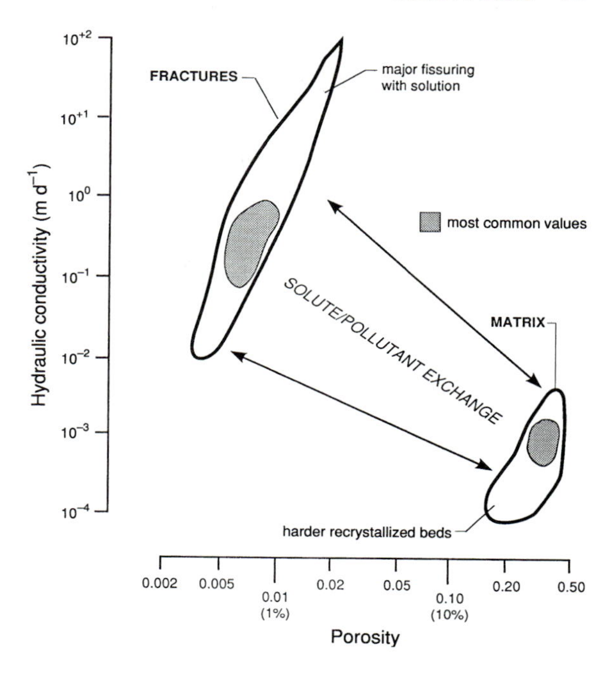

Fig 6.1. Dual-porosity components of the Chalk that enter into solute/pollutant exchange by diffusion in aqueous phase.

from adjacent Tertiary strata and flowing on to the Chalk outcrop, karstic solution features may develop with swallow-holes in the unsaturated zone and major solution conduits at, or just below, the water table (see, for example, Chapter 7; Atkinson and Smith 1974; Rodet 1978; Price *et al.* 1992).

Fracture–matrix contaminant exchange

The very small pore size of the Chalk matrix inhibits gravity drainage. Thus the matrix remains almost fully saturated even in the unsaturated zone (Foster 1975), except within the depth of direct influence of plant roots which can exert higher suctions and extract a proportion of the matrix pore-water (Wellings and Cooper 1983). The generally very low intergranular permeability means that, below the depth affected by plant roots, the rate of flow through the Chalk matrix is very slow and insignificant compared with that in fractures.

Exchange of solutes and contaminants between the (relatively) static matrix water and the mobile fracture water occurs (Fig. 6.1) as a result of

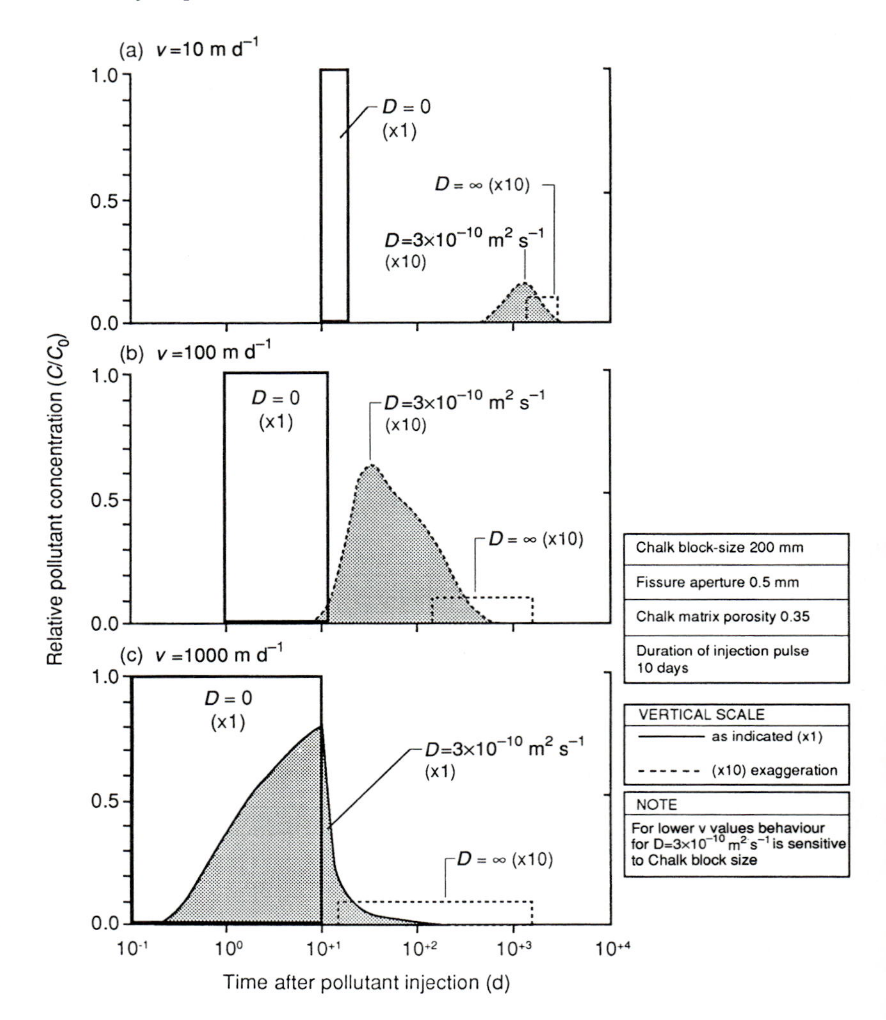

Fig 6.2. Simulated breakthrough curves for transport in the saturated zone, over a distance of 100 m, of a non-reactive pollutant of diffusion coefficient D in an idealized fissured porous Chalk aquifer showing sensitivity to fissure-flow velocity (v) (developed from Barker 1982).

aqueous diffusion, and according to the prevailing chemical concentration gradients as defined by Fick's second law, which in the one-dimensional case can be expressed as follows:

$$\frac{\partial C}{\partial t} = D^* \frac{\partial^2 C}{\partial x^2}$$

where C is the concentration of the solute and D^* is the apparent diffusion coefficient.

This is a very important process, especially in a dual-porosity medium like the Chalk (Foster 1975; Oakes 1977; Foster and Smith-Carington 1980; Barker and Foster 1981). It is capable of greatly retarding contaminant transport and can result in contaminants coming into more intimate contact with sorption sites on clay minerals and organic carbon disseminated within the rock matrix. It also provides enhanced opportunity for biodegradation of certain contaminants, since the bacteria responsible tend to form a biofilm on the walls of joints, fractures and fissures. Conversely, this mechanism also greatly prolongs the persistence of any pollution episode involving non-degradable contaminants and increases the difficulty of cleaning up pollution incidents and rehabilitating the aquifer once it is contaminated.

The degree of exchange and equilibration of a given contaminant or solute between fracture water and matrix water and thus the delay in breakthrough of pollution (Fig. 6.2), will depend upon the density of fractures, the velocity of fluid flow in the fractures, the aqueous diffusion coefficient of the contaminant or solute concerned (Barker and Foster 1981; Barker 1982; 1985*b*), and the extent to which any deposits present on the walls of fractures impede the diffusion process. The influence of the last factor remains largely unknown.

Mineralogical composition

The Chalk is for the most part a very pure limestone, composed largely of the debris of carbonate micro-fossils and recrystallized calcite (Chapter 2; Hancock 1975*b*). Other than in a few limited but persistent marl bands and stylolitic seams, its acid insoluble residue does not normally exceed 5 per cent. This residue is typically made up of montmorillonite, kaolinite, pyrite, apatite, and quartz (Chapter 2; Weir and Catt 1965); the first two provide significant ion-exchange capacity.

The proportion of organic matter in the Chalk is very low, generally about 0.05–0.2 per cent (Pacey 1989). Not nearly enough is known about the precise speciation of this carbon but it probably falls into three main groups (Whitelaw and Edwards 1980; Pacey 1989; Foster *et al.* 1991):

(1) fulvic and humic acids, immature polar material generally similar to much soil organic matter;

(2) bitumen-related materials of lower molecular weight, but more mature and less soluble;

(3) kerogen, which is insoluble organic material with a high molecular weight, and something of a graphitic character.

Only the first group is likely to be highly active in the absorption of non-polar organic compounds in aqueous solution.

Contaminant transport mechanisms

Unsaturated zone

Over extensive areas, the Chalk aquifer has an unsaturated zone more than 10 m thick, and at some locations more than 50 m thick. This zone represents the first line of natural defence against groundwater pollution. This is not only because of its strategic position between the land surface and the water table but also because it normally has a favourable environment for contaminant attenuation.

It is essential, therefore, that the unsaturated zone be fully considered in the evaluation of aquifer vulnerability (Foster 1987; Adams and Foster 1992). However, its behaviour can be complex and difficult to predict. It can perform a valuable role by totally eliminating some water pollutants, although the movement of persistent contaminants will be merely retarded, introducing a large time-lag before they reach the water table, and thus potentially concealing the occurrence of major pollution incidents for a long time. In the case of most other contaminants the degree of attenuation is very dependent upon the flow regime and the residence time in the unsaturated zone.

The flow velocity in the unsaturated zone is described by a modified form of Darcy's law, which in the vertical direction can be written:

$$\upsilon_\chi = \frac{K(\Theta)_\chi}{\Theta} \cdot \frac{\partial h}{\partial_\chi}$$

where $K(\Theta)_\chi$, the unsaturated vertical hydraulic conductivity, is a function of moisture content or matrix potential, Θ, and $\partial h/\partial_\chi$ is the hydraulic gradient. In the case of the Chalk the variation of the vertical hydraulic conductivity with saturation can be dramatic, since the permeability of fractures is orders of magnitude higher than that of the rock matrix, but the fractures only retain and conduct water during, and for limited periods after, major infiltration events.

The degree of longitudinal dispersion of a non-reactive solute or contaminant, with a concentration of C, is expressed through the hydrodynamic dispersion coefficient (D) in the one-dimensional version of the transport equation:

$$\frac{\partial C}{\partial t} = D\frac{\partial^2 C}{\partial x^2} - \upsilon_x \frac{\partial C}{\partial x}$$

It will vary widely with the types of fractures conducting water.

Fracture–matrix exchange by aqueous diffusion is considered to be an important mechanism of solute and contaminant transport in the unsaturated zone of most types of chalk, at least below the depth of direct influence of plant roots. It is a mechanism that can result in little dispersion (Fig. 6.2a) since almost total equilibration of concentrations between fracture water and matrix water is possible for a soluble ion in the unsaturated zone when film flow on fracture walls is slow (Foster 1975; Oakes 1977; Barker and Foster 1981; Barker 1982). Under these conditions the rate of downward movement of a non-reactive solute should be in the range 0.5–1.5 m a^{-1} and would vary with the average annual rainfall and the effective porosity of the Chalk involved. This was initially demonstrated and subsequently corroborated by the well-preserved and clearly defined thermonuclear tritium peak in unsaturated zone pore-waters (Fig. 6.3) at numerous sites (Smith and Richards 1972; Foster 1975; Foster and Smith-Carington 1980; Foster and Bath 1983).

On the other hand, with heavier hydraulic loading, as a result of very intense rainfall or effluent discharge, there will be a high probability of so-called preferential flow in the Chalk's unsaturated zone. This is taken to include all forms of more rapid downward movement, which do not allow sufficient time for complete fracture–matrix exchange, and in extreme cases result in the almost complete bypass of the matrix. Such a process can be shown to be physically feasible (Fig. 6.2b,c).

The potential for preferential flow in the Chalk, coupled with the existence of a very high fissure hydraulic conductivity in the unsaturated zone, has been proven at some sites (Wellings and Cooper 1983; Gardner *et al.* 1990), but fissure flow is difficult to observe under natural conditions. Indirect evidence comes from the analysis of rates of downward movement of tritium peaks (Smith and Richards 1972; Foster and Smith-Carington 1980; Geake and Foster 1989) and the fact that some unsaturated zone matrix pore-water profiles of environmental tritium show marked dispersion with strong 'forward tailing' of the 1963–5 peak concentrations (Fig. 6.3; Foster *et al.* 1986; 1991). This could only occur if the velocity of water movement had been high enough to prevent full interchange of solutes with the micro-porous matrix.

The conditions that trigger preferential flow cannot, as yet, be specified. In some types of Chalk

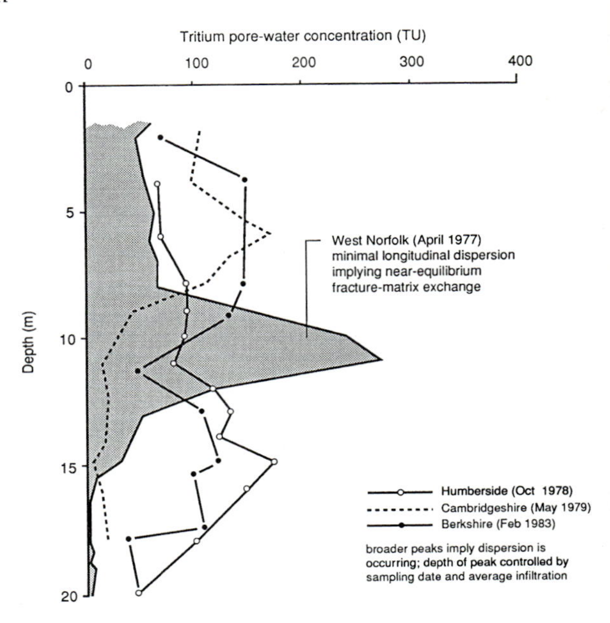

Fig. 6.3. Selected tritium profiles from the unsaturated zone of the Chalk showing departure from 'classic non-dispersive transport' (modified from Foster *et al.* 1986).

interflow through the soil zone is considered to occur when the rainfall infiltration rate exceeds around 2–5 mm d^{-1} and this would result in water flowing in fissures (Chapter 3; Downing *et al.* 1978).

The proportion of total downward water movement conducted by preferential flow is difficult to estimate, although figures of up to 30 per cent appear possible (Downing *et al.* 1978; Foster and Smith-Carington, 1980; Foster *et al.* 1991). It is equally difficult to estimate preferential flow velocities. Values of 1–10 m d^{-1} appear to be indicated (Foster and Crease, 1974; Downing *et al.* 1978) and would explain bacteria being detected in groundwater after transport through unsaturated zones more than 20 m thick.

Other than where subjected to a heavy organic load, the Chalk's unsaturated zone appears to be an aerobic environment, although it has not proven possible to make definitive measurement of dissolved oxygen or Eh on pore-waters.

The fact that the behaviour of a contaminant in the unsaturated zone can vary very widely with external hydraulic loading and type of chalk, cannot be overemphasized. It is also important to stress that the type of chalk in the unsaturated zone varies not only spatially with stratigraphical horizon, depositional facies and diagenetic history, but also vertically at any given location as a result of geomorphological history.

The acquisition of adequate data to model fully the transport of pollutants in the Chalk's unsaturated zone is inhibited by the practical difficulties and high cost of making appropriate measurements and monitoring the system. It is only in exceptional situations, at sites of major pollution incidents or hazards, that sufficient data can be collected for adequate calibration of unsaturated zone flow and pollutant transport models. Short of this, models can be used that assume a simple 'time-lag function' and non-dispersive conservative transport (Young *et al.* 1976b; Oakes 1982).

Saturated zone

The factors that control groundwater flow and pollutant transport in the saturated zone differ from those in the unsaturated zone. The entire rock mass, including all fractures and fissures, is fully saturated. The bulk of the horizontal groundwater flow is concentrated along major secondary fissures, where these are present. Under natural hydraulic gradients, flow velocities of the order of tens to hundreds of metres per day are common. Even higher velocities may occur close to pumping wells and where solution features are well developed.

While the same process of fracture–matrix exchange of solutes and contaminants by aqueous diffusion will tend to occur, flow rates in secondary fissures in the saturated zone can be so high that the process is much reduced (Figure 6.2b,c), and is even negligible for pathogenic micro-organisms and organic molecules with relatively low diffusion coefficients.

All contaminants transported in major secondary fissures will have very limited contact with inorganic and organic mineral species and with autochthonous bacteria, and thus are less likely to be significantly retarded or degraded. The only attenuation mechanism affecting pollutants may be dilution through hydrodynamic dispersion during flow in fissures. Hence lateral transport over distances of up to 1 km or more is possible. Evidence for this comes from viral tracer experiments in the Chalk of Humberside (Table 6.2) which showed very limited attenuation and dilution with increasing lateral travel distances, once the tracer had presumably found its way into major horizontal fissures.

The situation may be somewhat different in the confined parts of the saturated zone where hydraulic gradients, aquifer transmissivities, and horizontal

Table 6.2 Summary of results of viral tracer experiments in vicinity of Chalk production borehole in Humberside (derived from Skilton *et al.* 1985)

Tracer measurement*	Lateral travel distance**	
	100 m	200 m
Proportion tracer recovered (%)***	0.5	0.4
Peak dilution factor	5×10^8	5×10^9
First arrival time (min)	150	310
Peak arrival time (min)	380	750
Corresponding transport rate (m d^{-1})	380	380

* Viral tracer used was *Serratia marcescens*; other bacteriophages were used with broadly similar results.

** From injection borehole to production borehole in unconfined Chalk, latter pumping at a steady rate of around 5000 $m^3 d^{-1}$ throughout experiment.

*** Tests were not sufficiently long for tracer concentration to decline to zero; most tracer appeared to be held within the Chalk close to injection point, but the very small proportion which entered major horizontal fissures was transported rapidly with minimal attenuation.

groundwater flow rates are all generally much lower, allowing more time for exchange of solutes and contaminants by diffusion between fracture and matrix water.

While the unconfined aquifer appears to be predominantly aerobic, at some point along the groundwater flow path into the confined parts of the aquifer (horizontally or vertically, depending upon whether fully confined by Tertiary clays or semi-confined by Quaternary deposits), the dissolved oxygen in the fissure water is consumed, conditions become progressively anaerobic and bacteriologically catalysed denitrification and sulphate reduction occur (Foster *et al.* 1985*b*; Parker and James 1985; Edmunds *et al.* 1987; Chapter 3).

In principle, the modelling of contaminant transport in saturated fractured porous media, such as the Chalk, is straightforward, since adequate analytical equations and numerical codes exist (Chapter 4; Barker 1985*b*). Critical parameters for these models, such as the frequency, aperture and continuity of fractures and fissures, are, however, difficult to establish in the field. It will often be more realistic to use the stochastic approach, assigning probabilities to each of these variables and using the simplest possible model configuration to investigate the interaction of these probabilities and determine the likelihood of pollutant transport at critical concentrations over the distance under consideration (Price *et al.* 1992).

Not enough is known about the vertical flow component in the saturated zone of the Chalk, and the mechanisms of water transfer between well-developed horizontal flow horizons. This component is significant both regionally in groundwater flow from recharge to discharge areas and locally in the vicinity of pumping boreholes. Detailed investigations of the Chalk's hydraulic conductivity suggest that bulk values in excess of 10^{-6} m s^{-1} are present throughout most of the aquifer (Price *et al.* 1977), probably associated with primary fractures in the principal joint directions, although, of course, less frequent major vertical fractures do exist. Vertical groundwater movement along the (densely-spaced) primary fractures is likely to be slow. The evidence from pore-water profiles of the saturated zone for mobile solutes, such as chloride, sulphate and nitrate (Fig. 6.4), suggests that there is normally time for near-equilibrium exchange between fracture and matrix waters (Foster *et al.* 1982), in contrast to the situation during rapid horizontal flow in secondary fissures.

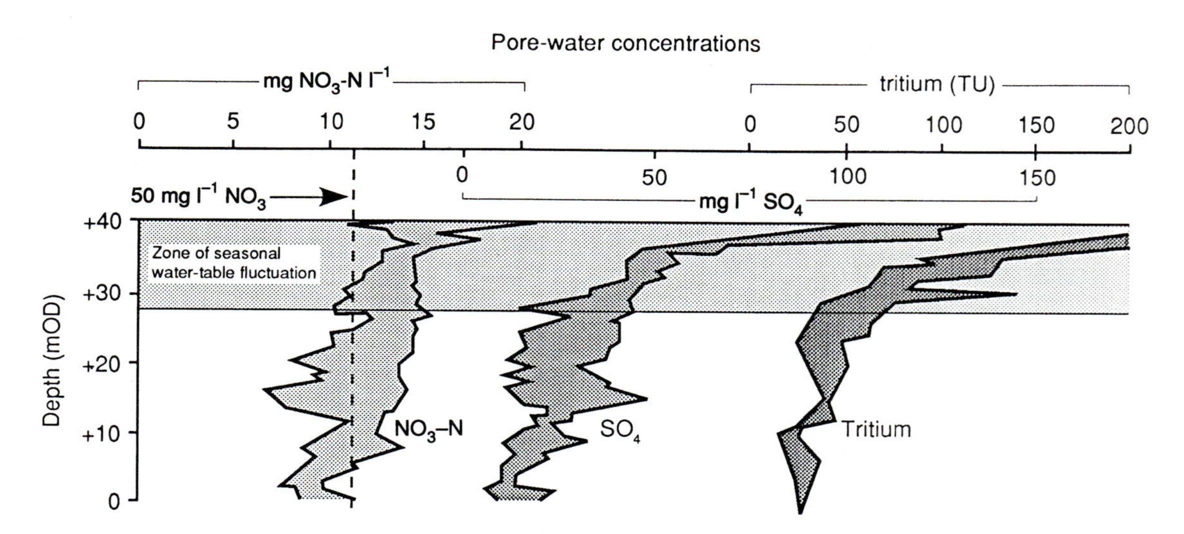

Fig 6.4. Pore-water profile envelopes in saturated zone of unconfined Chalk aquifer of West Norfolk (in an area of intensive arable cultivation and shallow groundwater table) in 1976–7, showing high level of fracture–matrix interchange with anthropogenic inputs and deteriorating quality of more recent recharge (modified after Foster *et al.* 1982).

Assessment of key pollution issues

Nutrient leaching from cultivated land

Although instances of individual Chalk wells producing groundwater with nitrate concentrations above 50 mg l^{-1} (the European Community's maximum acceptable concentration for drinking water) have been intermittently recorded throughout this century, the contamination could generally be assigned to nearby discrete sources, such as leaking sewers or farmyard drains (Foster and Young 1979; 1980). During the 1970s it became apparent (Greene and Walker 1970; Foster and Crease 1974; Robertson 1974; Person, 1978) that high and increasing nitrate concentrations were becoming widespread in groundwater in the Chalk where it is unconfined, suggesting a diffuse agricultural origin.

A relationship with changes in agricultural cultivation, particularly increasing regular cropping of cereals, oil seeds and root crops, sustained by greatly increased applications of inorganic nitrogenous fertilizers, was postulated, with the likelihood that the full effect of changes in the previous 25 years had not yet become apparent because of the very slow movement of pollutants through the unsaturated zone (Foster and Crease 1974; Young *et al.* 1976a). In subsequent years this interpretation has been substantiated, with values in groundwater in the Chalk exceeding 50 mg l^{-1} NO_3 over increasing areas of the unconfined aquifer in England and France (Foster *et al.* 1986; Parker *et al.* 1987; Roux 1991) and with levels continuing to rise at rates of 0.3–1.5 mg $l^{-1}NO_3$ per year (Foster 1976; Gray and Morgan-Jones 1980; Lawrence *et al.* 1983, Académie des Sciences 1991; Roux 1991).

As a result of this increase in nitrate concentration public water-supply sources have been abandoned and water-supply companies are making major capital investments to provide treatment and blending facilities. Well over 2 million people in England and France have been supplied with water from the Chalk containing more than 50 mg $l^{-1}NO_3$ for periods of many years.

Detailed hydrogeological research on the unsaturated zone of the Chalk has reached the following conclusions in relation to nitrate leaching (Foster and Young 1979; 1980; Foster *et al.* 1982; Oakes 1982; Foster and Bath 1983; Foster *et al* 1985*a*; 1986; Geake and Foster 1989; Foster 1989; Chilton and Foster 1991):

1. The intensification of arable cropping since the 1960s has resulted in major increases in leaching of nitrogen from typically less than 25 kg $ha^{-1} a^{-1}$ to 40–80 kg $ha^{-1} a^{-1}$, leading to pore-water concentrations in the unsaturated zone generally in excess of 50 mg $l^{-1}NO_3$ (Fig. 6.5).
2. The ploughing of permanent grassland to increase arable acreages during 1940–50 was another significant factor in some areas.
3. Leaching losses appear to be highest from the thinner permeable soils and in the drier regions, where lower infiltration rates also result in relatively higher concentrations.
4. There is evidence that certain cultivation practices have a disproportionate influence on leaching losses and need to be avoided in the interests of groundwater quality. These include the excess and superfluous application of fertilizers with the sowing of winter cereals and oil seeds, the untimely ploughing-in of grass leys and the over-fertilization of some crops such as sugar beet.
5. Practices such as the increasing cultivation of early-sown winter cereals, the use of minimal cultivation techniques, and the choice of appropriate crop rotations, can all lead to significant reductions in the leaching of nutrients, as should the close control of irrigation to meet more closely the requirements of the plant for moisture.
6. Unfertilized grassland and low-productivity pasture (receiving up to about 150 kg $ha^{-1} a^{-1}N$), gives minimal nitrate leachate (less than 10 kg $ha^{-1} a^{-1} N$), since, even in highly permeable soils, the presence of a mature grass root system has the effect of continuously maximizing the use of nutrient and produces conditions more favourable to soil denitrification.
7. Intensive grass production, with large fertilizer applications, leads, perhaps surprisingly, to major loss of nitrate through leaching, comparable to that below intensive cereal cultivation, particularly where high-density grazing or regular ploughing-in of grass leys are practised.

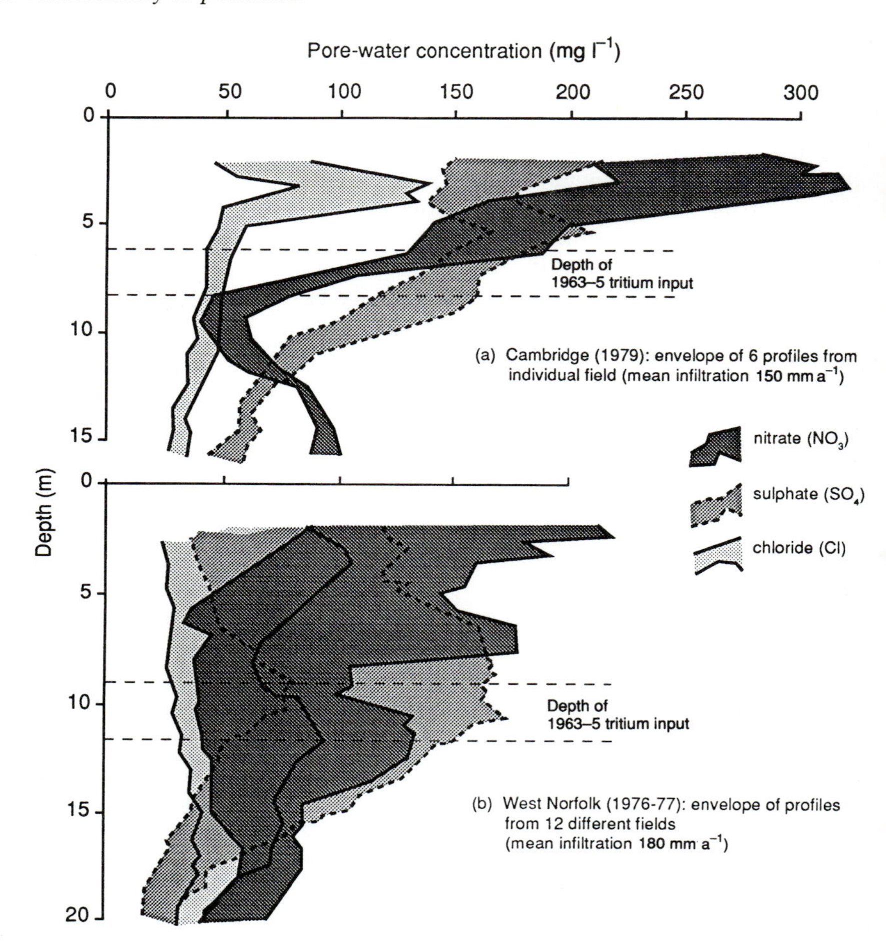

Fig. 6.5. Pore-water profiles in the unsaturated zone of the Chalk beneath fields under intensifying arable cultivation (after Foster *et al.* 1985a).

The possibility of nitrate attenuation through *in situ* denitrification within the unsaturated zone seems rather unlikely, especially under arable land (Foster and Young 1979; 1980). It cannot yet be discounted, however, since significant (insoluble) organic carbon concentrations and autochthonous nitrate-reducing bacteria have now been measured at depths well below the soil interface at some sites (Whitelaw and Edwards 1980; Whitelaw and Rees 1980).

Predictions of future trends in nitrate concentrations in Chalk groundwater are made difficult by the spatial variability of leaching, uncertainties about the proportion of rapid preferential (bypass) flow through the unsaturated zone, the degree of longitudinal dispersion affecting the slow component of downward movement of contaminants and the scale of any *in situ* denitrification in the unsaturated zone. It is, nevertheless, evident that concentrations in the parts of the Chalk outcrop which are extensively used for arable cultivation are likely to exceed 50 mg l^{-1} NO_3 (Department of the Environment 1986; Foster *et al.* 1986; Roux 1991), even assuming improved cropping practices to constrain the future leaching of nitrate.

Controls of fertilizer application and cropping practices are required to prevent the problems of excessive nitrate leaching spreading to areas without a long-term history of intensive arable monoculture.

There are indications of concomitant increases in the concentration of various soluble ions in pore-waters in the unsaturated zone of the Chalk beneath land under intensive arable cultivation. These include Cl, Ca, Na, Mg, K, Ba, Sr and B, and most notably SO_4, which is widely present at levels of 100–200 mg l^{-1}, indicating peak leaching rates as high as 250 kg ha^{-1} a^{-1} SO_4 in some areas (Foster *et al.* 1982; 1985*a*). This sulphate must be in part derived from fertilizer compounds, such as $(NH_4)_2SO_4$ and $Ca(H_2PO_4)_2$-$CaSO_4$, whose application has now decreased greatly, and also from aerial deposition associated with the acid-rain phenomenon. The latter is more pronounced under woodland plantations as a result both of increased capture of airborne pollutants and of reduced infiltration rates, and concentrations of 200–400 mg l^{-1} SO_4 have been widely observed (Neal *et al.* 1991).

Mechanisms of pesticide contamination

The use of pesticides across the Chalk outcrop of North-west Europe has increased considerably since the 1970s, and most rapidly in the case of herbicides and fungicides applied to autumn-sown cereals. Typical rates of application are in the range 0.2–5.0 kg ha^{-1} a^{-1} of active ingredients. There has also been increasing non-agricultural use of triazine compounds for general defoliation of railways, roads and airfields, forestry firebreaks, and fruit orchards, with unit rates of application much higher than for agricultural use.

The question of pesticide leaching has only recently begun to be addressed (Croll 1986; Lawrence and Foster 1987). All pesticide compounds pose a significant environmental health hazard since they are, to a greater or lesser degree, chemically tailored to be toxic and persistent. The availability of pesticide analyses of groundwaters is limited but the stringent maximum acceptable concentration, defined by the European Community, for pesticide compounds in drinking water of 0.1 μg l^{-1} (effectively surrogate zero) has already been considerably exceeded in many water-supply boreholes, although concentrations in excess of 1 μg l^{-1} have not yet been widely recorded (Croll 1991; Foster *et al.* 1991; Roux 1991). The compound most frequently recorded is the herbicide atrazine; to a lesser degree simazine, isoproturon, and mecoprop have also been consistently detected.

Most pesticide compounds have water solubilities in excess of 10 000 μg l^{-1}. The mode of application and action of the pesticide is important, since those targeted at weed roots, and at soil insects are much more mobile than those acting through leaves. The concentration in soil–water solutions required to achieve their agronomic objective is thousands of times greater than the corresponding maximum acceptable concentration of the EC, in contrast to nitrate, for which the corresponding ratio is 1.5–2.5 (Foster *et al.* 1991). Some compounds, however, may be neutralized in calcareous soils with the generation of less soluble residues.

Of greatest importance in considering susceptibility to leaching is the mobility in soil solutions, which is normally expressed by the corresponding partition coefficient for organic carbon, although polar adsorption on clay minerals and organic matter should also be considered.

Pesticide compounds will, to varying degrees, experience degradation as a result of bacteriological degradation or chemical hydrolysis, with half–lives in soil commonly in the range of 10 days to 10 years, but for more mobile compounds normally less than 50 days. However, given the timing of applications, they are sufficiently persistent to remain in the soil for significant periods at times when leaching may occur. In contrast to nitrate, the highest pesticide concentrations in soil–water are usually found after spraying in late winter or early spring, when soils are likely to be at 'field capacity' moisture content, and there is a high probability of infiltration following rainfall (Foster *et al.* 1991).

It must also be borne in mind that the listed adsorption and degradation characteristics of pesticide compounds relate to fertile organic clayey soil. They are not representative of the soils most widely developed on the Chalk outcrop and far less so of the Chalk's unsaturated zone, which contains a much smaller proportion of clay minerals and organic matter and a greatly reduced population of autochthonous bacteria (Lawrence and Foster 1987).

The limited monitoring of groundwater for pesticides means that pollution of groundwater by pesticides has to be considered from first principles. In the unsaturated zone, matrix transport rates of much less than 1 m a^{-1} are to be expected, given that some retardation of most pesticide compounds is likely (Table 6.3). Since few compounds have been in regular use for more than 10–20 years, pesticides leached from agricultural soils would be generally

Table 6.3 Estimated retardation factors (relative to a non–reactive pollutant) of widely used pesticides for matrix transport in the unsaturated zone of the Chalk

Pesticide compound	Partition coefficient* Koc (ml g^{-1})	Maximum retardation factor**
Simazine	50–100	1–2
Atrazine	100–200	1–3
Mecoprop	100–200	1–3
Isoproturon	100–200	1–3
Chloroturon	200–500	2–8
Lindane	500–1000	3–11
Paraquat	10,000+	2–100+

* Koc is the partition coefficient between organic carbon and water. The range given is approximate, reflecting the variation in published values.

** Based on Chalk porosity of 0.25–0.45, organic carbon content of 0.05–0.2% and assumption that all of latter will be active in pesticide adsorption.

expected to be still in the Chalk's unsaturated zone (Foster *et al.* 1991), if this were the only transport mechanism operating. Such a mechanism could delay the impact on the aquifer even more than that of agriculturally derived nitrates, but it is likely to be much more limited because of lower aqueous diffusion coefficients and the possibility of bacterial degradation or chemical hydrolysis. Very few analyses of pore-waters from the unsaturated zone have been made; those available have not proven pesticide concentrations greater than 1 $\mu g\, l^{-1}$ below a depth of 3 m.

Preferential flow is of major importance in the consideration of pesticide transport in the Chalk and will be more significant than for nitrates because proportionately much higher concentrations are likely to be involved (Foster *et al.* 1991). It provides a route for deep penetration of those pesticide compounds which are present in the soil in winter and readily leached by rainfall. It would be greatly favoured in situations where surface drainage to the ground is facilitated by soakaways and this route is probably responsible for the widespread penetration of triazine compounds (atrazine and simazine) used for non-agricultural defoliation.

Impact of hydrocarbon leakages and spillages

A serious threat to groundwater in the Chalk arises from the accidental leakage or spillage and casual discharge of hydrocarbons including petroleum products, phenols and chlorinated solvents, on or into the ground (Aspinwall 1974*b*; Caous *et al.* 1978*b*; Hunter-Blair 1978; Quaghebeur and Wulf 1978; Hunter-Blair 1980; Lawrence and Foster 1987; Roux 1991). The EC's maximum acceptable concentration for hydrocarbons in drinking water is less than 10 $\mu g\, l^{-1}$, but for many individual contaminants it is less than 1 $\mu g\, l^{-1}$. Small volumes of pollutant are thus capable of contaminating extremely large volumes of groundwater.

A large range of liquid hydrocarbons is widely transported, stored, handled and used, in human activities at all scales on the Chalk outcrop. Major foci of potential point pollution include airfields, fuel-storage facilities, industries involving metal working and modern light industries such as electronics, pharmaceuticals, photographic and dry cleaning.

Most hydrocarbons are relatively insoluble in water and are transported in the subsurface in the immiscible phase. Two broad groups of water pollutants are recognized: the so-called *floaters,* which are lighter than water and include both petrol and diesel, which are complex blends of many hundreds of petroleum-derived chemical compounds in the C_4–C_8 and C_{12}–C_{22} ranges, respectively; and the so-called *sinkers,* mainly halogenated hydrocarbons (Fig. 6.6), which are heavier than water and include the common industrial solvents (tetrachloroethylene, trichloroethylene, methyl chloroform, chloroform, carbon tetrachloride, methylene chloride, and so on). The latter were progressively introduced into more and more industrial processes from the 1920s, with volumetric consumption peaking in the 1970s and declining somewhat in recent years as a result of recycling and/or substitution.

The method by which the hydrocarbon enters the subsurface is a major factor in determining how a pollution plume develops and migrates in the Chalk. An instantaneous spillage or overnight leakage of, say, 5000 l of liquid chemical (in an immiscible

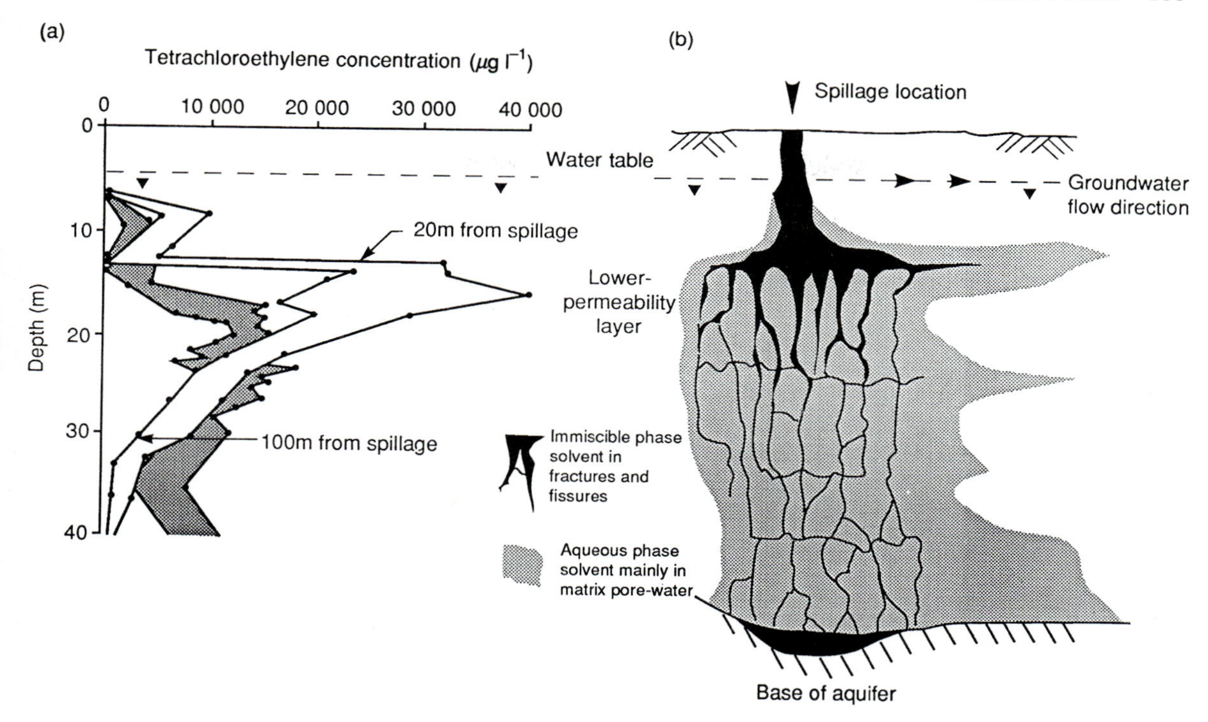

Fig. 6.6. Pollution of unconfined Chalk aquifer near Cambridge with tetrachloroethylene solvent from leather-processing industry:
(a) pore-water profiles from saturated zone determined by pentane extraction from cores;
(b) reconstruction of solvent distribution following major spillage (derived from Lawrence and Foster 1991).

phase) will produce a very different plume from that caused by contaminated drainage (in an aqueous phase) that is regularly infiltrating from a factory yard (Lawrence and Foster 1991).

The downward migration of the immiscible phase (if present) will be via primary joints (Fig. 6.6). Entry into the Chalk matrix is unlikely as a result of the very small pore sizes.

The halogenated solvents possess physico-chemical properties which make them more insidious groundwater pollutants (Lawrence and Foster 1987). Their immiscible phase is significantly denser (1.3–1.6 g cm^{-3}) and less viscous (0.3–0.6 mm^2 s^{-1}) than water, so that rapid, deep entry into the Chalk aquifer can be expected and evidence of penetration to depths of more than 40 m has been found at a site of a major discharge and spillage of tetrachloroethylene on the Middle Chalk of Cambridgeshire (Figure 6.6a).

If present in sufficient quantity, they could occupy all the fissures and/or fractures, thus dramatically reducing the effective permeability of the aquifer to the aqueous phase. The result is a decrease of the local groundwater flow and solvent–water contact, and thus a radical reduction in the rate of solution and dispersion of the pollutant.

Although the Chalk's matrix does not contribute significantly to groundwater movement, it can have a major influence on the rate of aqueous (immiscible) phase contaminant transport as a result of diffusion exchange, and concentrations in excess of 10 000 μg l^{-1} have been detected at the site mentioned above (Fig. 6.6a). This will result initially in some of the dissolved hydrocarbons being retained in the matrix, thus retarding their migration relative to the mean groundwater flow velocity, although transport over large distances will still be possible where major secondary fissures are present. Subsequently, concentration gradients will reverse and diffusion will promote movement from the matrix back into the major fissures. In effect, these processes combine to lengthen greatly the time these contaminants persist in the Chalk and to increase significantly the difficulty of cleaning up the aquifer once it has been contaminated.

Sorption on to organic matter and possibly also on to mineral surfaces will also retard hydrocarbon migration. Few data are available but preliminary laboratory work suggests that the sorption of tetrachloroethylene on to powdered chalk is low but may not be insignificant (Lawrence and Foster 1991).

The halogenated solvents are known to be chemically stable and resistant to biodegradation in subsurface environments, where volatile losses are also greatly suppressed. Any microbial degradation is likely to be cometabolic and require a considerable excess of more readily utilizable organic carbon as primary substrate. The detection of a contaminant plume of trichloroethylene, some 600 m in length with maximum concentrations in excess of 1000 $\mu g\ l^{-1}$, at the site of a former maintenance facility for military aircraft on the Upper Chalk in southern England, abandoned more than 25 years earlier with no subsequent industrial activity or waste tipping, testifies to the subsurface persistence of these compounds (Lawrence and Foster 1991).

The light hydrocarbon petroleum fuels and heating oils, in contrast, float on the water table, if they are discharged in sufficient quantity to exceed the sorption capacity of the unsaturated zone, which is probably of the order of 1–10 $l\ m^{-3}$. Here they will also dissolve in groundwater and then diffuse into the matrix pore-water, but they can also migrate substantial distances if major secondary fissures are present. They are, however, much more readily biodegradable and less persistent, their presence stimulating the growth of autochthonous bacteria. The limiting factor in natural *in situ* biodegradation appears normally to be the rate of oxygen supply through the water-bearing porous medium.

Attenuation of solid waste leachates from landfills

Although current European policy is towards total pollution containment and effluent treatment at new landfill waste-disposal facilities, there are many older landfill sites on the Chalk outcrop that continue to produce leachate.

The volume and composition of such leachate depends on numerous factors, including the age and construction of the landfill and especially the nature of the waste deposited, which may not be reliably known. Where disposal of industrial waste has occurred either alone, or together with domestic wastes, the leachate will often be strongly acidic, contain high concentrations of heavy metals, and a wide range of organic contaminants, frequently including phenols, cyanides, and petroleum-related and halogenated compounds. In some cases it may even include pesticide compounds, such as atrazine, which have been disposed of by landfill or used in industrial processes and incorporated with the wastes.

All leachates, except those derived from inert wastes, tend to have a high salinity (Cl, SO_4 and Na concentrations, for example, often above 1000 $mg\ l^{-1}$), a heavy organic carbon load with total organic carbon (TOC), usually in the range 200–5000 $mg\ l^{-1}$ with carboxylic acids predominating, elevated levels of ammonia (50–500 $mg\ l^{-1} NH_3$–N), and a significant concentration of phenols. These contaminants are generally, but not entirely, attenuated in the unsaturated zone of the Chalk, if one is present beneath the site.

Detailed research has been carried out on attenuation mechanisms in the Chalk at two sites of domestic and industrial codisposal: Ingham, on the Middle Chalk in eastern England; and Ewelme, on the Lower Chalk in southern England (Department of the Environment 1978; Towler *et al.* 1985; Blakey and Towler 1988; Williams *et al.* 1991). Both sites have unsaturated zones some 20 m thick (Fig. 6.7), but they are complicated somewhat by the presence of perched water tables on more clayey layers within or overlying the Chalk.

The following are the most prominent attenuation mechanisms:

1. The buffering of acidity to neutral pH, with the associated precipitation of heavy metals, even the most mobile species such as nickel and chromium showing limited mobility. At Ewelme a complex series of acid–base redox and precipitation reactions results in the formation of a metal-rich deposit on fissure walls in the unsaturated zone of the Chalk.

2. The oxidation of assimilable organic carbon by autochthonous bacteria as much as the availability of oxygen allows. The pH buffering prevents leachate acidity from eliminating autochthonous bacteria, although their population may be restricted if bactericidal compounds are present in the leachate. At Ewelme, TOC rapidly decreased from more than 2000 mg l^{-1} to below 100 mg l^{-1} (Fig. 6.7a), and fully aerobic conditions were established before the water table was reached. At Ingham biodegradation in the unsaturated zone was also a prominent process (Figure 6.7b).
3. The concentrations of TOC persisting below the water table at both sites (above 10 mg l^{-1}) suggest that a residual soluble, and presumably recalcitrant, organic fraction reaches the water table. Less than 20 per cent of the individual organic carbon species present have been positively identified, and the toxicity of this dilute leachate is thus unknown. It is likely that common community-derived and industrial chlorinated hydrocarbons will often be present.
4. Phenols also proved rather resistant to biodegradation and at Ingham, for example, they had only decreased to 11 mg l^{-1} at a depth of 15 m.
5. As soon as aerobic conditions begin to be established, oxidation of ammonia to nitrate commences (Figure 6.7a), although this will not always be complete within the unsaturated zone.
6. Gas-phase transport may also be significant at some sites and involve not only methane and carbon dioxide generated by anaerobic digestion and aerobic degradation of the organic load, respectively, but also a range of volatile organic compounds.

Current knowledge thus suggests that, even where there is a significant thickness of unsaturated chalk below a landfill, some residual contamination by persistent organic compounds, including phenols and halogenated aliphatics, and other unidentified dissolved organic carbon will occur. Increased salinity (notably Na, Cl, Ca, SO_4, K and B) and either high ammonia or nitrate concentrations will

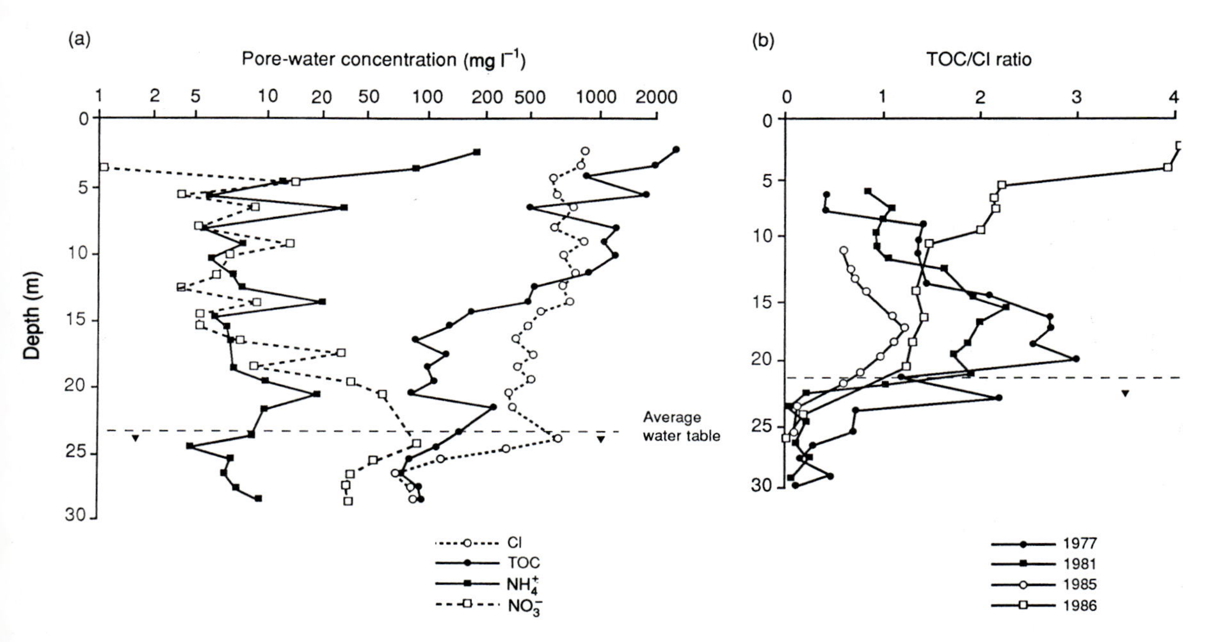

Fig. 6.7. Profiles of parameters indicating pollution below Chalk landfill codisposal sites (receiving both domestic and industrial wastes) showing major but incomplete attenuation of organic carbon load and partial oxidation of ammonium nitrate:
(a) Ewelme on Lower Chalk in Oxfordshire (modified after Department of Environment 1978);
(b) Ingham on Upper Chalk in Suffolk (derived from Williams *et al.* 1991).

also occur. The movement of this plume of polluted groundwater will depend on the prevailing groundwater flow regime, but it may be modified according to the density of the leachate.

Insufficient is known about the kinetics of leachate transformation in the unsaturated zone of the Chalk for it to be possible to indicate a 'typical unit attenuation capacity'. Moreover, where large leachate volumes are generated, preferential flow in the unsaturated zone may develop, allowing much less opportunity for attenuation.

If the landfill leachate is discharged directly to, or close to, the water table, far more severe groundwater contamination will occur, with the development of a significant volume of highly contaminated and strongly anaerobic water at the heart of the pollution plume.

Land application of effluents

The application of domestic sewage sludge and farmyard slurries to agricultural land does not appear to create such an increased risk of excessive nutrient leaching to groundwater, as the use of inorganic fertilizers, provided that guideline recommendations are followed (Joseph and Clark 1982). Indeed, it may help to suppress soil nitrification and thus reduce the leaching of nitrate.

In some hydrogeological conditions, however, it may give rise to microbiological contamination of abstracted Chalk groundwater if practised close to production boreholes. There is also likely to be a more general increase in NaCl and dissolved organic carbon (DOC) concentrations to above normal background (Edworthy *et al.* 1978), and the DOC may include traces of chlorinated aliphatic, benzene and phenol compounds that are in use as chemicals in the community.

There is somewhat conflicting evidence on the capacity of the unsaturated zone of the Chalk to attenuate bacteria. There are well-substantiated reports of intermittent groundwater pollution by coliform bacteria in areas with a thick unsaturated zone, when effluent application is followed by high-intensity rainfall. This presumably is a consequence of rapid, deep preferential flow.

On the other hand, the unsaturated zone of the Chalk has been shown to be highly effective for the microbiological purification of wastewater. Municipal wastewater disposal into the Chalk has been practised for many years at various locations in southern England (Table 6.4a). Recharge is achieved through soil drains, land spreading, over-irrigation of non-agricultural plots or lagoon infiltration, at rates in the range 0.02–0.40 m d^{-1}. Given the need for rotation of the infiltration sites, land requirements are appreciable (10–20 ha per 10000 $m^3 d^{-1}$ of wastewater recharged).

Table 6.4a Summary of sites on Chalk outcrop in southeastern England where effluent is recharged into the aquifer

Location		Depth to water table (m)	Effluent volume/ treatment method		Recharge	
Town	County		($\times 10^3 m^3 d^{-1}$)		Method	Rate ($mm\ d^{-1}$)
Winchester	Hampshire	5–35	11.0	Primary	Ditches/ Flood irrigation	25
Whitchurch	Hampshire	10–15	1.3	Secondary*	Soil drains	300
Alresford	Hampshire	25–35	0.6	Secondary	Soil drains*	–
Ludgershall	Hampshire	?	0.8	Secondary	Flood irrigation	400
Caddington	Bedfordshire	10–15	1.2	Secondary	Soil drains	200
Royston**	Hertfordshire	10–15	2.8	Primary	Lagoons	10

* Prior to 1981 only primary treatment and recharge via lagoons at rates of around 80 mm d^{-1} (see Table 6.4b)

** No longer operating, but formerly practised to dispose of mainly industrial and saline wastewater, unlike other sites which receive predominantly domestic effluent

Table 6.4b Summary of changes in wastewater quality following infiltration through the unsaturated zone of the Chalk at Whitchurch site (after Montgomery *et al.* 1984; Beard and Giles, 1990)

Quality parameter	Units	Recharge method: Unlined lagoons with primary effluent (pre-1981)		Soil drains with secondary effluent (post-1981)	
		Treated wastewater	Groundwater recharge	Treated wastewater	Groundwater recharge
NH_3–N	mg l^{-1}	17	14–20	0.3	<0.1
NO_3–N	mg l^{-1}	<1	<1	19	18–20
C1	mg l^{-1}	47	43	54	47–54
B	mg l^{-1}	1.1	0.6	1.0	0.9
DOC	mg l^{-1}	47	5–12	7	3–5
FC }	CFU per	8×10^6	1×10^3	2×10^5	1×10^3
RV }	100 m1	40	<1	?	?

CFU = Colony forming units
FC = Faecal coliforms
RV = Rotaviruses

Investigation and monitoring of these sites (Baxter *et al.* 1981; Montgomery *et al.* 1984; Beard and Giles 1990) have led to the following conclusions, which are summarized in Table 6.4b:

1. The soil and unsaturated zone are extremely effective for the removal of pathogenic microorganisms, and although contamination was observed in the saturated zone immediately below the recharge site in some instances, there was little evidence of any contamination, bacterial or viral, at distances of more than 200 m downstream of the site boundary.
2. There is also very effective attenuation of the dissolved organic carbon in the wastewater, but DOC levels significantly above background are often found in the groundwater below and downstream of the wastewater recharge facility, with the possible presence of toxic compounds, especially where the chlorination of wastewater was practised.
3. The total nitrogen content of wastewater is high. Groundwater contamination by ammonium is generally present below recharge facilities receiving primary effluent, but this is rapidly oxidized to nitrate and diluted in the regional groundwater flow system. Where secondary effluent is used, the oxidation takes place prior to recharge, and nitrate persists throughout the process.
4. There is no evidence of significant transport of heavy metals, but iron and manganese concentrations may increase locally.

The extent and intensity of any contamination will vary widely with the detailed hydrogeological conditions, the mode of application of the wastewater and the scale of the disposal process. Wastewater recharge may be justified in some circumstances as a means of discharging effluents indirectly to Chalk streams or of forming a hydraulic barrier against aquifer seawater intrusion, provided that no municipal borehole or wellfield, or other potable source, is threatened.

Concluding remarks

Vulnerability to pollution

The Chalk has some capacity for purifying water infiltrating through it, and for assimilating pollutants. Its very pure calcium carbonate composition results in a virtually unlimited buffering capacity for acidic pollution loads, and the naturally very aerobic character of its unsaturated zone promotes the breakdown of degradable contaminants.

Thus under favourable conditions the unsaturated zone of the Chalk is capable of significantly attenuating and effectively eliminating some contaminants. Pollutant transport, however, can be very sensitive to the hydraulic load and to relatively small changes in the bulk hydraulic characteristics of the rock mass. Difficulties arise in predicting pollutant movement and estimating the degree of protection offered by the unsaturated zone at a given location.

In absolute terms the unconfined Chalk aquifer must, therefore, always be regarded as vulnerable, and in some cases extremely vulnerable, to pollution, primarily as a result of its well-developed fracture systems, together with limited capacity for the adsorption of contaminants because of its very low clay mineral and organic carbon content. This vulnerability is compounded by the fact that once the saturated zone becomes polluted, it is highly probable that long-distance lateral transport of contaminants in major secondary fissures will occur. It will also be extremely difficult to clean up the aquifer because of the retention of large amounts of contaminant in the micro-porous matrix.

As the public debate about the pollution of groundwater in the Chalk intensifies, views of administrators and politicians tend to polarize towards the existence of either an 'imminent environmental disaster' or a 'minor localized problem'. Informed professional opinion is that the situation lies somewhere between these two extremes—a complex and potentially costly problem affecting northern European nations to differing degrees. Diffuse agricultural pollution is already serious and widespread, and policy decisions intended to rectify the situation are only now being implemented. Urban and industrial point sources of pollution are becoming more evident each year, although many are a legacy of inadequate control policies in the past.

Chalk aquifer protection policy

In consequence, groundwater in the Chalk should generally be afforded the protection measures summarized in Table 6.5, and a high priority should be given to systematic data collection and monitoring of existing sources of potential point pollution.

The establishment of protection areas around important municipal water-supply boreholes and wellfields, and other sensitive sources, provides a pragmatic basis on which to erect more rigorous pollution control measures where they are most needed (Table 6.5; Southern Water Authority 1985; Lallemande-Barrès and Roux 1989; Adams and Foster 1992). These should normally include the following (Adams & Foster 1992):

1. An outermost zone corresponding to the recharge area defined by the long-term average recharge and the authorized abstraction rates, which is in effect the capture zone for infinite pumping time. This should be defined using the best available hydrogeological data, aided by the use of numerical computer codes provided that the results are carefully reconciled with

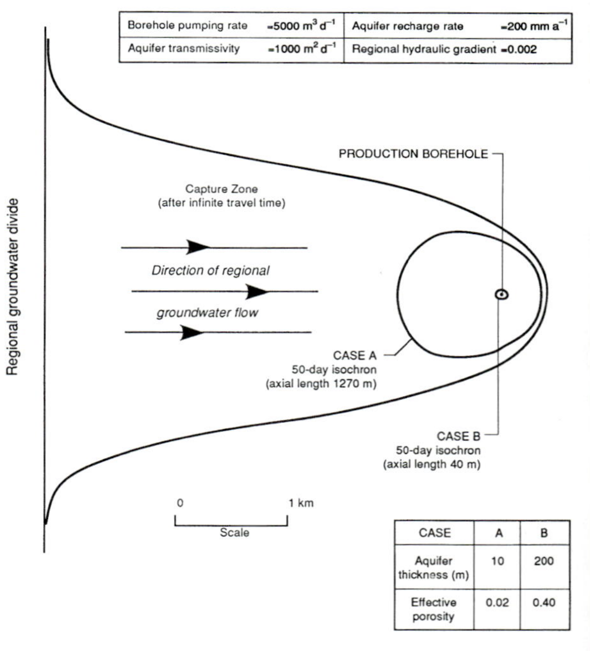

Fig. 6.8. Special protection areas around production borehole in idealized unconfined Chalk aquifer showing sensitivity of size of 50-day isochron to errors in effective aquifer thickness and porosity (modified from Adams and Foster 1992).

Table 6.5 Measures required to protect groundwater in the Chalk from pollution

Potential polluting activity	Level of control required		
	Source protection area		Aquifer outcrop/recharge area
	Microbiological zone (50-day transit time)	Remainder of capture zone	
Drainage and effluent percolation			
– roofs and minor roads	NC	NC	NC
– major roads	PR	SE	NC
– industrial and commercial site	PR	PR	SE
– septic tanks	PR	SE	NC
– cooling water	SE	NC	NC
– industrial effluents	PR	PR	SE
Effluent application to land			
– farmyard slurries	PR	NC	NC
– sewage sludge	PR	NC	NC
– sewage effluent	PR	SE	SE
– industrial effluents – organic	PR	SE	SE
– industrial effluents – others	PR	PR	SE
Solid waste landfill			
– industrial hazardous	PR	PR	PR
– municipal refuse	PR	SE	SE
– inert materials	PR	NC	NC
Storage and other facilities			
– chemicals and oils	PR	SE	SE
– farmyard wastes	PR	NC	NC
– industrial process effluents	PR	SE	SE
– graveyards	PR	SE	NC
– intensive livestock units	PR	SE	NC
– sewer lines	SE	NC	NC
Agricultural cultivation practices*			
– fertilizer application	NC(SE)	NC(SE)	NC
– pesticide application	NC(SE)	NC(SE)	NC

PR – prohibited as always unacceptable
SE – possible but subject to special evaluation and control
NC – permitted subject to normal controls

* Normal controls (that is, best agricultural management practice including consideration of groundwater quality degradation) will normally apply but many source protection areas will be subject to special evaluation with a view to encouraging additional controls over certain practices

available knowledge of local hydrogeological conditions.

2. An inner protection zone based on the distance equivalent of a horizontal flow time of 50 days. This value is considered a reasonable compromise between the maximum subsurface pathogen survival times of around 400 days and the maximum observed distance of significant pathogen transport which rarely exceeds that travelled by groundwater in 20 days.

The definition of this latter zone using analytical equations or numerical computer codes is very sensitive to errors in the estimation of aquifer porosity and effective saturated thickness (Fig. 6.8). This presents special problems in the Chalk aquifer, where there is danger of grossly overestimating these parameters. The computation of worst-case estimates or the use of a stochastic approach to modelling that allows for uncertainty in the values of parameters, should be considered.

Special problems arise with the definition of recharge capture areas in situations where the groundwater divide is at a great distance, the regional hydraulic gradient is very low, and the aquifer is very exploited.

Hydrogeological research requirements.

This review has identified certain critical gaps in current knowledge and understanding of groundwater flow and pollutant transport in the Chalk aquifer:

(1) the general scale of and threshold conditions for rapid preferential (bypass) flow in the unsaturated zone, which is of major importance in assessing the risk of contamination by pesticides and in predicting future trends in the concentration of nitrate in groundwater;

(2) the mechanisms and rates of chemical breakdown and/or bacterial degradation of the common pesticides in both the unsaturated and saturated zones;

(3) the distribution and availability of organic carbon that can act as a substrate for bacterial denitrification and the biodegradation of organic micropollutants;

(4) the extent of the transport of colloids through the aquifer system and its significance in terms of the adsorption of contaminants;

(5) the controls over the natural distribution of dissolved oxygen in the saturated zone and its rate of diffusion in the aquifer system;

(6) the speciation and significance of the increased levels of dissolved organic carbon in Chalk groundwater below some effluent and waste-disposal sites;

(7) the effect of natural linings on fissures on the rates of exchange of contaminants with the micro-porous matrix by aqueous diffusion;

(8) the mechanisms and rates of vertical groundwater flow in the saturated aquifer, both regionally and locally around production boreholes, and its significance in terms of the transport of pollutants.

Acknowledgements

This contribution is published by permission of the Director of the British Geological Survey, a component institute of the Natural Environment Research Council. The author is indebted to many British hydrogeologists for the interchange of data and ideas on the transport of pollutants in the Chalk, and especially to his colleagues John Barker and Adrian Lawrence.

7. France

N. Crampon, J. C. Roux, and P. Bracq, in collaboration with F. Delay, M. Lepiller, G. Mary, L. Rasplus, and G. Alcaydé

The geographical distribution of the Chalk

In France, the Chalk outcrop forms a wide belt around the Paris Basin, covering an area of about 70 000 km^2. For convenience, it can be sub-divided into four units: (I) Artois–Picardy; (II) Normandy; (III) Maine–Touraine; (IV) Gâtinais–Champagne (Fig. 7.1).

Artois–Picardy includes the French *départements* of the Nord (Dunkerque, Lille, Valenciennes), Pas-de-Calais (Calais, Boulogne, Arras) and also Picardy (Amiens, Saint Quentin), as far as the Oise in the south-east and the Bray anticlinal axis in the south-west. A dominant physical feature is the Artois Horst, which runs from north-west to south-east, at an average height of 140 m. It is dissected by numerous dry valleys and, in places along the Channel coast, by hanging valleys (*valleuses*). In the north-west this massif ends against the eroded core of the Boulonnais–Weald Anticline, from which the Chalk has been eroded. Picardy is also dissected by dry valleys, although it lies at a lower elevation.

One of the geomorphological characteristics of the area is the numerous 'steps', abrupt and sharp breaks of slope with the axes parallel to the thalwegs, several metres high, sometimes accentuated by the presence of a hedge. Although the presence of colluvium deposited behind the hedge suggests a man-made origin for the features, a structural element is clearly present (Lasne 1890; Leriche 1926; Roux 1963), at least for the 'steps' that are cut into the Chalk. Their trends are commonly similar to those of other geomorphological features of the area, which are themselves related to the regional fracture pattern.

The Chalk of the Artois Horst plunges towards the north-east below the Tertiary and Quaternary of the low-lying Flanders Plain, which is locally below sea level (the last transgression of the sea occurring in the tenth and eleventh centuries). Near the present shoreline, the maritime Flanders Plain is bordered on the south-west by the Sangatte fossil cliff. Other maritime plains, the 'Bas-Champs', extend to the base of ancient chalk cliffs of Eemian (Upper Pleistocene) age, in the area around the Somme, Authie and Canche estuaries. But the coast is still subject to active erosion near Calais, where 2–3 metres are eroded each year from Cap Blanc-Nez.

The eroded core of the north-west to south-east Bray Anticline forms a natural break in the Chalk and the second unit, Normandy, lies to the west of it. The unit includes the Pays de Caux (Dieppe and Le Havre) to the north-east of the Seine, and the Pays d'Ouche in the south-west. The southern boundary is an east–west line from Argentan to Dreux, drawn arbitrarily, but corresponding approximately to a boundary of hydrogeological significance. This area, of rather high average elevation, has a more marked relief than Artois–Picardy, particularly because of a dense network of deep dry valleys related to karstic features. In places the Pays de Caux cliffs are receding at about 1 m a^{-1}, although in other places there has been little erosion since the Flandrian (Rodet in Colloque Karst et Quarternaire de la Basse-Seine 1988; Ochietti, personal communication).

Maine–Touraine includes Perche, Maine (Le Mans), the east of Anjou (Angers), Touraine (Tours), and the north of Berry. The eastern boundary may be taken arbitrarily as the River Loire upstream of Orléans. The relief is rather moderate and in the form of a plateau; the altitude decreases from north to south, from about 320 m to 80 m, towards the boundary at Poitou.

Gâtinais–Champagne groups together Gâtinais (approximately the area between the Loire and the Yonne), Sénonais (Sens) between the Yonne and the Seine, and the Champagne region (Troyes,

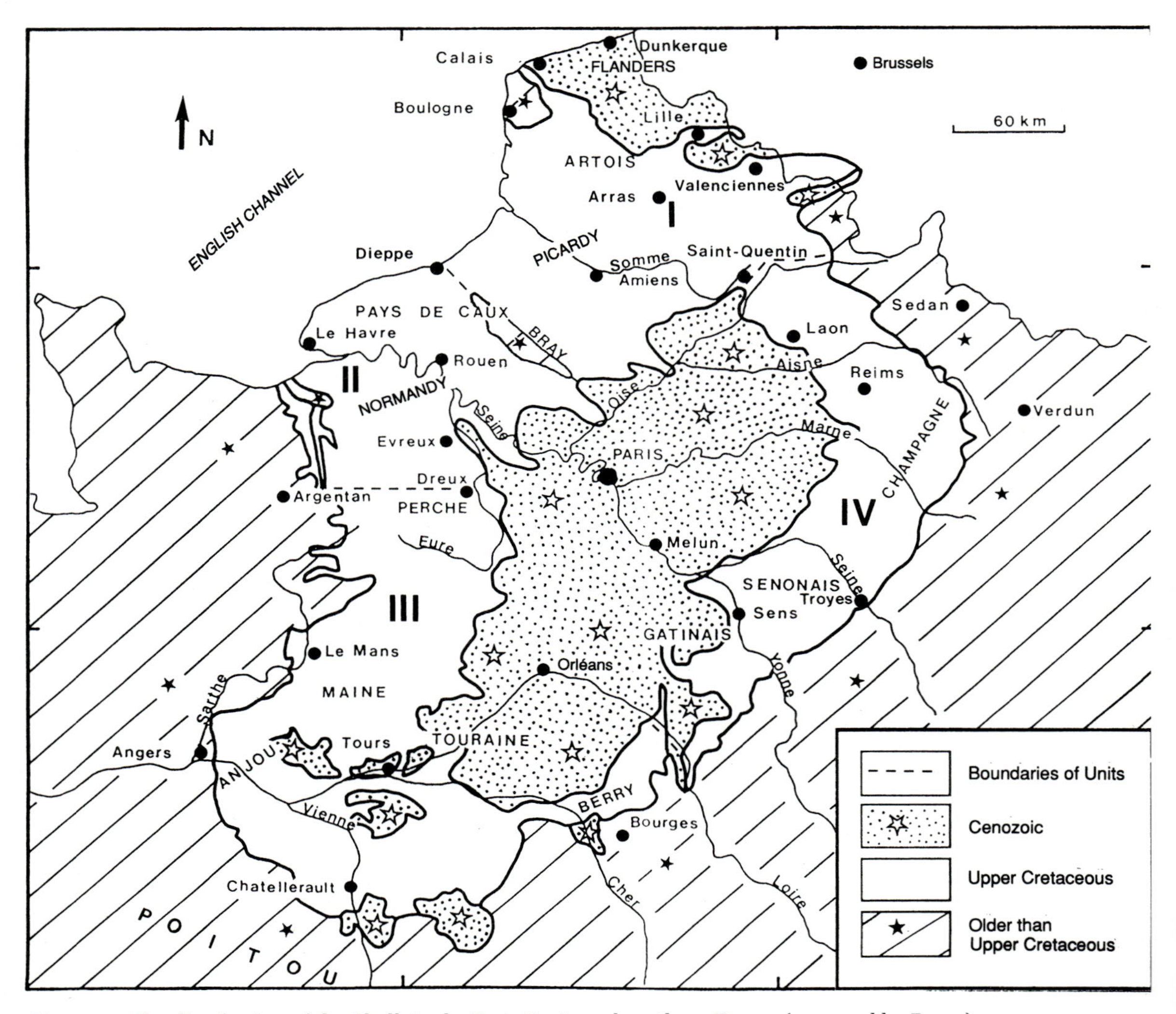

Fig. 7.1. The distribution of the Chalk in the Paris Basin and northern France (prepared by Bracq).

Châlons-sur-Marne, Reims) which extends as far as the Oise, thereby completing the girdle of chalk around the Paris Basin. Gâtinais–Sénonais is a region of uplands partially covered by Tertiary argillaceous-sandy formations, and dissected by the tributaries of the Seine. In the Champagne region, the average altitude is less than 200 m and the relief is subdued except in the Montagne de Reims.

Chalk has often been extracted for either building or agricultural purposes, and it also provides raw material for the cement industry, paper whitening or laying (412 000 t a^{-1}), lime for paper pulp (123 576 t in 1983), and for the paint and varnish industries (150 000 t a^{-1}) (Ricour 1985). Phosphatic chalk has been actively sought and exploited for fertilizer in Picardy, where it is particularly widespread. Underground quarries, now disused or not properly recorded, occur in many districts (particularly in Artois, Picardy, Normandy and Touraine) and pose problems for modern engineering work, particularly at urban sites. Sometimes, as in Artois, bottle-shaped underground quarries (*catiches*), possibly connected with galleries, replace the 'pillars and stalls' of classic quarries. Some of these former quarries are still used for mushroom cultivation. Between Picardy and Touraine, these artificial cavities, as well as some natural caves, served, and still serve locally, as cave dwellings and even as churches (for example Haute-Isle, between Mantes and Rouen), but now they are used mainly as cellars (wine cellars in Touraine), barns, or garages.

As far as agricultural activities are concerned, Artois, Picardy, Normandy (Pays d'Ouche), Touraine and Champagne are large cereal- or beetroot-growing areas. Where the soils are too thin for arable farming, meadows are used for sheep breeding (as in Champagne). Dairy farming occurs in Normandy and Avesnois (the eastern part of Artois–Picardy), and horse breeding in Perche. The vineyards of the Anjou–Touraine hills and Champagne have an excellent reputation.

General stratigraphy and structure

It was during the Lower Cretaceous period that the North Sea began to coalesce with the Mésogée (the palaeo-Mediterranean Sea): in the Albian a wide strait extended across the Paris Basin towards the Weald and Burgundy, and from the beginning of the Upper Cretaceous Cenomanian transgression it grew progressively wider. This marine transgression extended directly over the folded Palaeozoic rocks from Avesnois (at the western end of the Ardennais block) to Calais (that is, over the entire northern part of Artois–Picardy except for the Boulonnais region) and Anjou; it extended over Jurassic or Lower Cretaceous rocks in the remainder of the area. The lithology of the Upper Cretaceous deposits was influenced by the proximity and elevation of the nearby ancient massifs (the Ardennais block, the Armorican Massif, and the Central Massif) which provided detritus; it also depended on the distribution of shallow water, such as existed in the Pays de Bray–Pays de Caux, and on the presence of subsiding basins (Fig. 7.2). Consequently, the chalk facies is not distributed uniformly over the Paris Basin and in some districts alternations of chalk, clay, and sand occur which obviously influence the behaviour of the Chalk as an aquifer (Fig. 7.3).

In Artois–Picardy, above the basal conglomerate called 'tourtia', which overlies coal-bearing rocks and which decreases in age with the marine transgression, marls and marly chalk were deposited which are sometimes water-bearing from the base of the Middle Cenomanian (as at Cap Blanc-Nez). The deposits became more and more chalky during the Turonian period, but with episodes that were still very marly. By the Upper Turonian–Senonian, the deposits were mostly chalk. Pure chalk occurs earlier in the north-west than to the east of Lille, where it appears only during the Upper Turonian, above marly formations of the Middle Turonian called *dièves*. The presence of hard nodular chalk (referred to as *tun*), marks a temporary pause in the sedimentation sequence at the Turonian–Senonian boundary. Some calcium phosphate occurs in the Campanian Chalk in Picardy (in the Amiens area, Fig. 7.2). The Maastrichtian is not represented.

In Normandy, because of shallow seas and associated hardgrounds, the earliest chalk deposits that are preserved date from the Upper Cenomanian on the Pays de Bray, whereas to the south of the Seine (on the Norman Platform) many chalk facies follow one another—initially glauconitic chalks during the Lower and Middle Cenomanian, then chalks with more and more flint from the Upper Turonian. The famous Senonian chalk cliffs of Etretat show the regularity of the frequent flint layers, with the occasional layer oblique to the bedding testifying to their early origin (Pomerol 1975).

The 'Mancello–Tourangeau' Basin lies to the south of the Merlerault axis (Fig. 7.2), a structure that played an active role during the Cenomanian. In this basin, Cenomanian sedimentation consists mainly of detrital rocks (argillaceous, glauconitic, and later siliceous types), including the sands and sandstones of the Mans (Middle Cenomanian) and of Perche (Upper Cenomanian). During the Turonian three superimposed facies developed: argillaceous chalk with *Inoceramus*, followed by white micaceous chalk (white tuffeau) with branching flints, and finally a biocalcarenite (a yellow more or less sandy tuffeau) that passes into greensands towards Anjou and shelly sands towards Maine. In these provinces, the sandy and sandstone-like deposits continued during the Senonian, whereas in Touraine, the siliceous chalk facies continued during the Coniacian–Santonian period with the Villedieu Chalk changing near Blois and Chaumont into a white-chalk with flints no younger than Lower Campanian. Towards the south, in Berry, to the south of Orléans, the Senonian becomes clayey and sandy.

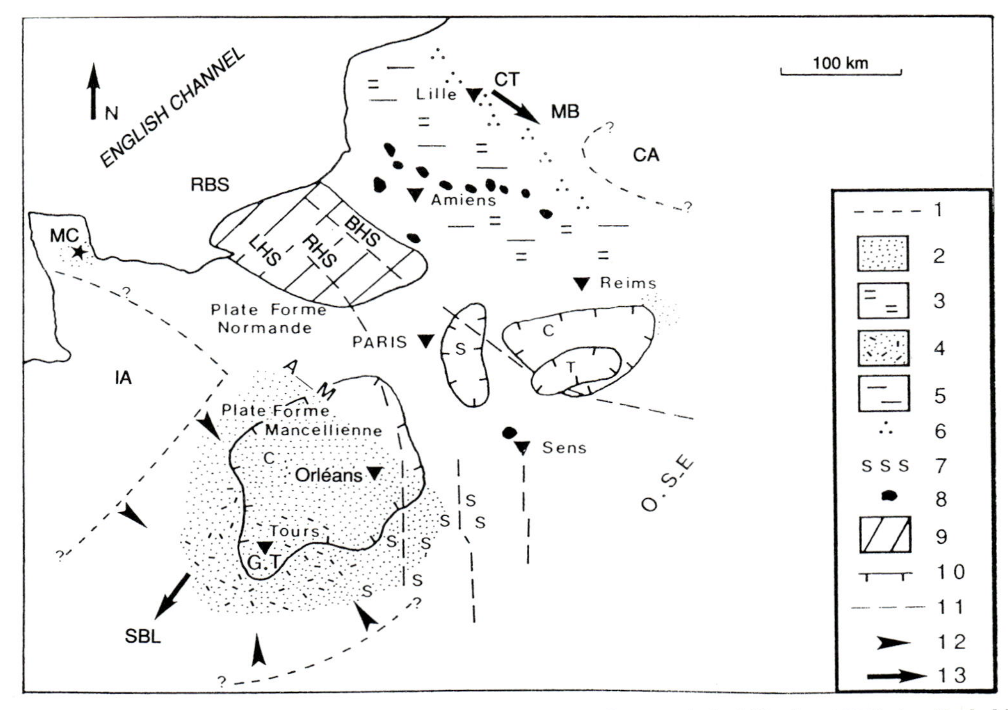

Fig. 7.2. Palaeogeography of the Upper Cretaceous (after Monciardini *et al.*, in Mégnien 1980). 1 – Probable limit of Upper Cretaceous. 2 – Detrital Cenomanian facies. 3 – Marly Cenomanian facies. 4 – Detrital Turonian and Senonian facies of Touraine. 5 – Marly Turonian facies. 6 – Cenomanian and Turonian 'Tourtia' (or basal conglomerate). 7 – Siliceous Turonian and Senonian facies of Berry. 8 – Phosphatic Senonian Chalk. 9 – Shallow water Upper Cretaceous. 10 – Areas of maximum subsidence during: Cenomanian (C), Turonian (T) and Senonian (S). 11 – Craton (or basement) faults. 12 – Direction of sediment input. 13 – Directions of transgressions. BHS: Bray, RHS: Rouen, and LHS Lillebonne areas of shallow-water sedimentation; RBS: inversion of Bray shallow-water area; MC: Maastrichtian of Cotentin, MB: Mons Basin, IA: Island of Armorique; SBL: Shallow sedimentation area of Basse-Loire; CT: Cenomanian transgression; CA: Craton of Ardennes; AM: Axis of Merlerault; OS–E: Opening to the south–east; GT: Gulf of Touraine.

To the east of Berry, the Cenomanian maintains a marly sandy facies in the Gâtinais, becoming chalky in the Sénonais (from the Middle Cenomanian), whereas in Champagne marl is again dominant, chalk appearing only in the Lower Turonian. The chalky facies continues up to the Campanian, with white-chalk developing basically from the Upper Turonian.

Some examples of typical average lithologies are given in Fig. 7.3, where the facies of hydrogeological significance have been emphasized, such as the marly horizons and the hard bands that separate the individual chalk aquifers. The scale of the figures does not indicate the true thicknesses.

The Cenomanian is less important to the north and north-west of the Paris Basin, being less than 50 m thick, but in Champagne, Sénonais, Gâtinais, and in the centre of Touraine it attains its maximum thickness of 75–150 m. The Turonian is less than 50 m thick over the major part of the periphery of the basin, except in Sénonais, Champagne, and in the middle of the basin, where it reaches 150–200 m. The Senonian is relatively thin in Anjou–Touraine (50–100 m), around Bray, in Picardy, and in the Nord, and its maximum thickness of 400 m is found to the east of Paris under Tertiary cover. The entire Upper Cretaceous exceeds 700 m in the sedimentation trough at Brie, and is of the order of 500 m in Paris (Fig. 7.4, Monciardini *et al.* in Mégnien 1980).

At the end of the Cretaceous, the sea retreated to the north and the Chalk was eroded. During the Palaeogene

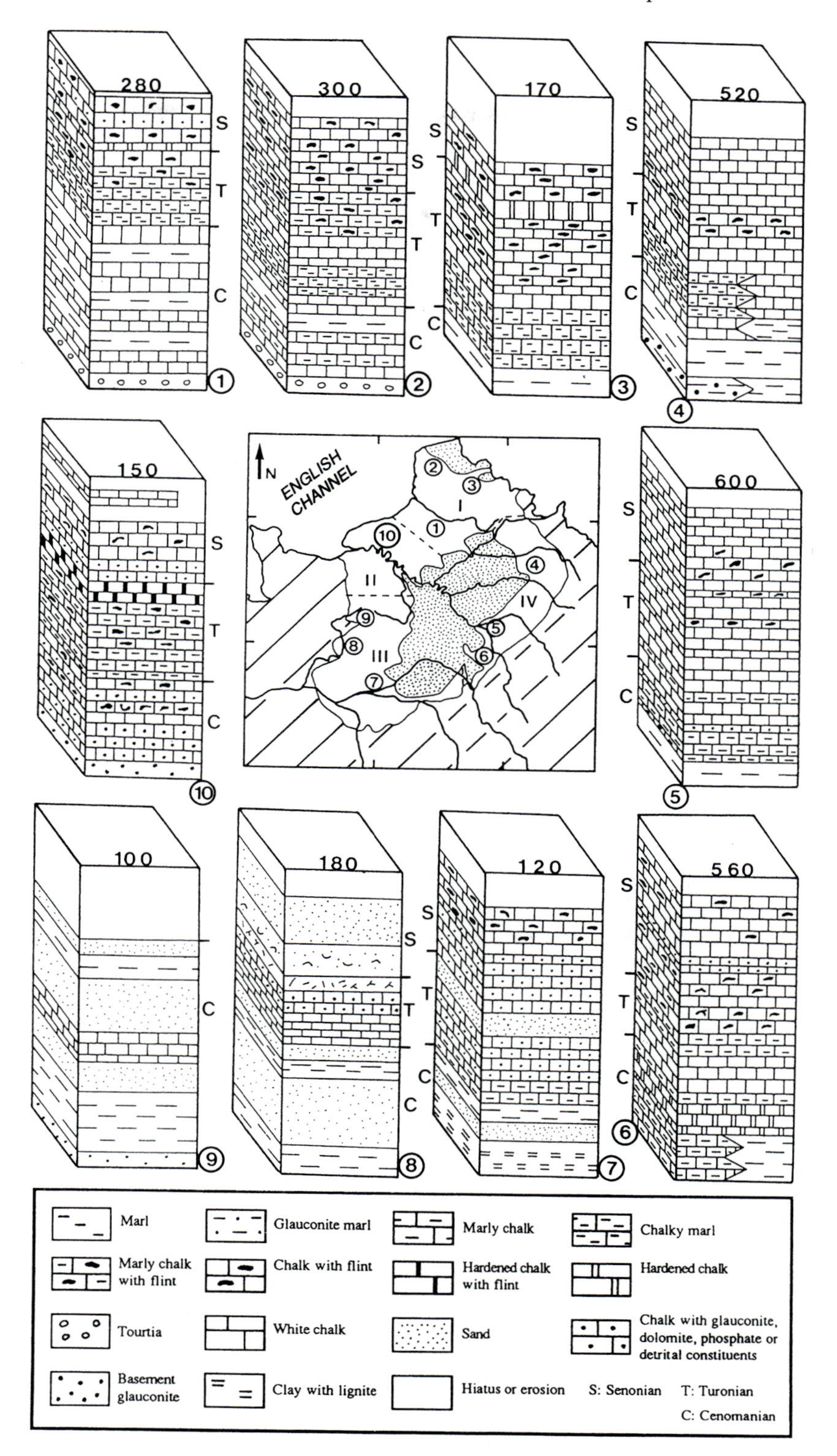

Fig. 7.3. Lithology of the Upper Cretaceous in the various units. Figures at head of columns indicate thicknesses in metres (prepared by Bracq).

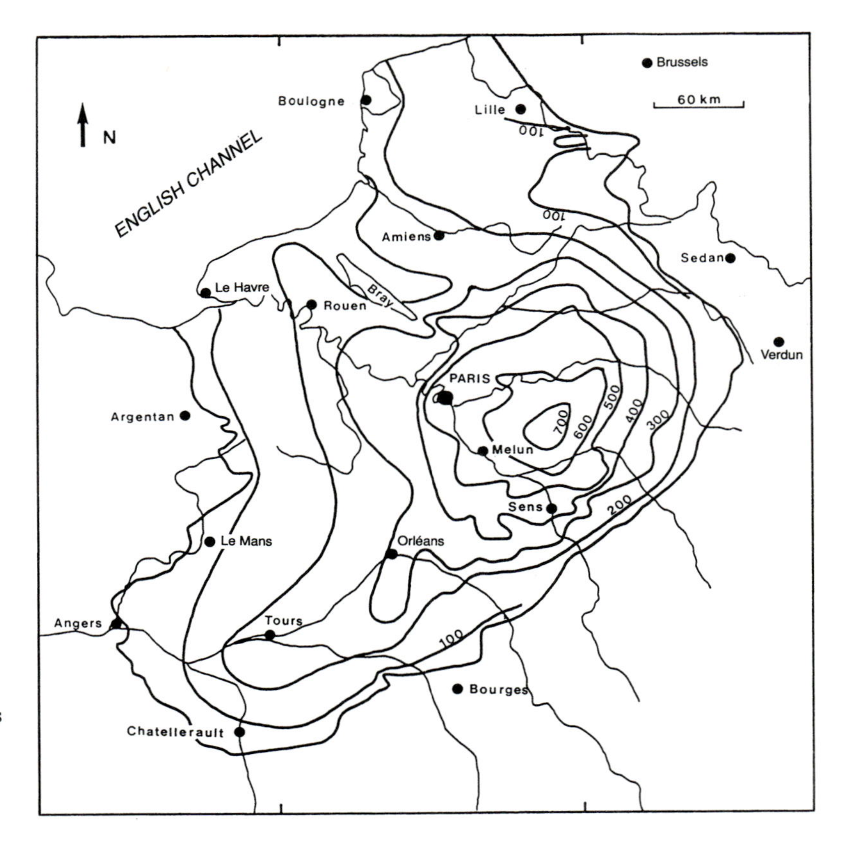

Fig. 7.4. Thickness of the Upper Cretaceous in the Paris Basin, in metres (after Monciardini *et al.*, in Mégnien 1980).

the sea again advanced over the Paris–Brussels Basin, and repeated transgressions and regressions deposited an interbedded marine and terrestrial sequence in the Paris Basin and southern parts of the Brussels Basin, which were beginning to become separate basins as a result of the early stages of uplift of the Artois Horst. These clays, sands and calcareous deposits have an important bearing on the development of the boundaries of the Chalk aquifer. The Tertiary formations overlying the Chalk may be either of marine or continental origin according to locality.

Except in Champagne and part of Artois where the Tertiary transgressions did not occur, the Chalk is covered with a mantle of Clay-with-flints that developed during the Tertiary and Quaternary through weathering and erosion of both the Chalk and the Tertiary formations. In Normandy and in Gâtinais-Sénonais, this clay mantle can attain thicknesses of 30 m. In Artois, Picardy and Champagne, the principal cover on the Chalk is loess from a few centimetres to a few metres thick.

Both at outcrop and below Tertiary cover, the Chalk of whatever age or facies, has been subjected to erosion, and has been altered or modified to variable depths, as revealed by:

(1) a variable thickness of fissured chalk (20–50 m thick along the edge of the Flanders Plain, for example);
(2) the presence of solifluction deposits;
(3) the existence of solution pockets that are funnel-shaped and isolated or interlinked, from a few metres to more than 30 m deep, with flint, clay and sand filling the spaces between residual chalk blocks;
(4) the development of deep valleys, sometimes with a meandering form;
(5) the presence of numerous dry valleys;
(6) the development of karstic networks such as those of the Pays de Caux, of the Pays d'Ouche or of Sénonais.

These features probably developed in Lower–Middle Pleistocene times (Sommé, personal communication).

The dip slope of the Chalk, as it appears today, has the general form of a basin with its centre to the east of Paris (Fig. 7.5). This basin began to develop, as a result of subsidence, during the Upper Cretaceous period and this increased in the Cenozoic era.

The basin contains numerous faults and small folds (Fig. 7.6), the result of multiple phases of post-Palaeozoic tectonic activity. This activity continued throughout the time when the Chalk was being deposited. Areas of local uplift and subsidence are revealed by evidence of ridges and shallow waters, such as those of the Pays de Bray, of Rouen and the Merlerault axis, and by the persistence of deep sedimentation troughs, such as can be recognized in Brie. However, subsidence was not always centred on the same locality from the Cenomanian to the Senonian (Fig. 7.2).

Groundwater levels and groundwater flow

The Upper Cretaceous of the Paris Basin covers about 110 000 km^2 (one-fifth of the total area of France) of which two-thirds is at outcrop and one-third under Tertiary cover. Although chalk is not the only facies in the Upper Cretaceous, it dominates most outcrops, except in the south-west of the basin (in Perche, Maine, Anjou, Touraine, and Berry). Where the Chalk is overlain by thick Tertiary cover (in the centre of the basin and in Flanders) it is generally not an aquifer: the formation is generally productive only at outcrop or within a short distance of the edge of the Tertiary cover.

The Chalk is the most important unconfined aquifer of the Paris Basin, both because of its extent and the size of its resources (equivalent to $11–12 \times 10^9 m^3 a^{-1}$). But can one speak of a single Chalk aquifer, over the entire Paris Basin or even locally? The answer, of course, depends on the scale of the investigation. Even at its maximum development, the thickness of the Chalk is only about one-thousandth of its north–south extent. This relatively thin aquifer is folded into a basin that contains subsidiary folds and faults. It is open to the English Channel to the north-west, and cut by valleys that play important roles for both drainage and recharge. It behaves hydraulically more as a system of separate aquifers than as a single aquifer, although each individual aquifer is not totally independent of its neighbours (Margat in Colloque Régional de Rouen 1978). Generally it is considered to be homogeneous, but not necessarily isotropic. The form of the water table, groundwater–river interconnections, effective infiltration, vertical water flow, and groundwater abstraction are treated as if it were homogeneous on a scale of, say 10–100 km. But this homogeneity is apparent only if one overlooks the factors that produce heterogeneity—for example, the following:

1. The superposition of several distinct aquifers—layers of chalk separated by marly interbeds.
2. The variable development of fissures with depth, which depends upon the exposure to climatic agents: fissured chalk may be a few metres to about 10 m thick under uplands, and sometimes several tens of metres under the valleys. This fissured zone (or effective thickness of the aquifer) is very irregular and thus difficult to estimate other than locally, but obviously it is the only zone suitable for aquifer development and management.
3. The preferential flow at the top and/or bottom of indurated chalk or flint bands.
4. The existence of permeable zones or 'drains' which have developed because of faults or joints, or along bedding planes.
5. Local karstification, which varies from the simple solution pocket to a multi-kilometre underground network, including dolines, collapse zones, and open pipes, some tens of centimetres in diameter.

The Chalk usually has a homogeneous, isotropic, very porous matrix, but with a low permeability, even when it does not have an argillaceous component. The various factors, referred to above, that produce heterogeneity produce an aquifer system with double porosity and multiple permeability, in which there is a slow, generalized flow but also very localized rapid flows.

On the local scale (of metres to several kilometres), it is the heterogeneities of the lithofacies, and the modification of them leading to fissuring, fracturing, and karstification, that present particular problems of water production and of quality protection that must be considered by hydrogeologists and engineers.

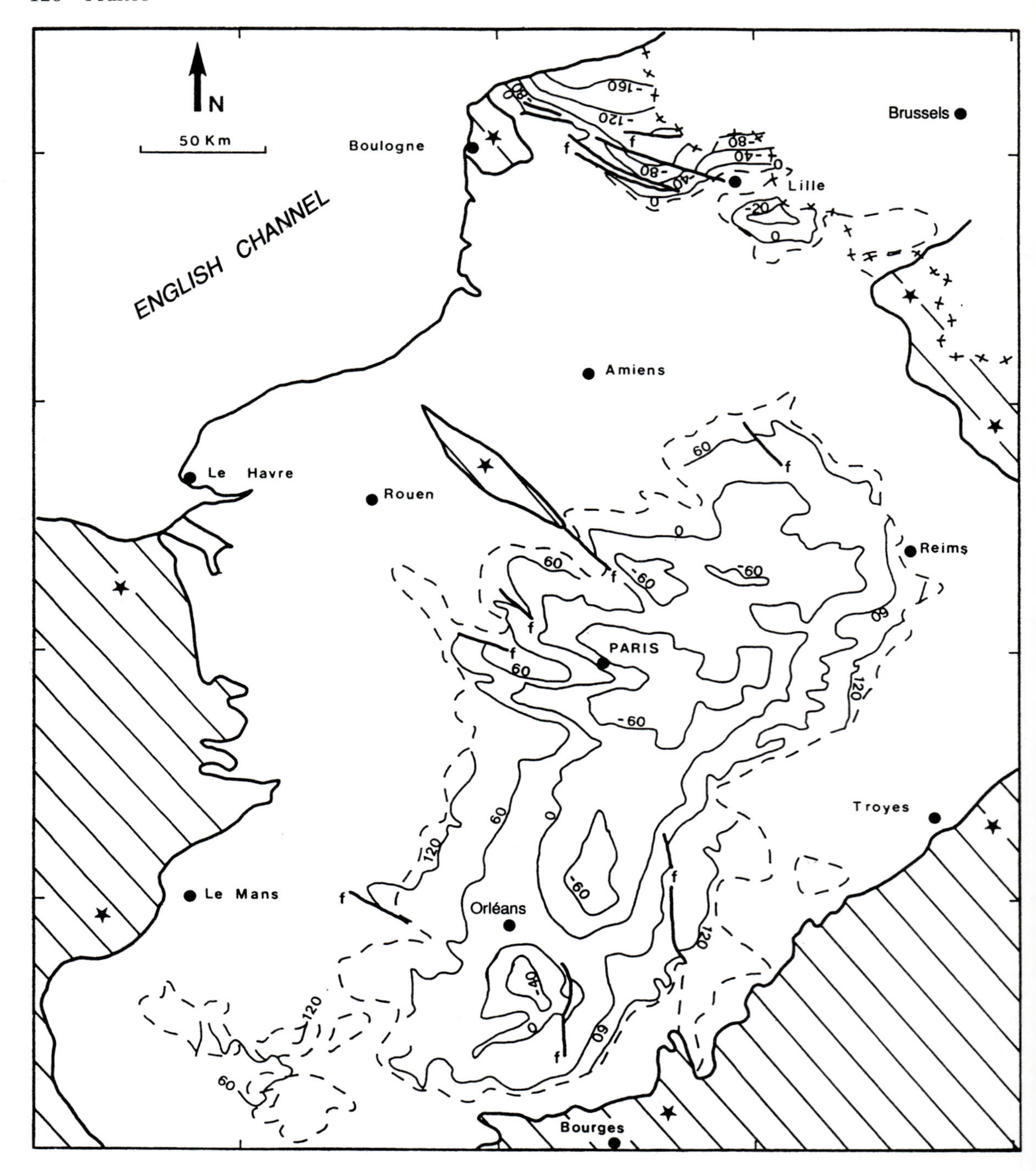

Fig. 7.5. Contours on the top of the Chalk in the Paris Basin in metres relative to sea level (after Monciardini *et al.*, in Mégnien 1980).

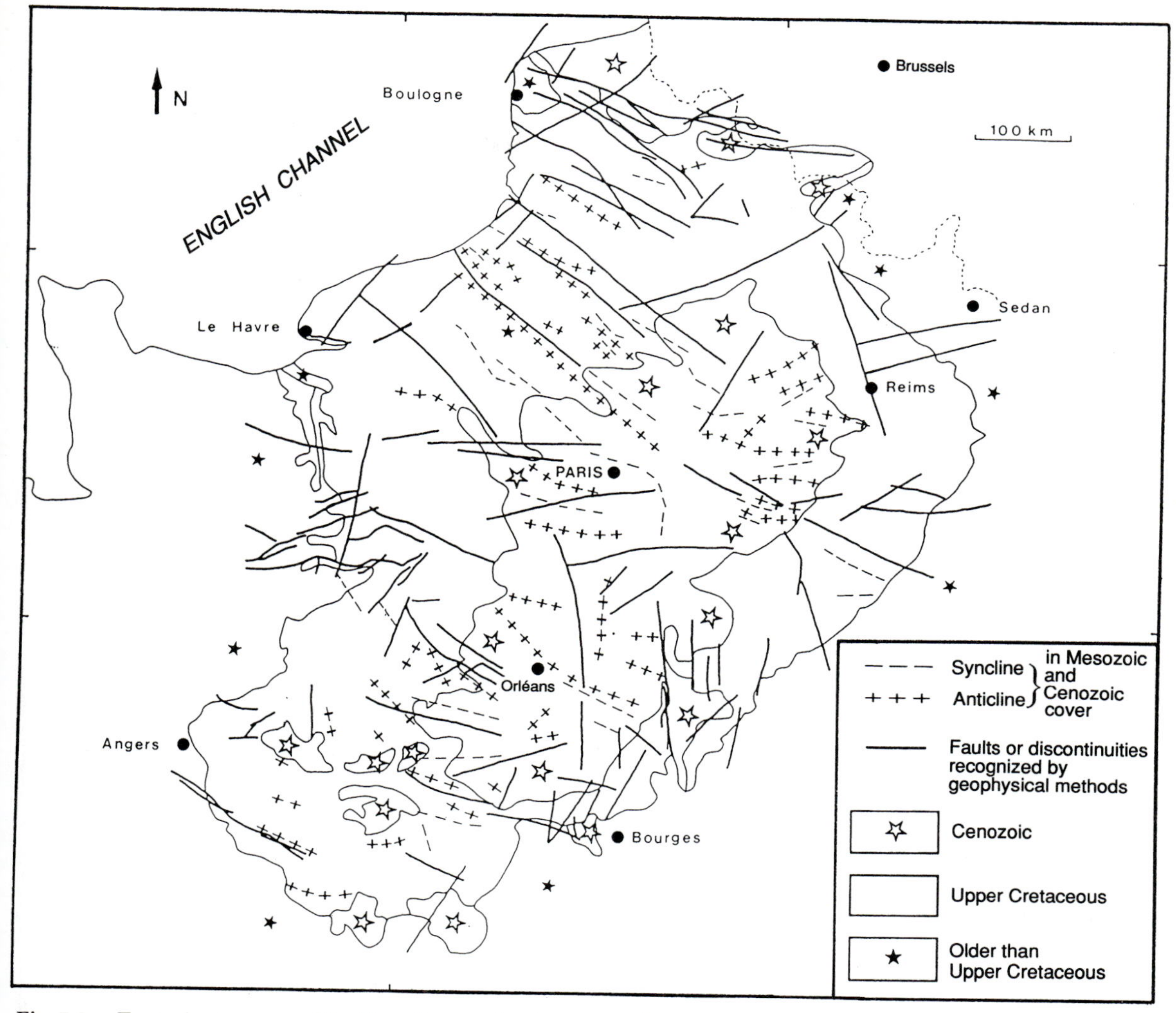

Fig. 7.6. Tectonic structure of the Paris Basin (after Autran *et al.*, in Mégnien 1980, modified by Bracq).

The nature of regional aquifers

Because of what has just been emphasized about the heterogeneity of the Chalk, the form and fluctuation of groundwater levels and the pattern of flow can change significantly from one region to another and even from one district to another. The variations of the facies of the Upper Cretaceous are a main cause of the differences in hydrodynamic behaviour.

At the northern extremity of Artois, near the coast at Calais and Cap Blanc-Nez, despite the presence of marly horizons and probably because of fractures, the Chalk from the Middle Cenomanian to the Lower Senonian can be considered to behave as a single aquifer (hydraulic continuity has been demonstrated by a tracing test at Escalles). The Cap Blanc-Nez cliff, at the Cran d'Escalles, shows perfectly, in section, the saturation level (water-table surface) in the Cenomanian Chalk as well as the numerous springs associated with a group of faults. Numerous overflow springs emerge from the base of the Cenomanian Chalk on contact with the Gault clays (Upper Albian). They originate at the bottom of the escarpment that encloses the exposed and eroded Boulonnais anticlinal fold.

To the east of the Boulonnais, in the upper valley

of the Lys (near Fruges), two superimposed aquifers, with distinct potentiometric levels (differing by no more than a few centimetres to a few decimetres), may be recognized in Upper Cenomanian marly chalk and in the Upper Turonian–Senonian. The marly chalk confines water under pressure and in hollows it is artesian. Locally, it is even possible to distinguish three aquifers, where fissured and significant chalk horizons are intercalated in the Middle Turonian marls.

Further east, near Lille and Valenciennes, the aquifer is only in the Upper Turonian–Senonian Chalk. This is also the case in northern Picardy, because of the widespread occurrence of the Turonian *dièves* (or marly horizons) (Fig. 7.2).

In the Pays de Bray, as along the Boulonnais anticline, groundwater discharges as overflow springs from the Chalk (Upper Cenomanian) on contact with the Gault clays in the core of the eroded Bray anticline.

In Normandy, in the Pays de Caux as well as in the Pays d'Ouche, the Chalk behaves as a single aquifer and is particularly well developed from the Middle Cenomanian (the Rouen chalk) to the Campanian (the Ailly chalk). The most extensive developments of karst occur in this region, with its marked relief, deep river valleys and high-level dry valleys.

In Maine, the Chalk aquifer (comprising basically the Middle and Lower Turonian) is no more than 70 m thick and generally much less than this. It is unconfined, but divided into units corresponding to individual, separate uplands. To the north-east of Le Mans, it is, for the most part, dewatered as a result of basal drainage by the Perche sands (Upper Cenomanian).

In Touraine it is generally unconfined, comprising the Turonian tuffeaux (sandy chalks), the nodular chalk of Villedieu, and the white flinty chalk of Blois and Chaumont. Some karstic flow systems are revealed by the way water emerges from the aquifer.

Beyond the detrital facies of Berry, in Gâtinais, the Chalk aquifer is Turonian in age, and displays marked contrasts in permeability; it is really productive only in fracture zones or karstic networks which develop in the east of Loiret. In Sénonais, large-scale abstractions are mainly from the karstic springs in chalk which is particularly thick and from Middle Cenomanian to Campanian in age. In Champagne,as in east Artois and in Picardy, the aquifer is developed only in the Upper Turonian and in the Senonian.

The marine and terrestrial Tertiary deposits act as aquifers that are in hydraulic continuity with the Chalk, or as aquitards that allow recharge of the Chalk and thereby increase the total ionic content of its groundwater (de la Querière 1972), or as aquicludes sufficiently thick and impermeable to protect the Chalk effectively and confine water under pressure, even artesian pressure.

Potentiometric surface

The map of groundwater levels in the Chalk of the Paris Basin (Fig. 7.7) has been drawn by Bracq and Delay for this chapter. It is based on the general map of the Paris Basin in Albinet (1967) and numerous regional maps prepared by the Bureau de Recherches Géologiques et Minières (BRGM) as well as university publications and unpublished papers: Cottez and Dassonville (1965), Caulier (1974), Beckelynck (1981), for the Nord and Pas-de-Calais; Roux and Tirat (1968), de la Querière (1972), Roux *et al.* (1978*a*; 1978*b*); Chemin and Holé (1980; 1981), Caous and Roux (1981), Caous *et al.* (1983) and Caous and Comon (1987) for Picardy and Normandy; Mary (1988) for Maine; Panetier (1966) and Lasne and Lepiller (1989) for Gâtinais; Panetier (1966) and Mégnien (1970; 1979; 1980) for Sénonais; Duermael *et al.* (1966) and Morfaux (1976) for Champagne; Mégnien (1970) for the Paris region; Castany and Mégnien (1974) for the Seine–Normandy Basin.

In spite of the fact that the Chalk is exposed along the English Channel from Calais to the Seine estuary, the outflow of groundwater to the sea is not significant except for baseflow to the river system draining to the Channel coast. There are coastal springs and even discharges of freshwater in the sea, but these are only of local interest. The synclinal structure of the Paris Basin does not lead to a converging pattern of groundwater flow towards the centre of the basin; groundwater does not flow as springs from the Chalk through Tertiary cover in the centre of the basin where the top of the formation is more than 60 m below sea level (Monciardini *et al.* in Mégnien 1980; Fig. 7.5), although groundwater is pumped from the Chalk in this region. In the north, the Chalk is also deeply

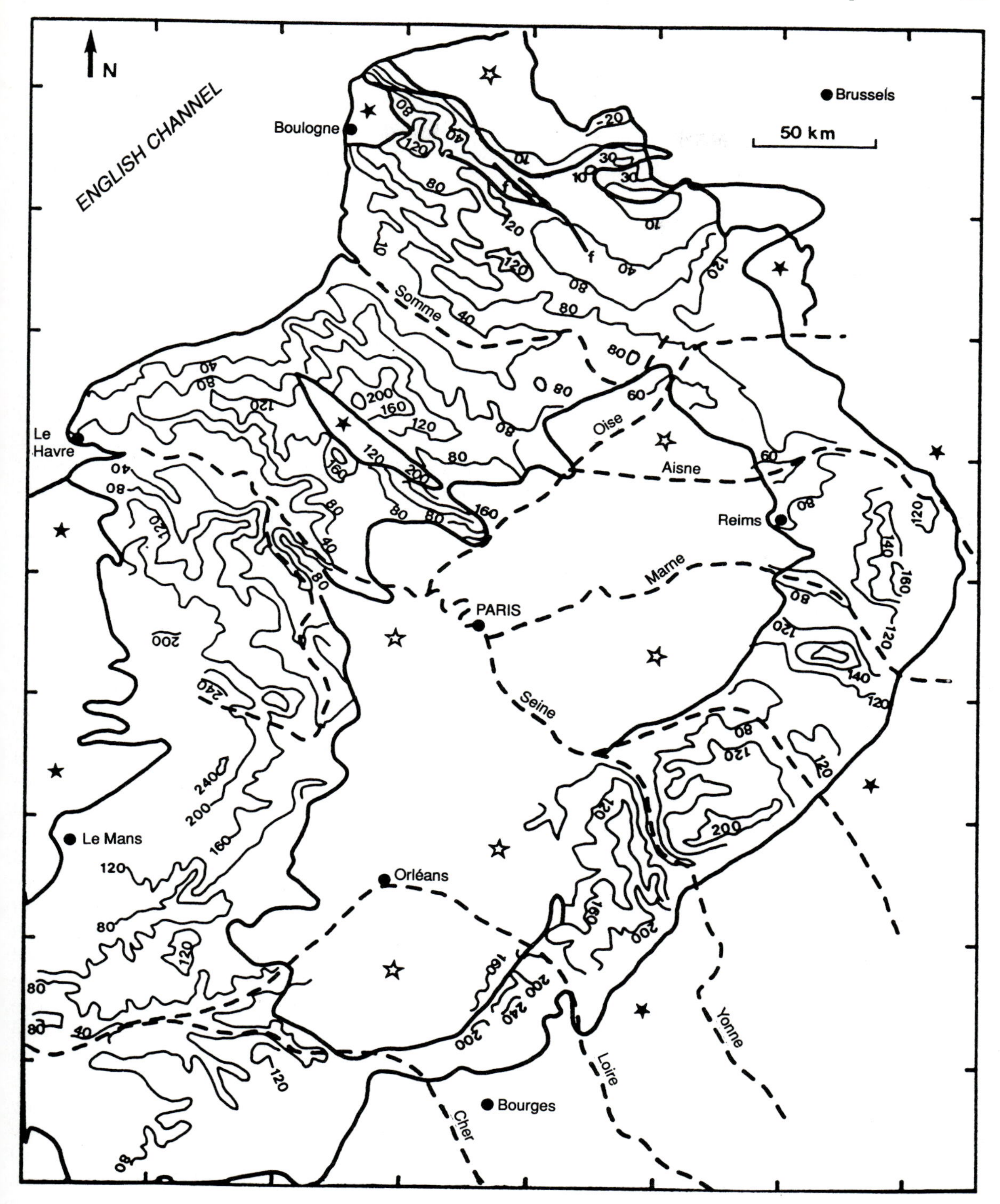

Fig. 7.7. Groundwater levels in the unconfined Chalk of the Paris Basin in metres relative to sea level (by Bracq and Delay).

buried beneath the Flanders Plain (at more than 160 m below sea level). Hence the outflow towards the North Sea can only be small, especially as the permeability decreases rapidly as the Tertiary cover thickens.

Taking a broad view of the Chalk of the Paris Basin, several features are evident:

1. Two large anticlinal axes, trending north-west to south-east, create separate groundwater provinces:
   (a) the Artois Horst, in the north, which separates the groundwater flow between Flanders and Picardy at progressively lower elevations towards the south-east;
   (b) the core of the Pays de Bray Anticline (Fig. 7.1), over which the Chalk has been removed, and which completely interrupts groundwater flow between Picardy and Normandy, except at the extremities of the structure.
2. The drainage role of the main rivers, evident in their upstream reaches as they cross the chalk outcrops of the eastern half of the Paris Basin (for example, the Oise, Aisne, Marne, Aube, Yonne, and Loire) and in a more definite manner as they cross the western half of the Chalk outcrop (the Canche, Authie, Somme, Seine downstream from Paris, and Loire in Touraine).
3. The limited outflow of groundwater to the sea along the entire Picardy and Normandy shoreline that faces the English Channel.
4. Areas where the groundwater levels are highest (for example, Berry, Perche, and Maine) are where the groundwater flow is limited and the resources are small.

However, generalizations reached from such a broad overview may not necessarily apply when the region is looked at more closely. For example, consider the coincidence between the boundaries of groundwater basins and river basins. In many areas, such as Artois, Picardy, Maine, Touraine, Sénonais, and Champagne, the groundwater divide noticeably coincides with the surface divide, although the irregularities are smoothed out. However, deviations of several kilometres exist between topographic and groundwater divides in the Somme Basin, in Picardy (Caous and Roux 1981). In Normandy, in the Pays de Caux, the groundwater divide between the flow systems drained by the Seine and those flowing to the English Channel is 3 km north of the topographic divide (Lepiller 1990). Real groundwater 'piracy' (made evident by tracer tests) occurs in that area, in particular in the Dun, Commerce, and probably the Oison basins, although the best example is at Yport where the area of the topographic basin is only about 18 km^2 but the groundwater basin covers 90 km^2. The karstic underground drainage system completely overshadows the surface drainage of the Chalk aquifer. In Gâtinais (in the east of Loiret), the situation is complicated by narrow, deep depressions in the potentiometric surface in the Chalk; major springs emerge from these groundwater drainage axes where the potentiometric surface meets the valley floor (Lepiller 1990). In this area, the karstic groundwater flowing (N 140°) between Clairis and Ouanne (to the east of Montargis) no longer follows the form of the topography and the difference between the topographic and groundwater divides is of the order of 1–1.5 km.

The drainage role of the main valleys, occupied by large rivers, has been emphasized above. Most perennial streams are supplied by 'depression springs'. These are often large and occur not only at the heads of valleys but also commonly along their courses, at the edge of and—in places—below the river deposits where these are sufficiently permeable. Where alluvial deposits are less permeable, artesian springs locally appear in the floors of valleys containing rivers. However, the valleys do not always serve as outlets for groundwater in the Chalk. Alluvial deposits in valley floors sometimes contain a local aquifer that is totally independent of the Chalk, as in Albert's zone of the Upper Somme Basin (Roux 1963). The dry valleys exist because the water table has fallen; they are favourable zones for the infiltration of rainfall, and for the recharge of any surface run off and effluent discharged by man to the environment. Some dry valleys still contain terraced river deposits, with a meandering old water channel that is completely dry, as in the Pihen ravine (near Wizernes) in the Aa valley in Artois. More generally, the dry valleys do not contain significant alluvial deposits. They represent distinct subsurface drainage axes in the groundwater basins, with rapid circulation through fissures or karstic openings, as in Normandy and Sénonais.

The water table is sometimes marked by 'highs'

or mounds which can be linked to the bedrock structure: for example, they coincide with domes and anticlines in underlying *diéves* of Picardy. East of Gâtinais and in Puisaye wide regular mounds in the potentiometric surface occur and where they intersect the ground surface marshy conditions may be present. These mounds occur where the Chalk has a low permeability because it is not fissured. In Touraine, local 'lows' or troughs in the groundwater surface, of some 20–25 m, result from drainage into the sandy Cenomanian aquifer (Rasplus and Alcaydé 1991).

The hydraulic gradient varies considerably. Under the uplands it is generally no more than a few units per thousand, but it can increase very markedly towards the valleys which intersect the uplands, reaching several per cent and sometimes 5 per cent. High gradients (possibly up to 10 per cent) are also found beneath the scarp slopes around eroded anticlines (for example, in the Boulonnais and Pays de Bray), along the Artois–Picardy fossil shoreline cliffs, and close to some faults. Gradients of several per cent are usual in the karstic systems of Normandy and of the Gâtinais–Sénonais. Lastly, there are similar high gradients (up to 5 per cent) along the edge of the Tertiary formations where they recharge the Chalk, as in the Montagne de Reims (Champagne).

In the valleys that carry a river or stream, the gradients are generally very low (from less than 1 per mille to a few per mille), except at the heads of the valleys where they can reach 1–2 per cent. In the dry valleys, values along the axes of the valleys are of the same order (0.2–2 per cent).

The major structural axes, particularly those orientated north-west to south-east, in the north-west of the Paris Basin, between Artois and Normandy, clearly influence the form of the groundwater surface, as previously mentioned in connection with the Artois Horst and the Bray Anticline. The valleys of Canche, Authie, Somme, Bresle, Béthune, and even the Lower Seine Valley, which are quite straight and parallel to each other, also indicate the role in the structural pattern of these large NW–SE lineaments. Fractures play an essential role in groundwater flow through the Chalk, and certain faults or fault zones obviously control the flow directions. This is particularly the case in numerous dry or flowing valleys. Thus in Rouen, the Aubette spring appears along a karstic development of a fault cutting the impermeable Turonian marly chalk. Similarly, in the topographic basin of Yport, to the south-west of Fécamp (in the Pays de Caux), the principal sink-holes form a line, trending N 160°, similar to the direction of the Fécamp fault. Sink-holes in the valley of the Sec-Iton, south-west of Evreux (Pays d'Ouche) also have a similar alignment. On the other hand, in Gâtinais and Sénonais, the general direction of the karstic groundwater flow does not correspond to the north–south regional orientation of the fractures.

In other situations,the major regional faults can interrupt groundwater flow. This occurs where faults with large vertical throws (more than 100 m) intersect the Chalk on the north-east edge of the Artois Horst. The water table is displaced up to 40–70 m along the Pernes and Marqueffles faults (Fig. 7.8). In Pays de Caux, the Lillebonne fault, acting as a drain, lowers the groundwater level by several metres.

Because of these factors, the depth to groundwater varies greatly from one area to another. The potentiometric surface intersects the ground surface in perennial valleys, at the bottoms of small hills or in hollows. Groundwater can discharge as artesian springs (for example, the marshes of Guines and of Ardres, on the edge of the Flanders Plain) or as depression springs or ponds (for example, the ponds of the Sensée to the south of Douai, and the marshes of the Souche to the north-east of Picardy, close to Laon). In upland areas, in contrast, the thickness of the unsaturated zone can be very great: frequently it is about 30 m in Cambrésis (eastern Artois), of the order of 40–100 m in the higher parts of Artois (particularly near Cap Blanc-Nez) and up to 70 m in Picardy. The maximum recorded depth to the water-table is 105 m in Normandy (Eberentz *et al.* in Colloque Régional de Rouen 1978).

Fluctuation of groundwater levels

Since 1967, when a groundwater-level measuring network was established for the Chalk,the lowest levels were recorded in 1973 or 1974, following (particularly in Picardy and Normandy) four years (from 1970 to 1973) without any recharge. Groundwater levels were also low in 1976, 1986, and 1990, although not so low as in 1974. In 1991 and in the first quarter of 1992, the levels were somewhat lower than in 1990, and by mid-1992

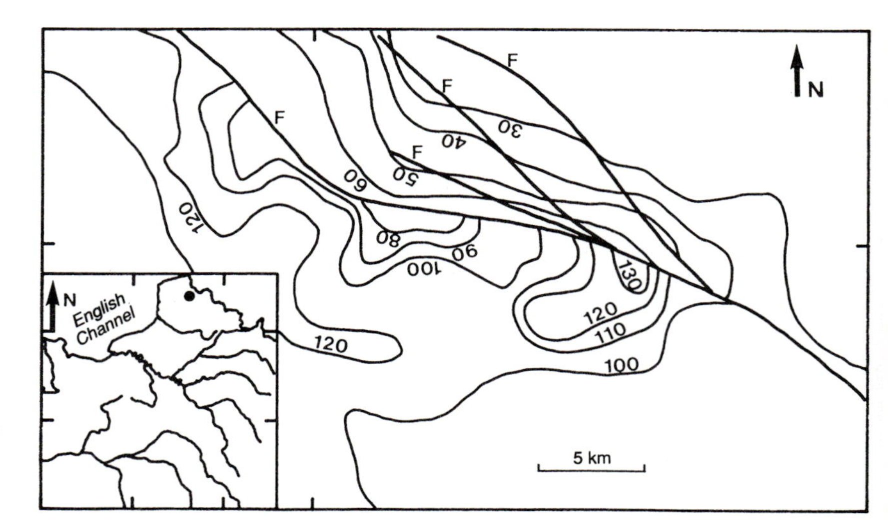

Fig. 7.8. An example of the influence of a fault (F) on groundwater levels (in metres above sea level) in Artois–Picardy (Beckelynck 1981).

were approaching those of 1973 and 1974 (Fig. 7.9). The highest levels were in 1966–7, 1968, 1970, 1981, and 1988. In some areas, as in Picardy, records go back to the beginning of the century: for example, 1902 at Nibas and 1904 at Breteuil. Within these longer time series, the lowest water levels in the Chalk were observed in 1902–4, 1921–2, 1934–5 and 1949–51, and the highest in 1926–8, 1937, 1940 and 1952. This long-term series of measurements stopped around 1955–6.

The magnitude of the water table fluctuations is generally inversely proportional to the degree of fissuring. The amplitude of the long-term range of variations is often large, of the order of 10–20 m (for example, up to 21.5 m in the Grandes-Loges in Champagne) in the upland areas where the water table is deep and the permeability is low. Individual seasonal variations are smaller; for example, a few metres in Artois–Picardy, and 6–17 m in the Grandes-Loges. On the other hand, seasonal variations in valleys are generally small, of the order of a metre in dry valleys, and less than a metre in flowing valleys, for example in Picardy (Caous and Roux 1981). In the basin of the Hallue, in Picardy, with a drainage area of 219 km^2, the average fluctuation registered in 65 wells over 10 years was about 2 m. In Normandy, the long-term maximum amplitude registered in 110 piezometers between 1968 and 1977 varied from 0.68 to 28.4 m (with 90 per cent varying from 1 to 3 m, the median amplitude being 9 m), according to Roux and Trémembert (Colloque Régional de Rouen 1978).

The response of the aquifer to precipitation depends very much on the location of the measurement, the geological context and the level of the drainage outlet. Three typical situations in Normandy (Chemin and Holé 1981) are as follows:

1. The aquifer reacts to each effective rainfall event but the minimum water level and hence the minimum storage do not vary significantly from one year to another. The water table is moderately deep (5–20 m), the aquifer relatively permeable and the observation point close to a spring. This is the most common case (Fig. 7.10a).
2. The water level responds only to heavy precipitation and the response is rapid. The minimum water level never falls below a base elevation. This behaviour indicates the presence of karstic conditions (Fig. 7.10b).
3. The water-level fluctuations are small, the aquifer is not very permeable and the groundwater is deep (35–85 m) below a thick semi-permeable cover (Fig. 7.10c).

In Normandy, groundwater levels are clearly affected by changes in the level of the Seine (due to variations in the volume of flow and tides), the changes decreasing with increasing distance from the river. At a distance of 1 km, along the margin of the terraces, there is no diurnal tidal fluctuation, but the effect of fortnightly cycles can still be seen. The fluctuations in potentiometric surface are not trans-

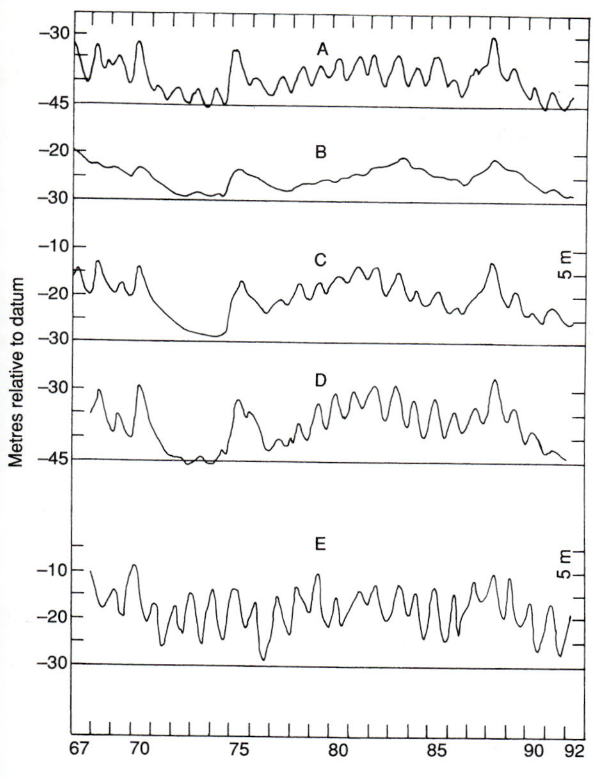

Fig. 7.9. Fluctuations of groundwater levels in three units between 1967 and 1990. Artois: A (well 24-8-5; Haravesnes), B (well 36-5-3; Barastre). Picardy: C (well 32-6-8; Feuquières-en-Vimeu). Normandy D (well 77-7-8; Catenay), Champagne E (well 108-6-11; Fresnes-les-Reims). Depths to the water table are on the left axis (data from BRGM) Note the low levels in 1974 (following 4 years without recharge), 1976, 1986 and 1990–92.

mitted instantaneously through the aquifer; for example, in the karstic system of Yport, a well situated 5 km from the coast shows the effect of the variation in sea level (fortnightly tides) as fluctuations of only a few tens of centimetres but with a delay of the order of 8 days.

In Champagne, in non-karstic chalk, very regular piezometric variations have been observed in Fresnes-lès-Reims (Fig. 7.9e), where the lowest levels were in 1976 and 1990. The behaviour seems to indicate a well-developed surface drainage network. The reduction in amplitude of fluctuations with distance from a river has also been observed in non-karstic zones in Touraine.

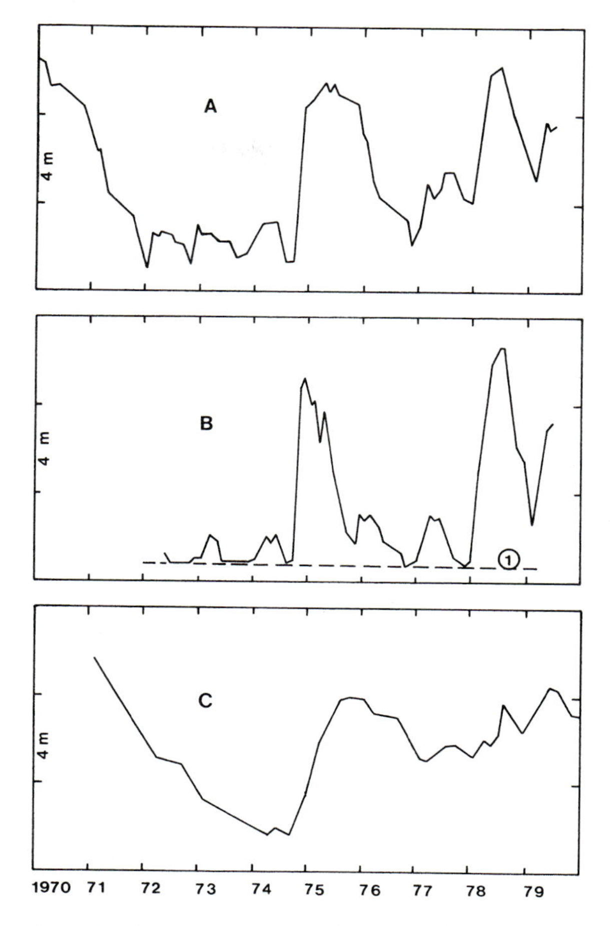

Fig. 7.10. Variations of groundwater levels in three chalk aquifers in Normandy with different physical characteristics.

(a) (well 76-5x-17; Touffreville-la-Corbeline): the aquifer reacts to effective rainfall, the water table is relatively deep (from 5 to 20 m below the surface), the aquifer is relatively permeable and the observation point is close to a spring, the minimum water level and storage does not vary very much from year to year. This is the most common situation.

(b) (well 59-4x-23; Baillolet): the water level responds rapidly to heavy rainfall. The minimum level is stabilized at a base elevation which does not vary (indicated by 1). This is typical of karstic zones.

(c) (well 58-3x-5; Tocqueville-en-Caux): water-level fluctuations are small. The groundwater levels are deep (35–85 m below the surface), the aquifer is not very permeable and it is below a thick semi-permeable cover (Data provided by BRGM).

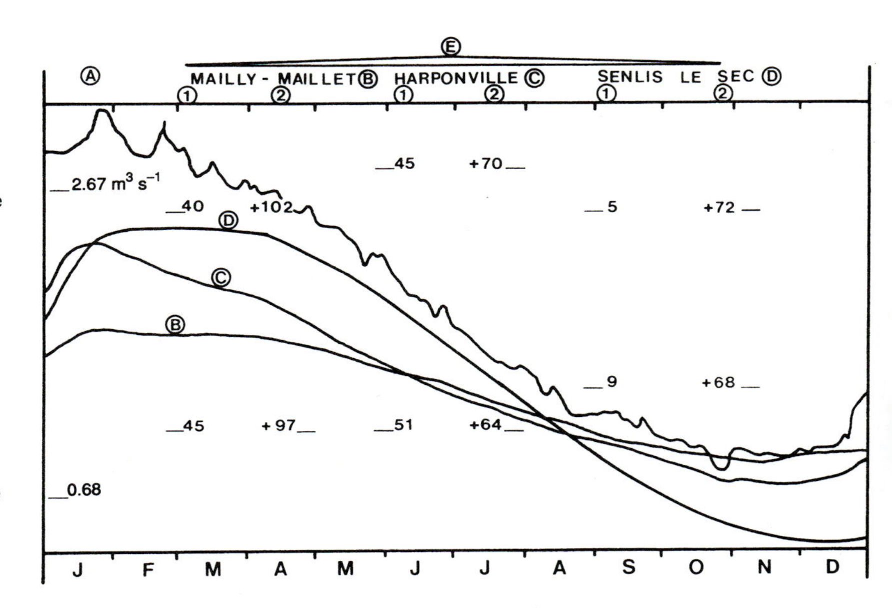

Fig. 7.11. Comparison of the hydrograph of the Hallue river and groundwater-level fluctuations in three wells in 1967.
A is the hydrograph of the Hallue at Daours. Flow rate is given on the vertical axis.
B, C and D are groundwater levels for wells at Mailly-Maillet, Harponville and Senlis le Sec, respectively. Figures under E, 1 and 2, give in each case the depth and elevation scales in metres. (after Caous and Roux 1981).

As a general rule, numerous observations and studies have shown an excellent correlation between the potentiometric levels in the Chalk and river discharge, particularly during months when water levels are low (Fig. 7.11).

Long-term changes in groundwater level can affect the position of springs. Displacement of springs has been observed along valleys, notably in Artois and Picardy. In Artois, the emergence of a new outlet of the Lys springs caused the collapse of a street in the village of Lisbourg in February 1984. In Picardy, in 1966, the springs of Hallue were 5 km upstream of their present perennial outlet, while in 1981, the springs of Trie appeared in the centre of the Village of Nibas (Picardy) where they flooded the streets.

Groundwater levels are obviously particularly sensitive to groundwater abstraction if the observation points are close to pumping wells and if the aquifer is confined. But the greatest impact is due to continuous intensive pumping, particularly in areas which are highly urbanized and industrialized, although, in recent years, economic recession has reduced or even stopped some industrial abstraction. In the industrial zones of the Seine valley in Rouen, abstraction reaching 250 000 $m^3 d^{-1}$ depressed the water levels in the Chalk and overlying alluvium by about 6 m in places. Since the beginning of the century, the Chalk aquifer has been intensely developed in the Nord and in the Pas de Calais, to the north of the Artois Horst, by mine dewatering and for industrial supplies. Here the water level has fallen by as much as 15 m in the west and 25 m in the east. Since 1955 the levels have started to recover because of the closure of some mines (Fig. 7.12). The levels have been rising more rapidly recently, as abstraction declined from $171 \times 10^6 m^3$ in 1975 to $130 \times 10^6 m^3$ in 1986. The recovery of levels was not evenly distributed in space and time. In combination with subsidence due to mining, the recovery of the water table has formed marshy areas.

Underground civil engineering works have also affected the water table in the Chalk: the construction of the north canal tunnel in Ruyaulcourt, Picardy, in 1962, led to a general fall of 9 m which reduced to 3.5 m after the tunnel was opened; the effect was perceptible at a distance of 4 km from the canal. Construction of the Channel Tunnel, on the other hand, is expected to cause only a slight rise in Chalk water levels south of Sangatte (Crampon *et al.* 1990).

Circulation in fissured and karstic chalk

The Chalk is an aquifer only because it contains a network of fissures produced by tectonism and physico-chemical alteration. In places a more rapid flow system is sometimes superimposed on the

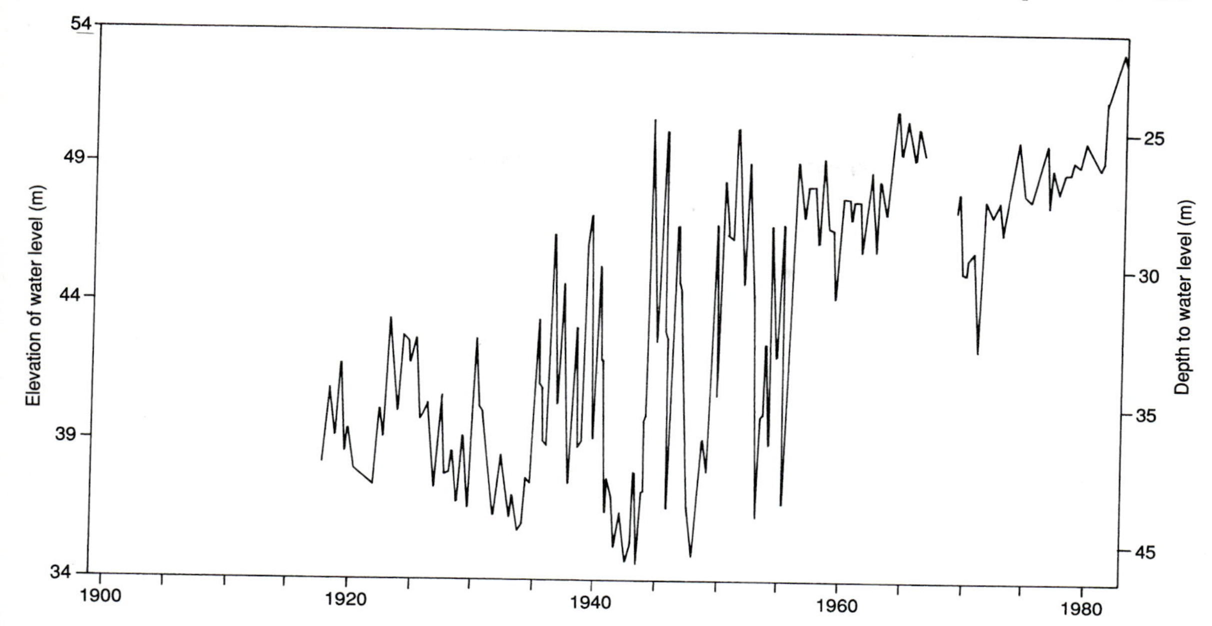

Fig. 7.12. Variation of groundwater level between 1916 and 1982 in Artois–Picardy in Well 19-5x-8 at Bruay-en-Artois (data provided by BRGM). The rise in the level after 1955 is due to the closure of coalmines.

more general fissure network, whose slower flow system is often treated as a relatively uniform porous medium for the purposes of estimating the size of the resource at the scale of the aquifer system (Colbeaux and Mania 1976). Flow occurs in fractures or along bedding planes that are sufficiently open to allow good drainage of the more finely fissured chalk mass. The transition to a karstic system, characterized by the presence of marked dissolution, is not always obvious. In addition to the characteristic morphology which results from this dissolution, it is the speed of groundwater movement and the flow regime of the discharge areas that distinguish a karstic flow system. The geographical distribution of karst regions and of springs (karstic or otherwise) in the Chalk is very uneven (Fig. 7.13).

In Artois–Picardy, locally important flow conduits have been recognized, for example: No. 11 shaft at Lens Colliery seems to have intersected, at a depth of 58 m, a fracture which is 0.4 m wide discharging 680 $m^3 h^{-1}$ (Gosselet 1904), and the borehole at the sugar refinery of Dompierre-en-Santerre (Roux 1963) crossed a void 0.55 m wide at a depth of 64 m.

Recent work for the Channel Tunnel (in the form of wells, galleries, and boreholes) has also intersected fractures or very open joints with appreciable water inflows. By interpreting pumping tests, Berkaloff (1970) drew attention to the frequent correspondence of zones of flow with bedding planes, especially at the point of contact with hard nodular chalk (or 'tun' beds). This area is usually considered to be devoid of active karstic systems, although outward expressions such as solution pockets and even some dolines can be seen. Lasne (1890) mentioned the existence, near Doullens, of cylindrical 'wells' with a flint and clayey-sand filling, which can be as deep as 20–35 m. Following engineering works for the TGV Nord, the ground surface recently collapsed along the upper part of the Sensée, where the river is apparently linked with an underground flow system. Dissolution pipes, now inactive, have also been found in other areas (Bracq *et al.* 1992). In the absence of direct observations, distinguishing between rapid circulation in fissures and karstic circulation is difficult. Tracer tests at Escalles (Cap Blanc-Nez) in 1989–90 indicated a maximum flow velocity of 53 $m h^{-1}$ over an apparent distance of 2225 m below a dry valley; this is very rapid, especially if infiltration through about 57 m of unsaturated chalk is taken into account.

In Normandy, karst is well developed both north

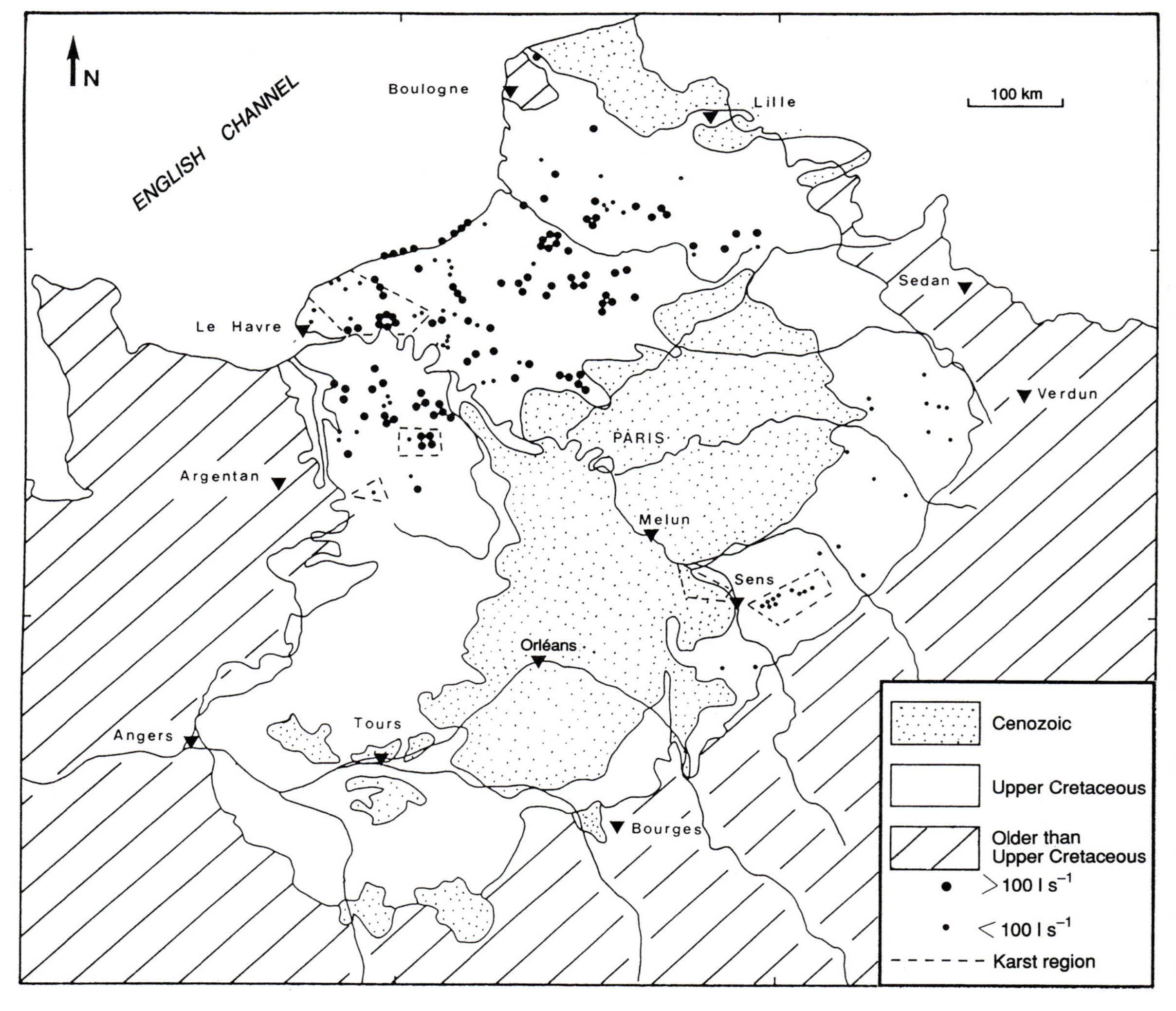

Fig. 7.13. Location of karst regions and principal springs (prepared by Bracq).

and south of the Seine. According to Rodet (in Colloque 1988) there are two kinds of karst: an active karst, consisting of quite open cracks that cannot be entered but containing water under pressure; and an ancient Pleistocene or Tertiary karst which generally does not carry water but consists of large passages which can be entered. The passages are over-sized in relation to the flows which still run through them and they play an apparently reduced role in the drainage because of a filling of clayey-calcareous sand. An example of the ancient karst is the caves of Caumont, on the left bank of the Seine near and downstream of Rouen, which contain more than 5000 m of explored galleries with numerous caverns (including the cave of La Jacqueline which is 800 m long), chimneys and an underground river and lake. A further example of a karstic network is the caves of Villequier, mid-way between Rouen and Le Havre (Fig. 7.14).

The Iton Valley, in the Evreux region, contains a very interesting karstic network (Roux, in Colloque Régional de Rouen 1978). As a result of recent tectonic movements, the stream profile has become convex along almost 40 km, leading to a fall of groundwater levels relative to the stream bed, and a loss of stream flow by seepage through the bed of the stream. When the flow is equal to or less than 1 $m^3 s^{-1}$, the stream dries up over almost 7 km. Ferray explored this karstic network in 1894 below the uplands adjacent to the valley and discovered several quite large caves as well as a water-filled channel, 3–4 m wide, flowing at 700–800 $l s^{-1}$.

According to Rodet (Colloque Régional de Rouen 1978), the karstic systems of Normandy comprise an upstream network of numerous fissures which join into a single channel before splitting into different channels near the outlet zone; this suggests that the outlet has moved over the years. Some galleries are drowned and nowadays may even be beneath sea level because of the Flandrian transgression. At Yport, part of the conduit has been inspected by diving, with a view to using onshore wells to develop the freshwater that the system contains.

The initial development of karst includes *bétoires* (local name for sink-holes) of which the largest reach several tens of metres in diameter and several metres deep; the biggest is La Fosse du Champ Guérin, in Evreux Forest, which has an elliptical shape, and is up to 200 m wide and 18 m deep. These *bétoires* become progressively filled by loose surface formations that slip sideways into them as they are deepened by dissolution. Substantial sudden collapses do occur, but less frequently than small subsidences that are nevertheless capable of disturbing small buildings. These sink-holes can follow a line for several kilometres and probably coincide with a fault, as has been pointed out previously in the case of Yport and the Sec–Iton Valley. According to Rodet (Colloque Régional de Rouen 1978), 89 per cent of known karstic features in the Chalk of France are found on the Pays de Caux shoreline or in the Seine valley.

Most karstic systems seem to develop along faults, joints (Fig. 7.14) and bedding planes, and the size of the outlets is of the order of a decimetre. In the Lillebonne area, a borehole intersected a few voids, up to 0.6 m wide at a depth of 21 m, at the base of the Turonian. One of the most remarkable karstic networks is in the Yport Basin, where the underground circulation paths, revealed by tracer experiments, do not coincide with the directions of the surface network. The tests indicated flow velocities between 7 and 270 $m h^{-1}$ over apparent distances that exceed 10 km. The network of the Sec–Iton valley also extends over about 10 km and inlet–outlet differences of 18 km have been demonstrated in the zone of the Avre springs to the southwest of Verneuil. During about 70 tracer tests, Roux (in Colloque Régional de Rouen 1978) noticed that the velocities ranged from 7 to 540 $m h^{-1}$(with a median of 120 $m h^{-1}$), of which half were between 50 and 250 $m h^{-1}$; speeds of less than 50 $m h^{-1}$ corresponded to tests where the tracer was injected into the unsaturated zone. The greatest distance covered was about 27 km, with an average speed of more than 200 $m h^{-1}$(from Glos-la-Ferrière to the fish farm at Beaumont, in the basin of the Risle). These high speeds are associated with a marked drop in elevation (sometimes more than 100 m) which generally exists between the inlet and the outlet of the system, and the consequent steep gradient. The average gradient is as much as 7 per cent.

In Maine the karst, although less spectacular, is present over distances of some hundreds of metres. In some valleys, surface run-off flows very rapidly into the ground. The majority of the springs exploited from the Turonian Chalk become turbid at times of heavy rain, and their flow rates increase almost tenfold, for example at Courdemanche (Mary 1990).

In Touraine, the karstic zones in the 'yellow sandy chalk' and in the chalk of Villedieu can extend over 1–2 km under the uplands (for example, at Pocé-sur-Cisse, Montrésor, Genillé, Saint Paterne-Racan, and Paulmy) but they can only rarely be entered (as, for example, at Montrésor). The waters become turbid at times of heavy rain. Flow speeds can reach 230 $m h^{-1}$ in the karstic tuffeau, as in the Senonian Chalk of Villedieu (where 234 $m h^{-1}$ over 1 km was recorded in Saint Paterne-Racan). During periods of low water level the karst systems dewater progressively and the groundwater outflow becomes regular (Rasplus and Alcaydé 1991).

In Gâtinais, karst is less well developed than further east in Sénonais and the Pays d'Othe because of the long-term protection afforded the Chalk by the slightly permeable Tertiary cover (Lepiller 1990). Observations in quarries and measurements with flowmeters in boreholes suggest that karstification develops along horizontal sedimentary discontinuities, and the resulting voids are of the order of decimetres. Larger voids occur, however, in association with underground rivers. For example,

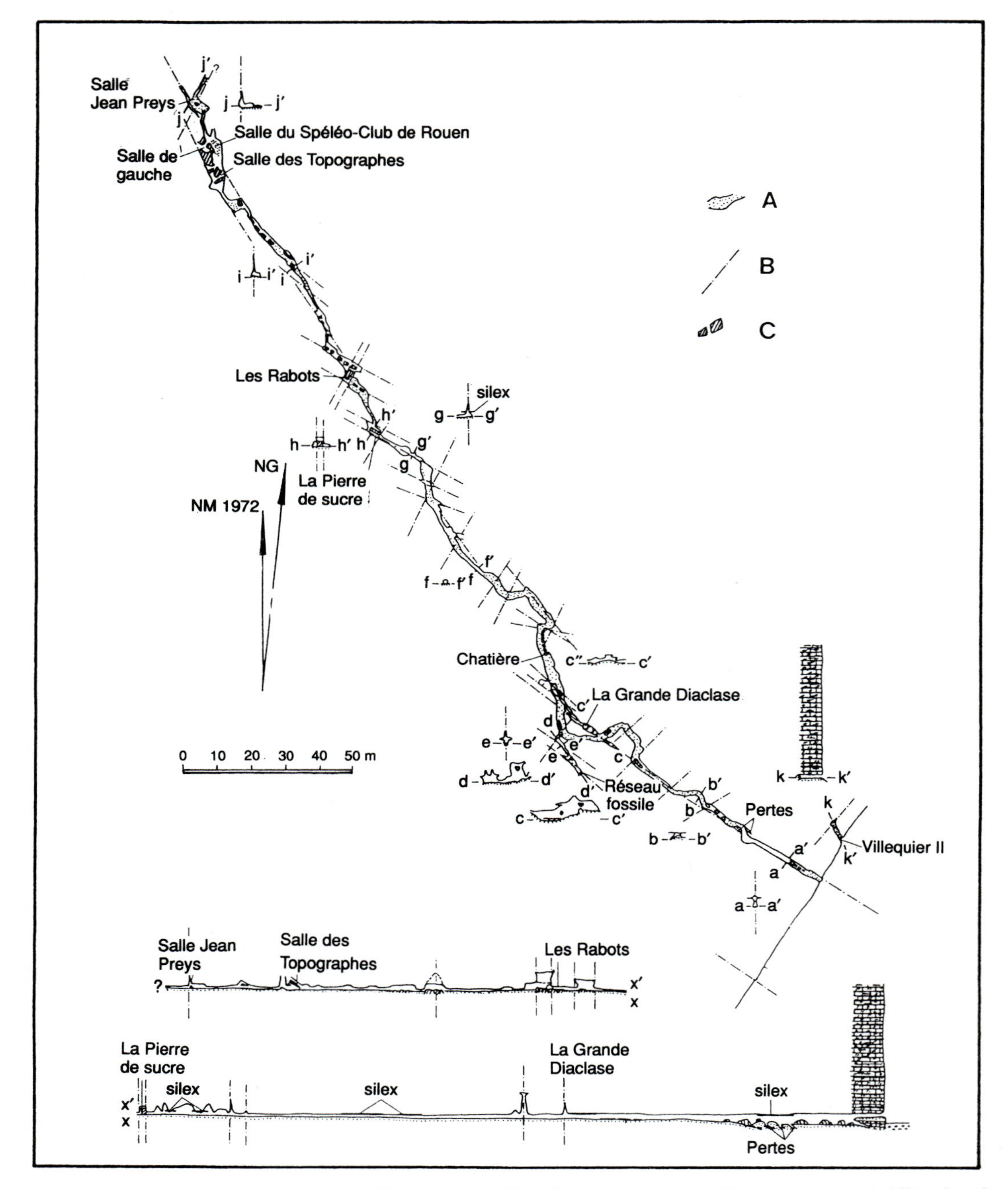

Fig. 7.14. An example of a karstic network: Les grottes de Villequier in Normandy. A represents infilling by clay and sand, B joints, and C fallen blocks (Lepiller 1975).

in the Puits-Bouillant system, the underground river is accessible for about 1500 m through a joint with a maximum width of 2 m but with a vault reaching to a height of 15 m (Blavoux and Panetier, in Colloque Régional de Rouen 1978). In the two main karstic systems, the Puits-la-Laude and the Trois-Fontaines, with areas of 100 and 50 km^2, respectively, to the east of Montargis, the general drainage direction (N 140°) follows neither the surface pattern, nor the north–south structural axes, and the flow velocities do not seem excessive.In the Trois-Fontaines system, 34 m h^{-1} has been recorded at times of low flow (Lepiller and Lasne 1990).

In Sénonais and in the Pays d'Othe, karstic features are revealed by outflows from the drainage systems of the Cochepie and Vanne, to the east of Sens. Sometimes these systems can be entered. Mégnien (1959) showed that the water levels in the karstic network and in the mass of the Chalk aquifer coincided approximately during normal hydrological conditions: the karstic network either loses or gains water according to whether the conduits are above or below the water table, and the mass of Chalk thus serves as a regulating reservoir. During periods of low water levels, the water table drops below the karstic network, and then only the finer fissures of the Chalk aquifer supply the springs. The apparent directions of underground flow sometimes coincide with the valley axes, and much less frequently with the orientation of subvertical fractures or the north–south Arces–Cerilly regional fault. In 35 tracer experiments, the average flow velocity was about 162 m h^{-1} over apparent distances of 6–14 km.

In Champagne, in the Montagne de Reims, karstic conduits have developed in the upper part of the Chalk, under and at the edge of Sparnacian sandstone and sandy deposits. Several networks have been explored for some hundreds of metres (600 m in the case of Trepail); sink-holes in the Chalk are about 30 m deep (for example, the Martin–Godard sink in Verzy and the chasm of Creusin in Villers-Marmery) but the infiltrating water can descend about a further 50 m (as in the cave of the Grande-Fontaine of Verzy).

Springs

The distribution of the principal springs in the Chalk is shown in Fig. 7.13. In Artois-Picardy the largest springs, which are 'depression springs', are often found at the heads of valleys and give rise to perennial streams. The Lys spring, in Artois, whose origin is not yet fully understood, has an average flow rate of 80 l s^{-1} and a maximum of 500 l s^{-1}. The overflowing springs of Rivierette, at the edge of the Flanders Plain near the Channel Tunnel terminal, are aligned along a 60° structural axis and can discharge more than 150 l s^{-1}. As previously mentioned, other overflowing springs appear at the base of the Chalk aquifer along the outcrop of the Albian argillaceous rocks, at the edge of the exposed and eroded anticlinal folds of Boulonnais and the Pays de Bray; some of them have high flow rates, for example 250 l s^{-1} at Ons-en-Bray. The plunge of the aquifer under the coastal plains of the Bas-Champs also gives rise to notable springs at the bottom of fossil cliffs (for example at Dannes). Some springs with rapid variations in flow rates (ranging from zero to 100 or 150 l s^{-1}) form a line at the bottom of escarpments emphasising the presence of faults, notably in the basin of Authie (d'Arcy 1969; Colloque Régional de Rouen 1978). At the margin of the Tertiary cuesta of South Picardy, the reduction in the permeability of the Chalk gives rise to overflowing springs in the valley of the Brèche and the marsh of Sacy. In Picardy, Caous and Roux (1981) have counted 42 springs with flows in excess of 100 l s^{-1}, of which two exceed 500 l s^{-1} (the Spring of Môle in Araines and the Spring of the Celle in Fontaine-Bonneleau).

As stated previously, artesian springs are also not rare; they are found in the bottoms of flowing valleys in Artois and Picardy, particularly the Canche, Authie, and Somme. The dip of the Turonian–Senonian Chalk aquifer under the Tertiary formations of Flanders gives rise locally to artesian conditions, as for example in the marshes of Guines and Ardres (Mania 1978). Artesian conditions can also occur because of less permeable horizons within the Chalk itself; in western Artois, the form of the outcrop of the Cenomanian aquifer gives rise to artesian springs, particularly in the Upper Lys Valley, between Saint-Omer and Fruges; some of the springs are natural, others are created artificially to supply watercress beds. The Chalk is confined under a bank of compact chalk in the valleys of the Ancre and of the Hallue, in Picardy, and the water rises up in a kind of vertical chimney called a *puits tournant*.

About 616 springs have been studied in the Seine-Maritime *département*, in Normandy, and of these 70 per cent have a flow rate of less than

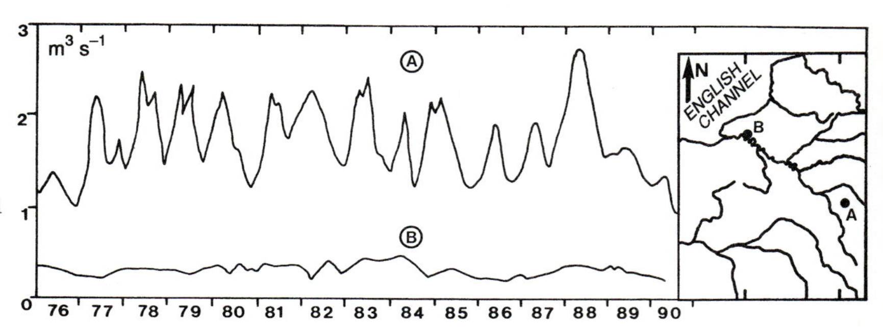

Fig. 7.15. Variation of flow rate of major springs. A: Vanne springs (Sénonais) and B: Saint-Laurent spring (Normandy) between 1976 and 1990 (data provided by BRGM, 1991).

$25\,l\,s^{-1}$ and 10 per cent more than $100\,l\,s^{-1}$. The largest are found in the Seine Valley, for example at Radicatel, Moulineaux ($350\,l\,s^{-1}$) and Elbeuf, and on the shoreline at Saint-Valery-en-Caux, Veules-les-Roses, and Yport, where the flow is as much as $1500\,l\,s^{-1}$. In the Eure *département*, 6 per cent of the known springs have an average flow of over $100\,l\,s^{-1}$, including the spring of the Bave ($1050\,l\,s^{-1}$), the group at Glisolles ($1060\,l\,s^{-1}$), Gaudreville ($1180\,l\,s^{-1}$), Bonneville ($1250\,l\,s^{-1}$), and those of the Avre. Most of the large springs have a karstic origin, as for example the Yport Spring. This drowned conduit has been partly explored by entering both the outlet and a well 2 km from the outlet. Many towns in this area derive part of their water supply from chalk springs including Le Havre, Rouen, Dieppe, Brionne, and even Paris. Springs in the valleys of the Eure, the Iton and the Avre provide $5400\,m^3\,h^{-1}$ for Paris. The flow rates of the springs generally vary from year to year by as much as a factor of 2 (Fig. 7.15). Artesian springs also occur in Normandy. For example, new artesian springs appeared in the side of the Risle Valley in 1978.

The Turonian Chalk springs in Maine have for the most part a flow rate of less than $10\,l\,s^{-1}$ at times of low water levels but they can increase almost tenfold after precipitation (for example, from 8 to $70\,l\,s^{-1}$ in the case of the spring of Courdemanche, already mentioned). In Touraine, high flow rates of $150–250\,l\,s^{-1}$ in Saint Paterne-Racan, near the Loire, are linked also to the karstic systems referred to above. Artesian conditions can appear in this latter area, in particular in the Turonian aquifer below Tertiary and Campanian cover, as in Montbazon, south of Tours (Rasplus and Alcaydé 1991).

Of 17 springs in Gâtinais between the Ouanne and the Yonne, to the west of Auxerre, four flow at more than $50\,l\,s^{-1}$ and one at over 100 or even $200\,l\,s^{-1}$ (Blavoux and Panetier, in Colloque Régional de Rouen 1978). Lepiller and Lasne (1990) registered eight springs of over $100\,m^3\,h^{-1}$ ($28\,l\,s^{-1}$) to the east of the Loiret, including those in Puy-la-Laude ($150\,l\,s^{-1}$) and in the karstic system of Les Trois-Fontaines ($150\,l\,s^{-1}$).

In Sénonais, the Vanne group of springs (see the previous section), which is used to supply Paris, discharged for 100 years (1876–1975) at a flow rate generally between 1.2 and $2.3\,m^3\,s^{-1}$ with an average of $1.8\,m^3\,s^{-1}$; the recent extremes were $1\,m^3\,s^{-1}$ (at the end of 1976) and $2.7\,m^3\,s^{-1}$ (in 1988) (Fig. 7.15). The springs of the Loing and of the Lunain are also used for supplying Paris at a rate of $1.35\,m^3\,s^{-1}$.

Regional aquifer characteristics

The ability of the Chalk to store and transmit water depends, as previously emphasized, on the distribution of the argillaceous and sandy facies, on the compactness of the chalk, on the degree of weathering (that produces cracks and fissures), on the extent of fracture systems and, above all, on the extent to which fractures and joints have opened in response to tectonic stresses, and their development by dissolution (karstification). All these factors mean that the hydrogeological characteristics of the Chalk vary from region to region, depending on depositional characteristics, on the geomorphological setting (whether in uplands or in dry or perennial valleys), the depth of the aquifer and the geology (for example, whether Tertiary cover exists, and the degree to which the area has been affected by tectonic disturbances).

Porosity

Chalk is a very porous rock. Typical porosity values are 30–45 per cent in Artois–Picardy, 23–43 per cent in Normandy, 37–44 per cent in Champagne, 15–40 per cent in Touraine, with values varying according to the facies. The pore-throat sizes are very small. In the white chalk of the Upper Turonian–Senonian of the Nord they range from 0.1 to 1 μm (with a modal value of 0.3 μm) (Henry, in ENPC 1989). Less than 2 per cent of pore-throats are over 1 μm (mode 0.4 μm) and the total porosity is 45 per cent in the Senonian chalk of Haubourdin. In Normandy various chalks have median values of 0.55–0.77 μm. Chalk from depth in a borehole at Grande-Paroisse in the Seine valley, downstream of Montereau (in Sénonais), and at the edge of the Tertiary in the south-east of the Paris Basin, had modal values for the lower Turonian and Coniacian, respectively, of 0.2 and 0.5 μm (Mégnien 1979). Samples from the Lower Senonian of Champagne gave a modal value of between 0.5 and 1 μm (Ballif 1978).

Hydraulic conductivity

With such fine pores and such a large porosity, non-fractured chalk has the characteristics of an aquiclude or aquitard. The hydraulic conductivity of the matrix is generally negligible in comparison with that created by fissures. In the wells and galleries of the 'Project GEOVEXIN' (in the Seine Valley, upstream of Mantes), the chalk at depth is almost impermeable; for example, the Turonian Chalk (155 m deep) has a hydraulic conductivity of 9.4×10^{-8} to 2.6×10^{-7} m s^{-1}, while that of the Santonian Chalk (75 m deep) is 3.2×10^{-8} to 1.5×10^{-7} m s^{-1},. In contrast, the flows in fissured chalk indicate a general hydraulic conductivity of the order of 2.5×10^{-6} m s^{-1}.

In some galleries, excavated to depths of 150 m, for underground storage in the Seine Valley near Rouen, there is no water inflow. Hydraulic conductivity values even lower than those mentioned above have been obtained in chalk penetrated by the borehole at the Grande-Paroisse, referred to above: Upper Campanian (90 m deep), 1×10^{-8} m s^{-1}; Coniacian (230 m deep), 8.5×10^{-9} m s^{-1}; Upper Turonian (390 m deep), 9×10^{-10} m s^{-1}; Cenomanian (540 m deep), $< 1 \times 10^{-10}$ m s^{-1}.

Thus chalk, even in its purest form (without clay minerals) remains an aquiclude until it is fissured; it *becomes an aquifer* only *if* it is sufficiently *interconnected by open fractures*. Because the development of fissuring is progressive, a range of values can be found at a site according to the depth and the argillaceous content. For example, the marly chalk, *craie bleue*, of the Blanc-Nez Lower Cenomanian, in which the Channel Tunnel has been excavated, has an average hydraulic conductivity of 2×10^{-7} m s^{-1}, whereas values increase steadily up to 10^{-5} m s^{-1} in the white-chalk of the Upper Cenomanian (Bertrand *et al.*, in ENPC 1989). In the Upper Turonian–Senonian Chalk intersected by drilling in the same area, the hydraulic conductivity ranges from 9×10^{-7} to 2×10^{-4} m s^{-1}.

Infiltration tests in wells and boreholes in Normandy in unsaturated chalk (de la Querière, in Colloque Régional de Rouen 1978) gave vertical hydraulic conductivities of the order of 10^{-4} m s^{-1} in dry valleys and 10^{-5} m s^{-1} in upland areas. In the latter situation, the values are higher in fissured chalk, with fissure openings 2–5 cm wide below the residual Clay-with-flints formation, than in the deeper and more compact chalk where the fissures are only millimetres wide (unless locally large karstic voids exist). Near Mantes, the hydraulic conductivity is of the order of 10^{-3} m s^{-1} in a dolomitic facies.

In Touraine, the hydraulic conductivity of the Chalk ranges generally from 10^{-6} to 10^{-4} m s^{-1}, reaching 5×10^{-3} m s^{-1} in the most productive zones. Similarly in the Sénonais, at the springs of Lunain, it is 4×10^{-3} m s^{-1}.

Transmissivity

The transmissivity depends on both the hydraulic conductivity and the saturated effective thickness of the aquifer. Values vary considerably, and typical examples are given in Tables 7.1 and 7.2.

The subjective nature of the statistical data should not be overlooked, for they are derived from wells that are preferentially sited in the most productive areas. Because there are few reliable measurements of transmissivity, many attempts to characterize the hydrodynamic nature of an area depend on using measurements of the specific capacity (flow rate per metre of drawdown). Values cover a very large range, of the order of 0.1 to several hundreds of cubic metres per hour per metre. In Picardy, results from more than 300

Table 7.1 Typical values of transmissivity

	Location or situation	Transmissivity ($m^2 s^{-1}$)	Notes
1.	Upland areas and small hills of Artois	10^{-5} to 10^{-3}	
2.	Dry valleys	10^{-3} to 10^{-2}	e.g. 1.4×10^{-2} $m^2 s^{-1}$ in Airon-Saint-Vaast, Artois
3.	Under Tertiary less than 5 km from edge of cover	10^{-3} to 10^{-2}	e.g. experimental site at Béthune, 20 m Tertiary: 2×10^{-3} $m^2 s^{-1}$
4.	Under Tertiary more than 5 km from edge of cover	10^{-5} to 10^{-3}	From Mania (Colloque Régional de Rouen 1978)
5.	Along the lower slopes of hills	10^{-2} to 10^{-1}	e.g. North flank of Artois Horst between Sangatte and Guines: 2×10^{-2} to 2.6×10^{-1} $m^2 s^{-1}$
6.	Valleys with perennial rivers:		
	(a) Seine at Montereau	1.5×10^{-2} to 4×10^{-2}	Based on 20 test wells
	(b) Seine at Aubergenville near limit of Tertiary cover	10^{-3} to 4×10^{-2}	In the valley
		10^{-4} to 10^{-3}	Under Tertiary cover
	(c) Seine at Rouen	10^{-1} to 3×10^{-1}	Along the axis of valley
		2×10^{-2} to 3×10^{-2}	Near small hills under Quaternary terrace
	(d) Seine at Lillebonne	9×10^{-3}	
7.	Confined Upper Cenomanian	6×10^{-3}	In upper valley of the Lys

Table 7.2 General regional values of transmissivity

Area	Transmissivity ($m^2 s^{-1}$)	Notes
Nord–Pas–de–Calais	1.5×10^{-2}	Average of 698 values: 90% less than 3×10^{-2} *
Somme Basin	1.3×10^{-3}	Average value**
Touraine	10^{-5} to 2.5×10^{-2}	10^{-2} in karstic sandy chalk (tuffeau)†

* from Caulier (1974)
** from Roux (1963)
† from Rasplus and Alcaydé (1991)

boreholes gave values of more than 10 $m^3 h^{-1} m^{-1}$ in 56 per cent of the cases, and over 50 $m^3 h^{-1} m^{-1}$ in 15 per cent of the cases; values in excess of 100 $m^3 h^{-1} m^{-1}$ were derived from perennial valleys and, more rarely, from dry valleys (Roux 1977). In the karstic zone of the Pays de Caux, values up to 1000 $m^3 h^{-1} m^{-1}$ have been measured near Yport. In Maine and Touraine specific capacities are of the order of 3 $m^3 h^{-1} m^{-1}$ or less in slightly fractured chalk, reaching 30–70 $m^3 h^{-1} m^{-1}$ where karst is developed (Mary 1990; Rasplus and Alcaydé 1991). In Gâtinais, results from 275 boreholes indicated values rose in steps from 3.5 to 150 $m^3 h^{-1} m^{-1}$, with an average of 14 $m^3 h^{-1} m^{-1}$. However, 80 per cent of the boreholes gave less than 18 $m^3 h^{-1} m^{-1}$ (Lepiller and Lasne 1990). Specific capacity values of 30–260 $m^3 h^{-1} m^{-1}$ have been attained in the valleys of Champagne (Vesles, Suippe, Marne), whereas at the edge of the Cenozoic cover or in upland areas values of 0.1–1 $m^3 h^{-1} m^{-1}$ are more typical.

Storage coefficient

Values of the storage coefficient calculated from pumping tests are not very common because a second borehole or well is necessary to provide appropriate data.

In the unconfined aquifer, values seem to be of the order of a few per cent under upland areas (0.5–1 per cent near Reims; an average of 2 per cent in the Sénonais and in the tuffeau of Touraine; 2–3 per cent in Artois) and from 2 to nearly 10 per cent in valleys (2 per cent for the Vanne; 5–6 per cent for the Hallue; 6–9 per cent for the Canche). In the confined aquifer, the typical range is about 10^{-4} to 10^{-3}, for example, 4×10^{-4} at the Béthune experimental site (where the Chalk is below 20 m of Tertiary cover) and $2.5–4.5 \times 10^{-4}$ in the confined Cenomanian aquifer of the Upper Lys valley (below 50 m of marly Turonian).

On a basin-wide scale where the Chalk crops out, the storage coefficient can be estimated from the ratio of the volume of water that has drained from storage by gravity flow (deduced from the recession curve) to the volume of aquifer drained (derived from the average water-level fluctuation in the aquifer of the corresponding groundwater basin). By this means Roux (1963) estimated the average storage coefficient of the Chalk in the Somme Basin as 3 per cent, but 5–6 per cent in the Hallue subcatchment.

Interpretation of pumping tests

The interpretation of pumping tests in the Chalk requires the use of a variety of methods. They range from the simplest techniques which assume the Chalk is isotropic, non-leaky and releases water instantaneously, to those in which it is treated as a multi-layered aquifer with delayed release from storage as described by Forkasiewicz (Colloque Régional de Rouen 1978) in an analysis of 39 selected tests in Normandy. It is not unusual for low storage coefficients to be determined where the aquifer is theoretically unconfined; for example, 1.54×10^{-3} in the new Sangatte boreholes near the entrance to the Channel Tunnel (Crampon *et al.* 1990). From pumping tests, it is possible to find apparent boundary effects well within the true limits of the aquifer, because the aquifer is fractured or stratified in a non-homogeneous manner, or to find apparent confined conditions in the absence of cover because of the confining effect of some sub-horizontal fractures. It is sometimes necessary to resort to fractured-media models, such as the one with vertical and horizontal fractures proposed by Bertrand and Gringarten (1978) precisely for application to the Chalk.

Productivity

The productivity of the Chalk is obviously related to its aquifer properties (T and S), but from a practical point of view it is expressed by the discharge rate of wells and boreholes.The regional distribution is similar to that presented for transmissivity. Caous and Roux (1981) estimated that, in the Somme, the average production rates are of the order of 10–35 $m^3 h^{-1}$ in upland areas, 20–60 $m^3 h^{-1}$ in dry valleys, and 75–270 $m^3 h^{-1}$ (and even 1000 $m^3 h^{-1}$) in valleys containing rivers. In Touraine, the rates range from a few cubic metres per hour to 250 $m^3 h^{-1}$ (Rasplus and Alcaydé 1991), while in Gâtinais the average flow rate for 275 agricultural boreholes was about 36 $m^3 h^{-1}$ with a maximum of about 240 $m^3 h^{-1}$ (Lepiller and Lasne 1990). Near Montereau, in the Seine Valley above Paris, the average yield from 20 test wells in chalk with an effective thickness of 30 m was 238 $m^3 h^{-1}$ for a drawdown of 4 m.

It is quite common to use acid treatment to increase the productivity of boreholes. This generally doubles the efficiency and may even triple it, as at Verchin, in the Upper Lys, in Artois (Caous *et al.* in Colloque Régional de Rouen 1978).

Effective rates of flow velocity

The effective rate of flow or velocity (that is, the distance divided by the time of travel by advection only) is usually determined by tracer tests of natural groundwater flows. This is a classic experiment in karstic networks, but tracer experiments in the Chalk in non-karstic areas are still very unusual. In the north of France, tracer tests have given effective velocities of the order of 0.5–1 m h^{-1} with a maximum (first arrival of the tracer) of the order of 1.3–4.9 m h^{-1} in tabular, well-bedded chalk. Velocities of the same order (0.125–1.25 m h^{-1}) have been measured, also by tracer tests, in the Seine Valley upstream of Montereau in alluvial deposits and chalk.

Below dry valleys, in areas considered to be nonkarstic, such as Artois, the velocity can be very much higher. An experiment at Escalles in 1989–90 gave multiple peaks of tracer at the outlet and an effective velocity of the order of 2 m h^{-1}, with a maximum of 53 m h^{-1}. This latter figure is close to values obtained in karstic environments elsewhere, when the tracer was not directly injected into the saturated zone. In the non-karstic tuffeau of Touraine, the average velocity of flow is only 0.3–30 m a^{-1} (Rasplus and Alcaydé 1991).

Hydrodispersive characteristics

The hydrodispersive characteristics of the Chalk are still poorly understood, for the following reasons. First, most of the tracer tests have been carried out in karstic systems, in which these parameters have little meaning. Second, when the aquifer is nonkarstic, it is still frequently heterogeneous with discrete fissure flows which give multiple answers from tracer tests (for example, experiments at Doignies, Avesnes-le-Sec, and Escalles, in Artois), so that only the maximum, modal (or multi-modal) and, possibly, the mean or median velocities can be determined. Third, the measured dispersivity tends to increase with distance in an environment that is not perfectly homogeneous.

Values obtained at the experimental site of Béthune are of the order of 0.5–1.8 m for longitudinal dispersivity and 1–6 per cent for kinematic porosity (for radial flow over distances of 10–27 m, in a semi-confined aquifer). In an unconfined aquifer near Sangatte, close to the Channel Tunnel, with combined two-dimensional and radial flows, the longitudinal dispersivity is about 9.7 m and the kinematic porosity of the order of 2–3 per cent over a distance of 39 m (the values for kinematic porosity agree well with regional values of specific yield, or unconfined storage coefficient). A two-dimensional natural flow model of tracer tests at Avesnes-le-Sec indicated a longitudinal dispersivity of the order of 25 m and a transverse dispersivity of about 4 m, over a distance of 700 m.

Hydrochemistry: Natural quality and pollution of the Chalk

Means of quality control

Measurement points for determining the quality of groundwater in the Chalk are basically springs and private and public wells and boreholes from which samples are taken periodically for physicochemical analysis. The oldest and most uniform network is that of the Ministry of Health, which monitors the quality of water supplied for human consumption. Until 1988, the number of constituents measured was not always adequate. Nevertheless, at some sites, such as the springs of the Vanne or of the Avre, which supply Paris, records date back to the beginning of the century. Since 1975 specific networks for monitoring the quality of water in the Chalk have been established over most parts of the aquifer. They include not only drinking-water supply wells, but industrial and agricultural supplies and springs.

Since 1985, a monitoring network and archive for groundwater quality has been progressively established in France following an initiative of the Ministry of Environment. It includes the identification of the aquifer, the geographical coordinates of the sampling points, as well as records of all physical, chemical and bacteriological analyses. The monitoring network offers considerable possibilities for periodical, historical and cartographical syntheses by aquifer system, basin or administrative district. It will cover the whole of the Chalk by 1993.

Mapping groundwater quality

A general impression of the quality of groundwater in the Chalk can be obtained from the map, produced by Landreau and Lemoine (1977) at the scale of 1:1 000 000, which adopts a classification of seven quality classes each corresponding to a type of use (for example, drinking water). Other more recent maps have been made for the Seine Basin (Agence de Bassin Seine-Normandie in 1988), for the Loire basin (Agence de Bassin Loire-Bretagne), and for the north of France which indicate physical and chemical parameters such as hardness, chlorides, and sulphates. Detailed maps illustrating different parameters and for different regions could be compiled from data provided by the quality-control network, for example for Aisne, Calvados, Seine-Maritime, and Essonne.

Natural quality of groundwater

Groundwater in the Chalk derives its basic natural chemical quality at the time that recharge infiltrates through the soil and the unsaturated zone. Research in the north of France, at Sainghin-en-Mélantois, has shown that the water obtains its chemical character in the first few metres of the unsaturated zone (Bernard 1979). The final composition depends on the mineralogical composition of the aquifer, on the type of water circulation, and on the residence time of water in the aquifer. Once established, the hydrochemical characteristics in an area remain stable if the aquifer is homogeneous and not contaminated.

As a general rule, the water in the Chalk of the Paris Basin and northern France is of a calcium bicarbonate type with small amounts of sodium, sulphate, and chloride; it is hard or fairly hard. The temperature is in the range 10–12°C; the pH is usually about 7–7.5 and at most 8; the specific conductivity is between 350 and 700 microsiemens per centimetre (μS cm^{-1}) (and more often between 400 and 550) and the hardness between 200 and 400 mg l^{-1} $CaCO_3$ (and generally between 250 and 350 mg l^{-1}). The natural nitrate content is usually less than 15 mg l^{-1} and sometimes—in some parts of Normandy and Artois—even less than 5 mg l^{-1}. The chloride (Fig. 7.16) and sulphate contents are generally less than 30 mg l^{-1}.

With the knowledge of the physical and chemical characteristics of Chalk groundwater currently available, there are no significant differences from one area to another across the Paris Basin. This is based upon a considerable body of data for some areas but on very few data for others. Some differences are apparent locally because of the flow conditions and the nature of the cover. Three categories of hydrochemical zones can be distinguished:

(1) the aquifer is unconfined and away from the main valleys (the most common situation);

(2) the aquifer is in a major valley with alluvial deposits, where chalk waters mix with waters in the alluvial deposits, and possibly with river water;

(3) the aquifer is confined under the Tertiary cover, in the centre of the Paris Basin and in the north of France.

In the areas where the Chalk crops out extensively and where the aquifer is well fissured and not too deep, the relatively rapid renewal of the groundwater leads to a low mineral content—especially of Cl^-, SO_4^{2-} and Mg^{2+}—and a conductivity of the order of 400 μS cm^{-1}. The mineralization increases where the reservoir is less well fissured.

The Tertiary rocks commonly contain pyrite and gypsum. Therefore, adjacent to or under Tertiary cover, runoff and leakage from the Tertiary aquifers often introduce a significant sulphate content into the Chalk groundwater. This occurs in the Seine Valley downstream of Paris (at Poissy and Boulogne), in the Thanetian of the Aisne Valley upstream of Soissons, and in the valley of the Vesle (Caous and Roux 1981). Under the Landenian sands and under the Louvil clay of the Orchies Basin (in the Nord, near Valenciennes), the sulphate content can reach 200–300 mg l^{-1} (Droz 1985). Invariably, when the aquifer passes under Tertiary cover, there is an increase in Mg^{2+}, Na^+, K^+, SO_4^{2-}, and a marked decrease in NO_3^-; in the Paris area, in deep chalk below thick Tertiary cover, the nitrate content is less than 5 mg l^{-1}. Iron enrichment can be observed locally as a result of the migration of water from the Tertiary aquifers, as in Artois–Picardy, Normandy, Touraine (1.5 mg l^{-1} in Montbazon) and in Sénonais (0.8 mg l^{-1} in Montereau). The increase in Na^+ and K^+ below the Tertiary cover is due to ion exchange. Eventually in a downgradient direction the groundwater becomes saline.

As a general rule, mineralization increases near and below valleys where alluvial deposits are in hydraulic communication with the Chalk; the conductivity can increase to about 650 μS cm^{-1}. Because of such interconnections iron enrichment has been observed in Artois–Picardy. On the other hand, abnormal manganese contents have not been observed in the Chalk below alluvial valleys (although they occur in other parts of France). Generally in Chalk groundwaters fluoride values do not exceed potable limits.

Underlying strata may also influence the chemical composition of groundwater in the Chalk. For

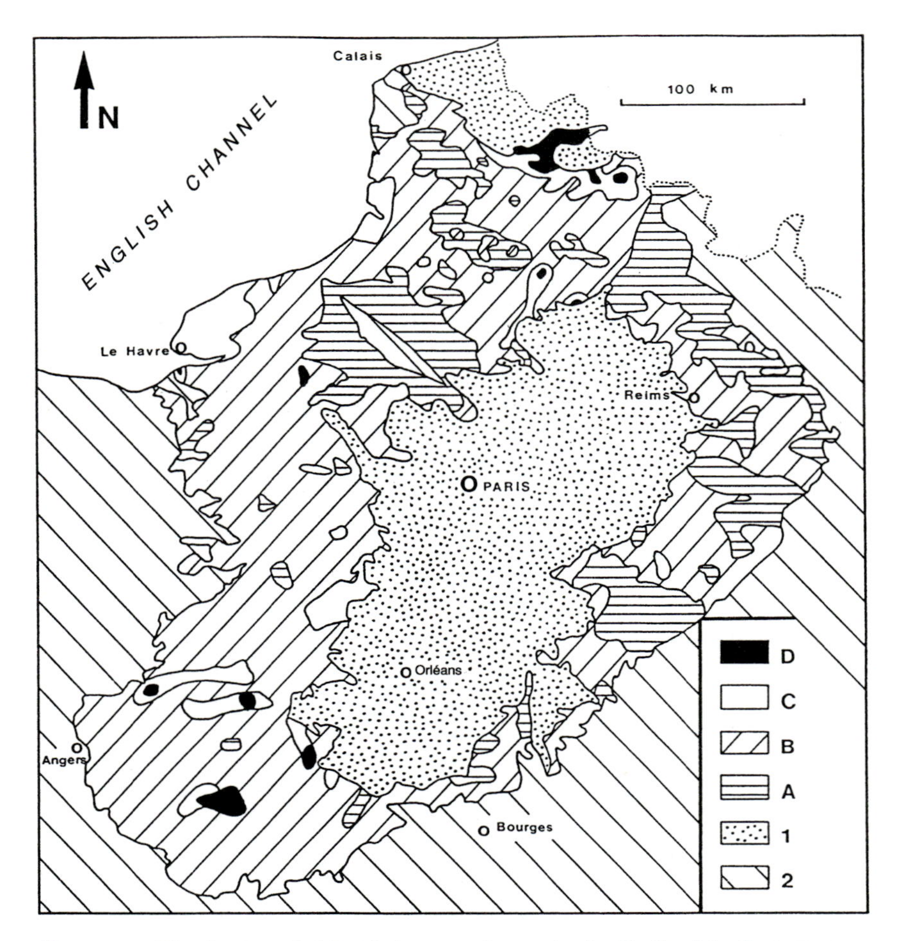

Fig. 7.16. Distribution of chloride in groundwater in the Chalk of the Paris Basin. (A < 15 mg l^{-1}; B 15–30 mg l^{-1}; C 30–50 mg l^{-1}; D > 50 mg l^{-1}; 1 is Cenozoic; 2 is pre-Upper Cretaceous rocks). (Prepared by Bracq.)

example, in the Orchies Basin upward movement of water (locally along a fault) increases the sulphate content; it can reach 600 mg l^{-1} in the Saint-Amand-les-Eaux area. The sulphate is derived from the dissolution of gypsum in the underlying Dinantian limestones. It has been distinguished from sulphate-rich water from the Landenian on the basis of the general hydrochemical facies and by studies of the sulphur isotopes (Droz 1985).

In coastal or estuarine areas, Chalk groundwaters can be contaminated by the intrusion of seawater, locally accentuated by pumping, as is the case at the Bas-Champs of Artois and Picardy (from Boulogne to Le Tréport). Chloride contents of 100 mg l^{-1} have been reached at Mers-les-Bains (Caous and Roux 1981), of 700 mg l^{-1} at Lillebonne, in the Seine Valley downstream from Rouen, and of 1–2 g l^{-1} along some parts of the Normandy coast.

At the edge of the Flanders maritime plain, between Calais and Cap Blanc-Nez, investigations for the Channel Tunnel have established the position of the saline interface. It is orientated east–west, parallel to the edge of the Flandrian Plain, before curving towards the south-west. The chloride contents are higher than 15 g l^{-1} in this area, where the salinity is related not to contemporary brackish water intrusion but to ancient seawater, still trapped beneath the former Flanders maritime plain; this is despite the regression of the sea. The saline water retained in the Chalk and the overlying cover is being slowly flushed out by the flow of fresh groundwater from inland areas of the Chalk. This flushing out of the saltwater is more rapid in the overlying marine Quaternary deposits which are more permeable than the Chalk.

The quality of Chalk groundwater varies with depth. Mineralization increases in deep zones where the circulation is slower and the residence times longer. The most characteristic feature is an increase in Mg^{2+}, which can rise to 20–30 mg l^{-1}. Many characteristics of these deep waters are similar to those found where the aquifer passes under thick Tertiary cover. Stratification of waters may sometimes be observed. In the Seine Valley downstream from Paris, Santonian–Campanian water has moderate contents of Cl^-, SO_4^{2-} and NO_3^-, but is particularly rich in Mg^{2+} (which ranges from 30 mg l^{-1} to about 50 mg l^{-1}), with a corresponding reduction in calcium (to 70 mg l^{-1}). At depths of more than 115 m, the inflows of water from the Turonian–Coniacian contain less magnesium, but more chloride, sulphate and especially nitrate (30 mg l^{-1}). It is clear that quite important differences can sometimes be observed in one locality.

Land use and the type of cultivation also affect groundwater quality. In the south-east of the Paris Basin, the groundwaters of the Loing basin are clearly more mineralized than those of the Vanne Basin, although they are less than 50 km apart. The conductivity is higher (480 as against 400 μS cm^{-1}), and the concentrations of Na^+ and K^+ are practically double; the Cl^- is 14 against 6 mg l^{-1}, sulphate 12 against 3 mg l^{-1}, and nitrate 30 against 15 mg l^{-1}. These differences can be related to the geology of the basins and their soils; the Loing Basin has extensive Tertiary deposits and well-developed agriculture, whereas the Vanne Basin is very forested.

Data collected over 20–30 years, in Normandy and in northern France in particular, have shown that the hydrochemical characteristics of Chalk groundwater, with the exception of nitrate, show little long-term variation.

Extent of pollution

Susceptibility to pollution

Groundwater may be considered to be polluted as soon as its hydrochemical composition differs significantly from the natural state; these changes are due to various land-use and human activities. The risk from pollution depends also on the vulnerability of the aquifer to pollution, particularly the protection afforded by the soil, the permeability and thickness of the unsaturated zone and the permeability of river beds and banks. As shown above, where there is no Tertiary cover, the vulnerability of the Chalk to pollution can vary considerably. For example, the loess covering the Chalk in Artois, Picardy, and Champagne is a protective layer although it is not effective everywhere because of its irregular and discontinuous character. In Normandy, loams or the Quaternary Clay-with-flints can provide impermeable barriers or natural filters, but their thicknesses and geographical distribution are very variable. Natural sink-holes (*bétoires*) in the Chalk are areas of maximum vulnerability. Furthermore, human activities facilitate the introduction of pollutants by developing quarries, and using injection wells and soakaways to dispose of runoff, agricultural drainage, and domestic wastewater.

A great part of the Chalk outcrop, particularly in the north of France and in the north-west of the Paris Basin, is covered by pollution vulnerability maps, at scales of 1:50 000, 1:100 000 and 1:250 000. These maps indicate the general vulnerability to pollution rather than give detailed information about local risks of contamination.

Pollution of the Chalk is found almost exclusively in zones where the aquifer is unconfined and particularly in urban areas or where agriculture and industry are developed. A convenient distinction can be made between diffuse pollution and local (or point-source) pollution, the latter often being found in valleys.

Diffuse pollution

This form of pollution is essentially caused by nitrates and pesticides and is almost entirely agricultural in origin. However, most highly-polluted nitrate zones generally correspond to old urban areas or villages with poor sanitation. In the Nord, 70 per cent of the drinking-water sources that have nitrate contents higher than normal do so because of domestic pollution. The treatment of roads and railways to prevent weed growth can lead to groundwater contamination by pesticides.

The steady increase in *nitrate* levels in the main unconfined aquifers in France, including the Chalk, over the last twenty years or so, has become a

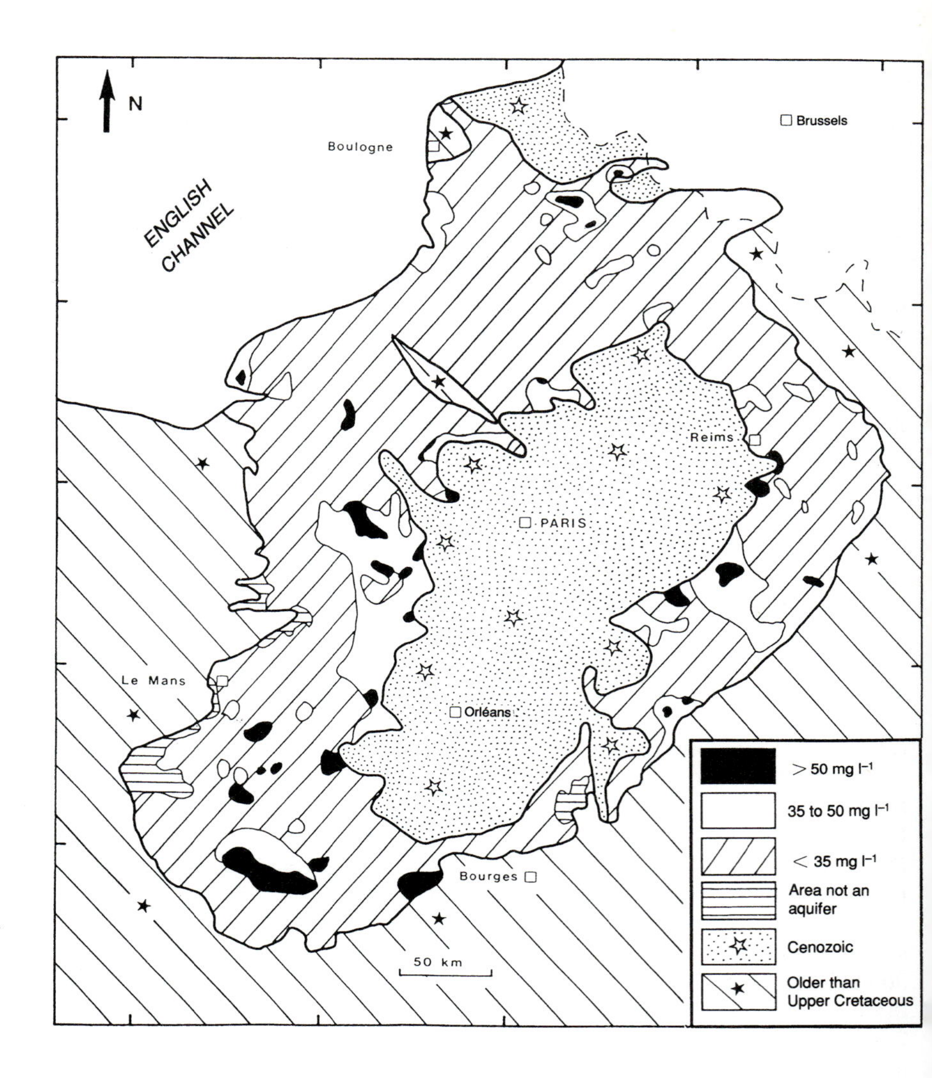

Fig. 7.17. Distribution of nitrate in groundwater in the Chalk of the Paris Basin (Prepared by Bracq.)

serious problem for the public services in charge of groundwater management and for the population as a whole. The last Ministry of Health inquiry, in 1987, showed that 2100 inhabitants use water with more than 100 mg l^{-1} of nitrate and 861 000 with more than the 50 mg l^{-1}, which is the accepted potability norm. In the Chalk it is estimated that at least 300 drinking-water sources have been closed for this reason.

A map at a scale of 1:1 500 000 made by Lallemand-Barrès and Landreau (1986) for the whole of France, indicates the average nitrate values in unconfined aquifers according to four concentration classes: <25 mg l^{-1}, 25–50 mg l^{-1} (sensitive zones), 50–100 mg l^{-1}, and >100 mg l^{-1} (critical zones). Part of the Chalk in the Nord, Touraine, Centre, and Sénonais is in the critical zone. A synthesis of the most recent data is given in Fig. 7.17.

Regionally, there are significant differences within a few tens of kilometres, as in Seine-Maritime, where concentrations are of the order of 15 mg l^{-1} in the east of the *département* and more than 50 mg l^{-1} in the western part of the Pays de Caux. Values can differ greatly even within a few kilometres. Thus, in the Pays d'Ouche, they range from 0.5 to 63 mg l^{-1}, with the highest concentrations occurring in the areas of extensive cereal cultivation.

In Nord-Pas-de-Calais, values range from 0.5 to 110 mg l^{-1}. The highest are in the mining basin and in urban and industrial areas, while the lowest

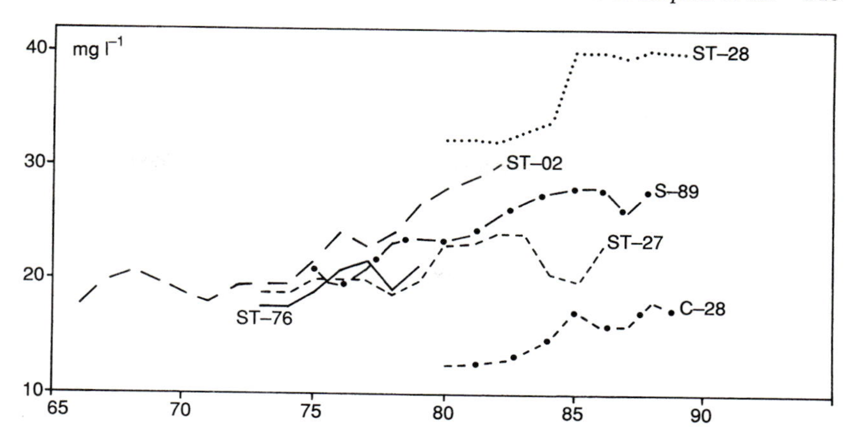

Fig. 7.18. Variation of nitrate concentration in groundwater in different Chalk aquifers between 1965 and 1990. Number on each curve indicates the French département, ST: Seno-Turonian, S: Senonian, C: Cenomanian (data provided by BRGM, and J. C. Roux).

(<20 mg l^{-1}) are found below Tertiary cover and in rural areas where the unsaturated zone is thick, as in the hills of western Artois (Crampon 1983). Under Tertiary cover, in the centre of the Paris Basin, values are less than 5 mg l^{-1}.

Nitrate concentration can change significantly even within a hundred metres. For example, between the axes of dry valleys and the valley sides, they may differ by between 25 and more than 50 mg l^{-1}, because preferential routes exist for the infiltration of rainfall (the unsaturated zone of the Chalk is not so thick and is more permeable below the valley floors than on the sides of valleys and below upland areas).

In the most polluted zones, individual sources show a steady increase in nitrate over the last 15 to 30 years (Fig. 7.18). Thus, in Normandy since 1975, the trend has been about 3 mg l^{-1} per year in the Eure, and about 1.8 mg l^{-1} per year in Seine-Maritime; in 1970 the concentrations were very low, ranging from 0.2 to 0.8 mg l^{-1}. If the average change in wells and springs in a single region is considered, the annual increase is from 1 to 2. 5 mg l^{-1} for the Chalk of the Cambrai area (Nord), 0.3 mg l^{-1} for 160 sources in Eure (Normandy), 1 mg l^{-1} for sources in Aisne (Picardy), 0.6 mg l^{-1} for the Cenomanian Chalk of Eure-et-Loire (Centre), and 1.1 mg l^{-1} for the Chalk of Gâtinais. In Seine-Normandy the highest increases (2 mg l^{-1} per year) are in zones which are already very contaminated by nitrate pollution (with values of more than 30 mg l^{-1}).

In northern France, tritium analyses indicate that groundwater in the Chalk at a depth of 30 m is over 25 years old. The nitrate infiltration front in the unsaturated zone in Connantre, Champagne, is moving downwards at a rate of 0.5–0.8 m a^{-1} (Fig. 7.19). Profiles made in boreholes 70 m deep in compact chalk in the unsaturated zone in Eure (Normandy), show that peaks of more than 50 mg l^{-1} of nitrate have reached a depth of only 25 m.

Given the low average velocity of nitrate migration in the unsaturated zone of the Chalk and the continuation of current agricultural practices, nitrate concentrations will continue to increase for a long time to come (Landreau and Roux, in Colloque AIDEC 1984). Model predictions of 15 sources in the Eure show that increases of 1 mg l^{-1} a year will continue for at least 10 years or so, and that, where

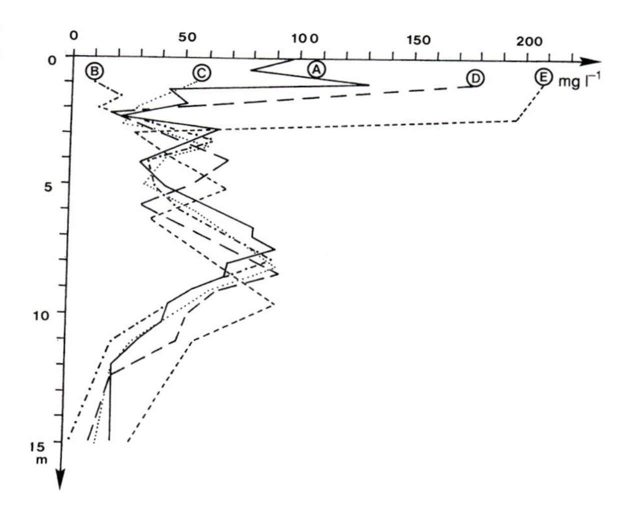

Fig. 7.19. Nitrate profiles in the unsaturated zone of the Chalk at Connantre (Champagne). A: June 1979; B: December 1979; C: March 1980; D: March 1981; E: March 1982. (data provided by BRGM).

the water table is deep, the nitrate front may not reach it for 30–40 years. A forecast made for the Eocene limestones in the centre of the Paris Basin (in Brie) indicated that concentrations will continue to rise for some 40 years, with actual concentrations doubling.

In the Seine–Normandy Basin, forecasts suggest that 25 per cent of the aquifer outcrop will eventually see concentrations of between 50 and 100 mg l^{-1} and 20 per cent will see values higher than 100 mg l^{-1}.

In the Nord, in the Douai and Lille regions, there is a rapid reduction of nitrate of 0.1 mg $l^{-1} m^{-1}$ N in the direction of groundwater flow, as the aquifer becomes confined under Tertiary clays. Thus, nitrate reduces from about 40 mg l^{-1} as NO_3^- in the unconfined aquifer near Béthune to less than 0.5 mg l^{-1} in the confined zone beneath 20 m of Landenian at the experimental site of the Institut Universitaire de Technologie at Béthune (IUT), 2 km from the Tertiary limit, in the Pas-de-Calais. Similar observations were made in Champagne, below the old argillaceous alluvium of the Marne valley where, in a few hundred metres, the nitrate content drops from 50–100 mg l^{-1} to 5 mg l^{-1}. Isotope measurements indicate that the decrease is due to denitrifying bacteria. There is an enrichment of nitrogen-15 as the NO_3^- concentration decreases. The increase in nitrogen-15 can only be caused by denitrification (Mariotti *et al.* 1988). An absence of dissolved oxygen in the water, and the presence of assimilable organic carbon and sulphides in the matrix of the Chalk provide favourable conditions for the process.

An experiment to rehabilitate groundwater with a high nitrate content was made in 1987 at Emmerin, near Lille. Pumped groundwater was treated at the surface in an aeration-infiltration basin to which starch was added. Just below the basin the nitrate content decreased from 100 to 40 mg l^{-1} and the process continued at depth in the unsaturated zone until values of 17 mg l^{-1} were attained.

Analysis of groundwater for *pesticides* was not required before January 1989. Since then systematic analyses made in the Seine basin have shown the presence of atrazine at 25–30 per cent of sampling points, and sometimes also lindane, but concentrations were lower than the EC guidelines, except in a few cases. Maximum values corresponded to the use of herbicides in the spring and early summer. It is still quite difficult to draw conclusions about trends or the relation of pesticide levels to land use and the type of agriculture, but potentially it is a serious problem now and for the future.

Point-source pollution

Whatever their nature (chemical or bacterial), point pollution sources are linked to rural housing and conurbations, and especially to industrial activities.

Bacterial pollution is more frequent in regions where the Chalk is extensively fissured or karstic, notably in Normandy and Sénonais, and it can sometimes lead to epidemics. In Pays de Caux, detailed studies have emphasized the risk of contamination of large springs (for example, at Radicatel) by rapid throughflow following heavy rainfall. Bacterial counts can reach 1200–2400 faecal coliform organisms per 100 ml, with peaks of 9300. The most recent incident occurred in the Le Havre area in November 1990, and was caused by an enterobacterium, *Shigella*, which led to a gastroenteritis epidemic affecting about 200 people.

With regard to *chemical pollution*, the most polluted areas are in the Lower Seine Valley, downstream from Paris (Roux 1977), and in the Lille area, in the Nord. In the industrial zone of Rouen, around chemical industries and former waste dumps, 'pollution centres' exist where groundwater contains more than 300 mg l^{-1} of chloride, 400 mg l^{-1} of sulphate, 270 mg l^{-1} of nitrate and 70 mg l^{-1} of nitrite, together with traces of heavy metals (Cr, Zn, Mn, Sr) and hydrocarbons.

A network of 32 representative sites in the Nord, with observations dating back to 1962–72, shows chloride contents that are below 40 mg l^{-1}, and sometimes 20 mg l^{-1}, in rural areas, and that values are not changing; but in industrial areas they vary from 50 to 80 mg l^{-1}. In the Lille area, failures in the urban sewerage systems have locally increased nitrate values to about 150 mg l^{-1}, and in large areas around Lille and in the Lens mining area, to over 80 mg l^{-1}. In this latter area, at one observation site, the sulphate content was 180 mg l^{-1} in 1983, increasing to 260 mg l^{-1} in 1985, although at nine out of ten of the other observed sites it was less than 35 mg l^{-1}.

Groundwater abstraction near streams that contain poor-quality water can cause pollution of the aquifer. Thus, again in the Lille area, water from the wellfield of Ansereuilles, along the Deule canal, has a sulphate content of the order of 150–250 mg l^{-1}. Water-table recovery, after pumping is stopped or reduced, remobilizes trapped pollutants in the unsaturated zone. In the Escrebieux Valley, in the Nord, from 1958 to 1982, a clear correlation was observed between increased nitrate levels and a rise of the water table.

With regard to trace elements, various heavy metals are found in Chalk groundwaters in industrial zones. The greatest pollution seems to be caused by hexavalent chromium (as in Picardy, with 800 $\mu g\ l^{-1}$ in Vimeu and 3.5 $mg\ l^{-1}$ in the Oise Valley). At Meulan in the Seine Valley, downstream from Paris, it has been necessary to build a plant to remove chromium from the water. Deposits in the banks of streams often filter or trap these pollutants. Thus, at Ansereuilles, in the Nord, metals are adsorbed by sludges in the Deule canal.

Contamination by organic or chlorinated solvents from industrial waste dumps or wastewater lagoons is common in the industrial regions already mentioned. Pollution by trichloroethylene was reported in the Seine Valley towards Mantes, in the Oise Valley (the Compiègne area) and in supply sources near Paris (the Dreux area), and by tar acids in the Nord mining basin.

Hydrocarbon pollution incidents, usually due to fractured pipes or tanks or to accidents during transport of fluids, are numerous. In 1978 the supplies to Saint-Quentin, in Picardy, had to be closed, as did those at Petit-Couronne and Grand Quevilly, near Rouen, in 1986, following pollution by domestic heating oil. Near the Rouen refinery in 1990, the same part of the aquifer was contaminated by more than 10 000 m^3 of heating oil and petrol. Other cases have been reported in Champagne, in Picardy and in Nord.

A full inventory of pollution incidents affecting the Chalk aquifer has not been produced because many are not reported and the contamination only becomes known when a well is affected. Many earlier incidents in industrial areas were also not recorded and the pollution may remain undetected for a long time.

Recharge, resources, and exploitation

Recharge of the Chalk by effective rainfall

The Chalk basically derives its recharge from effective rainfall on the outcrop, generally between the beginning of October and the end of March. The actual amount depends on the climate. The Nord and the Paris Basin have a temperate, oceanic climate. The dominant winds are from the southwest and the average long-term temperature is about 10°C. According to the primary meteorological network, the average annual rainfall (over 30 years) is between 570 mm (at Reims) and 700 mm (at Abbeville), but according to the secondary network the true range is greater, from 450 to 1000 mm. Normandy (Pays de Caux), the Picardy coast and the west of Artois are the wettest regions, and the Centre of Champagne, Beauce and Touraine the driest (Fig. 7.20a). The evapotranspiration is about 500–600 mm a^{-1} over the Paris Basin (Fig. 7.20b), except near the coast where it is 600–800 mm.

Calculation of the average, annual effective rainfall over a period of 30 years shows that the amount of water available for surface runoff and groundwater recharge is between 90 mm (at Chartres) and 300 mm (in the Pays de Caux and west of Artois). Regions where the effective rainfall is highest are Haute-Normandy and the west of Artois–Picardy, and where it is lowest, southern Normandy, Touraine, Gâtinais–Sénonais, and Champagne (Fig. 7.20c).

The ratio of effective rainfall to total rainfall is of the order of 25–30 per cent. According to Margat (1986), the average value for effective rainfall, over the basins of Artois–Picardy and Seine–Normandy (as defined for administrative purposes) is about 190 mm, which represents 25 per cent of the total rainfall of 775 mm. The proportion in the Nord-Pas-de-Calais (Beckelynck 1981) and in Picardy is similar. In this latter area, in the Hallue Basin, detailed measurements by BRGM between 1966 and 1975 (Auriol, in Colloque Régional de Rouen 1978) gave an average annual rainfall of 700 mm, of which 165 mm is available for runoff, implying evapotranspiration is 76 per cent and effective rainfall 24 per cent.

Surface runoff

The annual surface runoff is low, of the order of 5–10 per cent of effective rainfall. It varies according to the nature and thickness of Pleistocene cover overlying the Chalk. It is lowest in Artois–Picardy because the relief is flat or gentle and the Chalk is exposed or covered by thin, sandy-clay loams. In

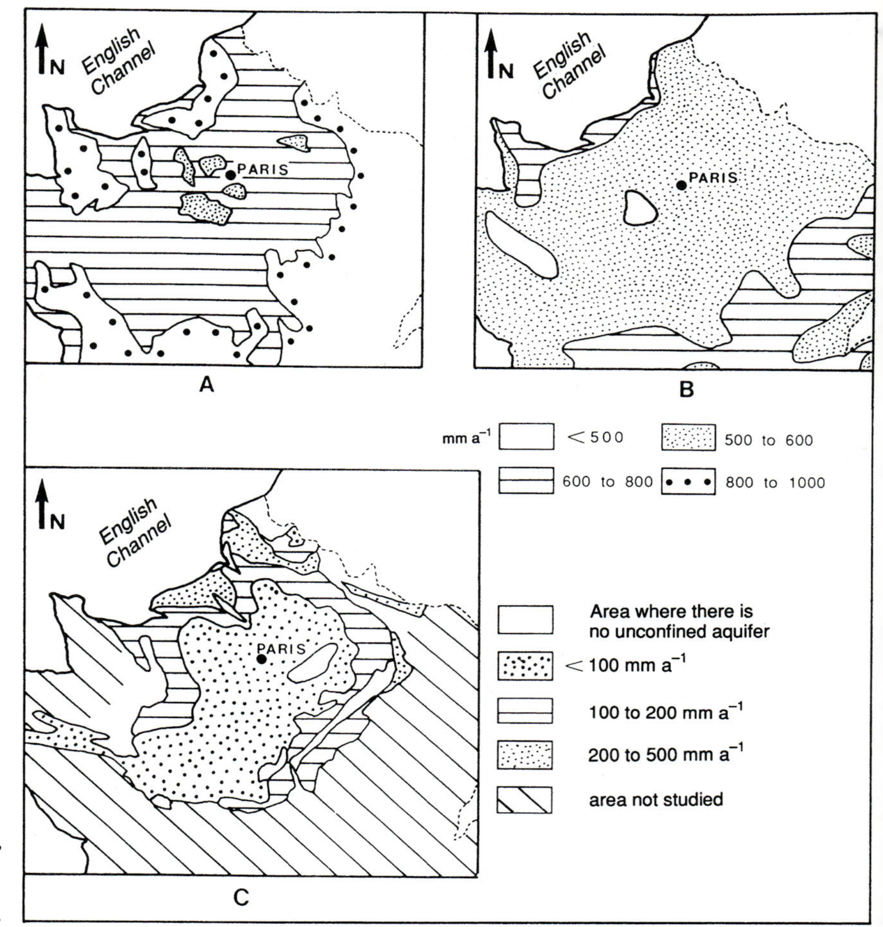

Fig. 7.20. Maps of (a) rainfall, (b) evapotranspiration, (c) effective rainfall in the Paris Basin (after Margat 1986).

Normandy, where the relief is more pronounced and the thickness of Clay-with-flints is significant, runoff coefficients are higher. Turbid spring discharges are a consequence of this runoff, which also causes considerable erosion of agricultural land, particularly in the Pays d'Ouche. A similar situation occurs in Touraine, because of superficial argillaceous formations.

In the Hallue experimental basin, the runoff is practically nil (ranging from 1 to 2 per cent of effective rainfall) except during heavy rain, as happened in December 1986 when 97 mm of rain fell in less than 10 days and the proportion of runoff reached 50 per cent over 2 days. However, this is quite exceptional. Hydrograph analysis for the Lunain Basin (in Gâtinais) estimated surface runoff as 22 per cent.

Infiltration modes and velocities

Even disregarding the differences in the permeability of materials covering the Chalk, infiltration modes and velocities in unsaturated chalk are extremely variable. Under uplands, where the chalk is, in general, quite compact or fissured to only shallow depths, infiltration velocities away from zones of preferential flow, such as faults and sinkholes, are of the order of 0.5–2 m a^{-1}, as has been shown by various detailed studies. Thus, in Nord, near Lille (at the Sainghin-en-Mélantois site), infiltration velocities of about 1 m a^{-1} were recorded over a depth of 8 m. Tritium measurements in the Cambrai area (Nord) gave values of the order of 1–1.8 m a^{-1} over 25–50 m (Arnoult 1981). In Normandy, in vertical fissured karstic chalk, in upland areas infiltration velocities were estimated to be 1.8–2 m a^{-1}; on the other hand, in the Eure, they were less than or equal to 0.5 m a^{-1} in very compact chalk. In Champagne, near Châlons-sur-Marne, the vertical downward movement established from oxygen-18 and tritium analyses was 0.45 m a^{-1} under cultivated ground and 0.75 m a^{-1} under natural vegetation, with an average velocity of 0.60 m a^{-1} (Vachier *et al.* 1987), and still in Champagne, in Connantre (Fig. 7.19), the vertical movement of nitrates is 0.5–0.8 m a^{-1} over a depth of 13 m. This explains why groundwater at a depth of 25 m can be more than 30 years old. Along the axes of dry valleys where the Chalk can be very fissured, vertical velocities can reach several metres per hour. Infiltration is, of course, very rapid in karstic zones, when surface water can flow directly into swallow holes and caverns.

In Normandy, tritium analyses have clearly shown how much infiltration velocities can vary: in two neighbouring basins in the south of the Pays de Caux (Fontenelle and Rançon), springs from one had an average of 10 TU, whereas the springs from the other varied between 2 and 44 TU. In the same area oxygen-18 and deuterium studies showed that a change in the oxygen-18 content occurred during flow, and that several recharge cycles could mix in the same aquifer.

In valleys where river deposits and the Chalk are in hydraulic connection, as in Montereau (in the Seine Valley, Sénonais), recharge of the Chalk can be almost immediate, especially in winter when rainfall intensity reaches a threshold value. In exceptional cases, the Chalk can be naturally recharged continuously by infiltration from streams, as for example at Iton, in Normandy, where the infiltration rate is up to 1 $m^3 s^{-1}$. On the other hand, induced recharge from streams or stretches of surface water or leakage from floodplain deposits or Tertiary aquifers, under the influence of pumping, is a frequent event, particularly in northern France and in the Seine Valley (see the section on over development).

Dating by tritium gives information on the age of groundwaters according to the depth of their occurrence, indicating stratification of waters in the aquifer. Below the Seine Valley in Gargenville (downstream from Paris), water from depths of up to 50 m was relatively recent (with 43 to 50 TU), from 50 to 95 m there was a mixed zone of recent and older waters (from 7 to 18 TU), and from 95 to 156 m the water was very old (with 3 TU).

Water balance and groundwater flow

Because there is so little surface runoff from the Chalk, measurements of stream flow from basins which are completely in the Chalk, and where the flow regime is not disturbed by large abstractions, equate precisely to the recharge of the Chalk and, consequently, to average infiltration (Roux, in Colloque Régional de Rouen 1978; Bodelle and Margat 1980).

Numerous measurements of stream discharges obtained over many years in various regions of northern France and of the Paris Basin give a good indication of specific flow rates (Table 7.3) which, over a period of 10 years or more, can be considered equal to the effective rainfall. These rates vary between 90 and 310 mm a^{-1}, confirming the values obtained by theoretical calculations (see section above on recharge).

In the Hallue Basin, in Picardy, the annual average water reserves have been estimated from recession curves to be between $34 \times 10^6 m^3$ (1966) and $4.6 \times 10^6 m^3$ (1974). In the Somme Basin, with an area of 5560 km^2, the average observed discharge from the Chalk, during the period 1963–90, was about $1040 \times 10^6 m^3 a^{-1}$, varying from $627.5 \times 10^6 m^3 a^{-1}$ in 1976 to $1542 \times 10^6 m^3 a^{-1}$ in 1988.

The average specific flow rates (flow rate divided by the surface area of a basin) from the Chalk oscillate, on average, between 3 and 7 l $s^{-1} km^{-2}$ according to the region, but can reach 10 l $s^{-1} km^{-2}$ along the Channel coast.

By adopting average values of about 161 mm, or 5.1 l $s^{-1} km^{-2}$, for the long-term annual groundwater flow for the whole of the Artois–Picardy and Seine–Normandy basins (Margat 1986), the average annual water resource for the entire Chalk aquifer (an area of 70 000–75 000 km^2) is of the order of $11–12 \times 10^9 m^3$.

Exploitation of the Chalk aquifer

The Chalk aquifer plays an important part in the economy of the Paris Basin and northern France. In several regions, such as the Nord, Picardy, Normandy and Champagne, almost all the towns and villages

Table 7.3 Specific flow rates from the Chalk of the Paris Basin (from Services Régionaux de l'Amenagement des Eaux (SRAE), Services Hydrologiques Centralisateurs (SHC) and Bureau de Recherches Géologiques et Minières (BRGM))

Region	Basin	Years	Flow rate over surface area of basin	
			($mm\ a^{-1}$)	($l\ s^{-1}\ km^{-2}$)
Artois–Picardy	East Picardy	1965–1975	150	2.3
	Coastal area and the west	1963–1990	338	10.7
	Somme	1963–1990	186	5.9
	Hallue	1966–1975	165	7.2
	Central Picardy	1980–1989	191	6.0
	South Picardy	1968–1990	199	6.3
Normandy	S. W. Normandy	1971–1990	118	3.7
	Central Normandy	1971–1990	260	3.3
	South Normandy	1971–1990	118	3.7
	Small basins	1964–1988	312	9.8
	Seine Valley	1965–1990	349	11.6
	Risle	1970–1990	201	6.4
Maine–Touraine	Thimerais	1965–1990	155	4.9
	Perche	1971–1990	220	6.8
	Gâtine	1973–1989	150	4.7
Sancerrois	Pays Fort	1970–1989	218	6.9
Gâtinais	Vanne	1969–1989	176	5.6
	Lunain	1962–1971	117	3.7
Sénonais	Vanne Springs	1962–1971	100	3.3
Auxerrois		1970–1990	208	6.6
Champagne		1970–1986	220	7.0

are supplied with water from the Chalk. It also supplies a large part of the requirements of industry—virtually all in the Nord-Pas-de-Calais. Where other resources are also available, the Chalk still contributes significantly to both drinking- and industrial water supply: for example, 55 per cent of the abstraction in Ile-de-France is from the Chalk.

Abstractions are listed by Basin Financial Agencies, which tax users according to the volume taken. Unfortunately, the source of the groundwater is not always precisely known and agricultural use is not recorded. Hence, it is often possible to give only a general estimate of use, by combining various sources of information.

It can be estimated that, apart from agricultural abstraction, the Chalk of the Paris Basin and of northern France supplies at least $10^9\,m^3$ of water a year (of the $7 \times 10^9\,m^3$ of groundwater abstracted annually in France), of which more than $0.7 \times 10^9\,m^3$ is for potable drinking supplies. Table 7.4, although incomplete, shows the distribution by region; the aquifer is most used in Nord-Pas-de-Calais

Table 7.4 Abstraction of groundwater from the Chalk in the Paris Basin ($\times 10^6 m^3 a^{-1}$)

Region	Year	Public water supply	Industry	Agri-culture	Total
Nord–Artois	1988	263	108	*	371
Picardy	1978	88	61	3	152
Normandy–Lower Seine	1979	154	46	0.4	200
Ile de France	1985	170	20	*	190
Maine–Anjou–Touraine-Centre	1989	31	*	*	31
Sénonais–Pays d'Othe					60
Champagne					50
					1054

* Value unknown

($371 \times 10^6 m^3 a^{-1}$), Normandy ($200 \times 10^6 m^3 a^{-1}$) and Ile de France ($189 \times 10^6 m^3 a^{-1}$). Within these regions, urban and industrial centres can each use more than $10 \times 10^6 m^3 a^{-1}$, for example the urban area of Lille (Nord), the Nord-Pas-de-Calais mining basin, Amiens (Picardy), Rouen and the lower Seine Valley (Normandy), and the Seine Valley in the Ile-de-France.

Abstraction for drinking-water supply

Where the Chalk is unconfined, towns and villages, or associations of villages, use it for drinking-water supplies; it is generally found to be satisfactory both in quantity and quality. Among the most important of these users are the following:

1. In the Nord, the Lille conurbation, with a total requirement of $95 \times 10^6 m^3 a^{-1}$, takes $8 \times 10^6 m^3 a^{-1}$ from wellfields at Emmerin and $23 \times 10^6 m^3 a^{-1}$ from Ansereuilles; Douai takes $14 \times 10^6 m^3 a^{-1}$; and Dunkirk, with its wellfield of 15 boreholes at Houle and Moule, takes $18 \times 10^6 m^3 a^{-1}$.
2. In Picardy, Amiens abstracts $17 \times 10^6 m^3 a^{-1}$.
3. In Normandy, the Rouen conurbation abstracts $30 \times 10^6 m^3 a^{-1}$, of which 70 per cent is from springs, and Le Havre takes $27 \times 10^6 m^3 a^{-1}$.
4. In Champagne, Reims takes $18 \times 10^6 m^3 a^{-1}$.

In general, abstraction is from wells or boreholes but in some regions, such as the Pays de Caux and the Pays d'Ouche (in Normandy), springs such as those at Rouen, Le Havre, Elbeuf, Dieppe, and Brionne have been used for many years.

Since the end of the nineteenth century, Paris has been supplied with drinking water from several chalk springs situated within a radius of 100 km, namely Avre to the west and Vanne, Lunain and Loing to the south. There are also two large wellfields of 35 wells at Croissy and Aubergenville, in the Seine Valley downstream of Paris (in alluvial deposits and chalk), which provide public supplies for the western suburbs of Paris. Aubergenville is the most important wellfield in France, yielding $45 \times 10^6 m^3 a^{-1}$ from the Chalk and alluvial deposits. Table 7.5 gives the details of these abstractions. On average, the Paris conurbation obtains 60 per cent of its supply from groundwater (Samson *et al.* 1986). Zones have been reserved in the Eure Valley and in the Upper Seine Valley (at Montereau) for future supplies for the capital; it was shown in 1965 that the latter area can supply between 300 000 and 600 000 $m^3 d^{-1}$.

As demand increases, and urban and industrial zones expand, it is necessary to look for drinking-water resources at greater and greater distances from demand centres. Thus, Le Havre has been obliged to abandon springs on its periphery because of pollution, and replace them by developing the Yport springs, 25 km away, to provide an anticipated flow rate of 50 000 $m^3 d^{-1}$.

Table 7.5 Abstraction of groundwater from the Chalk to supply the city of Paris ($\times 10^6 m^3 a^{-1}$). The number of the French *département* is given in brackets.

Location of site	Abstraction
Aubergenville (78)	45
Croissy (78)	32.5
Seine Valley:	
The Loing and the Lunain springs (77)	42.5
The Avre and the Eure springs (28)	47.5
The Vanne springs (89)	46.5
Total	214

Industrial abstraction

Industrial abstraction is also significant, particularly in the Nord, Picardy, and Normandy, where it represents 20–30 per cent of all abstractions. As a general rule, industrial use has noticeably decreased over the last twenty years or so. Thus, in the Nord, it has declined from 45 to 30 per cent of all abstraction between 1974 and 1986, particularly because of the closure of coalmines in the Nord-Pas-de-Calais mining basin. In the Seine Valley, near Rouen, the demand, which reached 324 000 $m^3 d^{-1}$ in 1964, was only about 84 000 $m^3 d^{-1}$ in 1976. These reductions have had a significant impact on water levels in the Chalk, leading to rises of the water table of more than 15 m in the Nord and of 6 m in the Seine Valley, with consequences for groundwater quality, the stability of buildings, and the flooding of low-lying areas. Industrial abstraction does remain very high in some places, often between 3 and 6 $\times 10^6 m^3 a^{-1}$, and sometimes, near Rouen, as much as 25 $\times 10^6 m^3 a^{-1}$. However, as a general rule, there has been a reduction of industrial abstraction to the advantage of domestic requirements.

Agricultural abstraction

This is poorly known because the amounts are not recorded. In 1978, the abstraction figures were believed to be 3 $\times 10^6 m^3$ in Picardy and 472 000 m^3 in Normandy, but these are certainly underestimates; moreover, irrigation use has increased considerably in the last 10–15 years.

Abstraction for energy production

The Chalk and adjacent river deposits are also used to supply water to water heat pumps, which heat houses or apartment blocks. As a general rule abstraction only occurs in valleys where the water level is less than 10 m below the surface. The volumes taken are not known, because they do not have to be declared. In theory, the law requires the water to be reinjected into the aquifer after use.

Overdevelopment and recharge

As referred to earlier, natural recharge of the Chalk by flow induced from streams or floodplains, under the influence of intensive pumping of groundwater, is common in northern France and in the Seine Valley. In the Seine Valley, at Lillebonne, near Le Havre, abstraction of 4000 $m^3 h^{-1}$ (in 1966) included 18 per cent induced recharge from the Seine. Isotopic analyses made in 1976 at Rouen in an industrial zone, demonstrated that seepage of riverwater amounted to 60–85 per cent of the abstraction flow rate of 13 500 $m^3 d^{-1}$. The groundwater in the Chalk had a tritium content of 2.5 to 6.0 TU; the Seine waters 153 TU; and the abstracted groundwater 11–110 TU depending on the borehole. In Aubergenville, downstream of Paris, the Seine contributes 50–70 per cent of the wellfield recharge. In the Nord, recharge induced by overexploitation of the Chalk consists of infiltration of surface water and leakage from alluvial deposits and from Tertiary aquifers. Thus, in the Deule–Scarpe area of 1105 km^3, Mania (Colloque Régional de Rouen 1978) estimated that of the total annual supply of 96 $\times 10^6 m^3$, 12 per cent comes from canals and 38 per cent from leakage, the remaining 50 per cent being derived from precipitation.

Overexploitation in the Nord is such that the form of the stream network has been considerably modified. The aquifer model used by the Lille water-supply authorities was used to reconstruct the state of the aquifer in 1910 from measurements made at that time. It indicated that all the rivers were flowing and the main outlet was the spring at Escaut. In 1973, all natural streams except the

Deudre had disappeared and the main outlet had become the Roubaix–Tourcoing pumping zone. It is estimated that between 4 and 45 per cent (12 per cent on average) of the present recharge of the Chalk in the Nord is from surface streams. On a smaller scale, abstraction at Reims, in Champagne, dries up the Vesle watercourse in some years.

Upward leakage can also occur as a result of overexploitation. It is estimated that the leakage from the underlying Carboniferous aquifer into the Chalk in the Roubaix–Tourcoing area amounts to $2–3\,l\,s^{-1}\,km^{-2}$. This can have a repercussion on water quality, as discussed previously for the Saint-Amand-les-Eaux area in the Orchies Basin of the Nord. In some cases temporary overexploitation is deliberate as in the Upper Lys Valley of Artois, where groundwater is discharged into the river to maintain flow at times of low flow and the water is abstracted downstream to supply Lille. This seasonal overexploitation is of the Cenomanian confined aquifer, which is hydraulically separate from the Turonian–Senonian aquifer in this area.

Finally, when the Chalk is unable to meet the demands, resources are supplemented by artificial recharge of the aquifer by infiltration of riverwater through basins, after appropriate treatment (including settlement after flocculation, and even filtration through artificial filters).

This practice is applied at three wellfields:

1. At Houle and Moule in Artois, where a group of 15 boreholes produces $18 \times 10^6\,m^3\,a^{-1}$ for the Water Association of Dunkerque. The source is supplemented with two recharge basins with surface infiltration areas of $20\,000\,m^2$, covered with 0.80 m of dune sand. The recharge capacity is $50\,000\,m^3\,d^{-1}$. The injected volumes have increased from $1 \times 10^6\,m^3$ in 1971 to $8 \times 10^6\,m^3$ in 1985.
2. At Croissy-sur-Seine at a wellfield for the western suburbs of Paris, the volume artificially recharged in 1985 amounted to $21.5 \times 10^6\,m^3$ out of a total production of $32.5 \times 10^6\,m^3$. The recharge capacity is $150\,000\,m^3\,d^{-1}$. The area of the recharge basins is $150\,000\,m^2$ and water from the Seine is treated and then remains in the infiltration basins for 2–5 days.
3. At Aubergenville, also for the western suburbs of Paris, recharge was $2.3 \times 10^6\,m^3$ in 1985 while the annual withdrawal was $45 \times 10^6\,m^3$.

Conclusions

The Chalk of the Paris Basin covers nearly one-fifth of France, although cropping out over only about one-eighth in the northern half of the country. The wide expanse of the Paris Basin joins the Mons Basin in Belgium in the north.

The development of the chalk facies in the Upper Cretaceous period is not uniform throughout the country: lateral variations in facies produce a multi-layered aquifer which, with the effects of folds, faults and of the drainage axes of the main valleys, divide it into almost independent systems.

The main feature of this carbonate rock is its double porosity: the high matrix porosity of the rock itself provides very little permeability, but the fissured and karstic networks, although of low porosity, have sufficiently high permeability to allow exploitation of considerable resources. It is of vital importance for water supply, providing an average resource of more than $10 \times 10^9\,m^3\,a^{-1}$, of which $1 \times 10^9\,m^3\,a^{-1}$ is abstracted; in some regions, it supplies all the drinking water.

But it is a vulnerable aquifer. When it is not protected by a sufficiently thick and continuous cover of low-permeability, the Chalk is becoming more and more contaminated, particularly by nitrates, even though the slow infiltration rate gives a false impression that the high-quality water at depth is not at risk. There are few zones where the groundwater is of excellent quality but the future maintenance of quality depends on obtaining precise, reliable knowledge of the recharge mechanisms and of the local flow conditions which will allow essential measures of protection, monitoring and development to be taken within the context of overall rational management of water resources at a national level.

The analysis of the data and documents available for writing this chapter has served to illustrate the shortcomings of the information available and the variability of scientific knowledge, indicating that there is still a long way to go to attain the objective advocated.

Acknowledgements

This chapter has been prepared from a synthesis of hydrogeological research on the Chalk of northern

France and the Paris Basin, published over more than 100 years, and from an analysis of documents kindly provided by the following organizations and individuals to whom we are very grateful and wish to express thanks:

- Bureau de Recherches Géologiques et Minières (BRGM): J. Margat, C. Mégnien and O. Rouzeau (Orléans), J. Y. Caous and J. Ricour (Nord-Pas-de-Calais), Ph. Roussel (Picardie), D. Baudry and P. de la Querière (Normandie), P. Morfaux (Champagne);
- Agences Financières de Bassin (Agences de l'Eau): A. Samson (Seine-Normandie), P. Billault, F. Casal, J. Gilbert, and A. Sauter (Loire-Bretagne), D. Bernard (Artois-Picardie);
- CEMAGREF: J. M. Panetier;
- Direction Départementale de l'Action Sanitaire et Sociale du Loiret: P. Peigner;
- SAGEP; A. Montiel;
- Services Régionaux de l'Aménagement des Eaux (SRAE): Ph. Vannier and D. Denninger (Bourgogne), M. Ghio (Centre), M. Dahy (Champagne-Ardenne), G. Deuss (Haute-Normandie), A. Couchot (Ile-de-France), E. Reynaud (Nord-Pas-de-Calais), C. Lecarpentier (Picardie), and the Pays de Loire;
- Services Centralisateurs de Bassin (SHC): M. Lang (Artois-Picardie) and the Seine-Normandie;
- professors and research workers of the Universities of Caen (Professor P. Juignet) and Reims (Dr M. Laurain).

8. Belgium

A. Dassargues and A. Monjoie

Geographical distribution of the Chalk

The Chalk in Belgium was deposited mainly during Cretaceous times. The only exception is in the Mons Basin, where chalks of early Tertiary age (Danian and Montian) occur. The formation is found in the northern part of the country, but it crops out in only two areas—near Mons in the west and near Liège in the east (Fig. 8.1). During the Cretaceous, a major transgression led to marine sedimentation over almost the entire country. The transgression was not uniform and all regions were not affected simultaneously.

Cretaceous marine sedimentation, comprising calcareous and glauconitic sands and sandstones, glauconitic marls and limestones and conglomerates, began in the west of the country during Albian times. Near the city of Mons, regional transgression along a north–south line can be recognized as extending east from the sedimentation centre of the Paris Basin (Fig. 8.2). During Albian and Cenomanian times, many transgressions and regressions of the sea occurred. In the Late Cenomanian, a major advance covered the western part of the country. The sediments deposited were not chalk but green, clayey marls, the lateral equivalents of the chalky marls of the Pas de Calais. This transgression continued during Turonian times when pure chalk was deposited for the first time in the western part of Belgium. During the early Cretaceous, Cenomanian and Turonian, the northern and north-eastern parts of the country were not submerged by the Cretaceous sea.

Sedimentation continued near Mons and in eastern Flanders (Fig. 8.1) in Coniacian and Santonian times, but it was not until Campanian times that transgressive seas, advancing from the west, extended over the Brabant Massif and the area between the Sambre and Meuse rivers (Fig. 8.2).

The first marine transgression in the north-east of Belgium was also in Campanian times. In the Late Campanian, a major advance of the sea from the north-east deposited chalk over the Pays de Herve and the Hesbaye region, and at the end of the Campanian the sea probably covered most of Belgium. The boundary between the southern sea advancing from the Paris Basin and the north-eastern sea advancing from the Netherlands was along the eastern side of the Brabant Massif, between Lonzee and the western part of Hesbaye (Fig. 8.1). Chalk sedimentation continued into Maastrichtian times over north-east Belgium, and into Montian (Early Eocene) times in the Mons Basin.

The distribution of the various stages of the Chalk is shown in Fig. 8.2. Over most of the country, the outcrop, or subcrop below the Tertiary cover, is of Late Campanian to Maastrichtian age. The total thickness increases away from the Brabant Massif, exceeding 250 m in the north-east; in the outcrop areas it is generally up to between 50 and 100 m thick, although in the centre of the Mons Basin as much as 400 m occurs (Fig. 8.3). The top of the formation slopes to the north and is over 600 m below sea level in the north-east of Belgium (Fig. 8.4).

Stratigraphy

In Belgium, the Chalk is mainly represented by sediments ranging in age from Turonian to Maastrichtian. As a result of the geological history, summarized in the previous section, the nature of the sequence in different districts varies in detail. The differences can be recognized particularly in and near the outcrops.

The various types of chalk, of Turonian to Maastrichtian ages in the Mons and Tournai basins, are described in Table 8.1.

In north-eastern Belgium, in the Pays de Herve, Hesbaye, the Campine Basin and northern Brabant, the Chalk is Campanian to Maastrichtian in age. Lateral changes in lithology and facies are common,

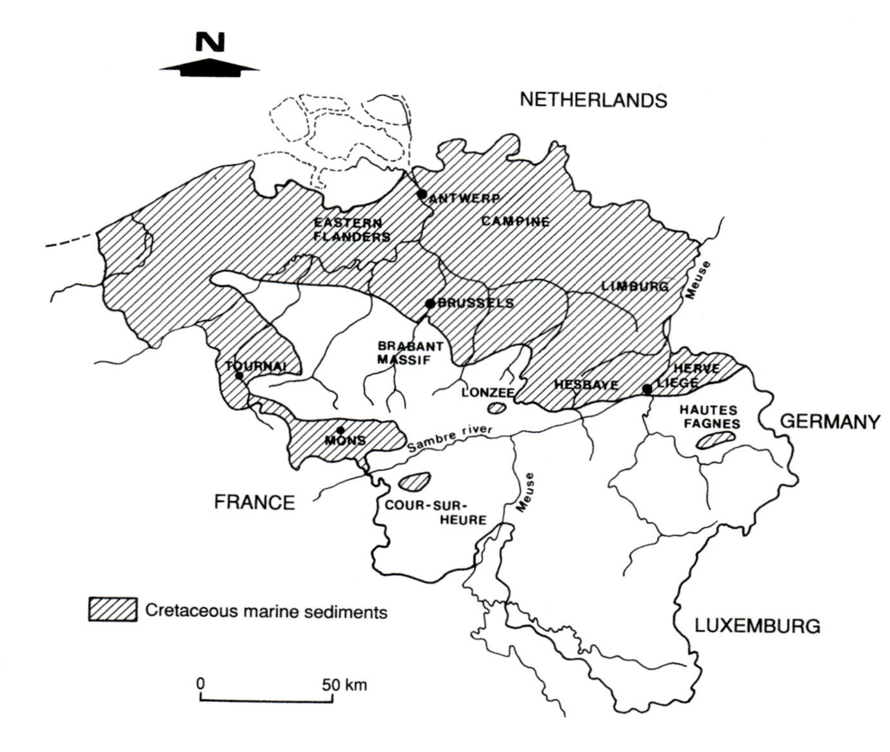

Fig. 8.1. Distribution of Cretaceous marine sediments in Belgium (after Legrand 1951).

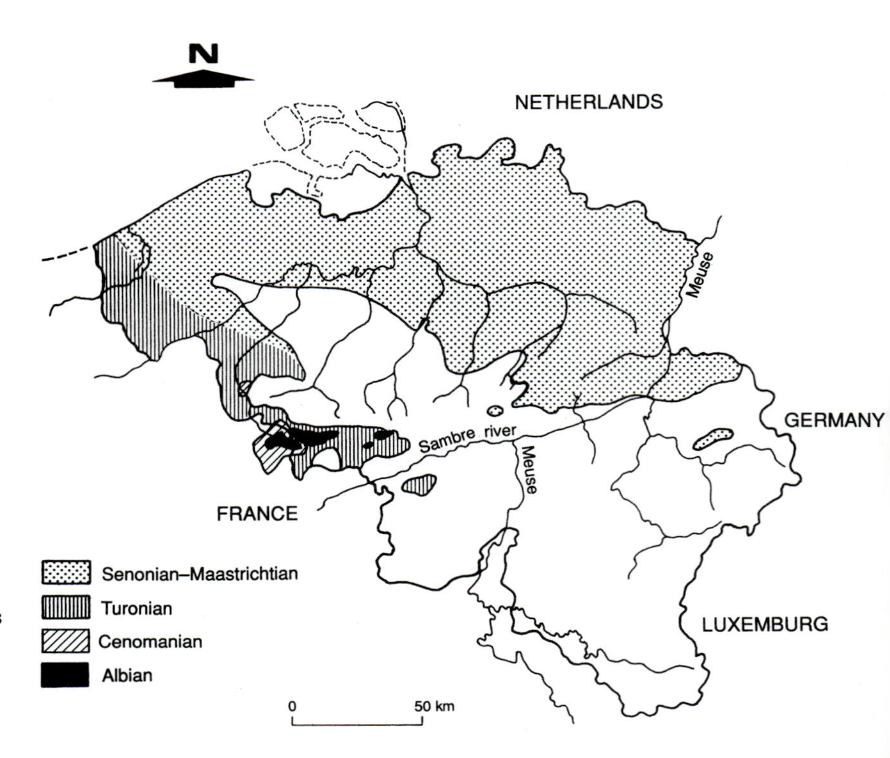

Fig. 8.2. Distribution of Albian, Cenomanian, Turonian and Senonian–Maastrichtian deposits in Belgium, The chalk facies occurs from the Upper Turonian (in the Mons Basin) to the Maastrichtian. Limited areas of Tertiary chalks occur in the Mons Basin.

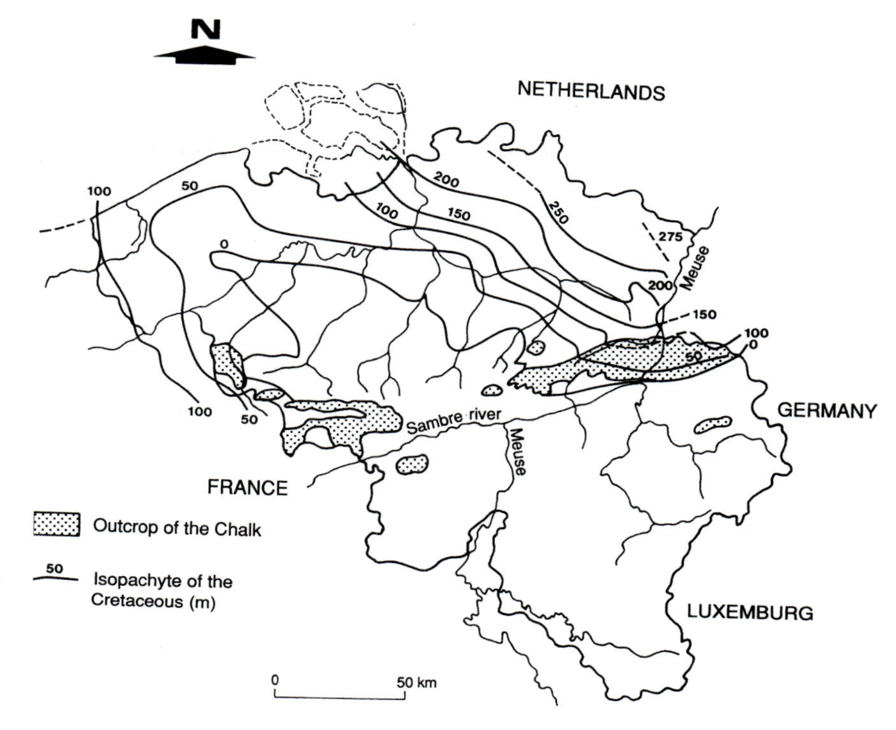

Fig. 8.3. Isopachytes (in metres) of Cretaceous marine sediments in Belgium. In the centre of the Mons Basin the thickness reaches 400 m but the isopachytes are not drawn because of the scale of the map. Most of the Cretaceous sequence is chalk (after Legrand 1951).

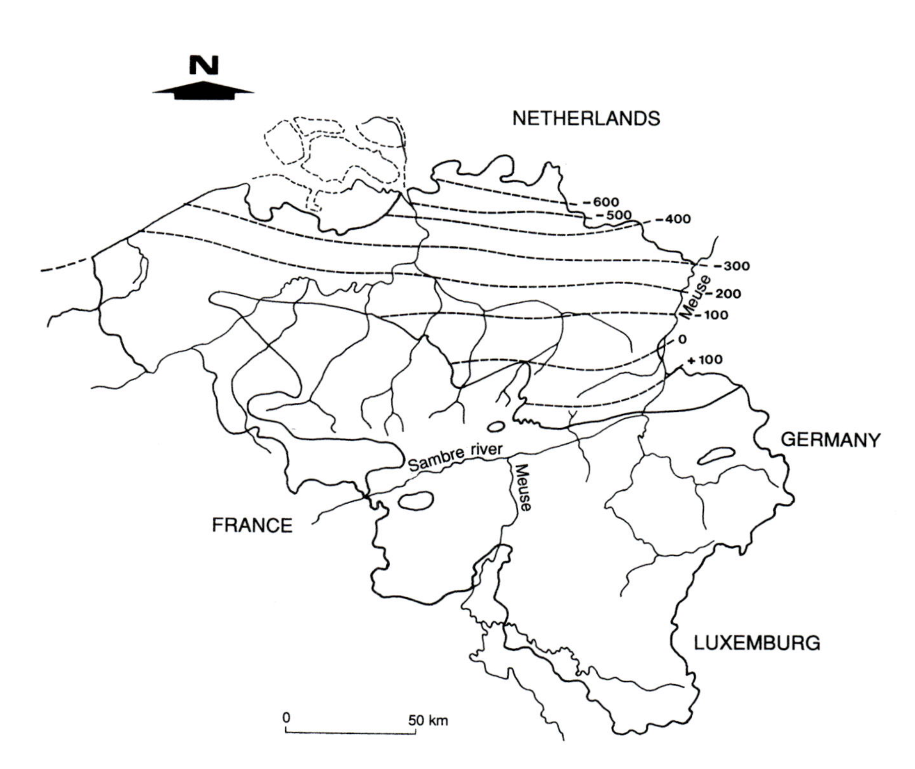

Fig. 8.4. Contours of the top of the Chalk (in metres relative to sea level).

Table 8.1 The nature of the Chalk in the Mons and Tournai basins

Stage	*Local name*	*Nature*	*Thickness (m)*
Maastrichtian	Ciply chalk	Brown or white-chalk, moderately hard, many phosphatic nodules which decrease downwards in the centre of the Mons Basin. At the margins of the basin the base is marked by a phosphatic conglomerate.	20–75
	Spiennes chalk	White coarse-grained chalk. Many large flints at the base with a poorly developed phosphatic conglomerate. Very good aquifer.	*c*.50
Campanian	Nouvelles chalk	Pure, fine white chalk without flints. Similar to Obourg chalk except for absence of flints.	13–20
	Obourg chalk	Fine white chalk; in the north of Mons Basin contains dark flints. Phosphatic conglomerate at base.	*c*.30
	Trivières chalk	White to grey marly chalk, without flints. Contains ferruginous and phosphatic beds. Phosphatic conglomerate at base. Separated from Saint-Vaast chalk by very short stratigraphic break.	*c*.80
Santonian and Coniacian	Saint-Vaast chalk	White-chalk with grey or dark flints. Glauconitic at base.	15–50
Upper Turonian	Maisières chalk	Very homogeneous over entire area. Uneven, highly friable very glauconitic chalk, with phosphatic nodules in places.	1.5–4.0
	Rabots	Uneven, white coarse-grained chalk with glauconite. Dark brown bedding plane flints. Locally the chalk is replaced by siliceous grit.	*c*.20–25

particularly to the west near Gembloux and to the north in the Campine. The Chalk overlies glauconitic marls and sands that are locally replaced by a hard calcareous clay called the Smectite de Herve. The Campanian is represented by chalk with glauconite and grey flints. It becomes uneven and coarse-grained in the upper part before grading up into white, pure chalk without flints; flints occur again at the top of the formation. Westwards in the valleys of the Mehaigne and Petite Gette rivers, the

Campanian Chalk is replaced by sandy tuffeau, an uneven coarse-grained limestone. An organic, detrital arenaceous facies prevails rather than chalk in the Campine Basin.

The overlying Maastrichtian can be divided into lower and upper divisions. The lower division is the Lanaye chalk, an uneven, white chalk containing virtually no phosphate but interbedded with continuous layers of brown flints. The upper part of the Lanaye chalk becomes more uneven and coarse-grained, and passes up into the tufaceous chalk (or tuffeau) that is well known in Limburg, in the Netherlands, where it is called the Kunrade chalk (see Chapter 9). In the north-eastern part of the Campine Basin, the chalk of this part of the sequence is replaced by silty and sandy marls.

The Chalk is overlain in north-east Belgium by a yellowish, organic, detrital, coarse-grained calcareous rock containing many flints and cherts called the Maastricht tuffeau. Where erosion has been considerable, as to the south of Hesbaye, Tertiary loess and loam overlie the Chalk.

The chalk sequences of the outliers of Lonzee and Cour-sur-Heure massif (Fig. 8.1) are similar to that of the Mons Basin, and the outlier of the Hautes Fagnes resembles the deposits of north-east Belgium. Over most of northern Belgium, the Chalk is overlain by thick Tertiary deposits, which are mainly uncemented sands, silts and clays that are complexly interlayered.

Structure

The overall distribution and form of the Chalk in Belgium is strongly influenced by the Brabant Massif. As already described, this restricted sedimentation, and the thickness of the formation increases away from the structure (Fig. 8.3). Subsequently, the Chalk's surface tilted towards the north as increasing thicknesses of Tertiary deposits covered the aquifer.

An important hydrogeological basin occurs in the south-west of Belgium, around the city of Mons. The Mons Basin (Figs. 8.1 and 8.3) is a large syncline with a general east–west axis and an average westerly pitch of 1 in 20. The decreasing thicknesses of the various deposits towards the east and at the margins of the basin indicate synsedimentary subsidence. However, more rapid and dramatic variations in thickness were caused by halokinetic structures in Palaeozoic anhydrite, underlying the Cretaceous (Fig. 8.5).

The Chalk crops out in the Hesbaye and Herve areas but towards Campine the thickness of the Tertiary and Quaternary cover gradually increases. The thickness of cover rocks is strongly influenced by faulting (Fig. 8.6). In the north-east of Belgium, the Chalk is characterized by regular horizontal or sub-horizontal layers dipping slightly towards the north.

Stratigraphic non-sequences occur in the Chalk in both the Mons Basin and north-east Belgium. South of the Dutch–Belgian border, in the northern Campine, the irregular base of the Cretaceous, indicated by seismic reflection surveys, illustrates the scale of the unconformity at the base of the Upper Cretaceous. The topography at the beginning of the Cretaceous marine transgression was influenced by the composition and nature of the Palaeozoic rocks.

Although the extents of the Cretaceous transgressions were linked to general rises of sea level, they were also influenced by local subsidence which determined some of the local variations in thickness of the deposits. The principal areas of subsidence were the Mons Basin and north-east Belgium. In the latter area, subsidence was controlled by faults extending from the Limburg area of the Netherlands. In general, the zones of subsidence reflect Palaeozoic tectonic features. At the end of Cretaceous times, almost the entire country was raised above sea level before early Tertiary lacustrine lagoonal sediments were deposited.

Groundwater in the Chalk

About $127 \times 10^6 m^3$ of groundwater are pumped annually from the Chalk, of which more than 90 per cent is used as drinking water. The aquifer provides about 20 per cent of the total volume of groundwater used for industrial and drinking purposes in Belgium. Overexploitation of the resources is not a problem as yet.

The main centres of abstraction are the outcrop areas across the centre of the country—the Mons Basin, the Hesbaye region and the Pays de Herve—and the confined areas in Limburg in the east, and in Brabant near and to the south of Brussels. Near Brussels, some $9 \times 10^6 m^3$ are abstracted each year

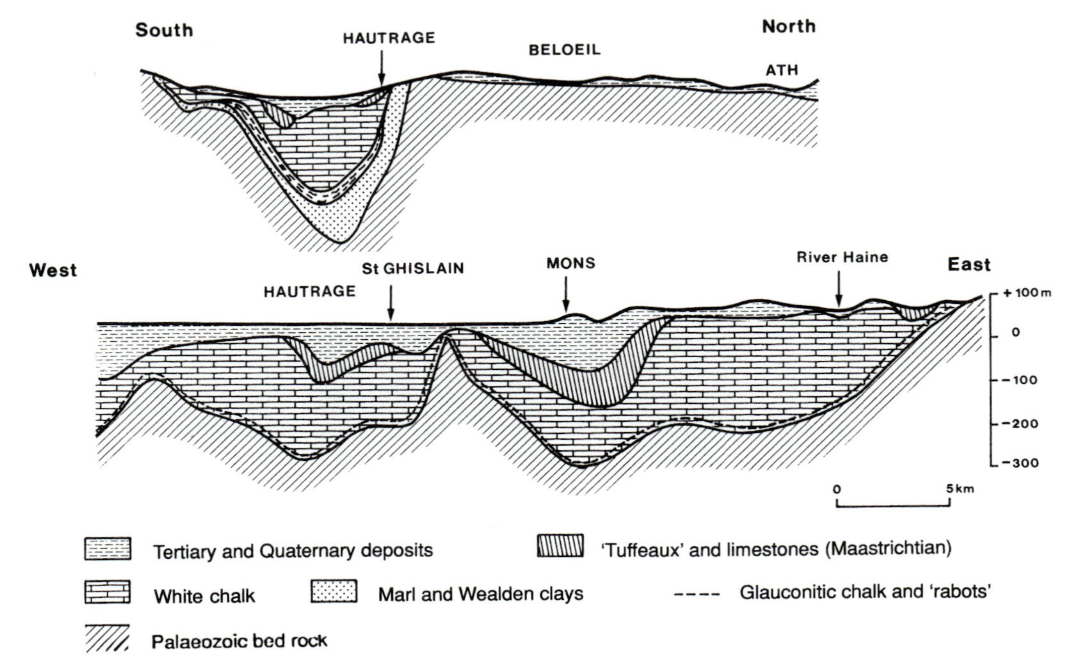

Fig. 8.5. Transverse (S–N) and longitudinal (W–E) cross-sections of the Mons Basin (after Gulinck 1966).

but the city itself is partly supplied by pipeline from the Chalk of the Mons Basin. Intensive pumping near Brussels has led to large drawdowns of water level and as a consequence saline waters in the north of the country have migrated south, the 500 mg l^{-1} isochlor now lying about 20 km north of Brussels. In Brabant province, especially to the south of Brussels, the thickness and lateral extent of the aquifer are limited and recharge by infiltration is, of course, very low. The Turonian chalk is ex-

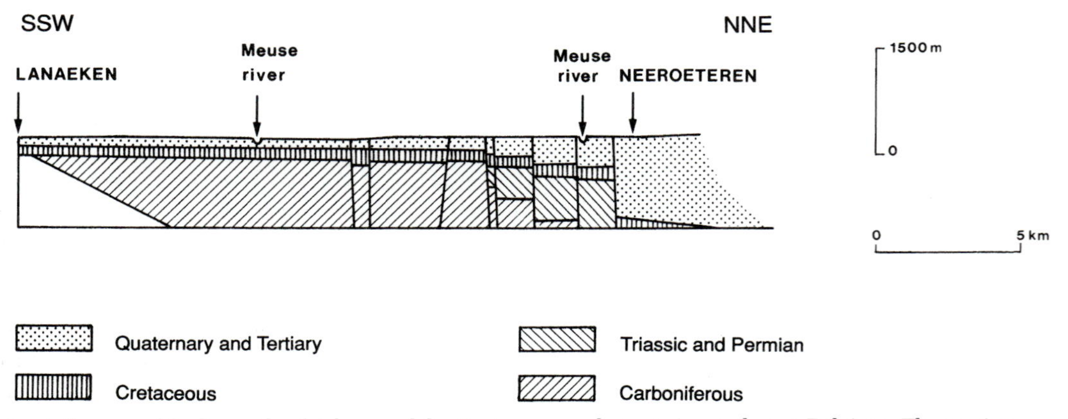

Fig. 8.6. Influence of faults on the thickness of the Cretaceous sediments in north-east Belgium. The section illustrates the horst and graben structure. The section is nearly parallel to the River Meuse in north-east Belgium, near the Dutch border.

ploited in the provinces of Flanders but the wells also penetrate the fissured aquifers of the underlying Palaeozoic.

Unconfined conditions exist in the outcrop areas of central Belgium except in the limited areas where alluvial deposits and, locally, thin Quaternary and Tertiary deposits create confined conditions. In the east, in the Hesbaye and the Pays de Herve aquifers, water levels range from about 100 m to over 250 m above sea level (asl). Further west, in the Mons Basin, levels fall from some 70–100 m asl in the east of the basin, below the higher ground, to less than 20 m asl in the lower parts of the basin in the extreme west. Throughout the northern part of the country, groundwater in the Chalk is confined by the Tertiary sediments and the piezometric surface is generally less than 20 m asl (Fig. 8.7). The most detailed information is available for the three main Chalk aquifers—the Hesbaye aquifer, the Mons Basin and the confined aquifer of northern Belgium, which are each discussed below.

Hesbaye aquifer

This aquifer is represented by a Chalk outcrop of 350 km^2 lying to the north of the River Meuse near Liège (Fig. 8.8). It provides about 60 000 $m^3 d^{-1}$ of potable water for Liège and its suburbs. An interesting feature of the Hesbaye aquifer is that it is developed by about 45 km of galleries driven in the lower part of the Chalk, with as much as 15 km constructed as recently as the 1970s. The cross-section of the galleries is about 2–2.5 m^2. Water drains into them from the Chalk and flows by gravity to supply Liège and its suburbs. Neither pumps nor filters are used. It is the only example in Belgium of large-scale groundwater development using galleries.

Recently, a complete set of data relating to the geology, hydrology, geomorphology, and geophysics has been collected in order to build a numerical three-dimensional finite-element model of the aquifer. This model has been developed by the University of Liège to forecast changes in the form of the water table and to provide additional hydrogeological interpretation, particularly about the main drainage axes in the aquifer.

The geological sequence (Fig. 8.9) may be summarized as follow:

- Recent alluvial and colluvial deposits; up to 5 m thick.
- Quaternary and Tertiary—sands and loess; 2–20 m thick.
- Residual conglomerate; 2–15 m thick.
- Maastrichtian Chalk—locally referred to as 'upper chalk'. It has been exposed to weathering and hence it is fractured; 10–15 m thick.
- Thin (less than 1 m) layer of hardened Upper Campanian Chalk (or hardground).
- Campanian Chalk—compact massive white-chalk, referred to as 'lower chalk', with many fracture zones providing preferred flow routes for groundwater; 20–40 m thick.
- Smectite de Herve—a layer of hardened calcareous clay of Campanian age, about 10 m thick, that forms the impermeable base of the Chalk aquifer.

Maps and cross-sections (for example, Fig. 8.9) have been drawn using geological and groundwater-level data from more than 500 boreholes.

Recharge to the aquifer is by infiltration through the overlying loess and the residual conglomerate and is estimated to be between 175 and 275 mm a^{-1}. The hydrological balance for the period 1952–66 has been assessed (Monjoie 1967) as follows:

rainfall (R)	740 mm a^{-1}
evapotranspiration (E)	525 mm a^{-1}
effective infiltration	175–275 mm a^{-1}
flow of the R. Geer (Ru)	$52 \times 10^6 m^3 a^{-1}$, equivalent to 120 mm a^{-1}
groundwater abstraction (G)	60 000 $m^3 d^{-1}$ equivalent to 65 mm a^{-1}
mean storage (S)	15 mm a^{-1}

giving a balance of:

$$
\begin{array}{lllllll}
R = E & + Ru & + G & + S & + \text{losses} \\
740 = 525 & + 120 & + 65 & + 15 & + 15\text{ mm}
\end{array}
$$

The 'losses' term includes flow across the boundaries of the aquifer when water levels are very high, as well as water flowing below the River Geer in response to high pumping rates north of the river.

Values for the aquifer properties have been obtained from 150 pumping tests in the different lithological units (Table 8.2). Because water-table

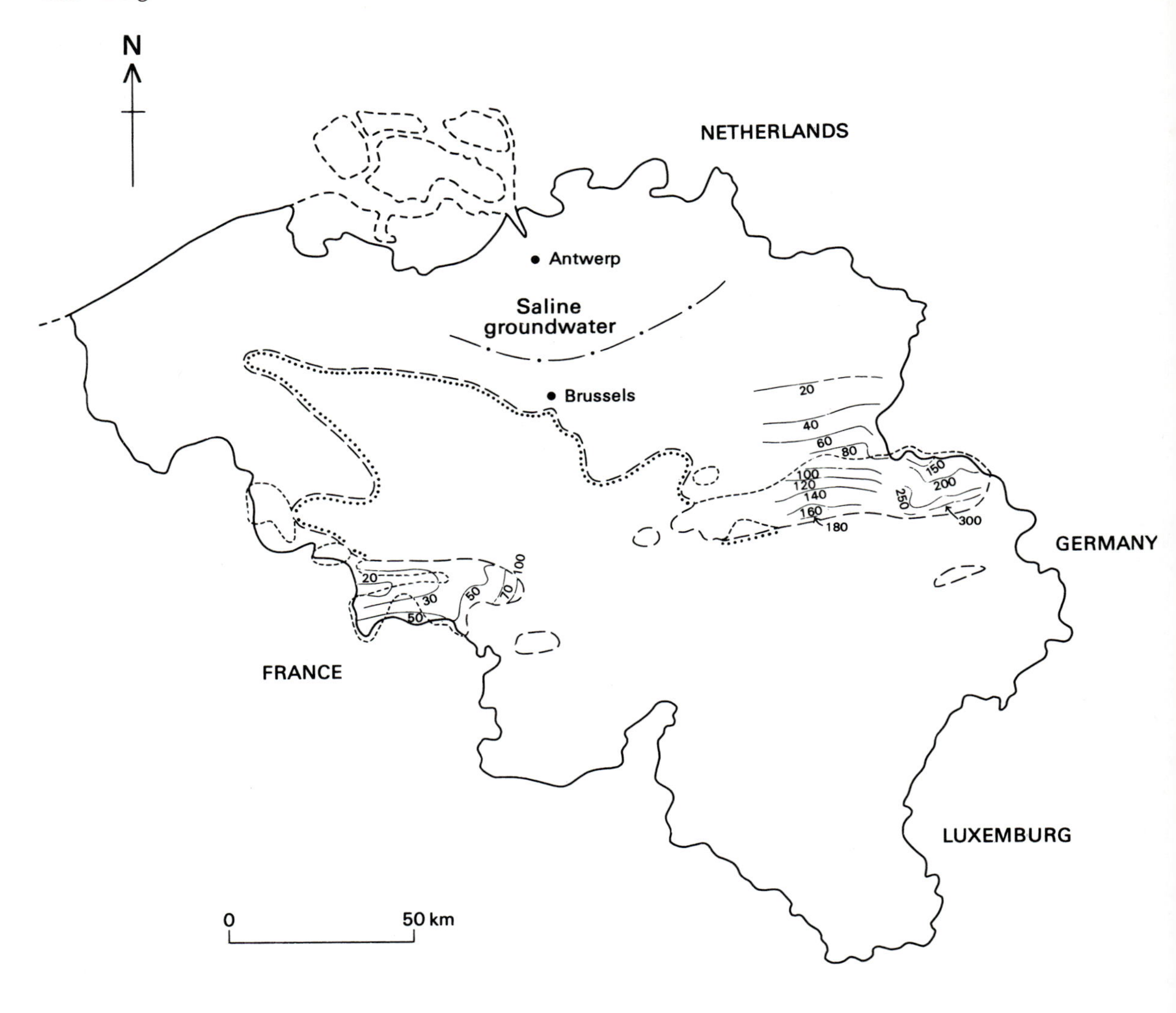

250 Groundwater level (metres above sea level)

Limit of the Chalk

Limit of Chalk outcrop

Limit of the Chalk and limit of the Chalk outcrop where they coincide

500 mg l^{-1} isochlor in confined aquifer

Fig. 8.7. Groundwater levels in the Chalk.

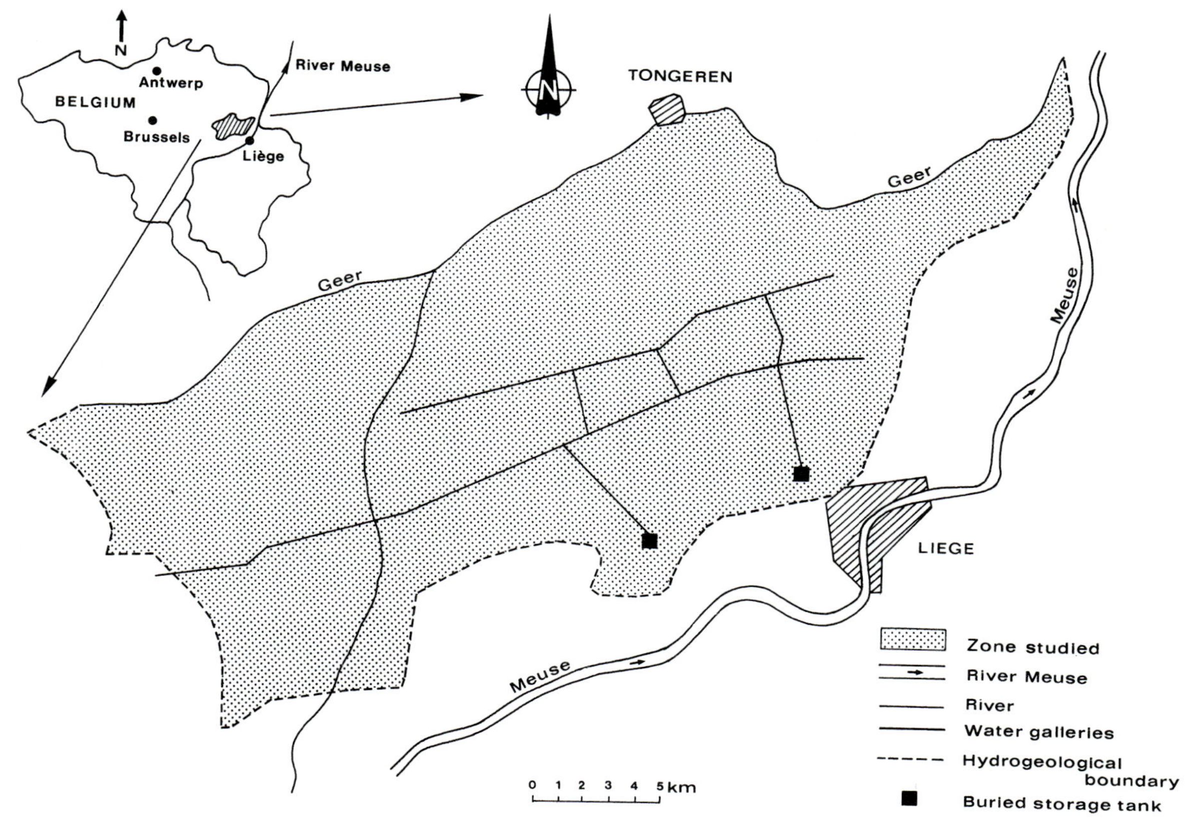

Fig. 8.8. Location of the Hesbaye aquifer.

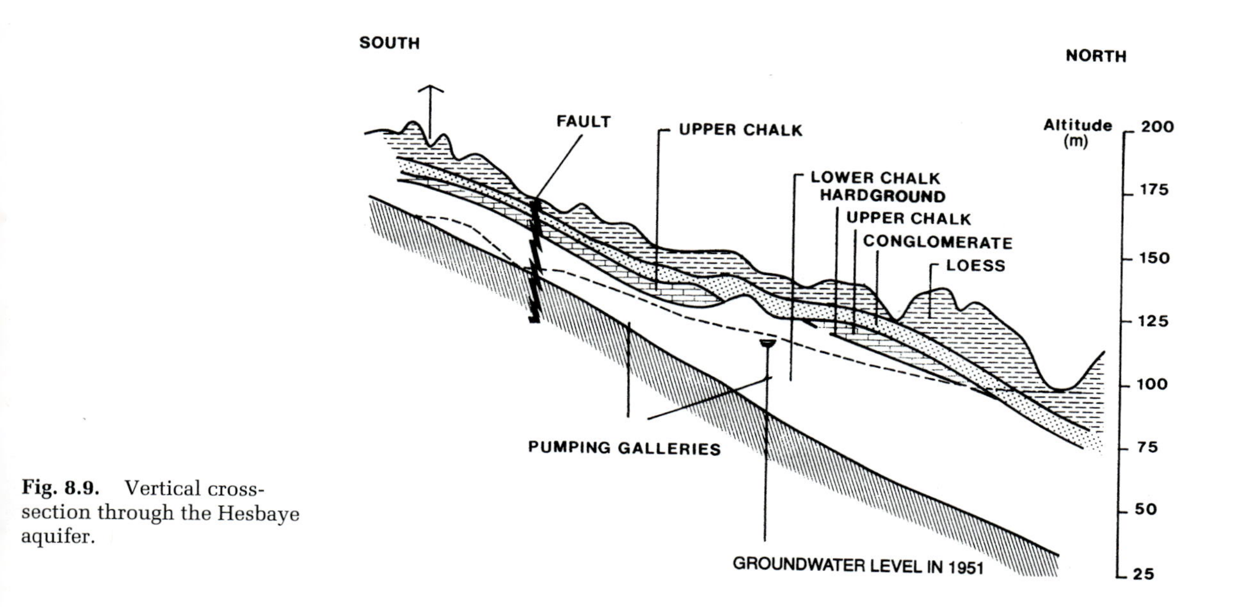

Fig. 8.9. Vertical cross-section through the Hesbaye aquifer.

Table 8.2 Aquifer properties of the Chalk in the Hesbaye aquifer

	Hydraulic conductivity (m s^{-1})	Storage coefficient
Loess	10^{-9} to 2×10^{-7}	–
Residual conglomerate	10^{-5} to 8×10^{-3}	0.05 to 0.1
Maastrichtian ('upper chalk')	2×10^{-4} to 5×10^{-3}	0.07 to 0.15
Campanian ('lower chalk')	10^{-5} to 5×10^{-4}	0.05 to 0.15

conditions exist, the storage coefficient is the effective porosity (or specific yield) and values are very high because it includes both fracture and matrix porosities.

The main drainage axes of the aquifer are below dry valleys in the Chalk, and are characterized by high hydraulic conductivities because of the presence of fractures and karstic features (Fig. 8.10). The direction of flow is towards the north-northwest. Because of the high permeability, surface drainage is very limited.

Groundwater levels have been carefully measured since 1951 and annual water level maps and cross-sections prepared. At specific points in the aquifer the change of elevation with time is known reliably. The flow of the River Geer, abstraction of groundwater from galleries, and the recharge of the aquifer by rainfall also give reliable time-distributed data.

The aquifer has been modelled by the finite-element method, which is well adapted for such a geometrically complex and heterogeneous system because the finite elements can readily follow the limits of the different layers or boundaries (Dassargues *et al.* 1988). Elements representing the porous medium, infiltration, and the abstraction galleries can be combined into a comprehensive model of the aquifer (Figs. 8.11 and 8.12). The transient form of the water-table surface has been modelled with a fixed-mesh network using a storage relationship dependent upon the groundwater elevation. The model has been carefully calibrated against historical data and tested against variations in permeability and storage coefficient. Figure 8.13 shows maps of groundwater flow for the 'upper' and 'lower' chalk aquifers as derived from the model. The flow is concentrated along a few preferred flow directions (corresponding to the axes of dry valleys) in the 'lower' aquifer because of the large contrast between values for hydraulic conductivity of frac-

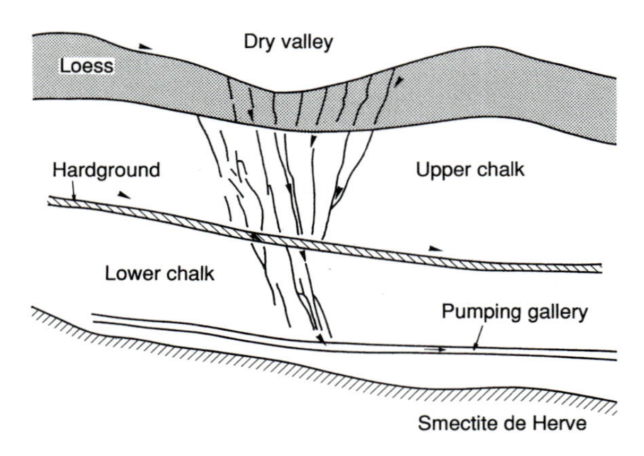

Fig. 8.10. Dry valley corresponding to a principal subsurface drainage axis through fractured chalk.

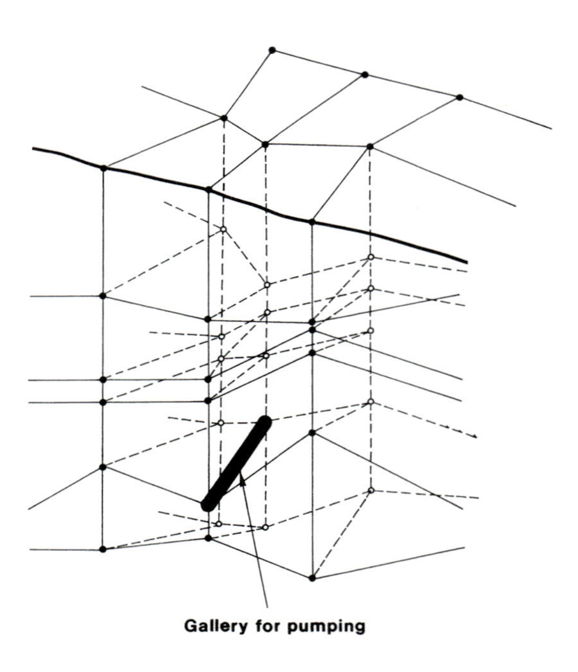

Fig. 8.11. One-dimensional 'pipe' elements representing galleries have been introduced into the mesh of the finite-element model (after Dassargues *et al.* 1988). These elements are given infinite permeability.

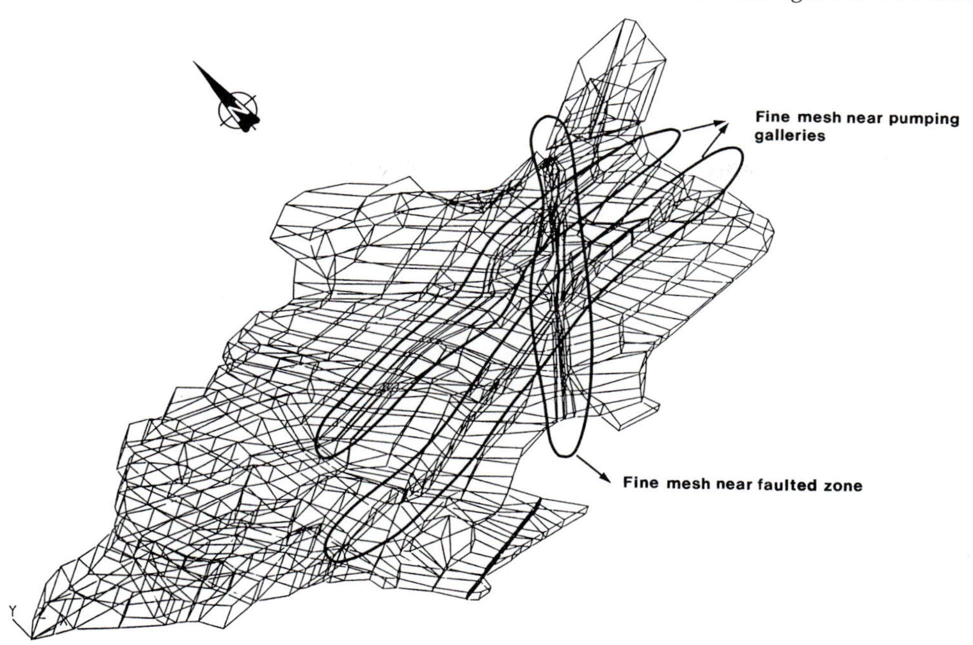

Fig. 8.12. The fourth of five layers of the finite-element model depicting the Hesbaye aquifer (after Dassargues 1991). The three-dimensional finite-element network incorporates a large body of geological and hydrogeological data including pumping wells and galleries, faults, and the boundaries of the basin or geological units. The total mesh is composed of 3300 8-node brick elements and 3600 nodes. The figure shows the complexity of the mesh network for one of the layers.

tured chalk below the dry valleys and those of the aquifer away from these axes. While these preferred directions can still be recognized in the 'upper' aquifer, the flow is more evenly distributed as the permeability values throughout this aquifer are of the same order of magnitude. A computed groundwater-level map for the Hesbaye aquifer for 1984 is shown in Fig. 8.14.

The model can be used to predict future conditions in the aquifer because groundwater levels, hydraulic gradients and volume of flow can be computed for any year as a function of infiltration and abstraction rates. The model also provides a basic framework for the simulation of contamination of the aquifer by pollutants as it incorporates advection and dispersion. With regard to pollution, the concentration of nitrites and nitrates (Figs. 8.15 and 8.16) is particularly relevant since nitrate concentrations exceed 100 mg l^{-1} over extensive parts of the upper aquifer. So far concentrations are less than 50 mg l^{-1} in the lower aquifer, where the galleries are located.

Mons Basin

The Chalk aquifer in the Mons Basin is in the form of an east–west syncline, covering an area of some 400 km^2. Drainage is by the Haine river, flowing in a westerly direction along the axis of the structure. The estimate of the mean annual infiltration, $80 \times 10^6 m^3$, is based on rainfall measurements and evapotranspiration calculations using the Thornthwaite and Turc formulae. The total annual abstraction of groundwater is estimated to be between 50 and $60 \times 10^6 m^3$. The water is supplied to Mons and its surrounding area as well as providing part of the requirements of Flanders and Brussels.

The lower part of the aquifer is formed by the Upper Turonian Chalk which overlies Cenomanian marls. The aquifer is, however, mainly in the Senonian and Maastrichtian Chalk and locally in the chalk and calcarenites of the Early Tertiary (Danian and Montian).

Water-table conditions exist over an area of about 240 km^2 but confined conditions prevail over some

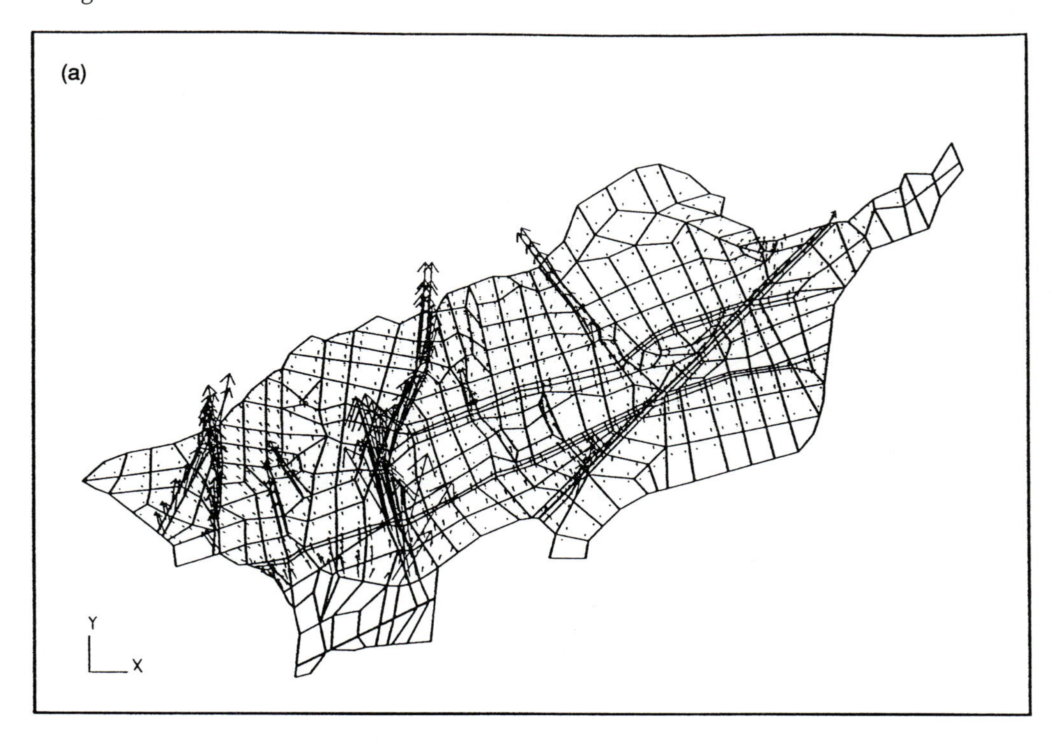

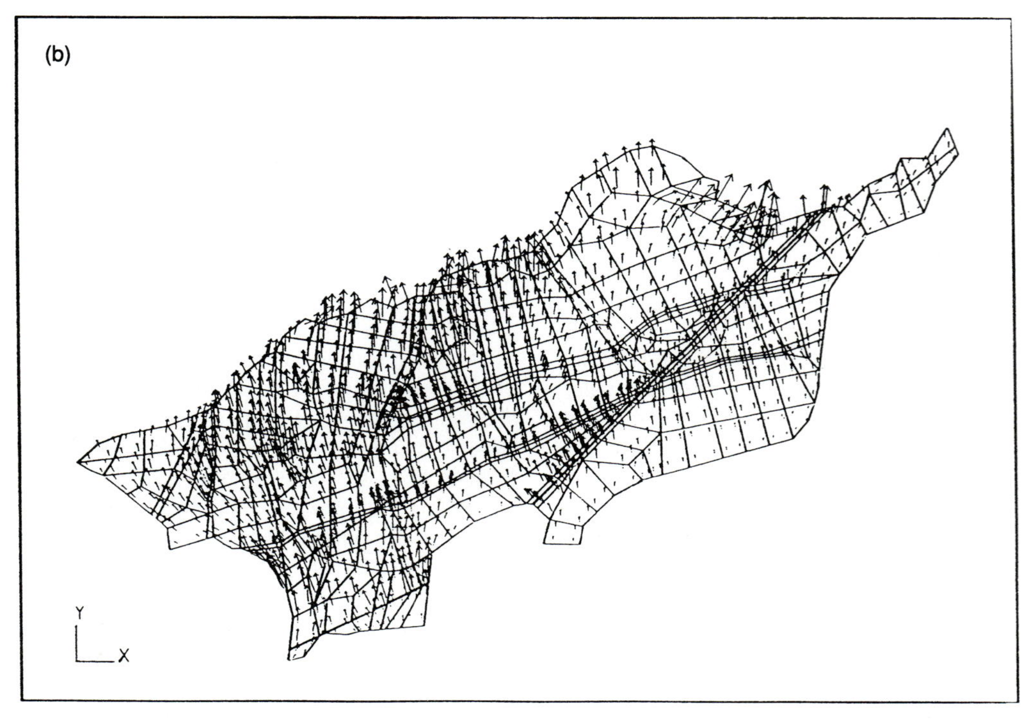

Fig. 8.13. Computed flows in (a) the 'lower chalk' and (b) the 'upper chalk' of the Hesbaye aquifer, in 1966.

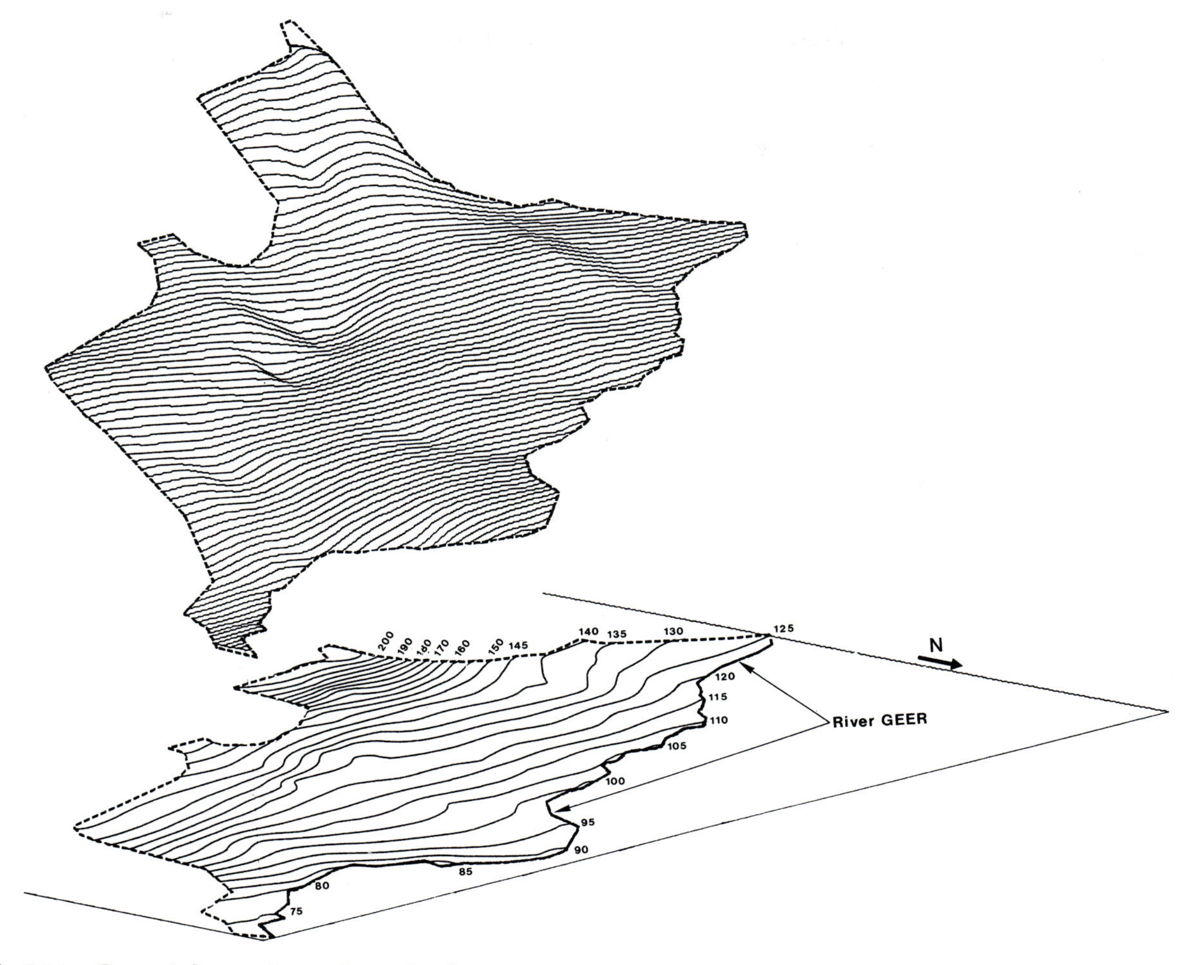

Fig. 8.14. Computed map of groundwater levels in the Hesbaye aquifer for 1984.

100 km^2 as a result of the presence of clays of the Ypresian, overlain by the silty and sandy formations of the Landenian (Fig. 8.17). The drawdown of groundwater levels due to abstraction has induced land subsidence where the aquifer is confined by Tertiary and alluvial clays and peats. In the Haine valley, drainage of peat lenses has induced severe differential settlement. Cumulative values for land subsidence due to water abstraction can attain 0.7–1.2 m.

The structure and thickness of the aquifer have been modified by synsedimentary subsidence during the Cretaceous period and by halokinetic migration in the underlying Palaeozoic rocks. As a consequence, the aquifer is relatively thin near its margins but in places it is as much as 200–400 m thick, especially near the centre of the basin. The vertical movements have led to the fracturing of the aquifer and the formation of peat in low-lying areas. Palaeokarst has developed in the fracture zones at the top of the Chalk and local uplifts of the Palaeozoic rocks have created barriers to groundwater flow within the Chalk, for example the Jemappes Sill, the Hornu Uplift and the Hensies Sill. The uplifts of impermeable bedrock divide the aquifer into compartments, and modify the general regional westerly flow (Calembert and Monjoie, 1975*a*). In 1907, the principal flow direction was towards the west, along the axis of the basin, but drawdowns of the groundwater level due to intensive pumping have accentuated the influence of the uplifts in the Palaeozoic basement on the flow

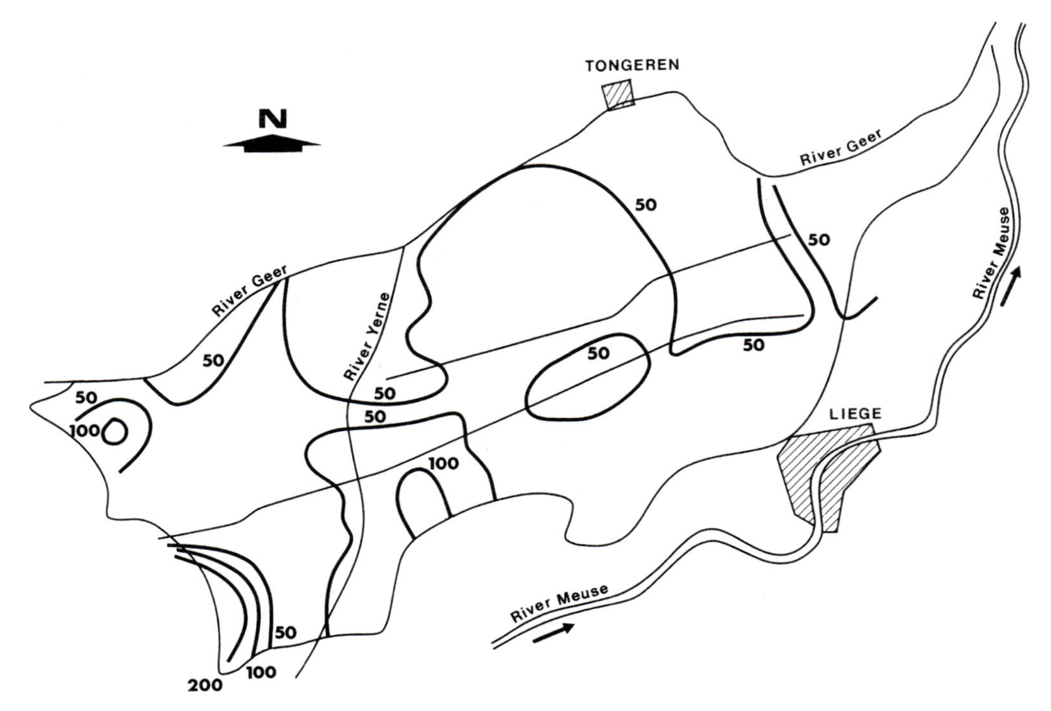

Fig. 8.15. Isoconcentration lines (in milligrams per litre) for the nitrite ion *at the top* of the Hesbaye aquifer in 1984.

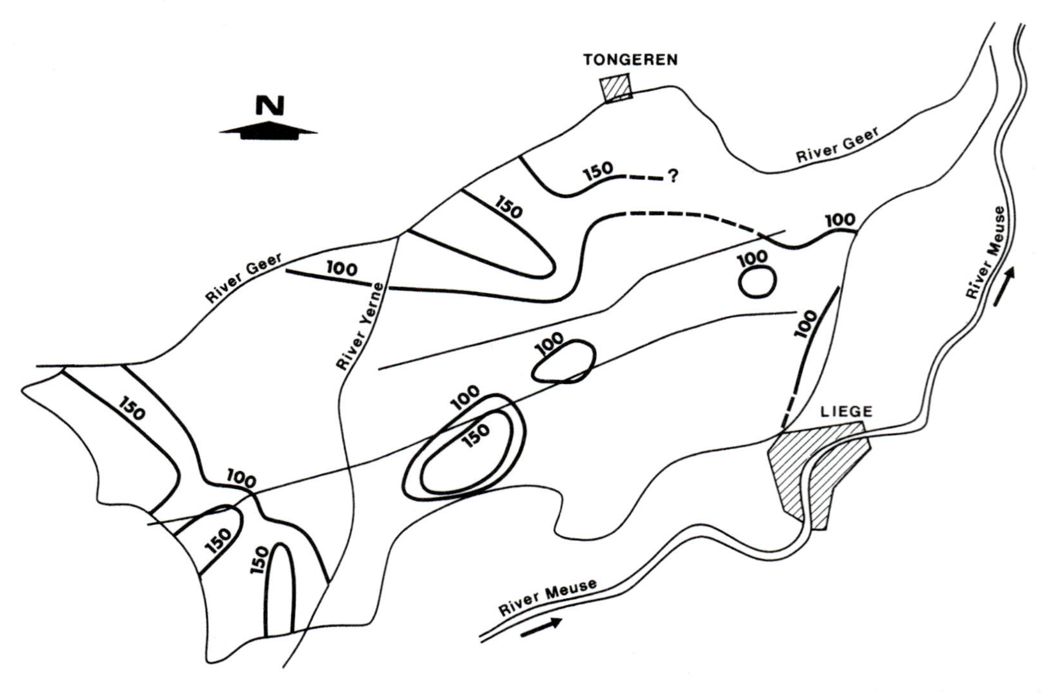

Fig. 8.16. Isoconcentration lines (in milligrams per litre) for the nitrate ion *at the top* of the Hesbaye aquifer in 1984.

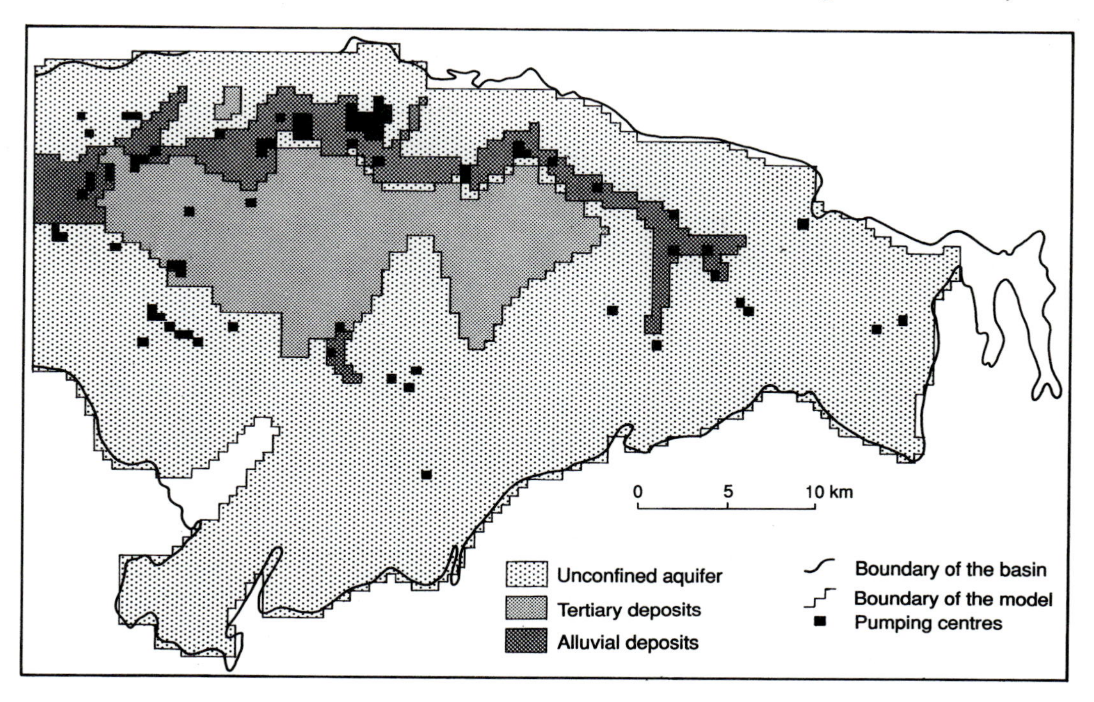

Fig. 8.17. The Chalk aquifer of the Mons Basin showing the factors that affect the aquifer's behaviour (after Rorive 1987).

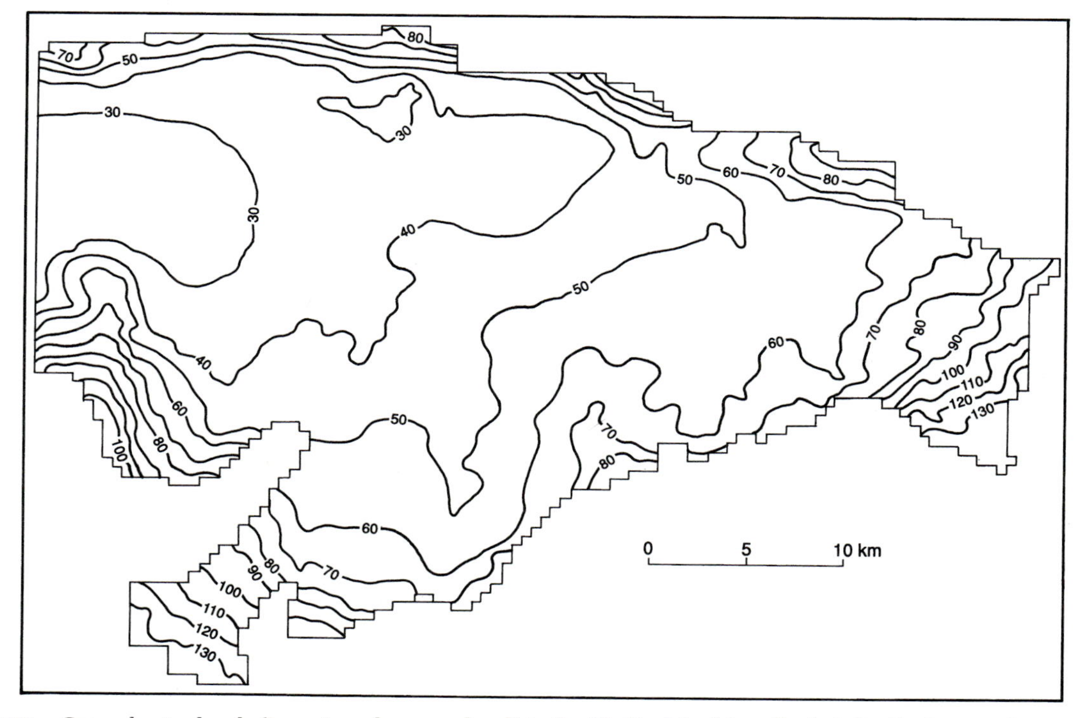

Fig. 8.18. Groundwater levels (in metres above sea level) in the Chalk of the Mons Basin (after Rorive 1987).

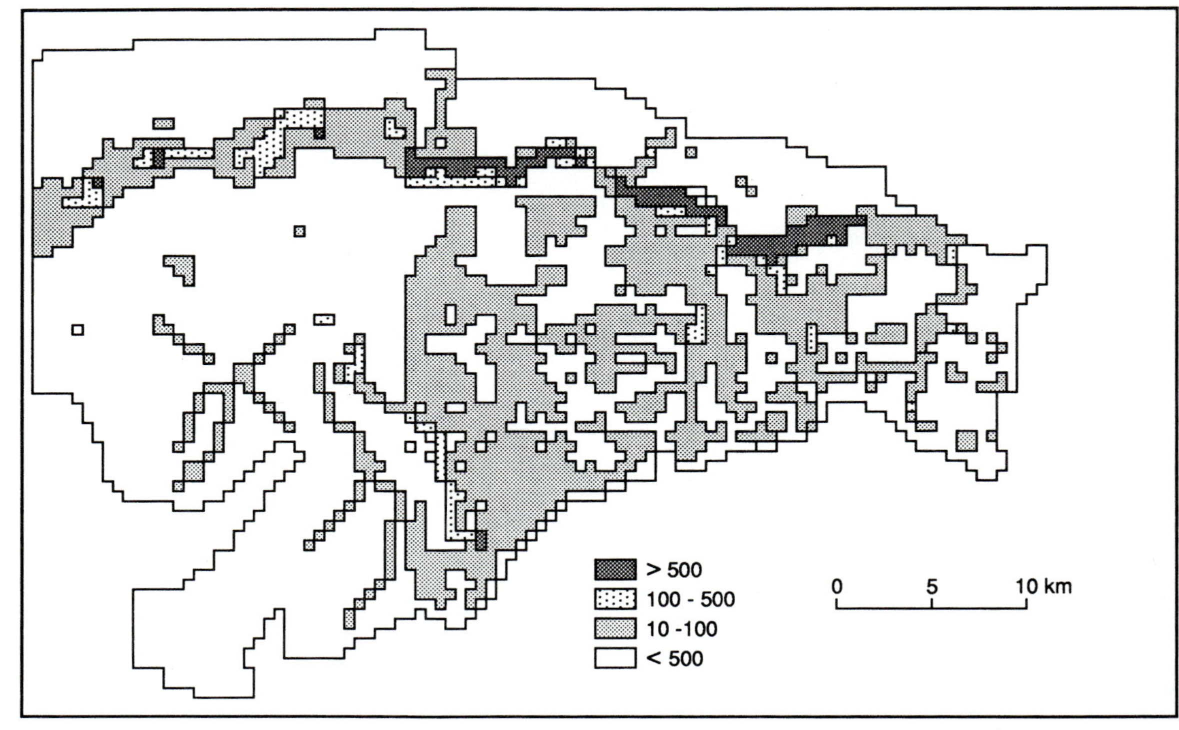

Fig. 8.19. The hydraulic conductivity of the Chalk in the Mons Basin deduced by calibrating a numerical model (after Rorive 1987). Units are $10^{-6}\,ms^{-1}$.

pattern between the different parts of the aquifer (Calembert and Monjoie 1975*b*)

The hydrogeological characteristics of the different lithological formations in the Mons Basin have been determined from about 100 pumping tests, and the mean values (Godfriaux and Rorive 1987) are as follows:

Peat	10^{-6} to $10^{-5}\,m\,s^{-1}$
Alluvial deposits of R. Haine	10^{-6}
Ypresian clay and clayey sand	10^{-9} to 10^{-8}
Landenian glauconitic or quartzitic sands	about 10^{-6}
Very fissured, palaeokarstic chalk	about 10^{-2}
Coarse-grained, fissured chalk	10^{-5} to 10^{-4}
Fine-grained chalk with virtually no fissures	5×10^{-10} to 5×10^{-8}

Because the saturated thickness of the Chalk is very variable, the transmissivity ranges from 6×10^{-4} to $8 \times 10^{-2}\,m^2\,s^{-1}$, while the storage coefficient is between 2.4×10^{-4} and 3×10^{-2}.

Since 1975, many studies of the nature of the Chalk aquifer have been made, and its behaviour has been simulated with a numerical model developed by the Polytechnical Faculty of Mons (Godfriaux and Rorive 1987; Derijcke 1978). The model is based on the finite-difference method and uses the Trepila program of the US Geological Survey. It is a two-dimensional model with a cell dimension representing about 250 m. Transient simulation takes into account unconfined, semi-confined and confined conditions (Fig 8.17). Monthly (and sometimes annual) measurements of water level at 300 locations, together with historical data for about 2000 sites, have been used in the model.

It has been calibrated by trial and error using historical groundwater levels while ensuring that the overall total flow through the aquifer is correctly balanced. Computed and measured water levels agree within 1 m in the unconfined and confined zones and within 2–3 m in the semi-confined zone (Fig. 8.18). The spatial distribution of hydraulic conductivity of the Chalk, as deduced from the model, is shown in Fig. 8.19; the high values along the axis of the basin are very evident as well as the generally higher values in the eastern part of the basin.

The model allows the computation, in each cell, of the recharge, abstraction by pumping, and lateral transfer and leakage between cells. It is a vital tool for the optimum management of the water resource of the Mons Basin and it can take into account the important aspect of land-surface settlement induced by the fall in groundwater levels and the consequent consolidation of the superficial peaty layers.

Confined aquifer in northern Belgium

In the north of Belgium, the Cretaceous, particularly the Maastrichtian, is covered by Tertiary deposits and a large confined aquifer exists, covering an area of 824 km^2 in Flanders, the Campine and part of Limburg. Some $5 \times 10^6 m^3$ of water are pumped annually to supply the Kempen area (Derijcke 1978), with over $9 \times 10^6 m^3$ near Brussels and $12 \times 10^6 m^3$ at Wavre, Jodoigne and Perwez in Brabant. The mean hydraulic conductivity of the Chalk is estimated to be about $10^{-4} m s^{-1}$. The geological sequence is:

- Tertiary—silts, sands and clays, 300–600 m thick, with a marl at least 20 m thick at the base which forms the top of the confined Chalk aquifer.
- Cretaceous—chalks with a thickness of 200–300 m.
- Herve Formation (Lower Campanian)—a marl, 20–50 m thick, that passes laterally into sand and clayey sand of the Aachen Formation.
- Permo-Triassic—occurs locally, mainly in the north-east.
- Carboniferous—shales and sandstones with interbedded coals.

These deposits dip between 1 in 50 and 1 in 10 to the north. They are intersected by many sub-vertical faults orientated NW–SE and NNW–SSE that have created a horst and graben structural pattern (Fig. 8.6). There is also a conjugate NE–SW fault system. Coal has been extensively mined in the past and some mines are still operating in the Kempen area. Because of mining, settlement and subsidence have occurred in the lower part of the Chalk, producing preferential flow channels between the Chalk and underlying Carboniferous aquifer. Intensive pumping is necessary for mine-drainage purposes and as a consequence large volumes of groundwater flow from the Chalk into the Carboniferous. When the coalmines are finally closed there will be a risk that the direction of flow will reverse and acid waters, with high iron and sulphate contents, will flow from the Carboniferous into the Chalk.

Chemical composition of the groundwater

Where the Chalk crops out, the groundwater in the aquifer is of the calcium bicarbonate type and it has a high calcium hardness. Below the Tertiary cover, hard waters persist initially but ion exchange gradually converts the water into a sodium bicarbonate type. In the north of the country, north of Brussels, and in Flanders and around Antwerp, the water is saline (Fig. 8.7).

Steadily rising contamination by nitrate is a feature of the Chalk's outcrop, particularly in the upper part of the aquifer. Water-supply distribution systems are now designed so that the nitrate content can be reduced by mixing groundwater with water from surface reservoirs. Although few data are currently available, pesticides are likely to become an increasing problem in the future. At present one of the main causes of groundwater contamination is leakage of oil and its by-products from literally thousands of storage tanks. This problem is being compounded by more and more accidents involving oil-carrying road tankers.

9. The Netherlands

P. van Rooijen

Geographical distribution of the Chalk

At various times during the Late Cretaceous, chalk was deposited over the Netherlands. A major transgression in Late Cretaceous times almost completely submerged the country and led to the extensive deposition of chalk and marly chalk.

At the end of the Early Cretaceous, a structural high extended from Zandvoort in the north-west towards Krefeld in the south-east (Fig. 9.1). This structure comprised the Zandvoort Ridge, the Maasbommel High, and the Krefeld High. On both sides of the structure, thick sequences of Lower Cretaceous sediments accumulated in depositional basins (Heybroek 1974). A transgression, which had become general by Aptian–Albian times, submerged these structural elements from the north and formed thick sequences of chalk in the northern and central parts of the country during Late Cretaceous times. In the central and northern Netherlands, chalk deposition began during the Cenomanian, but along the margins of the depositional basin, in the south-eastern part of the country, a supply of clastic sediments prevented the formation of chalk during the early part of the Late Cretaceous. Instead, argillaceous sediments and glauconitic sands were deposited during the Santonian and Early Campanian, to be replaced by carbonates only in Late Campanian times.

In the middle of the Late Cretaceous, Sub-Hercynian tectonic activity produced an inversion of the earlier tectonic pattern. Basins on both sides of the Mid-Netherlands Ridge were uplifted and the chalk was completely eroded, while the former ridge retained substantial thicknesses of the older chalk due to a relative subsidence. In the north-eastern part of the country and in the south-west, just south of Rotterdam, the deposition of chalk was more or less uninterrupted by tectonic activity and these areas contain the most complete sequences and maximum thicknesses. By the end of the Late Cretaceous, the Laramide tectonic phase resulted in a minor uplift over most of the area. Tectonic activity at this stage did not produce major deformation, but was responsible for erosion of the uppermost part of the Chalk. It was only in the very south-eastern part of the country, in the province of Limburg, that chalk sedimentation was apparently not affected by Laramide tectonics during the Late Cretaceous, as indicated by a seemingly uninterrupted transition from Late Cretaceous to Early Palaeocene Chalk. In the north-eastern part of the area, where the Laramide tectonic influence was relatively limited, some Palaeocene Chalk was possibly deposited, but removed again by subsequent erosion (Letsch and Sissingh 1983). The occurrence of Palaeocene Chalk is now mainly restricted to a WNW–ESE trending zone in the southern part of the country, concentrated around the Breda–Venlo line and extending from the latter town into Germany and south towards southern Limburg (Fig. 9.1).

The question may arise as to whether or not Upper Cretaceous Chalk was ever deposited in the south-eastern part of the Roer Valley Graben. It is quite possible that the Sub-Hercynian inversion uplifted this part of the former basin before the Late Cretaceous transgression was able to deposit sediments of a chalky facies that far south. This part of the graben now shows a relatively thin chalk sequence of Palaeocene age only. This also applies to the area north and north-east of Venlo, where no Cretaceous sediments were found in boreholes at Arcen and Wachtendonk and where a very thin cover of Palaeocene sandy chalk directly overlies Triassic and Carboniferous sediments, respectively.

The distribution of the Chalk in the Netherlands is shown in Fig. 9.2, and Chalk thicknesses in Fig. 9.3.

Early in Cenozoic times, the Netherlands became the south-eastern extremity of a major depositional basin in which substantial thicknesses of Tertiary sediments were concentrated. This deposition involved a corresponding subsidence of the

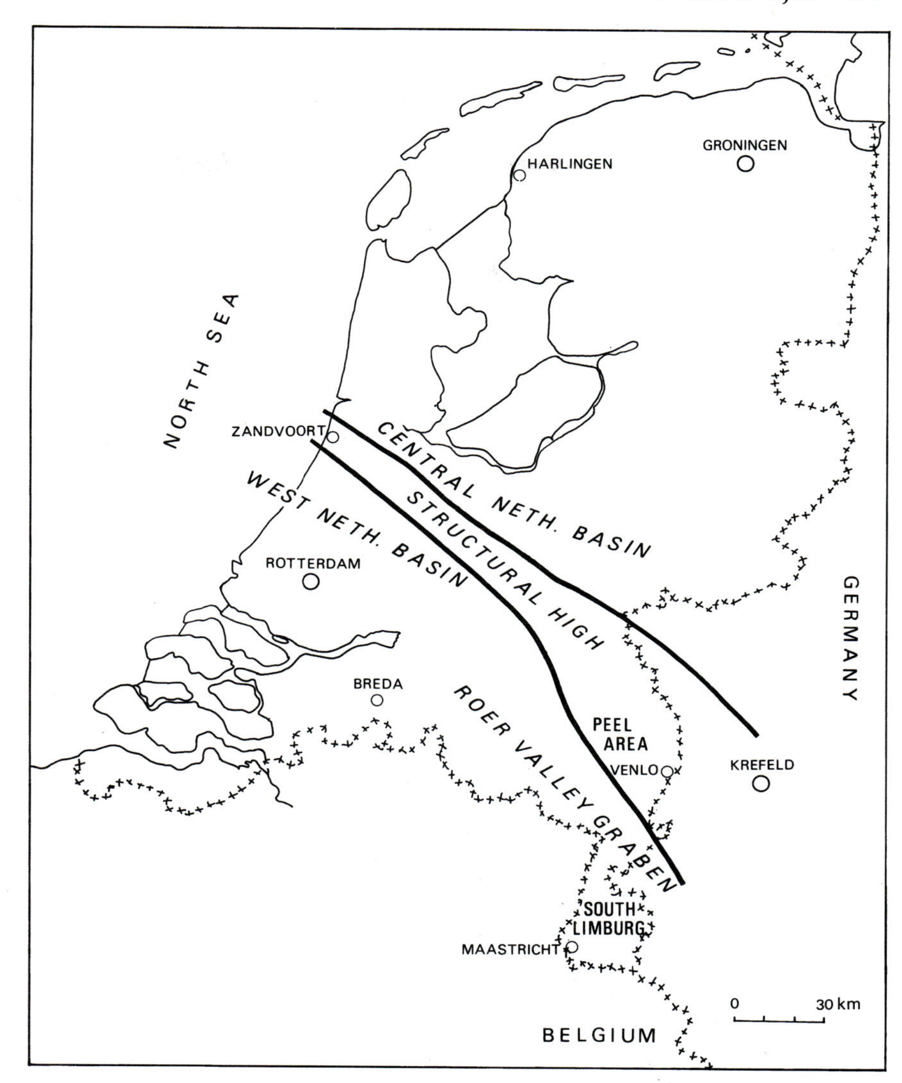

Fig. 9.1. Structural elements in the Netherlands at the end of the Early Cretaceous.

underlying Chalk, locally to a depth of more than 1000 m (Fig. 9.4). In the Roer Valley Graben, where especially rapid subsidence started in the Oligocene, the Chalk ended up at a maximum depth of about 1900 m. Conversely, in southern Limburg, which is tectonically attached to the uplifted Ardenno-Rhenisch Massif, both Upper Cretaceous and Palaeocene Chalks crop out or have only a thin Quaternary cover. It is in the latter area that the Chalk is extensively used as an aquifer.

General stratigraphy and lithology

The lithological nature and stratigraphical position of the Chalk Group has been described by the Nederlandse Aardolie Maatschappij BV and the Rijks Geologische Dienst (NAM and RGD 1980). In the northern and western Netherlands, away from the basin edge, chalk deposition commenced in the Cenomanian with the Texel Chalk (Fig. 9.5), a formation composed of light grey chalks and marly

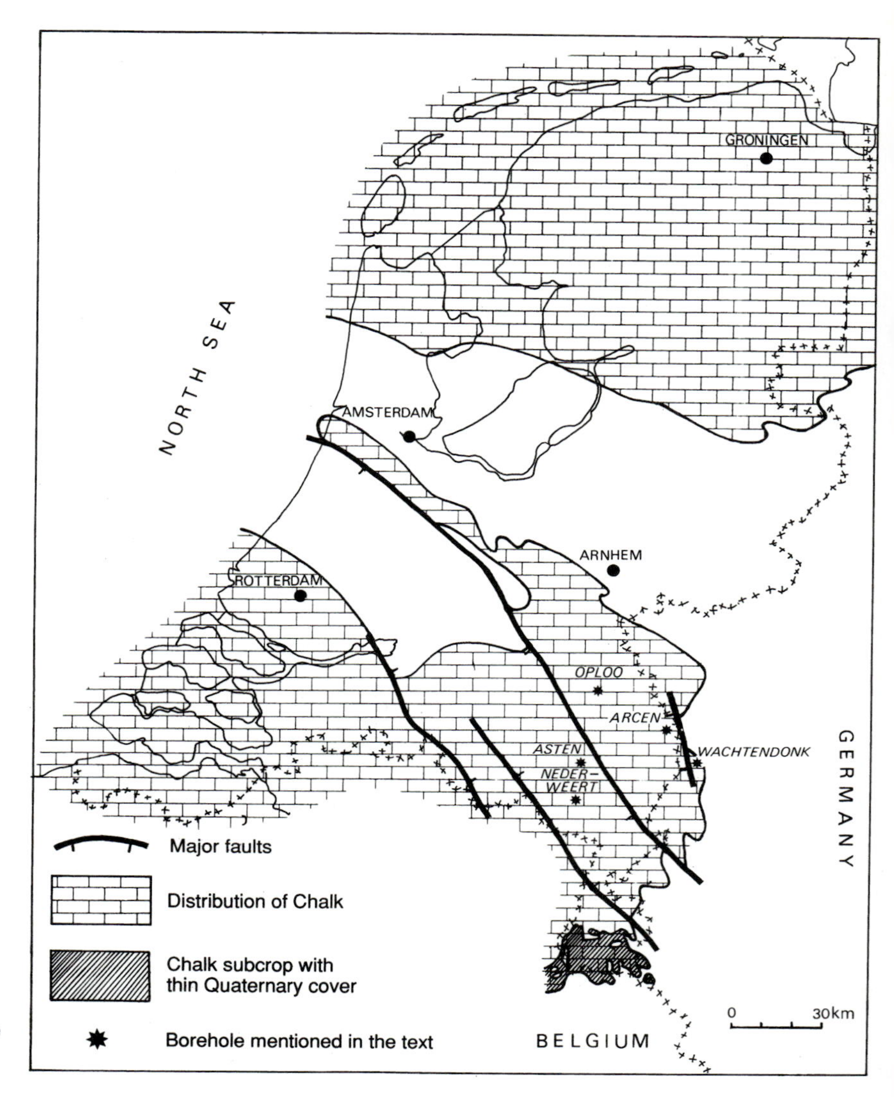

Fig. 9.2. Distribution of the Chalk in the Netherlands.

chalks with some marl intercalations. The top of this formation is marked by a shale bed, the equivalent of the Plenus Marl Formation of the United Kingdom. Where present, the Texel Chalk directly overlies the essentially argillaceous sediments of the Early Cretaceous Rijnland Group or older deposits.

Overlying the Texel Chalk, the Ommelanden Chalk ranges in age from Turonian to Early Palaeocene (Danian–Montian). This formation comprises a sequence of white to light grey chalks and marly chalks. Evidence of a terrigenous input in these northern and western parts of the Netherlands is limited. The Chalk is characterized by a low content of clastic material and only locally do marly chalk intercalations occur in the basal part of the formation. Flint concretions, often neatly arranged in layers parallel to the bedding, typically lace the upper part of the sequence. Depending on the local tectonic history, hiatuses may occur at various places and levels in the sequence. The Ommelanden Chalk is overlain by sandy, marly and clayey sediments of the Palaeogene Lower North Sea Group. The porous chalk rock is mainly composed of

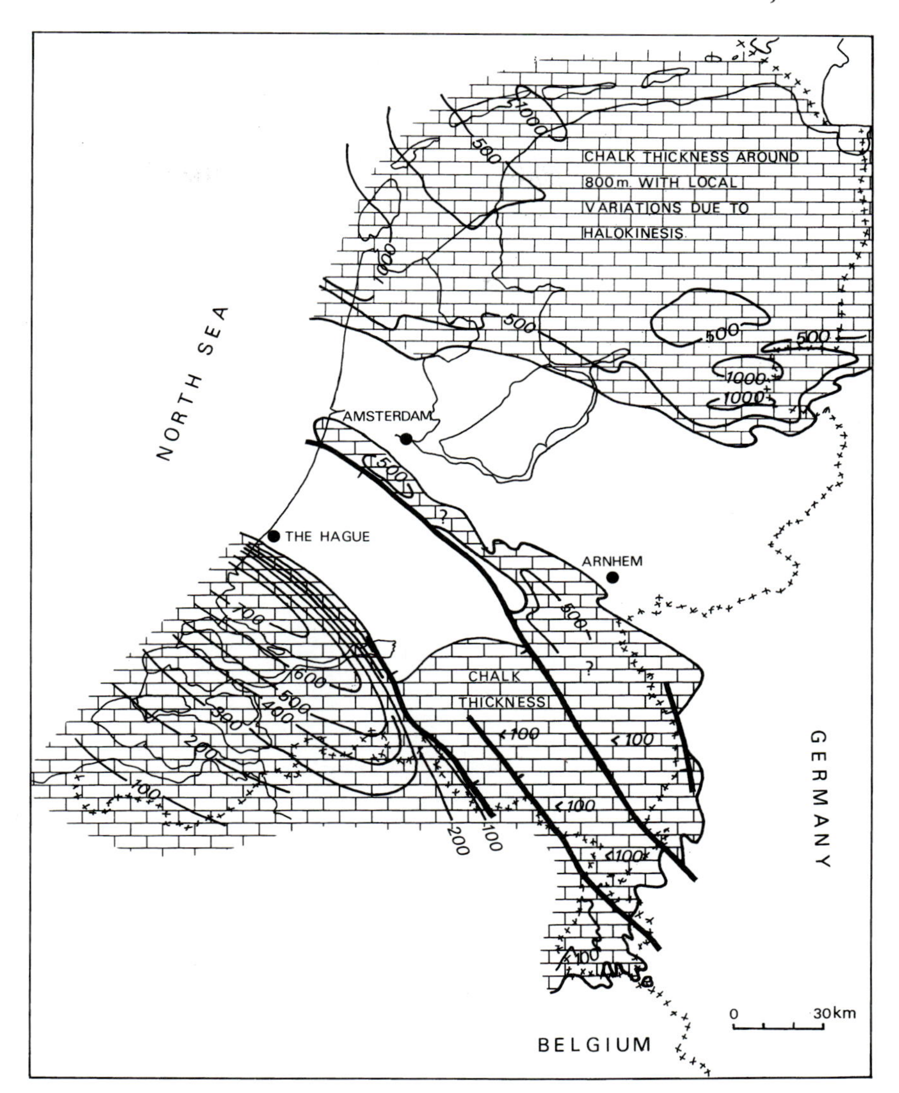

Fig. 9.3. Isopachytes of Chalk in the Netherlands in metres (northern part after Heybroek 1974).

pelagic calcareous fossils, like coccoliths, but also contains, *inter alia*, benthonic foraminifera, bryozoa, sponges, and lamellibranchs.

The south-eastern part of the Netherlands is, like the adjacent areas in Germany and Belgium, situated near the basin edge and shows a stratigraphical sequence quite different from that mentioned above. A detailed subdivision for this area has been described by Felder (1975).

The Late Cretaceous transgression, which brought the Chalk sea into the Netherlands, did not always cover this south-eastern part of the country. It was probably in Santonian times that the first Upper Cretaceous sediments were deposited in this marginal part of the basin. These lagoonal or nearshore sediments have been assigned to the Aken Formation which mainly comprises fine sands, with clayey and loamy intercalations in its basal parts. Early in the Campanian, the shallow marine Vaals Formation followed with fine sandy to loamy, glauconitic sediments. It was only from Late Campanian times onwards that sediments were deposited as a chalky facies. Although mainly subdivided on a chronological basis by means of foraminiferal

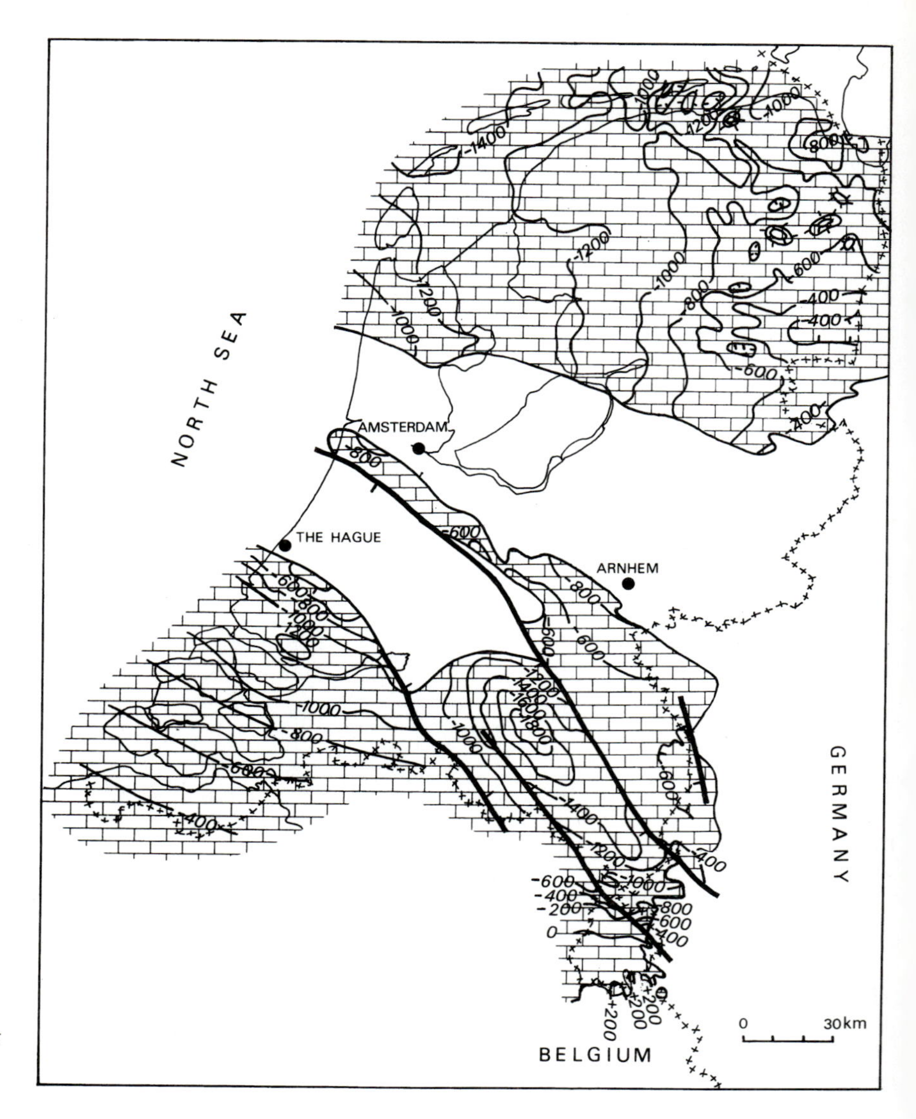

Fig. 9.4. Contours of the top of the Chalk in the Netherlands in metres below sea level.

associations, the Chalk sequence does show some lithological variations. The Gulpen Formation, at the base of the Chalk sequence in this area, is composed of fine-grained limestone with a mudstone texture. It is often built up of white to light grey coccolith ooze, and as such conforms to the description of the typical Chalk in the United Kingdom. The overlying Maastricht Formation is generally a soft, friable, light grey to light yellowish-grey limestone with a grain size grading from fine at the base to coarse at the top. The upper part of this sequence is intersected by a number of hardgrounds and layers or lenses of fossil fragments. Because of the Sub-Hercynian tectonic inversion in the Late Cretaceous, the original Roer Valley Graben and associated tectonic elements were uplifted and became a source of clastic material during the Maastrichtian. This resulted in the formation of the Kunrade facies of the Maastricht Formation, adjacent to these uplifted zones, with an alternation of hard and soft limestone layers. The supply of detrital material from the nearby tectonic highs is demonstrated by the high content of silica, especially in the soft layers of the Kunrade limestone,

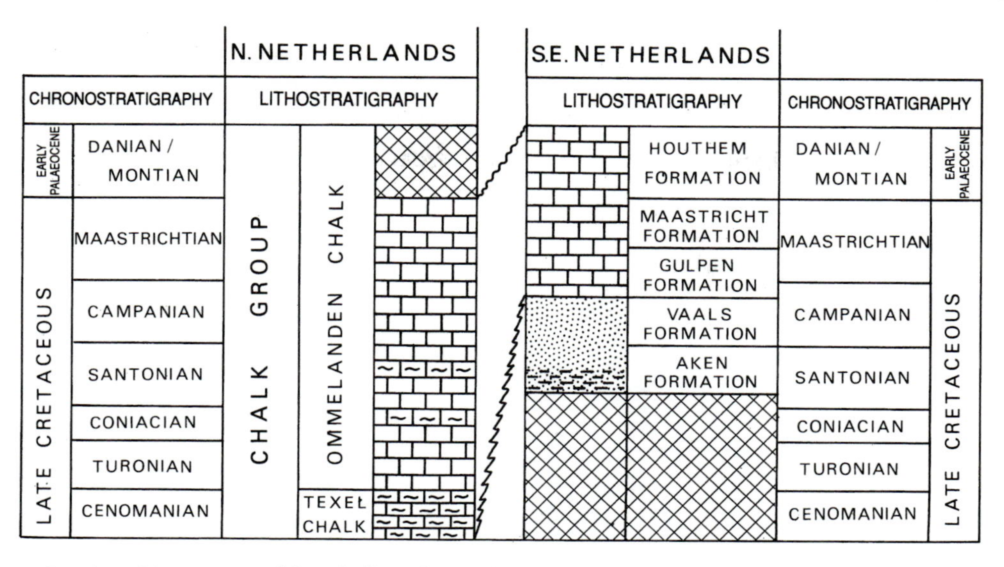

Fig. 9.5. Stratigraphic position of the Chalk in the northern and south-eastern Netherlands.

and by the presence of Carboniferous coal fragments, up to 10 cm in diameter, in this unit. The lithology of the Palaeocene Houthem Formation is very similar to that described for the Maastricht Formation away from the tectonic highs.

The approximate content of clastic material and flint for the different units that crop out in southern Limburg is given in Table 9.1 (after Felder 1983).

Groundwater levels and flow paths

Information on groundwater levels and aquifer properties of the Chalk in the Netherlands is virtually limited to the extreme southern part of the country, where the Chalk occurs at or near the surface. It is only in southern Limburg, that the piezometric head of the groundwater in the Chalk is closely monitored. Figure 9.6 shows the regional pattern of groundwater levels in the Chalk aquifer, slightly modified after the survey by Groundwater-survey TNO for October 1983. The pattern is based on information from more than 100 observation wells and, where appropriate, has been adjusted to the levels of creeks and rivers which are clearly draining the Chalk in the southern part of this hilly area. Hydraulic gradients generally range from 0.003 to 0.040.

The Chalk in southern Limburg dips gently in northerly and north-westerly directions. In the southern part of this area, it constitutes an un-

Table 9.1 Clastic and flint contents in the Chalk units in southern Limburg

	Gulpen Formation		Maastricht Formation				Houthem Formation
	Lower part	Upper part	Lower part	Upper part	Kunrade facies		
					Hard	Soft	
Clastics (wt%)	10–30	2–6	2–7	1–3	<10	20–50	1–4
Flint (vol%)	<1	15–20	5–10	<1	<1	<1	<1

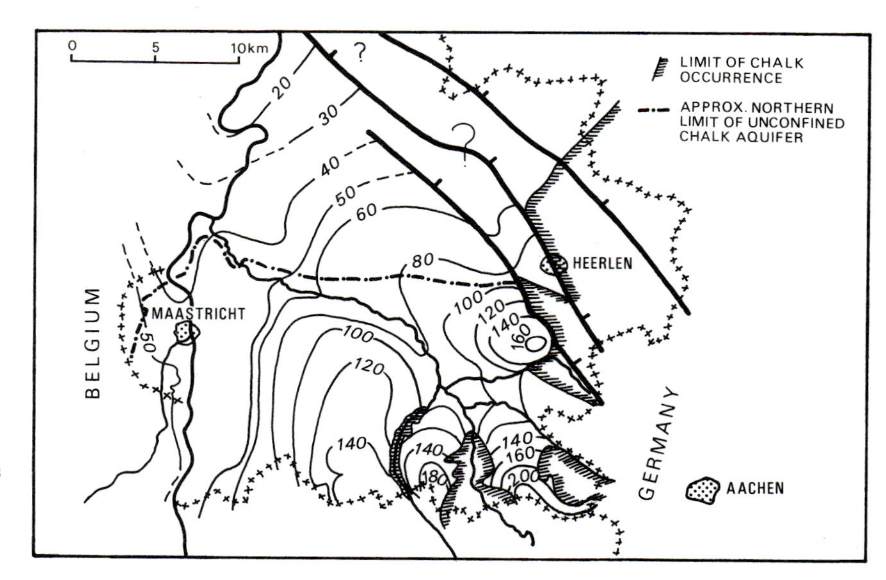

Fig. 9.6. Groundwater levels in the Chalk in southern Limburg in metres above sea level.

confined aquifer, but towards the north a cover of clayey Oligocene and younger sediments make it confined. Where the aquifer is unconfined, water levels vary in depth below the surface from about 2 m in major valleys to more than 60 m on the higher plateaux. In the extreme south, the groundwater in the Chalk is replenished by infiltration. On the other hand, in the confined Chalk aquifer, the groundwater has to be considered to be a mixture of water infiltrating the outcrop and water entering the aquifer by way of overlying or underlying strata. An increased salt content in the aquifer near the River Meuse clearly suggests a component of ascending groundwater from Carboniferous strata in this area north of Maastricht. The transition from fresh to brackish groundwater below the river is encountered near a pumping station some 8 km NNE of Maastricht at a depth of about 30 m below sea level, whereas freshwater appears to be present in the Chalk just 2 km to the east, away from the Meuse Valley, at a depth of 80 m or more below sea level. Other indications of the influence of the Meuse as a drain for deeper groundwater flow systems can be found south of Maastricht, where an old dug well just beside the river yields relatively warm and brackish water, and where some ponds in the Meuse Valley seldom freeze in winter. Carboniferous strata occur here either directly below Quaternary Meuse deposits or are separated from them by a relatively thin sequence of Upper Cretaceous sediments.

An interesting phenomenon in southern Limburg is the existence of different piezometric heads within the Chalk sequence in some localities. Apparently, the fractures and fissures, which provide the bulk of the permeability of the Chalk, do not prevent the existence of local hydraulic regimes. Relatively high fluctuations of water levels are found in some areas, especially below the plateaux, where the groundwater may be as deep as 60 m below the surface; fluctuations of up to 14 m have been recorded over the past years.

Elsewhere in the Netherlands information about groundwater in the Chalk is limited. Although many boreholes have traversed the formation during exploration for oil and gas, drilling methods never allow the observation of hydraulic heads in the Chalk. The only location which yielded a hydraulic head of Chalk water outside southern Limburg is in Oploo, approximately 35 km north-west of Venlo, where a piezometric head of about 6 m above sea level was found in a 1986 exploratory well for balneological applications. The local land surface is 18 m above sea level and the top of the Chalk was found at a depth of almost 500 m. Intriguingly, there has been an apparent decline in hydraulic head since 1912, when, during a mineral survey, the formation water 'blew high into the drilling rig' when the Chalk was penetrated at a site less than 100 m away. A reliable explanation for this phenomenon is not easily provided. No extraction of this highly saline formation water is known, and any association with mining activities in adjacent Germany seems unlikely.

Regional aquifer properties and karst features

Since the Chalk is used as an aquifer only in the southern part of the province of Limburg, it is this area which has provided most of the available information about the hydrogeological character of the formation. Experience with production wells for public water supply, but also some hydrogeological investigations in or near existing chalk quarries in this area, have contributed to the understanding of the Chalk as an aquifer. The aquifer properties described below are thus typical for this area in southern Limburg, and data on the Chalk elsewhere in the country are only incidentally mentioned where available.

Porosity

The Chalk can be regarded as a dual-porosity system with a very high primary, intergranular porosity and a relatively insignificant secondary porosity which can be attributed to fissures and fractures. Porosity values in the Chalk as measured in quarries in southern Limburg generally vary in the range 30–48 per cent. In the hard Chalk bands of the Kunrade facies of the Maastricht Formation, however, porosity values may be as low as 8 per cent, due to a high degree of cementation.

Chalk retains its original porosity relatively well during early diagenesis and, subsequently, upon being covered by substantial thicknesses of younger sediments. Matrix porosities as measured in the Chalk at a depth of more than 1000 m in the Harlingen Gas Field, in the northern Netherlands, lie in the range 24–38 per cent with average values around 30 per cent (van den Bosch 1983). In the Roer Valley Graben, the Chalk found at a depth of nearly 1600 m in the Nederweert exploration well has a porosity of about 20 per cent, whereas values south of Breda may be as high as 35–38 per cent at a depth of 800–900 m.

Porosity values attributed to fissures, fractures and dissolution phenomena are generally very small compared to the matrix porosity and amount to only a few per cent of the rock volume.

Permeability

For an adequate understanding of the permeability of the Chalk, one has to distinguish between primary and secondary permeability. The intergranular pores provide the primary permeability, whereas the secondary permeability is the result of fissuring, fracturing and dissolution phenomena. Unlike the primary porosity, the associated primary (matrix) permeability is in general very low. It is the permeability derived from fracturing and dissolution that makes the Chalk an effective or even a good aquifer. Permeability measurements on cores of undisturbed Chalk in the ENCI quarry, south of Maastricht (Fig. 9.7), yielded values averaging between 1.7×10^{-7} and 6×10^{-7} m s^{-1} (de Wit 1988). The higher values of 3.5×10^{-7} to 6×10^{-7} m s^{-1} were found in grey, original chalk, whereas the lower permeabilities were measured in yellowish, oxidized chalk in which the pore throats, already small after initial diagenesis, may have been further reduced by deposition of iron hydroxides. Local values as low as $1–3 \times 10^{-8}$ m s^{-1} were given for the matrix permeability in this quarry.

If presenting reliable values for the matrix permeability of the Chalk is difficult, quoting sensible figures for the mass permeability as a whole is mere speculation. This is mainly due to the wide variation in the amount of fissuring, fracturing and solution the rock has experienced since its formation.

The role of fissures and fractures in groundwater flow in Chalk is clearly indicated in quarries intersecting the natural water table. Fissures and joints normally occur in systems. In the ENCI quarry, these systems tend to be concentrated around directions N 120°E and N 40°E. The continuity of the sub-vertical joints and fissures is generally low and their frequency and width decrease with depth. Calculations and microscopic studies yield aperture values varying from 0.025 to 0.25 mm. Faults and major fractures generally penetrate relatively deeply into the Chalk formation. Fault zones normally comprise a concentration of smaller cracks and fractures and thus constitute a zone of increased permeability by which the Chalk can be either drained or fed, depending on the prevailing hydrological situation. Apart from these sub-vertical fissures, fractures and faults, the Chalk is also frequently intersected by discontinuities parallel to the bedding planes. Openings along the bedding planes but also nodular beds of fractured flints

Fig. 9.7. The ENCI chalk quarry, south of Maastricht.

(Fig. 9.8) and layers of shell fragments may form preferential flow paths for groundwater and thus have a positive effect on the permeability.

Experimental values for the total permeability of the Chalk in southern Limburg, based on the results of capacity tests and pumping tests, vary from 3×10^{-6} to $2 \times 10^{-4}\,\mathrm{m\,s^{-1}}$. However, these values were essentially derived from wells for public water supply and thus are representative only of the more productive parts of the aquifer.

Measurements in the ENCI quarry, often well below the natural water table, give lower values. De Wit (1988) carried out rising-head tests and calculated permeabilities varying between 6×10^{-7} and $3.5 \times 10^{-6}\,\mathrm{m\,s^{-1}}$. Slug tests, carried out by Waverijn (1990) in the Gulpen Formation, yielded permeab-

Fig. 9.8. Parallel beds of nodular flints in Chalk at the ENCI quarry near Maastricht.

ilities ranging from 1.4×10^{-7} m s^{-1} to more than 2×10^{-5} m s^{-1}. He also used pumping tests in the quarry just below the natural water table and found values of 6.9×10^{-6} and 7.4×10^{-6} m s^{-1} in the same area with different well configurations.

Summarizing the available information on the Chalk permeability in southern Limburg, a clear picture emerges of a relatively high permeability of several metres per day ($2–3 \times 10^{-5}$ m s^{-1}), or locally even several tens of metres per day ($3–4 \times 10^{-4}$ m s^{-1}), in the well-fissured, upper part of the aquifer, rapidly decreasing with depth to values of a few centimetres per day or less in the lower, unfissured parts. An exponential decrease of permeability with depth is generally confirmed by the results of flow logging in Chalk wells (Fig. 9.9).

A phenomenon associated with the occurrence of preferential orientations in the fissure and joint systems is the hydraulic anisotropy of the Chalk. Groundwater flows more easily through fractures than through the unfissured parts of the rock. Consequently, permeability figures may be expected to be highest in directions parallel to the preferred orientation of these sub-vertical discontinuities. Pumping tests, carried out by Waverijn (1990) in the ENCI quarry, confirm the anisotropic character of the rock. He calculated anisotropy ratios for permeability of about 1:3, with maximum values in directions between N 120°E and N 130°E, thus matching the preferred orientation of fissures, joints and fractures. The occurrence of horizontal discontinuities, such as those mentioned previously, may locally reduce the extent to which the preferred orientation of sub-vertical fractures causes anisotropy in the aquifer.

Information on the permeability of the Chalk outside southern Limburg is extremely scarce. During investigations of the hydrocarbon potential of the Harlingen Field in the northern Netherlands, van den Bosch (1983) found permeability values ranging from 8 mD in the fractured upper part of the Chalk, at a depth of just over 1000 m, to 0.7 mD at 20 m and more below the top of this unit. High permeabilities in the upper part of the Chalk are to be expected south of Breda, where a value of 263 mD was measured in a borehole at Meer, just south of the Belgian border, at a depth of about 820 m.

Based on the results of a pumping test in a balneological exploratory boring at Oploo, north-west of Venlo, Glasbergen (1987) calculated a

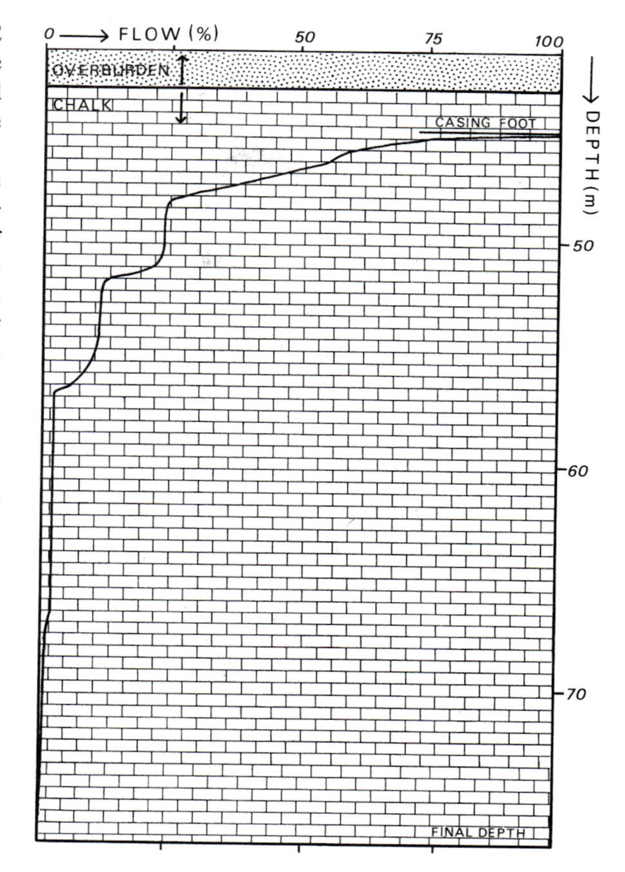

Fig. 9.9. Results of flow logging in a typical Chalk well in southern Limburg.

permeability of 1.56×10^{-4} m s^{-1} for the upper 25 m of the local Chalk sequence belonging to the Gulpen Formation and which is at a depth of 500 m.

Storage

Notwithstanding the very high porosity of the Chalk, the storage coefficient is generally very low. In the unconfined aquifer of southern Limburg, values in the range 0.02–0.05 are thought to be realistic. However, Waverijn (1990) mentions an average storage coefficient of about 0.0015 based on the results of pumping tests of short duration in the ENCI quarry. Pumping tests in the Chalk under semi-confined conditions, close to the limits of the confining clay bed, have yielded coefficients of 0.0005 and 0.0065 (De Wit 1988).

Fig. 9.10. Exposed surface of the Chalk ('karrenveld') at the ENCI quarry, near Maastricht.

Karst features

Since Chalk is generally composed of more than 90 per cent calcium carbonate, it is very sensitive to solution by CO_2-bearing water. Although a mature karst development is not known to exist in the Chalk in the Netherlands, karst features are quite common.

Solution of the Chalk mainly occurs near or above the water table. Below the water table, solution processes are restricted by a drastic reduction of the solution capacity and flow velocity of the groundwater (Dreybrodt 1990). The permeability is mainly contributed by an interconnected network of joints and fractures. Consequently, it is along these discontinuities that solution processes tend to be concentrated. In southern Limburg, solution of the top section of the Chalk often introduced a residual cover of loamy material in which a concentration of flints may occur. Where overlain by Oligocene sediments, the Chalk surface is sometimes characterized by solution along joints and fractures, giving it a rough appearance often called 'karrenveld' (Fig. 9.10). In some quarries, the occurrence of narrow, vertical solution pipes, locally called 'organ pipes' (Fig. 9.11), is apparent in the unsaturated zone. These pipes may be several tens of metres long and often extend to the water table. On plateaux, well above the water table, solution cavities high in the loess-covered Chalk sequence locally formed concentrations of sink-holes or dolines to the extent that one in three boreholes appear to be situated in such a feature. In southern Limburg the largest dolines are known to be almost 100 m in diameter. Major solution caves do not exist in the Chalk in southern Limburg but man-made caves do occur as a result of chalk and flint production, which is known to have occurred from Roman times.

The zone of groundwater-level fluctuation in particular is known to contain many solution features. The hilly area of the unconfined Chalk aquifer includes many dry valleys, beneath which a concentrated network of solution openings along joints, fractures and horizontal discontinuities is believed to provide drainage for all of the percolating rainfall. These zones of enhanced permeability below the dry valleys form preferential flow paths for the groundwater which may thus reach extremely high flow velocities. Evidence of the occurrence of exceptionally high permeabilities underneath dry valleys was also found in a borehole south-east of Maastricht. About one day after a heavy thunderstorm, groundwater could be heard gurgling through the rock at the bottom of the borehole and, with the help of artificial light, could be seen rushing through. It is believed that many dry valleys in southern Limburg are tectonic in origin and developed their high permeabilities by dissolution of chalk along fractures and fissures.

Solution and karst formation must have started immediately after the deposition of the Chalk. In the north-western part of south Limburg, where the Chalk is overlain by clayey sediments of Late Palaeocene and younger ages, karst features have been preserved in the upper part of the sequence, now at a depth of about 140 m below the surface. Boreholes through the Chalk here indicated the

Fig. 9.11. Solution pipes in the Chalk filled with Quaternary overburden at Halembaye, Belgium.

presence of openings in the upper part of the aquifer of a significant size, with apertures of at least 1 cm. Similar solution features along fractures may be interpreted in the northern Limburg Peel area, where increased secondary permeabilities were found in a number of old boreholes in the upper section of the Chalk at a depth of 500–600 m.

Groundwater resources and abstraction: Typical well yields

It is only in southern Limburg that extraction of potable water is possible from the Chalk. Elsewhere in the Netherlands, the deeply buried aquifer may only provide saline groundwater for balneological and geothermal applications and the like.

As an aquifer containing fresh, potable groundwater, the Chalk covers an area of some 400 km^2 in southern Limburg. In the northern half of this area the aquifer is confined by a cover of clayey Tertiary sediments. In the southern part the Chalk forms an unconfined aquifer from which most of the abstraction for public water supply takes place. Formerly, most of the domestic water supply in this area was the responsibility of local community water works, but now almost all the domestic water supply is extracted by the Waterleiding Maatschappij Limburg, which has its head office in Maastricht.

The annual abstraction of groundwater from the Chalk in southern Limburg for domestic water supply amounts to about $30 \times 10^6 m^3$, of which some $20 \times 10^6 m^3$ is derived from the unconfined Chalk. In the Netherlands as a whole, groundwater production for domestic use amounts to more than $400 \times 10^6 m^3 a^{-1}$, whereas surface water contributes some $200 \times 10^6 m^3 a^{-1}$. Drinking water in southern Limburg is almost exclusively extracted from the Chalk, mainly by means of some 90 production wells sited at 15 pumping stations. At two locations, at the foot of a plateau, the groundwater is collected by means of galleries dug into the Chalk. These galleries measure some 2×1.5 m. Production wells almost invariably consist of an open hole in the Chalk with a diameter of 400–500 mm. The overburden of loam and gravel is sealed off by a steel or PVC casing which is firmly cemented into the top of the Chalk.

If it were not for the secondary permeability, caused by jointing and fracturing in the upper part of the Chalk sequence and the associated dissolution phenomena, typical abstraction rates of 50–100 $m^3 h^{-1}$ would not be attainable from these production wells. Because of the hydraulic heterogeneity of the Chalk both yield and drawdown values show very wide variations. The relationship between yield and drawdown for some 120 pro-

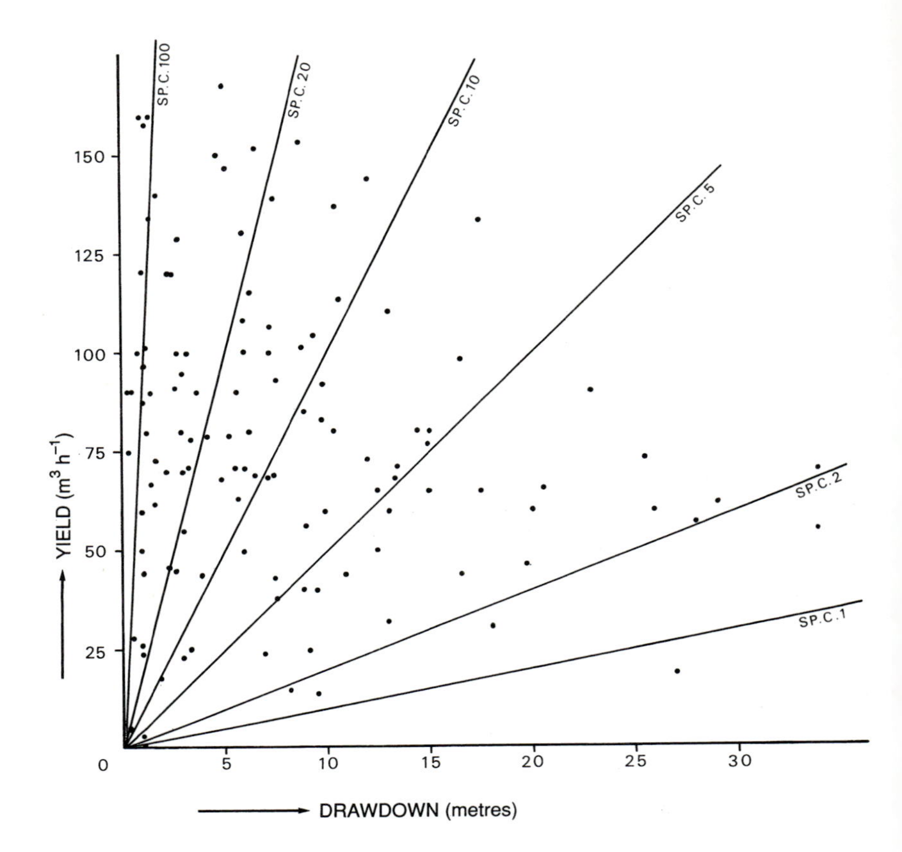

Fig. 9.12. Variation in specific capacity of Chalk wells in southern Limburg.

duction and observation wells in southern Limburg is given in Fig. 9.12. The wide scatter of this population and the substantial differences in specific capacity, defined as the yield in cubic metres per hour per metre drawdown, is apparent. The yield–drawdown relationship is dependent on a number of geological and hydrological conditions, among which the proximity to fracture zones and the piezometric level in the Chalk are well known.

The highest specific capacities were attained in a major fault zone. Productions of more than $150\,m^3 h^{-1}$ with a drawdown of slightly more than 1 m were measured in this fault zone which is at the foot of a Chalk plateau, where the groundwater appears to be dammed against clayey Oligocene sediments on the other side of the fault. At the other extreme, very poor results, with specific capacities near $1\,m^3 h^{-1} m^{-1}$, were encountered at higher locations, away from major fracture zones and with a substantial part of the Chalk sequence unsaturated. In these cases, the groundwater is released by the deeper, and thus poorly fissured and fractured, part of the Chalk, where a secondary permeability is not sufficiently well developed.

However, a specific capacity, once measured, is not necessarily a permanent characteristic for a particular well. An increased abstraction rate does not always show a corresponding increment of drawdown. In fact, Fig. 9.13 shows a disproportionate increase of drawdown with higher abstraction rates for most of the wells indicated. In a number of unconfined wells, this decrease of specific capacity with increasing abstraction rate can be explained by the cessation of flow from the upper production levels in the Chalk as the water table is lowered, and thus by a reduced transmissivity of the Chalk sequence.

The optimum depth of a production well has been a controversial topic in southern Limburg for several years. Traditionally, wells were bored deep

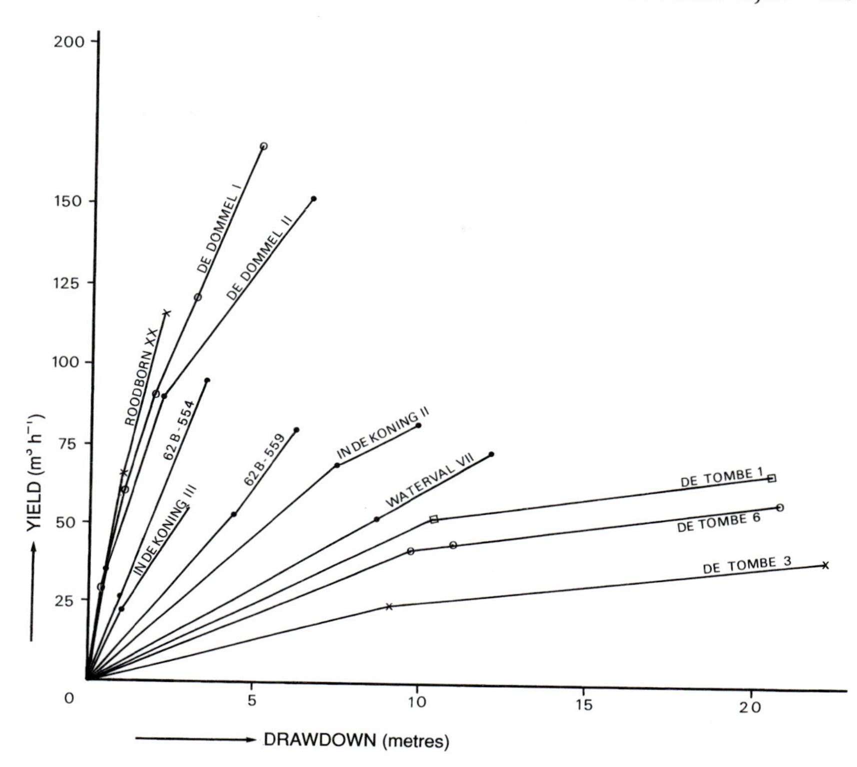

Fig. 9.13. Changes in specific capacity with abstraction rate for some typical Chalk wells in southern Limburg.

into the Chalk to guarantee good production rates. Penetration depths of 70 m into the aquifer were common. The argument that, because of the drastic decrease of permeability with depth, only the upper part of the Chalk sequence should be considered productive, was only hesitatingly accepted by the water authorities. But the concept of a heterogeneous, anisotropic aquifer, unlike a sand or a gravel, with a dual porosity and permeability, is winning ground. The results of flow-logging that are now available for deeper Chalk wells (see Fig. 9.9), indicate that one may safely assume that, in general, the upper 10–20 m of the Chalk sequence will yield the bulk of the groundwater. Only exceptionally, for instance in major fracture and fault zones, may additional water production be expected from the deeper levels of the Chalk.

The use of acid treatment in wells with a disappointing yield, so common in the United Kingdom, is not yet routinely practised in the Netherlands. An enhancement of the production from new wells and a successful restoration of wells with declining yields might, nevertheless, be expected by applying this technique.

The hilly country of southern Limburg experiences an annual precipitation of about 800 mm. Direct infiltration to the Chalk aquifer is only possible in the southern 240 km^2 of this area, where the clayey cover of Oligocene sediments is not present. In the northern part of the area, recharge of the aquifer through the clayey cover is considered to be negligible and from the Belgian area in the south no inflow of groundwater through the Chalk may be expected (see Fig. 9.6). With an estimated infiltration rate of 35 per cent of the precipitation, the annual recharge of the aquifer would amount to some 65×10^6 m^3. In order to avoid a steady decline of the piezometric head and damage to surface runoff and valuable ecological features only about half of this may in practice be considered as a resource for groundwater production. Together with local abstractions for industrial and agricultural

Table 9.2 Chemical composition of Chalk groundwater in southern Limburg

EC ($\mu S\ cm^{-1}$)	600–700	Na ($mg\ l^{-1}$)	6–10
CO_2 ($mg\ l^{-1}$)	40–50	K ($mg\ l^{-1}$)	1.5–3
HCO_3 ($mg\ l^{-1}$)	300–400	Ca ($mg\ l^{-1}$)	120–130
Cl ($mg\ l^{-1}$)	15–40	Mg ($mg\ l^{-1}$)	8–12
SO_4 ($mg\ l^{-1}$)	30–50	Total hardness (mmol)	3.1–3.7

purposes, the total use of the groundwater in the Chalk thus appears to be at its limit. Accordingly, the Provincial Water Authority recently put a limit on further expansion of abstraction, only allowing some additional use of the aquifer in areas with resources not yet fully developed.

Hydrochemistry

During the deposition of the Chalk, seawater was trapped in the formation. Erosion in Eocene times, and later from the Pliocene onwards, brought the Chalk to the surface again in southern Limburg and the saline formation water was gradually replaced by freshwater derived from rainfall.

The groundwater in the Chalk in southern Limburg is invariably of the calcium bicarbonate type. Because of chalk dissolution by CO_2-bearing infiltration water, both Ca and HCO_3 values are high compared with concentrations of other ions. Table 9.2 gives current averages for the major constituents of Chalk groundwater, extracted for domestic water supply in southern Limburg.

The NO_3 content varies considerably. In the well protected, confined part of the aquifer, values do not normally exceed $0.5\ mg\ l^{-1}$, but in the unconfined part values already exceed in some places the EC 'guideline' standard of $25\ mg\ l^{-1}$. The iron content is generally very low, in the range 0.03–$0.06\ mg\ l^{-1}$, but tends to increase towards the north-east, where values up to $1\ mg\ l^{-1}$ have been measured in confined Kunrade Chalk.

Analyses of groundwater in the Chalk in the northern part of southern Limburg, where the aquifer is covered by several hundred metres of Tertiary overburden, show a slight increase of the total dissolved solid values and a relative enrichment of Na and K contents as compared to Ca values. Apparently the process of ion exchange is active at these depths (Kimpe 1960).

Groundwater in the deeply buried Chalk outside southern Limburg is invariably saline and unsuitable for public water supply. Although fresh groundwater with chloride concentrations of less than $150\ mg\ l^{-1}$ may occur to depths of 500 m and more in Neogene sediments in the south-eastern part of the Roer Valley Graben, the water in the underlying Chalk is saline. The transition from fresh to saline water in the subsurface of the uplifted Peel area, north of the Roer Valley Graben, lies at a few hundred metres, again in Neogene sediments, but here also the Chalk water is highly saline. Observations in old boreholes in the Peel area show abrupt increases in chloride and sodium contents in the groundwater at a depth of 500–600 m, generally not far above the top of the Carboniferous strata. Above this level, present chloride values amount to about $2000\ mg\ l^{-1}$, but below this depth values rise rapidly to equal and exceed seawater composition (Kimpe 1963).

North of the Peel area, in the Oploo boring northwest of Venlo, a high chloride content of $12\,300\ mgl^{-1}$ was measured in the groundwater of the Chalk of the Gulpen Formation, at a depth of 500 m. The Houthem Formation Chalk in the deeper part of the Roer Valley Graben near Asten, 30 km west of Venlo, was found to contain groundwater with $27\,600\ mg\ l^{-1}$ of chloride at a depth of about 1600 m. Analyses of the Chalk groundwater found in boreholes at Oploo and Asten are given in Table 9.3.

The analyses of additional parameters such as carbon-14, oxygen-18 and deuterium in the Oploo groundwater showed it is 35 000–40 000 years old and mainly derived from infiltrated precipitation (Glasbergen 1987). Both in Asten and Oploo, the composition of the Chalk groundwater is believed to be influenced by water rising from greater depths.

Extremely saline groundwater may be expected in the Chalk in the vicinity of diapiric structures in rock salt in the most north-easterly part of the Netherlands.

The Chalk aquifer in southern Limburg in general yields good-quality drinking water. However, the substantial increase in the nitrate content, due to a liberal use of both fertilizers and manure, may constitute a threat to the use of the unconfined

Table 9.3 Chemical composition of Chalk groundwaters in boreholes at Oploo and Asten. Units are mg l^{-1} except where indicated

Parameter	Oploo Depth 500 m (Glasbergen 1987)	Asten Depth 1600 m (Heederik *et al.* 1989)
EC (μS cm^{-1}, T = 20°C)	33000	44700
Cl	12300	27600
HCO_3	510	273
SO_4		8.4
Na	7070	15950
K	153	191
Ca	550	1665
Mg	270	616
Fe	0.5	7
I	2.1	13
Br	45.5	105
F	1.15	1.8

aquifer for public water supply in some areas. In a small area, at the southern margin of the confined part of the aquifer, sulphate pollution of the groundwater occurs. Sulphate contents as high as 300 mg l^{-1} have recently been measured and are thought to be attributable to mobilization of sulphate ions as a result of groundwater-level fluctuations in the pyrite-bearing Oligocene overburden.

Protection zones for domestic water supply are designed either on the basis of a calculation of the travel time of the groundwater (for example, 60 days, 10 years, 25 years) or by means of a general rule of thumb (0.5 km downstream of the wells, 1 km laterally and 2 km upstream). For both methods, however, the Chalk is inevitably considered as a homogeneous, isotropic medium. The concept of the Chalk as a dual-porosity medium in which the bulk of the permeability is provided by fractures and joints, is bound to have its effect on the way protection zones are delineated. The possibility of rapid penetration and propagation of polluting agents into the aquifer along these discontinuities will often, especially in the case of the unconfined Chalk aquifer, necessitate protection zones elongated in known directions of preferred groundwater flow. The present lack of detailed knowledge of these directions of preferred flow means that the water authorities are forced to take a conservative approach in designing protection zones.

Acknowledgements

This chapter is published with the permission of the Geological Survey of the Netherlands (Rijks Geologische Dienst). The author is also indebted to the Eerste Nederlandse Cement Industrie (ENCI) for permission to publish the results of the geological and hydrogeological investigations carried out in their chalk quarry near Maastricht. Furthermore, the author wishes to thank the Director of Complan BV in Eindhoven and the Director of the Institute for Groundwater and Geo-Energy (IGG-TNO, formerly Groundwatersurvey TNO) in Delft for permission to use the exploration data from the boreholes at Oploo and Asten II. The results of recent investigations in production wells by the Waterleiding Maatschappij Limburg have also contributed significantly to the understanding of the hydrological behaviour of the Chalk. Finally, the author gratefully acknowledges the valuable comments on this chapter by Mr G. Remmelts of the Geological Survey of the Netherlands and Mrs M.H.A. Juhász-Holterman of the Waterleiding Maatschappij Limburg.

10. Denmark

E. Nygaard

Geographical distribution of the Chalk

A thick Chalk sequence is present beneath most of Denmark (Fig. 10.1). It comprises Upper Cretaceous and Palaeocene limestones, composed of coccolitic micrites (chalk proper), bryozoan limestones, calcarenitic greensands and marls. The Upper Palaeocene greensands and marls will be discussed only briefly.

Chalk is assumed to have been originally deposited throughout the Danish area. However, over the Scandinavian high, north-east of a major tectonic feature, the Tornquist Line (Fig. 10.2), it has been eroded and is preserved only locally where it has been downfaulted. Such occurrences are found on the Island of Bornholm.

Elsewhere in Denmark and its continental shelf the depth and thickness of the Chalk are influenced by the regional structural pattern, with basement highs separated by basins, such as the Danish Basin and the North Sea Central Trough (Figs. 10.1 and 10.2). In these depositional centres the Chalk may be locally more than 2000 m thick. The thickness gradually decreases away from these centres, and over the broad Ringkøbing–Fyn High (RFH) it is of the order of 500 m (Fig. 10.1). The Chalk succession is thinner than this only locally, for example over some salt diapirs.

General structure and stratigraphy

Knowledge of the Chalk succession (that is,the Chalk Group in the North Sea terminology of Deegan and Scull 1977) is derived from outcrops, wells, and seismic reflection data. Information from shallow wells and outcrops is particularly abundant in that part of the Danish land area, comprising roughly one-third of the country, or 15 000 km^2, where groundwater from the Chalk aquifer is used for water supply (Fig. 10.2).

Outside this area, onshore information is derived mainly from deep wells drilled in the search for hydrocarbons in formations below the Chalk. Relatively few hydrocarbon exploration wells have been drilled in the continental shelf as a whole. Except for construction sites, the only offshore area with a high density of drilling is the Danish sector of the Central Trough in the North Sea. The density of seismic reflection coverage coincides with that of the deep wells. Thus the greatest density of data available about the Chalk is for the North Sea Central Trough and those onshore areas where the uppermost part of the Chalk is used for water supply.

The area of most importance for the hydrogeology of chalk in Denmark is, therefore, an area extending east–west in the northern part of Jutland, and continuing south-eastwards into the peninsula of Djursland. In addition, the larger islands in the Danish archipelago (Møn, Falster, the northern part of Lolland, the eastern part of Fyn, and northern, southern and eastern Zealand) are also of major importance (Fig. 10.2). The areas developed for groundwater thus reflect the depth to the Chalk and occur around the Danish Basin (Fig. 10.3).

In what follows the emphasis will be placed on the regions where chalk is important as a source of potable water. Where the Chalk is directly overlain by the mainly glacial Quaternary deposits, its surface has been eroded to reveal horizons ranging in age from Santonian to Danian, but, with the exceptions of a local occurrence of Campanian chalk on the Island of Møn and of Lower Coniacian limestone on the eastern island of Bornholm (Tröger and Christensen 1991), beds older than the Maastrichtian are not exposed.

Within the Danish Basin, the oldest beds subcropping on the pre-Quaternary surface are generally located in the northern and eastern extremities of the basin. This generalization depends to some extent on the depth of the erosion surface, whereby isolated hills and cliffs of Danian lime-

Fig. 10.1. Isopachytes of the Upper Cretaceous–Danian Chalk Group within the Danish land and offshore area, and also the area of Danish sovereignty extending from the Bornholm region in the east to the Central Trough in the west. Modified from C. Andersen, personal communication; and Japsen and Langtofte, 1991.

stone have been selectively preserved (Fig. 10.4). On the Island of Møn, huge glacially dislocated blocks of Maastrichtian chalk and till form a hilly topography as well as a prominent sea cliff.

The stratigraphy of the Danish Chalk is well displayed at the type locality of the Danian stage at Stevns (Fig. 10.5). Here soft and friable chalk of Late Maastrichtian age is, through a thin succession of transition beds, overlain by a relatively hard limestone of Danian age. This limestone consists of banks of wacke and packstone and is termed 'bryozoan limestone' after its dominant constituent. The interrelationship of the transition beds is complicated, and they include a thin, dark clay bed, the so-called 'Fish Clay', which constitutes the basal part of the Danian succession. The lithological change from a coccolitic micrite to a coarser and harder limestone is characteristic of the chalk succession in a transitional setting between the coastal areas in southern Sweden (Surlyk and Christensen 1974) and the basinal deposits in the central North Sea. Westward, the exposed beds of Danian age become gradually more micritic and, in other parts of Zealand, beds of wackestone and micrite may alternate. In western Jutland the lowermost beds of Danian age consist of coccolitic micrite, which grades upwards into bryozoan limestone. Further west in the North Sea, all the beds of Danian age consist of more or less lithified and marly chalk. In a similar but much less pronounced way, the Cretaceous Chalk becomes richer in micrite away from the structural highs and towards the centre of the North Sea. Locally in northern Zealand, calcarenites dominate the Danian beds, while at Faxe a reef complex is centred around banks of coralline boundstone.

In the areas important for water supply, all the dominant rock types in the Chalk sequence occur and the bedded nature of the rock is emphasized by nodular and platey chert layers, occasional contemporaneous hardgrounds, and marly bands, which on average increase in frequency towards the top of the succession. The visibility of the bedding is further enhanced by open fractures, at 50–100 cm intervals, in the uppermost tens of metres of the sequence. In the eastern coastal part of Zealand, in the municipality of Frederiksberg, within the Copenhagen region, and in southern Stevns, one or two highly

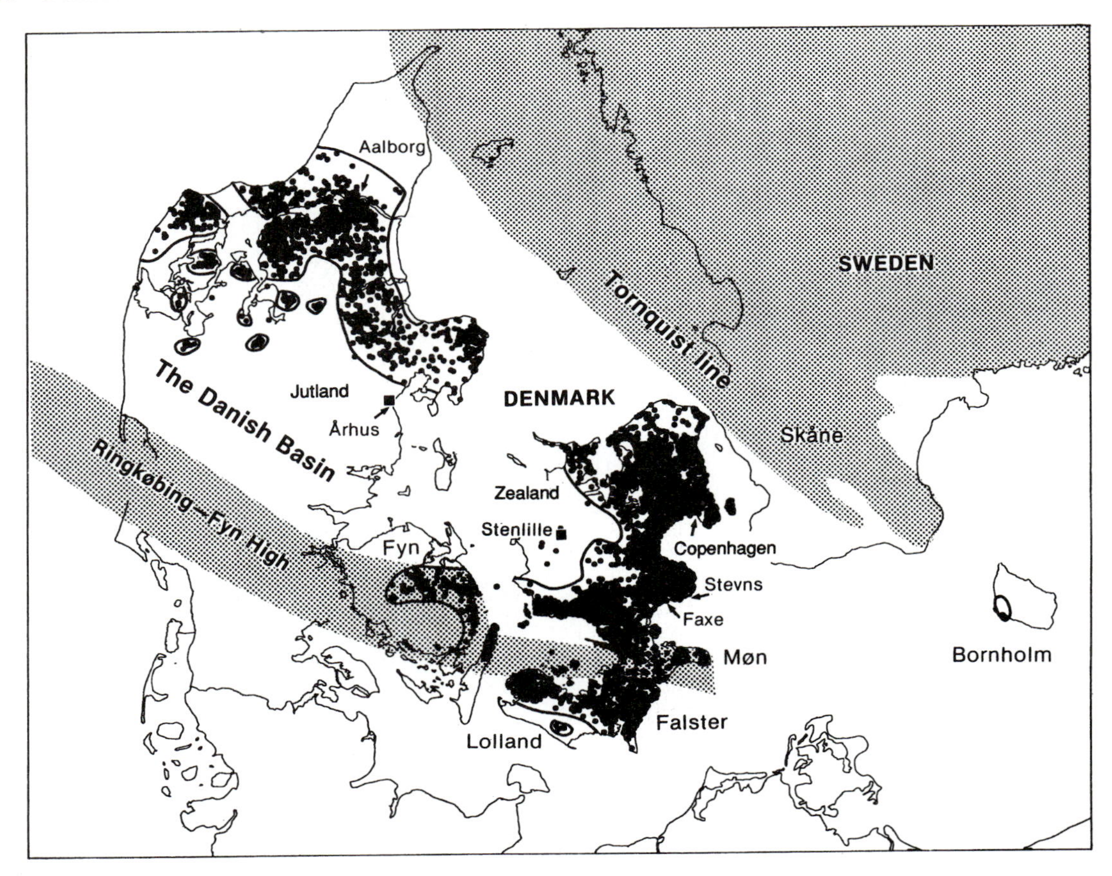

Fig. 10.2. The distribution of the Chalk as a freshwater aquifer highlighted by the locations (dots) of water wells producing from chalk. The main structural elements and geographical names used in the text are also indicated. Cities referred to are indicated by arrows. Partly based on Bækgaard and Knudsen (1979).

permeable zones have been found at depths of between 30 and 70 m below the ground surface. These horizons are assumed to consist either of an unusual lithology or open horizontal fractures arising from glacial compression and subsequent unloading upon melting of the ice. It is from such horizontal fractures and the top surfaces of the other bedding features that groundwater frequently seeps at outcrop.

The flint beds and sparse depositional hardgrounds have been used for local correlation of the Maastrichtian Chalk (Surlyk 1971). In recent years the stratigraphical correlation, especially of the deeply buried chalk, has been greatly improved by access to wireline logs (see, for example, Sorgenfrei and Buch 1964; Nielsen and Japsen 1992). Wireline logs from shallow wells are so far available only for a few local investigations. The Chalk succession in northern Jutland has recently been mapped using seismic data (Japsen and Langtofte 1991).

Throughout Denmark selected wells and field exposures have been studied to produce logs of a number of elements and isotopes (Jørgensen 1975; 1986) which supplement and largely confirm the wireline log correlations.

The time-stratigraphical correlation studies in Denmark have, in many instances, been concerned with beds of either Late Cretaceous or Danian age. This is partly because of better access to the Danian beds and partly due to the faunal variation in the different lithological facies in the beds from the two periods.

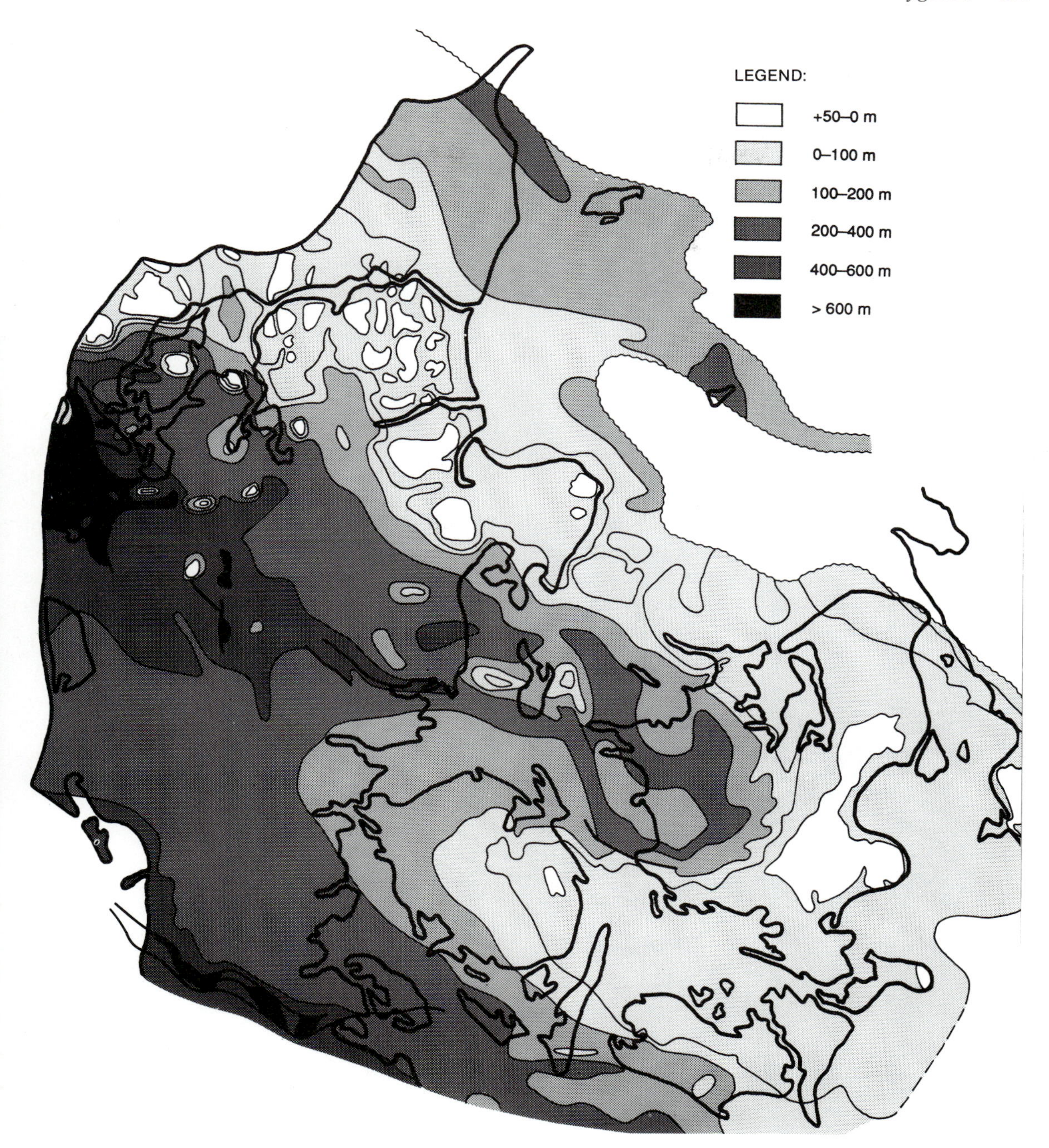

Fig. 10.3. Contours on the top of the Chalk in Denmark, excluding the Bornholm area. Regions with depths near to zero indicate the limit of the main post-Cretaceous area of subsidence around the Danish basin. The contour for 75 m below sea level is also shown. Modified from Ter-Borch (1991).

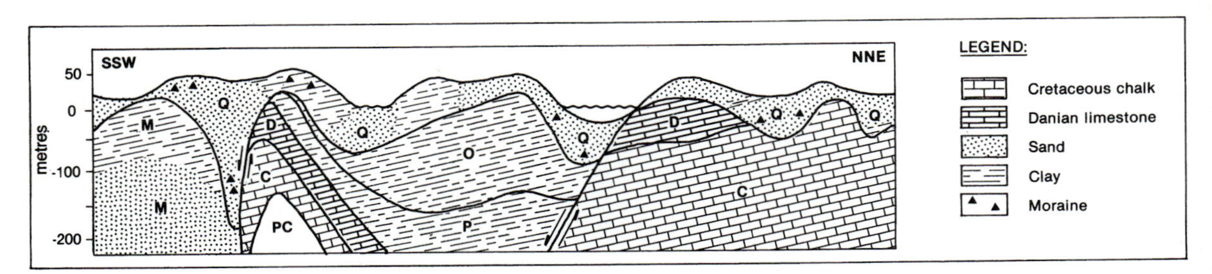

Fig. 10.4. Section illustrating the Chalk subcropping against Quaternary beds. The section is an SSW–NNE profile through northern Jutland. Glacial erosion left hills capped by relatively hard bryozoan limestone in the northern part of the region. Q, Quaternary; M, Miocene; O, Oligocene; P, Upper Paleocene; D, Danian; C, Cretaceous; PC, pre-Cretaceous, mainly Zechstein salt. In many locations in the region Eocene beds of clay and volcanic ash are also present. Faulting is indicated along the salt piercement dome and the regional Fjerritslev fault zone, which is a collateral of the Tornquist Line.

Groundwater levels and flowpaths

In Denmark, except on the Island of Bornholm, the Chalk represents the oldest aquifer containing freshwater. The salinity of pore-water in the Chalk increases with depth in the lower part of the aquifer and, in coastal zones, decreases with distance from the shoreline. In general, abstraction does not take place from chalk that is deeper than about 150 m below mean sea level.

The geography of Denmark means that none of the Chalk areas is very far from the coast. It follows that the groundwater flow systems, even those discharging to the sea, are only local in scale.

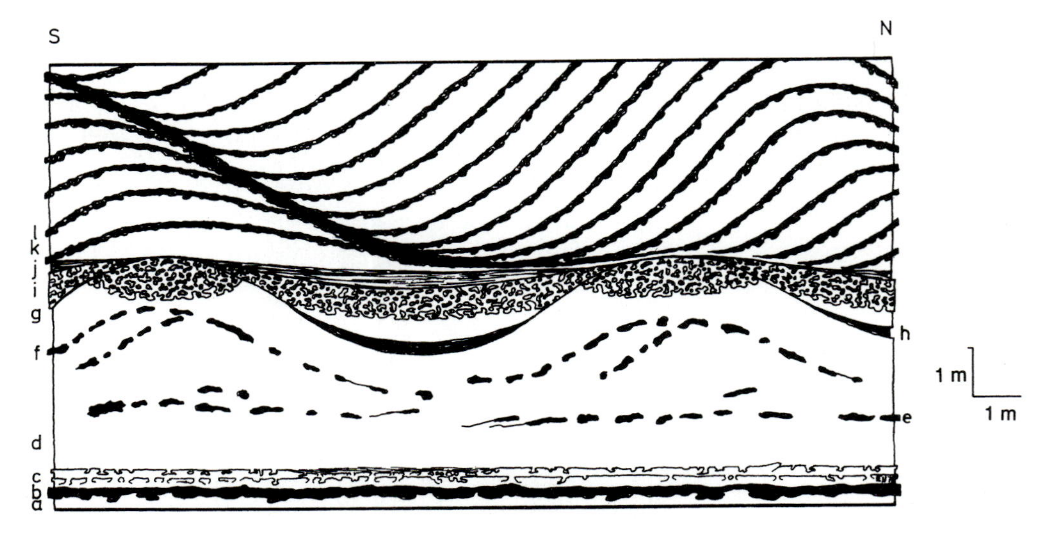

Fig. 10.5. Type locality of the Danian Stage at Stevns Cliff. The lithology in the 50 m high exposure is illustrated with a five times vertical exaggeration. a, Cretaceous Chalk; d, from this level upwards the Cretaceous Chalk is of a greyish variety; b, e, f, and l, flint in bands and nodules; c, two incipient hardgrounds; g, the mound-shaped upper surface of the Cretaceous Chalk; h, the lowermost Danian bed, termed 'Fishclay', only present in depressions in the surface of the Cretaceous beds, grades upwards into an arenitic limestone (named after the gastropod Cerithium), which created a level topography; i, successive erosion of both the Cerithium Limestone and the protruding Cretaceous mounds was accompanied by the development of a thick hardground; j, selective compaction in areas underlain by 'Fishclay' produced shallow basins within which calcarenitic material agglomerated; k, Bryozoan Limestone developed as banks, which overstep in a southerly direction.

Groundwater contours (Fig. 10.6) reflect this general pattern (Kelstrup *et al.* 1982). Locally the contour pattern is known to be influenced by cross-formation leakage, but this has not been distinguished from the topographic influence on a regional scale.

Major pumping stations supplying drinking water to the larger towns, and pumping from construction sites or major excavations, have an obvious influence on the water table, but, in the Chalk area in general, the effects of abstraction cannot be distinguished from climatically-induced fluctuations in the water table.

Regional aquifer properties and karst features

Measurements of the permeability of the matrix of the Danish Chalk have been infrequent and only a few have been made in connection with water-supply investigations. In the early 1980s the hydraulic properties of the 500 m thick chalk section overlying the Mors salt dome were investigated in detail. This was to evaluate the sealing capacity of the Chalk in connection with a study of a proposal to store radioactive waste in the underlying salt. The investigation showed that the permeable zone was

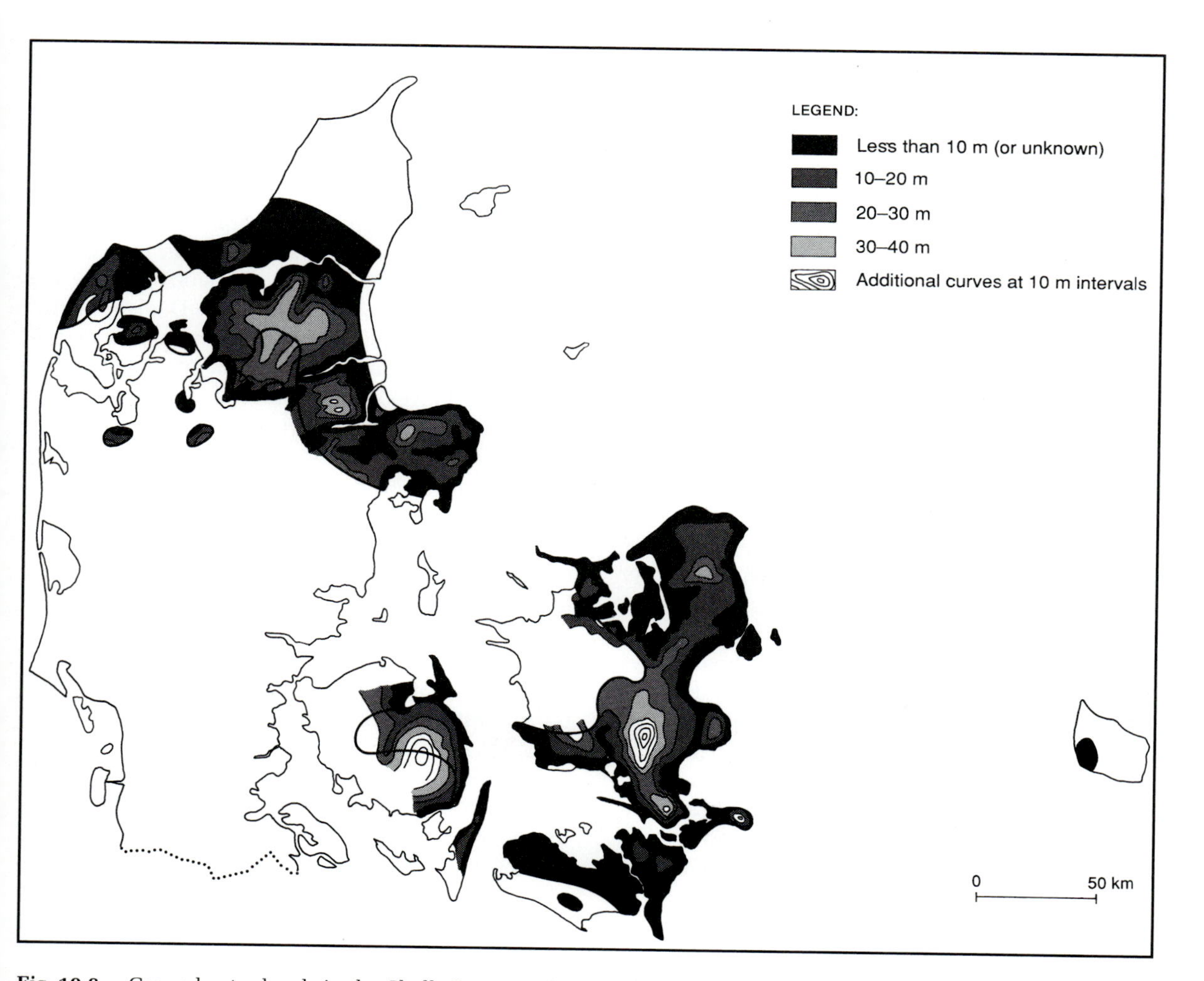

Fig. 10.6. Groundwater levels in the Chalk. In areas where Quaternary aquifers are in direct hydraulic contact with the Chalk, the levels extend into the Quaternary aquifers. Based on Kelstrup *et al.* (1982).

limited to the uppermost 30 m of rubbly and fractured chalk; at deeper levels, only the matrix contributed to the permeability (with values of 0.01–2 mD, decreasing with depth), the fractures adding very little to the permeability (Gosk *et al.* 1981).

In Stenlille, on Zealand, natural gas is stored in a deep Jurassic sand. The lower part of the overlying Chalk (the Turonian) at a depth of approximately 1.2 km contributes to the sealing of the reservoir with a permeability of the order of 0.01 mD, (Springer 1989).

Because of the very low permeability of the matrix of the Chalk, its usefulness as an aquifer depends almost entirely on the presence of permeable fractures. Apart from the open fractures parallel to the bedding, mentioned earlier, steeply inclined or open vertical fractures are also common. Such fractures are more closely spaced in the upper part of the Chalk that is *in situ*, on top of which a bed of disturbed, rubbly chalk is often present, especially in depressions in the pre-Quaternary surface. At depth, only a few of the steep fractures persist, and most of them are closed because of the relatively soft and friable nature of the Cretaceous Chalk. Only in tectonically disturbed areas and zones, such as around the flanks of salt diapirs, is water flow possible through deep-seated faults.

In combination, these factors mean that the Chalk has its highest transmissivity in tectonogenic fracture zones and where severe glacial disturbance has taken place. During the Weichselian glaciation the sea level relative to the land surface was approximately 25 m lower than today (Petersen 1989; 1991). The highly permeable zone along the upper surface of the Chalk therefore frequently continues offshore.

The upper fractured part of the Chalk both allows drainage from the overlying Quaternary deposits into the aquifer as recharge, and serves as a possible route for the intrusion of saline water into the aquifer. To a large extent it is the overlying beds that determine, among other things, the head conditions in the fractured chalk. Most groundwater in Denmark is actually produced from the unconfined upper aquifers or deeper confined aquifers in the Quaternary deposits. In the regions where the Chalk aquifer outcrops and is productive, such as in northern Jutland, it is often in direct hydraulic continuity with the unconfined Quaternary aquifers. Recharge to the Chalk may take place through fractures during intense precipitation, as referred to in Chapter 3. The chemical composition of groundwater in the Chalk may therefore differ between the fractures and the matrix.

Regionally, the hydraulic characteristics of the Chalk vary considerably. This depends partly on lithology and partly on the extent of erosion and strain on the aquifer, particularly during glacial episodes. The shallow Cretaceous Chalk is soft and friable, with a total porosity in the range 35–50 per cent and it is only weakly bedded. Therefore this rock is hydraulically comparatively homogeneous within the zone where freshwater occurs, unless secondary permeability is present. Such secondary permeability occurs as a result of fracturing or karstification. Both effects are most prominent in northern Jutland, where Cretaceous Chalk is overlain directly by Quaternary deposits. In this area many sizeable karstic cavities filled with siliciclastic Quaternary material have been discovered on excavating the Chalk. The high productivity of the Chalk in the area stems from a combination of these structures in the solid chalk and the apparently modest or gentle glacial abrasion during the Weichselian. On the south-eastern Danish islands the same rock type has suffered severe late glacial abrasion, so that any comparatively productive surficial layers have been stripped off. A few preserved karstic cavities in the upper parts of huge, glacially dislocated and stacked chalk blocks (or *olistoliths*) on the Island of Møn support this explanation.

Whereas the fractures in Upper Cretaceous chalk may be rehealed, they tend to remain open in the relatively more lithified and heterogeneous Danian limestone. The total porosity of this limestone is typically between 10 and 35 per cent. It is much more distinctly bedded than the Cretaceous Chalk, and alternations between hard and soft beds are typical. Flint, which may constitute up to one quarter of the Danian limestone, is for practical purposes non-porous. Fracture spacing varies considerably. The horizontal fractures, which are often present in the uppermost part of the succession, tend to be located between flint beds, while other fractures often cross-cut the sequence in an *en échelon* manner. Open fractures associated with faults do, however, cut across the entire sequence, and again appear generally to be relatively more permeable in the northern part of Jutland. The fractures, and voids in the rubbly weathered surficial limestone, usually remain open in the Danian limestone because of its hardness and heterogeneity. High yields from these fracture systems are well known, especially from the shallow aquifers in Jutland where karstification has also taken place, but also from Zealand (Sorgenfrei 1945). The Upper Palaeocene layers are sometimes even more distinctly bedded than the Danian, but fractures appear to be less frequent.

Because transmissivity is very dependent upon fracturing, and these in turn are most common in regional fault zones, it does vary considerably within short distances. For example, it is commonly found in the south-eastern part of the country that wells producing $10\,m^3\,d^{-1}$ may be located close to wells producing $100\,m^3\,d^{-1}$.

The Chalk is generally most productive in its upper part because of the shallow fracture system. On close inspection, however, the uppermost part of the glacially abraded chalk rubble, which in many places constitutes the uppermost chalk bed, has been compacted and weathered, and the secondary pore space filled with glacial material. Within the rubbly bed the permeability therefore sometimes increases downwards (Jacobsen 1991).

Hydrochemistry

General composition of groundwater from the Chalk

The hardness and bicarbonate content of groundwater from the Danish Chalk is generally in equilibrium with the partial pressure of carbon dioxide. Therefore the lowest concentrations (2–5 meq $l^{-1}HCO_3$ and 40–100 mg l^{-1}Ca) occur in the most shallow and unconfined groundwater. Conversely, the highest concentrations (5–10 meq $l^{-1}HCO_3$ and 100–200 mg l^{-1}Ca) are found in the deep confined aquifers. Roughly 10 per cent of the hardness is caused by the presence of magnesium. These concentrations of bicarbonate also apply to groundwater in the marly Palaeocene beds, where the concentrations of calcium and magnesium ions are often much lower. This is because of ion exchange of magnesium and calcium for sodium and potassium from interbedded marine clays. An example of chemical mapping in Denmark is given in Fig. 10.7. Selected aspects of the hydrogeology of part of the same area, in south-western Zealand, are illustrated in Fig. 10.8. The classical work on types of groundwater in Denmark was published by Ødum and Christensen (1936).

The deeper groundwater in the Chalk aquifers generally contains high levels of chloride and fluoride; an example is given later. Marine intrusion of saline water is especially prominent in coastal regions, as for example south of Copenhagen, where the groundwater level has fallen because of abstraction (Fig. 10.9).

Indirectly, the Chalk has a much greater influence on the composition of groundwater in Denmark than may be envisaged from the distribution of the Chalk aquifer. This is because of the erosion of the Chalk by glaciers, which included in their debris load a substantial amount of chalk. The amount of this disseminated chalk decreases in a westerly and south-westerly direction, partly as a result of the more extensive weathering in these areas and partly because of the distance from the source areas. In the country as a whole, it is the distribution of the Quaternary formations that influences the composition of drinking water. As shown in Figs. 10.10 and 10.11, the concentration of the chalk-derived constituents calcium and fluorine, in Danish drinking water from groundwater sources, decreases with distance from the easterly chalk-source areas. The expression of this tendency is further enhanced by a low concentration of these constituents in groundwater from the deeper Miocene aquifers in south-western Jutland.

High chloride content

Central Zealand has an abundance of lakes and rivers, and this region is important both for local water supply and for supplying water to distant Copenhagen. The Chalk aquifer in this region, taken to include the entire Cretaceous and Palaeocene sequence, is known for its high yields but also for relatively high chloride contents at some locations (Fig. 10.9). Over the years, the water table has fallen and, because there is some potential for further chloride contamination, aspects of production from the Chalk aquifer have been repeatedly studied. In a major study within the area, Jacobsen and Kelstrup (1981) have summarized the early data:

Zealand
0
5 km
Signatures
Wellsite
Ground elevation
and Reservoir lithology
Anions in %
Cations in %
meqv./l
Ca
Mg
Na
K
Fe
NO3
Cl
SO4
HCO3

Fig. 10.7. Chemical composition of groundwater in part of south-west Zealand (after Villumsen and Binzer 1979). The considerable variation in the total ionic content, as well as in the composition, reflects the different character of the various aquifers. In fractured Danian or Late Palaeocene aquifers, and also in Quaternary aquifers that are in hydraulic connection, sodium and chloride are dominant and the total ionic content is high. Where bicarbonate is dominant, the sodium and chloride contents are often not stoichiometrically equivalent because of cation exchange in clayey beds. It is noticeable that in the Quaternary and Cretaceous aquifers, which are often either not fractured or fractured to only a limited extent, the ionic content is comparatively low.
The aquifer lithologies refer to the Cretaceous Chalk, the Danian limestone, the Late Palaeocene limestone and siliciclastic Quaternary deposits; the aquifer signatures are given in Fig. 10.8. Well identification numbers within the map sheets 210, 211, 215, 216, 220, and 221 are indicated.

Within this territory the deepest aquifer studied is of Danian age, and consists of soft micritic limestone, rich in flint beds. The Danian beds are overlain by a variable succession of Upper Palaeocene deposits of glauconitic calcarenitic limestone, glauconitic calcareous sand and marls. Gradations between these lithologies commonly occur, and marls become more abundant up the section and towards the west. The succession as a whole is gently dipping toward the west. The central part of Zealand is traversed by at least two sets of regional fault zones, striking WNW–ESE and NNE–SSW; there are also both local accessory and glacially induced fractures. The valleys eroded in the pre-Quaternary (Chalk) surface are assumed to follow the fault trends and are also reflected in the Quaternary surface. It is in and along these valleys that most production wells are located.

The head conditions are generally artesian with separate pressure levels in the individual members of the aquifer. Cross-flow between individual permeable intervals of the aquifer is largely restricted to the valleys, where the bed continuity is broken, and to unscreened wells that unintentionally interconnect individual members of the aquifer.

The groundwater in both the upper arenitic and the lower micritic aquifer intervals as a rule contains between 15 and 25 mg l^{-1}Cl corresponding to that in the infiltrating water. The water in some of the wells, mainly the major producing ones, has a higher chloride content, so that the average value in the area is about 100 mg l^{-1}Cl. In individual wells almost 4000 mg l^{-1}Cl has been measured. The chloride is assumed to cone up into the producing wells from saline formation water that is present in deeper-seated and older beds. The saline water may be naturally discharged from a fracture zone at the easterly extension of the Palaeocene members of the aquifer, where the greatest width of the outcrop occurs, and may thus be encountered at relatively shallow depths.

The explanation is that the Upper Palaeocene, with alternating beds of high and low permeability, dips gently towards the west (that is, towards the centre of the basin). Towards the east they terminate, more or less, along a fault and fracture zone. The head increases with depth and saline groundwater from the deeper-seated beds in the basin discharge where impermeable, sealing beds in the Upper Palaeocene are broken by fractures. The greatest opportunity for the discharge of saline water is where there is no cover of low-permeability strata—that is, where the Upper Palaeocene attains its maximum easterly extension and where there is a fault zone.

In localities close to this studied area, salinities of up to 10 000 mg l^{-1}Cl are found. A current reinvestigation of the problem (Andersen 1992) has led to a reinterpretation of the data. The heads in the observation wells in the region have, over the past 50 years, equilibrated to the level of the lowest head, which corresponds to the water level in the rivers and the lakes. This equilibration apparently commenced decades before the abstraction increased (Fig. 10.12). Flow logs run in a number of observation wells with long screens (and which were not pumped) have documented vertical groundwater flow. Bearing in mind the high transmissivity and the cross-flow across the sequence, it is now considered that the declining heads more accurately reflect an early penetration of the sealing beds between the aquifers by unlined wells, in combination with relatively stagnant saline water in the deeper parts of the wells. More recent analyses of the groundwater in combination with repeated conductivity logging apparently cast doubt on the general assumption that there has been an increase in the upward flow of deep-seated saline water during the last 49 years. A halophilic flora on a nearby moor (Andersen and Ødum 1923) and other occurrences in the region (Andersen 1930) further indicate that the expulsion of saline water, apparently driven by compaction, has taken place over a very long period. The quality problems in the region could, therefore, possibly be reduced by plugging the deeper parts of the wells and thus preventing cross-formational flow.

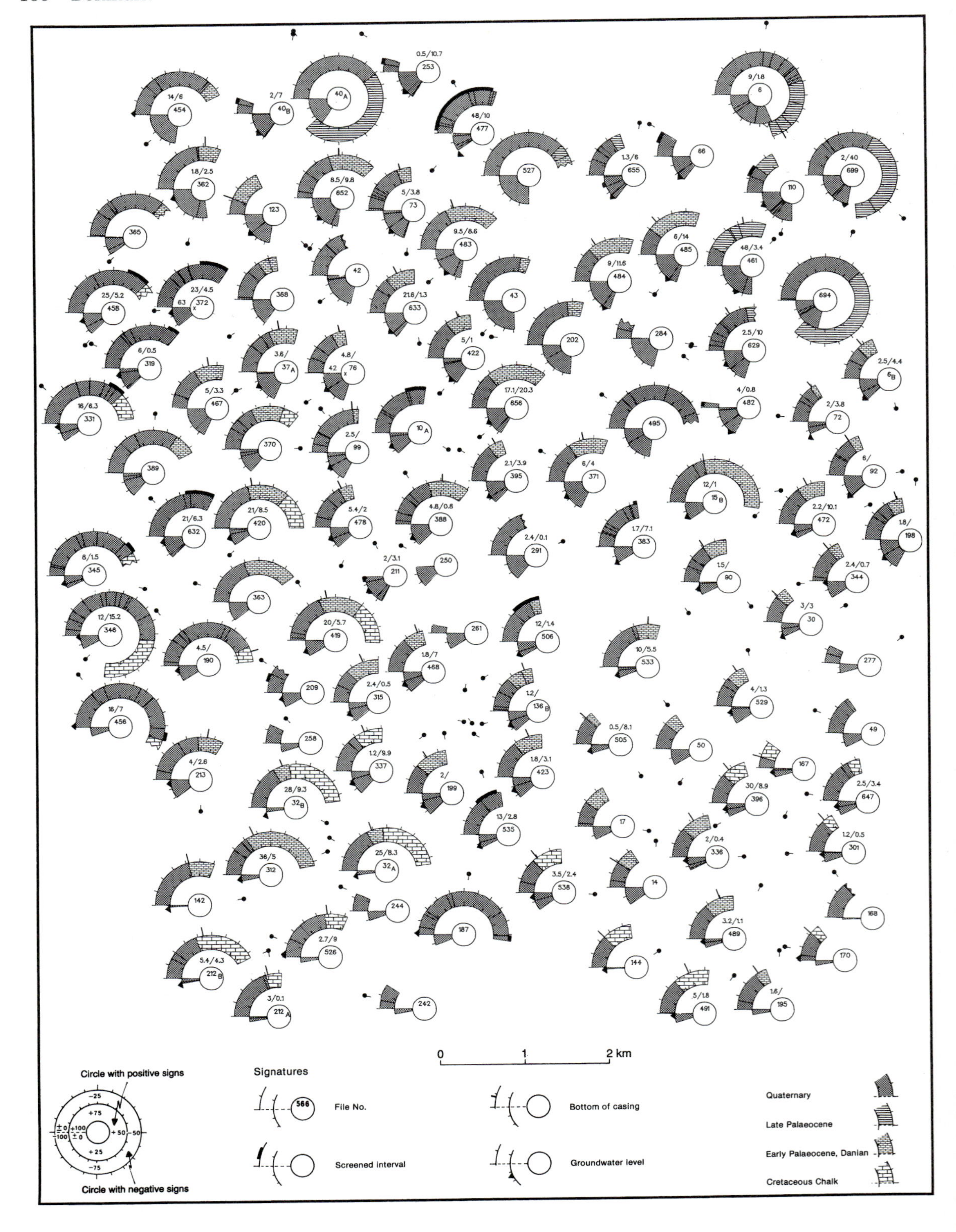
0 1 2 km
Circle with positive signs
Circle with negative signs
Signatures
File No.
Bottom of casing
Screened interval
Groundwater level
Quaternary
Late Palaeocene
Early Palaeocene, Danian
Cretaceous Chalk

Fig. 10.8. An example of cyclogram mapping depicting hydrogeological and well data for part of south-west Zealand (the area represented is the south-western part of that shown in Fig. 10.7).
Depth increases in a clockwise direction. The stratigraphy of the aquifer zones is highlighted by hatching and other subordinate bed boundaries are also indicated. In this area the age of the uppermost pre-Quaternary aquifer becomes younger towards the north; the surface of the aquifer is very irregular. The well file numbers refer to the map sheets 215 and 220, and the 'tadpoles' point to well sites. The numbers given for some wells indicate yield in cubic metres per hour per metre of drawdown.

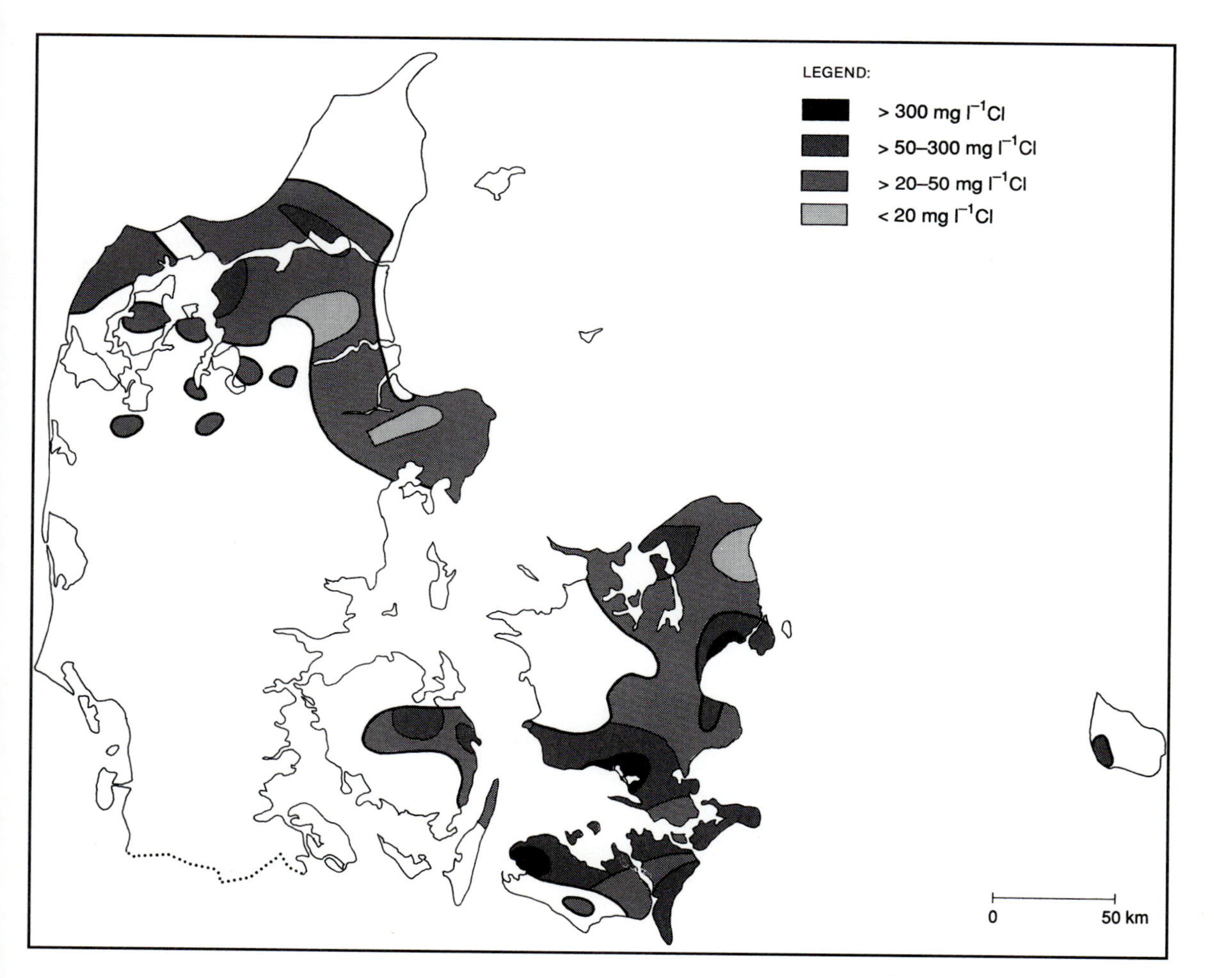

Fig. 10.9. Salinity of drinking water produced and supplied within the Danish Chalk aquifer region. The contours are based on estimated average values of all groundwater produced in the region. Chalk aquifers, however, are the dominant sources for saline groundwater. Based on Gosk *et al.* (1990).

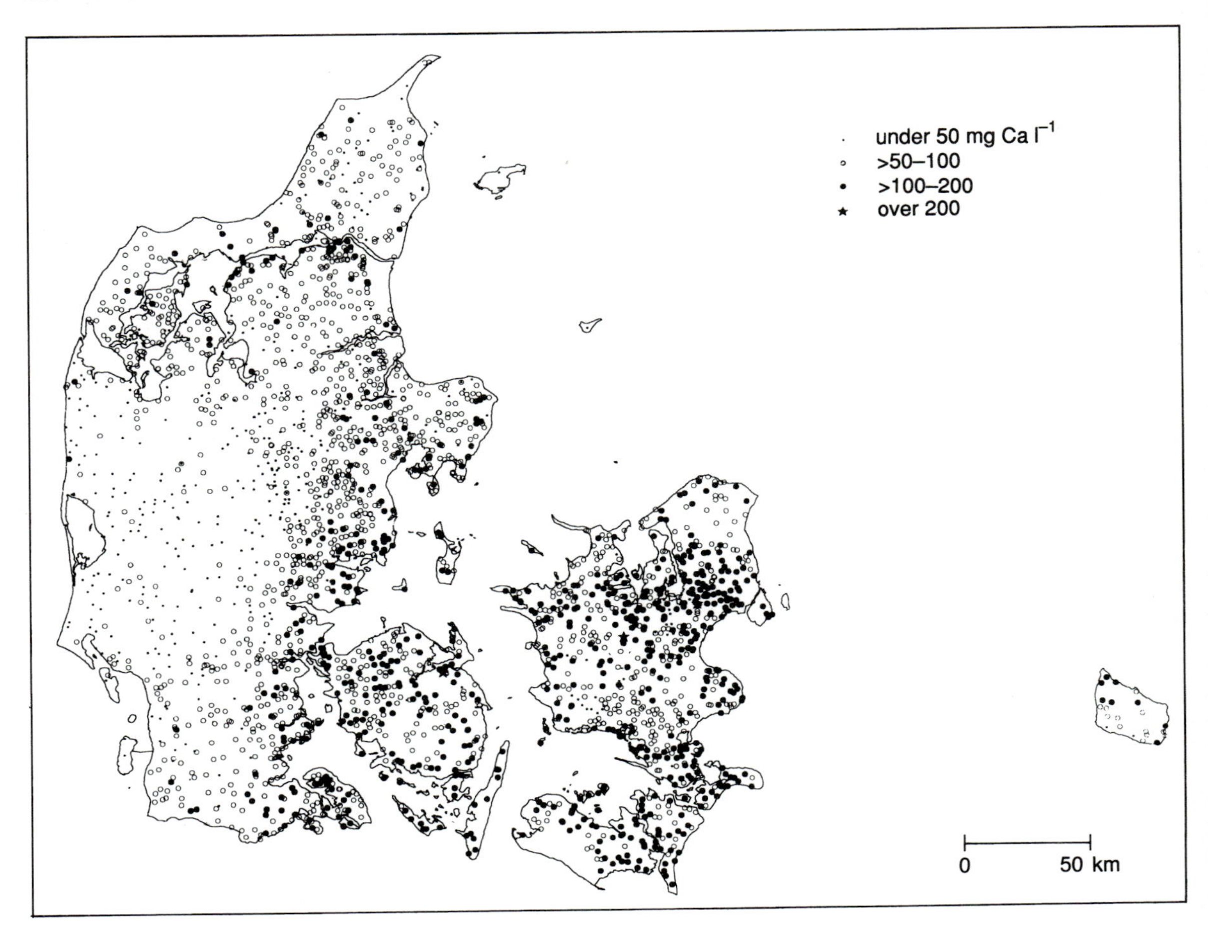

Fig. 10.10. Calcium concentration in Danish drinking water. Compare with Fig. 10.3 to see the relation to the protruding chalk surface within the Chalk aquifer region and the decreasing content in westerly and southerly directions. From Gosk *et al.* (1990).

General pollution problems

In unconfined chalk aquifers the groundwater often contains nitrate in excess of 50 mg l^{-1}, the highest concentration permitted in drinking water. This results from contamination of recharge by fertilizers, and is a problem particularly in Jutland. It may be aggravated by the short residence time of the recharge water in the thin and sandy glacial surface strata. In some areas, where fertilizers are not used, the nitrate-reducing processes in the top-soil have not been completed by the time the re-charge water leaves the zone of biological nitrate reduction and enters the Chalk, which itself has virtually no reduction capacity.

As conventional processes for groundwater treatment do not reduce the nitrate content, the only method in common use for achieving this is to mix groundwater from different sources. Therefore the higher nitrate content in drinking water (Fig. 10.13) highlights locations where this method of treatment has not been possible; this is the case in the Chalk in Jutland and the region immediately to the south. In the south-east of this chalk region in Jutland, there is, in addition to the excess of nitrate in the shallow groundwater, an increasing amount of fluoride in the deeper groundwater. Locally, therefore, the zone of exploitable groundwater may be only 10 m thick (Thorling *et al.* 1990).

In areas where the water table has been lowered

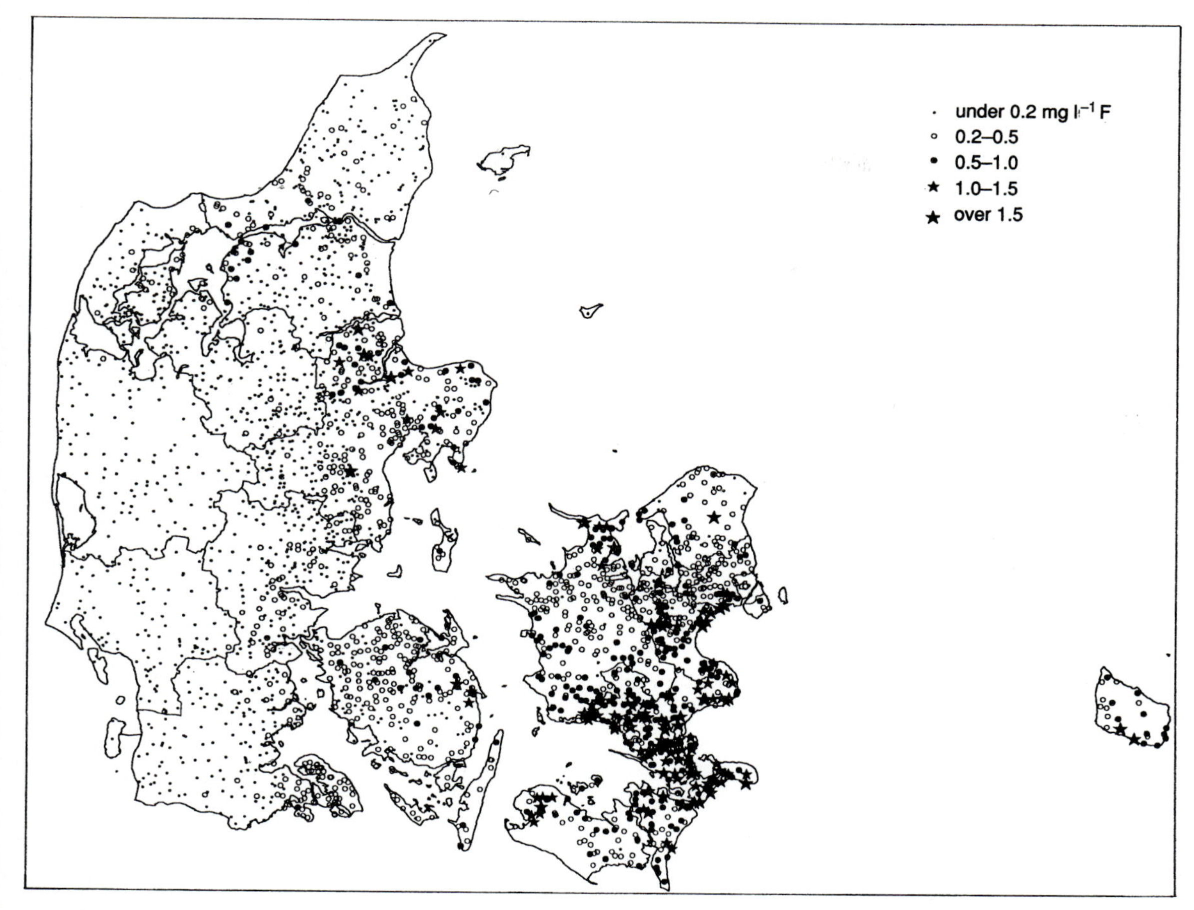

Fig. 10.11. Fluoride concentration in Danish drinking water. As for calcium (Fig. 10.10), the distribution of fluoride correlates with Chalk aquifers and calcareous Quaternary aquifers. From Gosk *et al.* (1990).

by over-abstraction, oxidation of pyrite has frequently resulted in an increase of both sulphate and permanent hardness; the latter can be relatively high. This is a characteristic of the area south-west of Copenhagen; it is also a feature of the Quaternary aquifers.

Pollution in the Copenhagen area

Throughout the last century, the major urban region of Copenhagen has been the focus of the main industrial development in Denmark. In the course of time this development has led to incidents of severe pollution caused by inadequate handling of waste.

The independent municipality of Frederiksberg is located in the centre of Copenhagen. Within Frederiksberg the only significant groundwater production is the $1.5 \times 10^6 m^3 a^{-1}$ produced from three wells at two locations in fractured Danian limestone. Following the discovery of small concentrations of chlorinated solvents in the groundwater from these wells, a major investigation was carried out with the objective of producing a detailed map of the groundwater head, identifying the pollution sources, predicting the implications of the pollution, and proposing priorities for remedying the problem (Markussen *et al.* 1991).

These aims were complicated by the fractured nature of the aquifer and the unevenly distributed superficial cover of Quaternary sand, till and fill

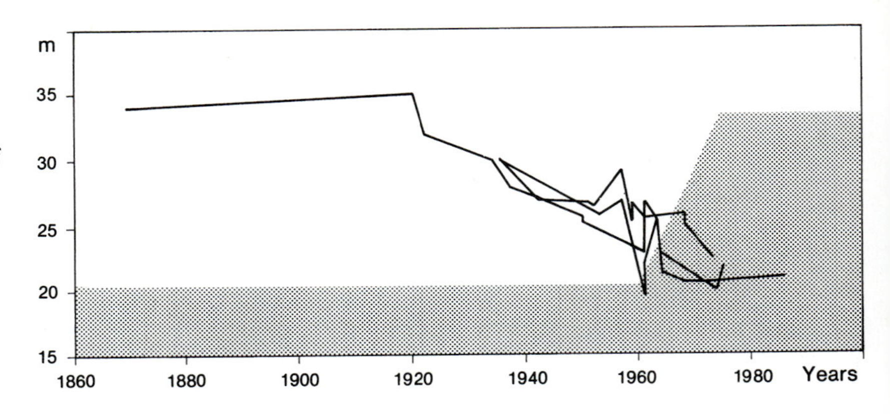

Fig. 10.12. Groundwater levels in an area in the central part of Zealand. With time more measurement locations are included. After a period of apparent stability a general decline in the water table commenced in 1920. This is many years before the major increase in abstraction during the late 1950s and early 1960s, as indicated by the hatched area. Based on Andersen (1992).

deposits. Based on the head distribution (Fig. 10.14) the greatest drawdown caused by abstraction clearly occurs within the 16 km^2 municipality of Frederiksberg, but the entire area within which groundwater flows towards the central wells is 35 km^2. Based on the indicated flow pattern and a large number of chemical analyses made at points defined both areally and stratigraphically, the plumes of pollution were located and related to proven and suspected sources (Fig. 10.14). The high transmissivity of the aquifer around the producing wells (15×10^{-3} m^2 s^{-1}, compared to the average for the entire area of 5×10^{-3} m^2 s^{-1}) stems from the prominent NNW–SSE trending fault system that intersects the area and from the surficial fractures of glacial origin. Along the fault zone, the Danian beds were downfaulted towards the east prior to the levelling of the land by glaciation. Based on the discoveries of chlorinated solvents together with the measured tritium content, recent water is assumed to constitute an increasing proportion of the groundwater that is now in the major flow conduits down to depths of 20 m below the surface of the Danian limestone. The flow and the concentration of the contaminant were modelled for a dual-porosity system, where a quarter of the 20 per cent effective porosity is attributed to fractures and the rest to the matrix. It was calculated that the combined dilution and dispersion in the amalgamated pore systems would cause the concentration of the pollutant in the abstracted groundwater to increase for 25 years, assuming no change in the strength of the source. Precisely where the concentrations presently found lie on the curve of increasing concentration of the pollutant is, however, not clear.

In an attempt to prevent further contamination of the Chalk aquifer, suspected surficial threats of pollution are being investigated, and local scavenging pumping within the plumes is being carried out. The polluted groundwater that is abstracted is reinjected into the aquifer after purification and treatment. These procedures are being carried out with the objective of securing a future supply of water that satisfies the legal requirement for quality within the municipality.

Groundwater resources and abstraction

Hydrometeorology

Precipitation within the Chalk areas range from 500 to 750 mm a^{-1}, and evapotranspiration is generally between 360 and 420 mm a^{-1} (Thomsen 1987). The highest precipitation occurs in the west and in the interior, and the highest evapotranspiration in the east of the country. In Denmark, practically all the precipitation not lost by evapotranspiration infiltrates and is potentially available to recharge groundwater, since overland flow is negligible. However, a large proportion of the infiltrating water is lost to the dense drainage system in the clayey cultivated areas. It is estimated that roughly between 100 and 250 mm of water percolates annually to the zone below this drainage level. The amount of water available for abstraction in the Chalk areas is roughly half of this because, among other things, there are regulatory demands for groundwater discharge into rivers and streams. In fact, at present, abstraction within Chalk areas amounts on average to slightly more than half of the annual recharge, and the Chalk aquifers in the Copenhagen and Aalborg regions are severely overexploited.

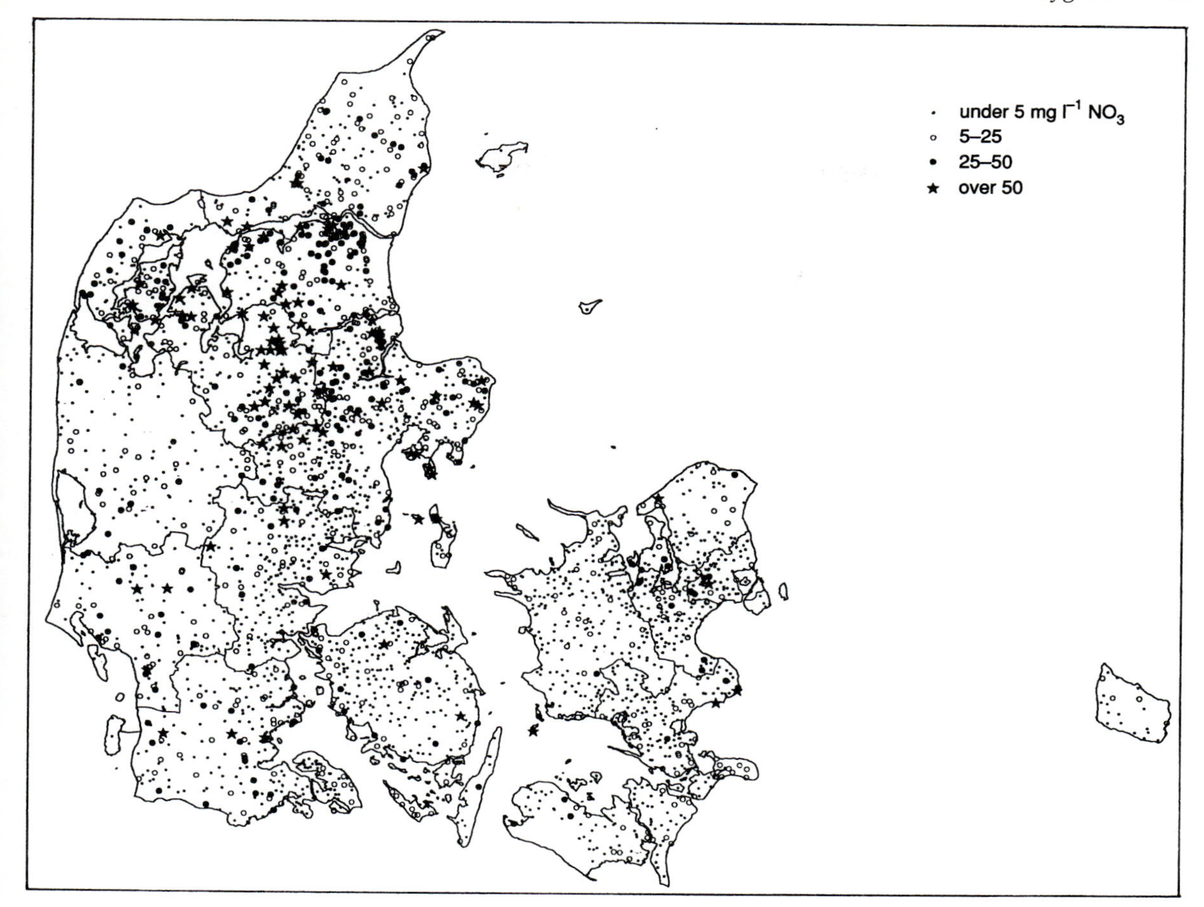

Fig. 10.13. Nitrate concentration in Danish drinking water. Nitrate as an areal contaminant is one of the main concerns of a major ongoing water environmental study programme. Concentrations above 25 mg l^{-1} in the main illustrate locations where there are insufficient alternative groundwater sources to the susceptible shallow and generally unconfined aquifers in present use, so that the possibility for remedial action is limited. The Chalk region in northern Jutland, with only thin Quaternary cover, is particularly evident. From Gosk *et al.* (1990).

Recharge and discharge areas

With a few local exceptions, the whole of Denmark is covered by deposits of glacial or post-glacial origin. However, these surface layers may be very thin, especially in northern Jutland where they are draped over the Chalk surface. The infiltration capacity of the Quaternary sediments is generally high enough to preclude overland flow except after rare storm events. Therefore, as previously mentioned, all precipitation in excess of that required to make up evapotranspiration losses infiltrates. Discharge areas, taken to be areas where the hydrostatic level is above the ground surface, are located along streams and lakes (Fig. 10.15; Kelstrup *et al.* 1982). Discharge from the Chalk into the Quaternary deposits has been investigated and mapped locally, as in the extreme north-east of Zealand (Andersen *et al.* 1975). Submarine discharges are known, but data are limited and scattered.

Groundwater resources

Information on groundwater resources in Denmark is collected regionally but not for individual aquifers. It is, therefore, not possible to give precise

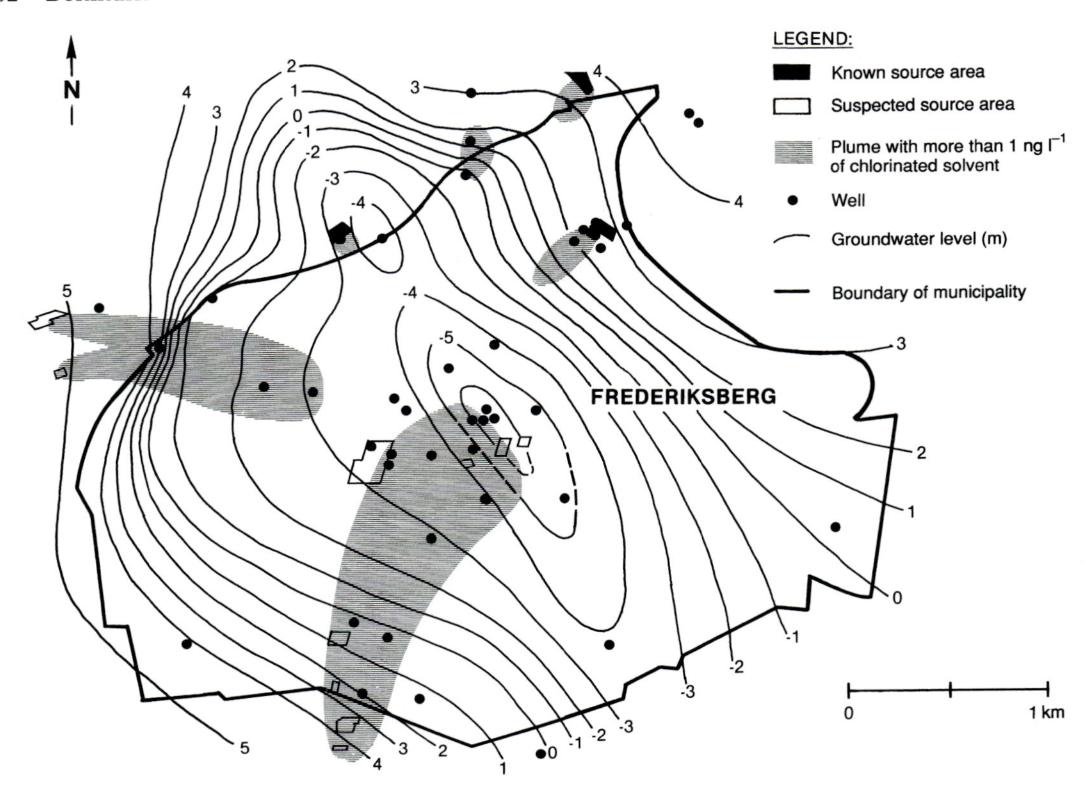

Fig. 10.14. Plumes of groundwater contaminated with chlorinated solvents within the drawdown area of the pumping wells in the centre of the municipality of Frederiksberg. Frederiksberg is located within Copenhagen. (After Markussen *et al.* 1991, by courtesy Rambøll and Hannemann A/S).

details of the resources of the Chalk. In general terms, the total resource of freshwater in the Chalk aquifer is probably $800 \times 10^6 \, m^3 \, a^{-1}$ (Kelstrup *et al.* 1982). At present the total annual permitted abstraction is only half this amount.

Local authorities have published information about the threat to the quality of groundwater reserves in the Chalk (Nygaard 1991).

Groundwater abstraction

In terms of the areal distribution of groundwater, the Quaternary aquifers are the most important in Denmark, yielding approximately 50 per cent of the total groundwater abstraction. The combined Chalk aquifers supply approximately 35 per cent of the total, of which 10 per cent is from Cretaceous Chalk and 10 per cent and 15 per cent from the Danian and Upper Palaeocene limestones, respectively (Frederiksen *et al.* 1991). Within the Chalk regions on the islands, the aquifers in the Quaternary deposits and the Chalk are in general equally important. But in Jutland and locally on the islands, in areas with freshwater head above sea level, the Chalk is often the main or the only producing aquifer. The topographically low-lying parts of the Chalk region in Jutland were flooded by the sea during early post-glacial (Neolithic) times. In these areas the recent marine deposits and their connate saline water usually affect the quality of groundwater in the underlying aquifers including the Chalk. The Chalk aquifer in Jutland therefore generally contains groundwater suitable for abstraction only in the areas that were islands and peninsulas during the early post-glacial period. This

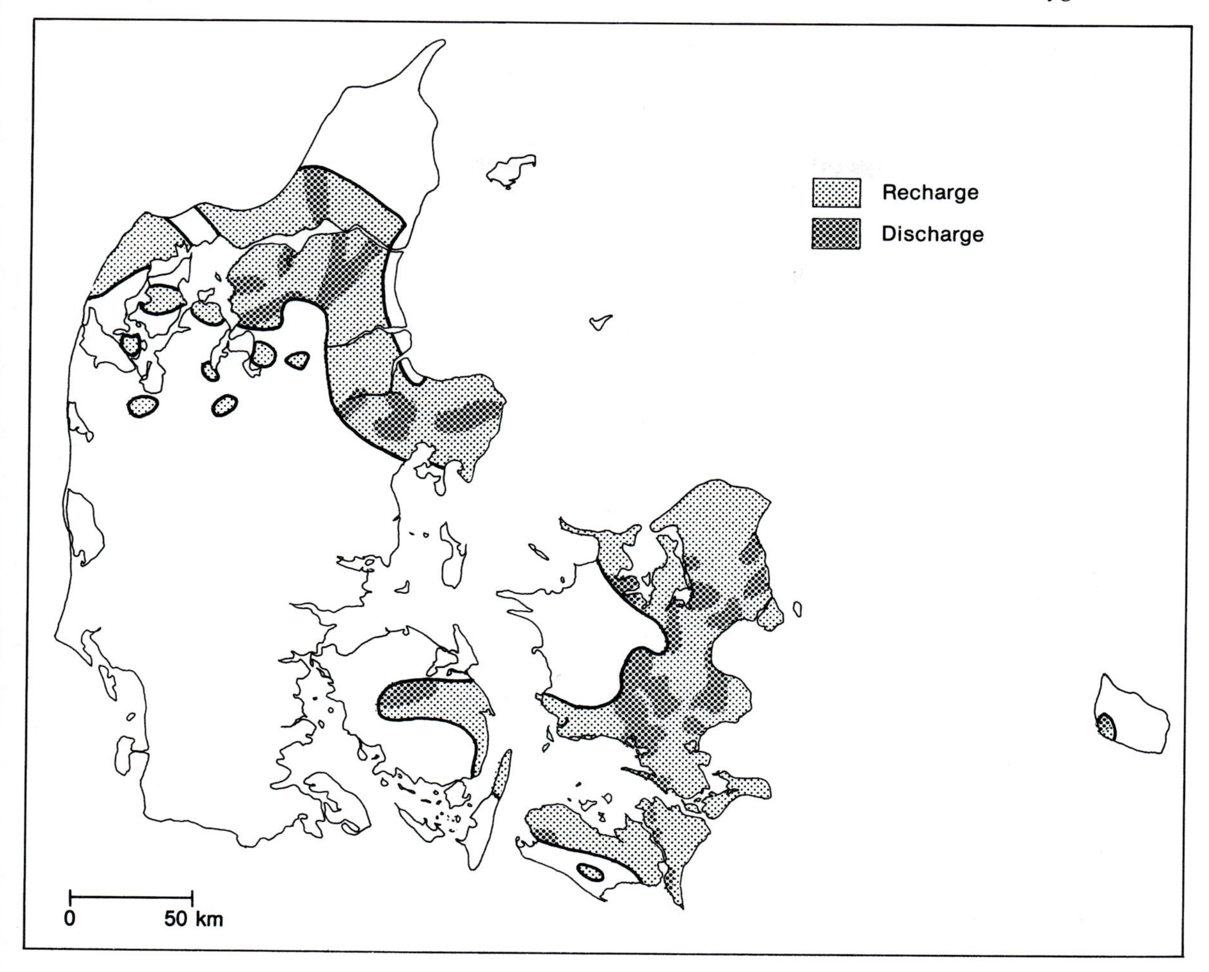

Fig. 10.15. Recharge and discharge areas within the Chalk aquifer region of Denmark. Modified from Kelstrup *et al.* (1982).

is mainly the case south of a line running east–west immediately north of Aalborg. In this region, the Chalk is known to be quite highly productive, with karstic features and many high-yielding springs in the valleys eroded into the aquifer. However, this high permeability is locally combined with a susceptibility to pollution by nitrate.

The municipal water supply for the major town of Aalborg is taken from the Chalk, which is locally karstified and fractured. Transmissivities are so high that well-drilling contractors are said to purchase pumps with the desired capacity prior to drilling the well, in the secure knowledge that the required yield will be obtained (Berthelsen 1987). The Aalborg urban area extends over both sides of the local fjord. The confined Chalk aquifer is continuous from the southern shore under the fjord and into the area to the north. Pumping for domestic supply and industrial purposes has lowered the groundwater level in an area below the fjord and the central part of the town to several metres below sea level with a resultant increase in the salinity of the groundwater. Pumping from the area to the north of the fjord has been reduced and the groundwater head has risen again, so that the main depression in the potentiometric surface is now south of the fjord.

On the islands the main effects of abstraction occur in the urban areas around Copenhagen, where in many places the groundwater level has been lowered by more than 10 m. The highest yielding wells are located in valleys in the Chalk surface which are eroded fault and fracture zones. The dangerous combination of lowered groundwater levels and high production from fractured zones has led to an increase in the salinity of the groundwater because of saline intrusion, especially in the coastal areas south of Copenhagen.

On the southern island of Lolland, a phase of relatively late glacial erosion seems to have cut deeply into the soft Cretaceous Chalk, removing any zone of highly permeable chalk that may have been present. On this topographically low-lying island the thin zone of chalk that contains freshwater does not, therefore, produce very high yields. Major abstraction of groundwater to allow construction and excavation works has so far been only of short duration in chalk areas. However, the need for permanent pumping is anticipated in connection with the planned bridge from Denmark to Sweden.

Typical well yields

Because of the various factors mentioned above, well yields within any chalk area are very variable. In northern Jutland, wells producing from Cretaceous Chalk typically yield 10–15 $l\,s^{-1}$, but yields as low as 1 $l\,s^{-1}$ and as high as 25 $l\,s^{-1}$ do occur. In eastern Zealand, yields from Cretaceous Chalk range from 0.75 to 7 $l\,s^{-1}$ but decrease southward from 3–7 $l\,s^{-1}$ in the north to typically 3–4 $l\,s^{-1}$ in southern Zealand. On the island of Lolland, the Cretaceous rocks yield only 1–3 $l\,s^{-1}$. The Danian limestone typically yields 5–8 $l\,s^{-1}$ in northern Jutland but ranges from 0.25–15 $l\,s^{-1}$. On the peninsula of Djursland, the Danian aquifer is typically more productive, with a range of 0.5–22 $l\,s^{-1}$ and a typical yield of around 10–12 $l\,s^{-1}$. On Zealand, the bryozoan limestone yields very much the same as in northern Jutland. For all these yields the drawdown is commonly 5–10 m.

The Upper Palaeocene glauconitic calcarenite most frequently yields 1–3 $l\,s^{-1}$ but yields as low as 0.25 $l\,s^{-1}$ or as high as 12 $l\,s^{-1}$ are not uncommon. In the centre of Zealand, specific capacities in the range 30–40 $l\,s^{-1}\,m^{-1}$ are common.

Cyclogram mapping

Since the establishment, in 1926, of a requirement to report data on water wells for inclusion in the archives of the Geological Survey of Denmark, an increasing number of such data points have been collected. At present there are 150 000 registered water wells. The difficulty in presenting and interpreting these data on maps led to the development of a graphical solution by Andersen (1973). On the basis that the drilled depths of these wells rarely exceeds 200 m and that the elevation of the ground surface in Denmark is only greater than 100 m over 5 per cent of the area, the depth scale was illustrated by concentric circles (Fig. 10.8). Each circle is intended to illustrate 100 m of drilled section with increasing depths in the clockwise direction and the index levels +100, 0, and -100 m pointing west. The space between the depth-scale bands is reserved for lithological information. On the circular scale information about screened intervals as well as casing and water levels may be indicated. The centre and other open space is reserved for the well-file number, reference to any supplementary information, and the yield in cubic metres per hour. Finally the well site is indicated on the map and the cyclogram located alongside.

Among the advantages of this form of presentation is the high density and diversity of data which may be presented. Also, correlation between wells is possible and can be understood by observing the angular position of the particular aspect under consideration. Furthermore, data are presented almost at the correct geographical location on the map. irrespective of well depth.

Cyclogram maps have been produced for three-quarters of the Danish land area and for two-thirds of the Chalk region. An example showing part of the Chalk region in south-western Zealand is given in Fig. 10.8; further explanation is given in the caption. For comparison, the hydrochemistry of the same area is included in Fig. 10.7.

Groundwater monitoring areas

Increasing awareness of the susceptibility of groundwater to man-made changes in quality led to the establishment in the late 1980s of a nation-wide groundwater monitoring system (Andersen 1990). This system is but one of the aspects embraced by a comprehensive water environment plan that is mainly concerned with the problem of reducing pollution by nitrate and phosphate in rivers, lakes, coastal waters, and groundwater. Groundwater is monitored with the aim of providing an early indication of quality problems, in time for management initiatives to prevent irreparable damage.

Groundwater is partly monitored in 67 monitoring areas and partly by regular chemical analysis of groundwater from all wells used for drinking-water supply. The water from all the producing wells is sampled directly at the well, prior to any treatment. The monitoring areas are preferentially located in agricultural regions which are not affected by point pollution sources and each constitutes the catchment area of a production well. The sizes of the individual monitoring areas therefore vary considerably from 1 km^2 to more than 50 km^2. Within these areas an average of 15 wells, many of which have been constructed exclusively for this purpose, are available for sampling groundwater.

Sampling sites and depths are chosen to provide water samples from the entire range of subsurface water. It is thus intended to follow the possible variations in the content of constituents in the water from infiltration into local aquifers until arrival at a producing well.

The groups of constituents analyzed include the major ions, inorganic trace elements, and organic trace compounds, including chlorinated solvents and pesticides. The initial period of the study was to establish the base levels of concentration of natural constituents and is now largely completed.

Of the monitoring areas, 23 are located where the Chalk is the main aquifer. As an example, a profile through one of these areas (from Nygaard 1991) is given in Fig. 10.16. Comparison of the nitrate distribution and concentration within these monitoring areas has shown that the concentration does not always decrease gradually with depth, irrespective of whether or not the aquifer is confined. This is presumably because of the dual-porosity nature of the Chalk, combined with the increased water movement in the aquifer due to abstraction. Groundwater flows and mixes readily in open fractures, and infiltrates into the unconfined areas and enters the confined aquifer through such fractures. From the point of view of pollution, the fractured chalk aquifers may be classified on the basis of chemical data as de facto confined or unconfined, but the confining beds cannot be relied upon to provide protection against pollution.

Relation of porosity and depth in the North Sea

The Upper Cretaceous–Danian Chalk Group (Deegan and Scull 1977) in the Danish sector of the North Sea Central Trough has been explored intensively for hydrocarbons over the last 30 years. Approximately 100 exploration wells have been drilled and 15 oil- or gasfields have been found in Chalk reservoirs. Most of these wells are located on structural highs at the Top Chalk level, but in recent years an increasing number of wells have had stratigraphical targets below this Chalk horizon. Many of the later wells have been drilled off the Chalk structures so that the top of the Chalk has been encountered throughout a depth range from less than 2 to 3 km. In general, the thickness of the Chalk increases as the depth of the top of the Chalk increases. The wells, which are usually water-bearing throughout most of the Chalk interval, have been logged for porosity, and in this way a substantial regional database for the entire Chalk interval has been assembled.

In a regional study of the northern part of the Danish North Sea Central Trough (Nygaard *et al.* 1990) the well data were interpreted using the concept of sequence stratigraphy with biostratigraphic and seismic data. An additional aspect of this study examined the porosity distribution with depth, within the six regional intra-Chalk sequences (Clausen *et al.* 1990). The purpose of adding this aspect was to establish a pre-drilling framework for anticipating and predicting the reservoir potential.

For each of the six regional intra-Chalk sequences the highest porosity measured in the wells, excluding isolated extreme values, has been taken. Together these values indicate the best porosity at any depth so far encountered, and suggest the position of very high porosities which can be applied for predicting conditions (Fig. 10.17). Conversely, the lowest porosity has also been extracted and presented on the figure as an indication of the sealing capacity of the rock. In the figure the envelopes through the extreme values thus indicate the poten-

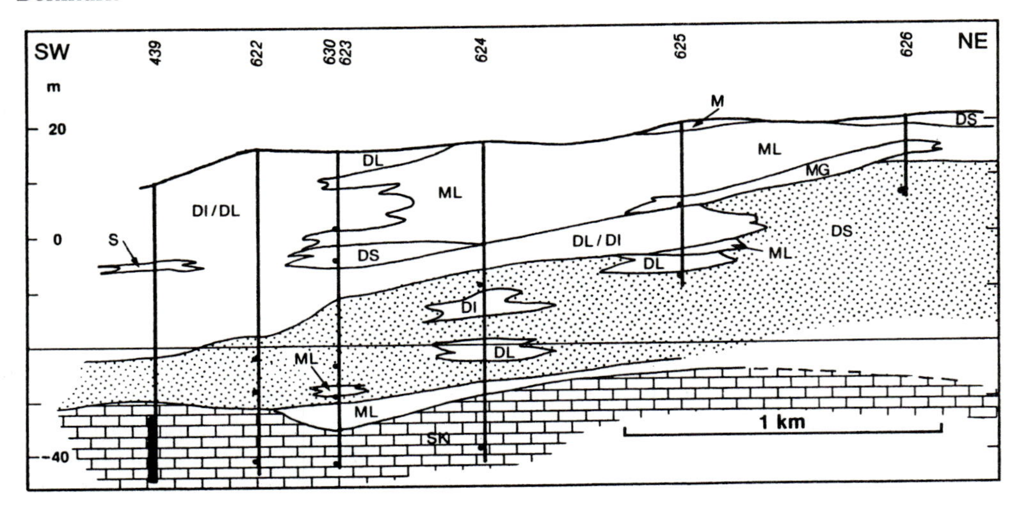

Fig. 10.16. Profile on the island of Falster through one of the 67 Groundwater Monitoring Areas in Denmark. Wells and well numbers are given, and screened intervals are indicated. The two main aquifers are hatched. Lithologies are: SK, Cretaceous Chalk; M and ML, clayey till; MG, gravelly till; S and DS, glacial sand; DI, glacial silt; DL, glacial clay. From Nygaard (1991).

tially highest and lowest porosities to be expected, at any depth in the study area, based on current data.

It must be kept in mind that, despite the smooth trend in the relationship between porosity and depth shown in Fig. 10.17, the average porosities do vary between the different sequences. In particular, beds with low porosity, and therefore a high sealing capacity, are present in the third sequence from the base, which was deposited during the Early Campanian.

The Danish data may be interpreted in terms of a parallel displacement of the average porosity–depth trends presented by Scholle (1977). From this it becomes apparent that the beds with the highest porosity are situated on the same trend of porosity with depth as are chalk sequences at much shallower depths. The porosity of these highly porous beds, therefore, appears to be preserved at the level appropriate prior to burial at depths of up to 3 km beneath younger Tertiary and Quaternary beds in

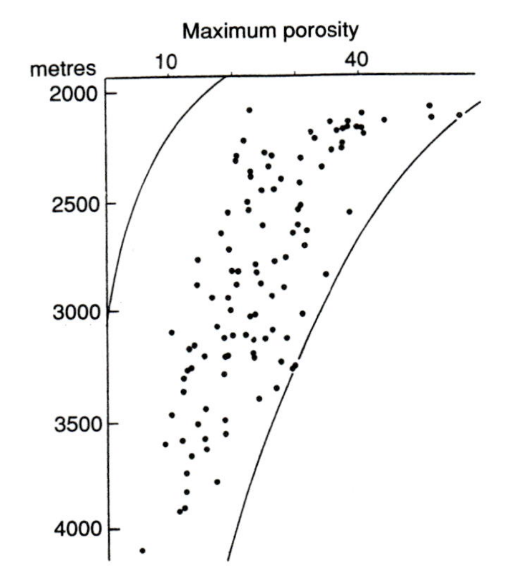

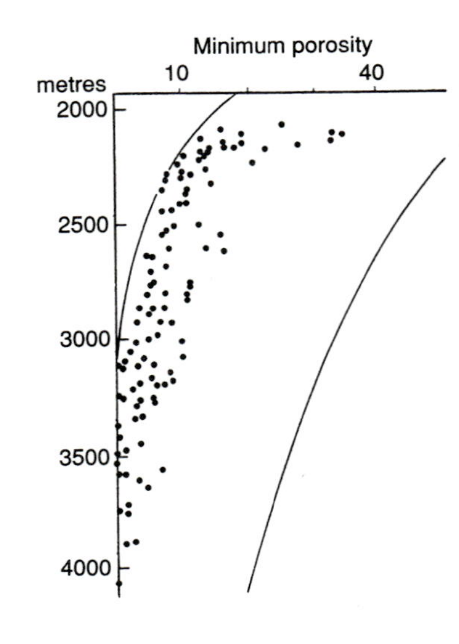

Fig. 10.17. The maximum (left) and minimum (right) porosities within the Chalk sequences in the Danish sector of the North Sea Central Trough. The enveloping curves given on both graphs illustrate, respectively, the best sealing capacity and reservoir porosity conditions to be anticipated at any depth. Based on Clausen *et al.* (1990).

the Central Trough. The beds with low porosity follow the world-wide average inverse relationship of porosity with depth for buried carbonates.

It has been demonstrated (D'Heur 1984) that the variation in thickness of the Chalk Group over local structures sometimes directly reflects the selective preservation of porosity caused by the presence of hydrocarbons in combination with the general loss of porosity with depth. On the regional scale, and indeed in many Danish fields, this effect cannot alone account for the major differences in thickness, so the porosity variations are also of depositional and/or diagenetic origin.

The Danish investigation of the porosity has indicated the value of considering the Chalk as two main diagenetic facies. One is resistant to a reduction in porosity, possibly because of over-pressure and/or hydrocarbon content, and the other follows the same porosity trend with depth as do other carbonate rocks.

Acknowledgements

The author is grateful to many colleagues at the Geological Survey of Denmark for their willing assistance and comments in the preparation of this chapter. Essential advice in the selection and preparation of the figures was given by P. Nyegaard. The manuscript was greatly improved by additions from L.J. Andersen, G. Andersen, O. Bertelsen, H. Kristiansen, and from H.-M. Moeller of R. and H. Consul.

11. Sweden

O. Gustafsson

Geographical distribution of the Chalk

Upper Cretaceous and Tertiary rocks are found in four areas in the most southern provinces of Sweden (Fig. 11.1). These sediments were deposited during many marine transgressions and regressions, often at the margin of the sea. For that reason they are very variable, consisting of conglomerates and sandstones, or sands, as well as marlstones and limestones. The limestones often contain argillaceous or sandy intercalations. Chalk was deposited mainly during Late Maastrichtian times in the south-western part of the province of Skåne. However, in the context of this chapter, chalk is taken in the wider sense to include all the Upper Cretaceous and Danian limestones.

The most extensive Upper Cretaceous and Tertiary deposits are found in south-west Skåne, where they cover an area of almost 1500 km^2 (Fig. 11.1). The Vomb Basin extends over about 400 km^2 and is situated in the depression between the Romeleåsen crystalline horst to the west and the Cambro-Silurian plateau to the east. The usually rather flat Kristianstad Plain covers an area of about 1000 km^2, but in the north about ten low residual mountains break the usually even topography. The Båstad area covers about 40 km^2 and is a flat coastal plain.

The Upper Cretaceous and Tertiary rocks (referred to as 'bedrock') are everywhere overlain by Quaternary deposits, and recharge of the groundwater resources of the bedrock aquifers is principally from infiltration through these younger deposits. The volume of recharge can be increased by lowering the water level in the bedrock aquifers to induce leakage from the Quaternary deposits.

The upper part of the bedrock sequence is well known from a large number of shallow wells, especially in south-west Skåne and on the Kristianstad Plain. Most of the area is covered by hydrogeological maps, and detailed groundwater investigations have also been carried out around some of the municipal wells. The deeper parts of the sequence are known from seismic investigations and drilling for oil and geothermal purposes. There are about 30 boreholes in south-west Skåne and the Vomb Basin which penetrate all or most of the Upper Cretaceous sequence.

General structure and stratigraphy

South-west Skåne

South-west Skåne is situated immediately south-west of the Fennoscandian Border Zone, the tectonic buffer zone between the Baltic Shield and Danish–Polish Trough. Along the north-eastern part of the area the displacement of the boundary fault zone is about 2000 m. The dominant tectonic feature south-west of the Romeleåsen Horst is the Svedala Fault running from Trelleborg in an approximately north–south direction. A characteristic tectonic unit in south-west Skåne is the Alnarp Valley, a depression in the Upper Cretaceous surface crossing the area from north-west to south-east. The Alnarp Valley is 4–6 km wide, usually 30–40 m deep and entirely filled with Quaternary deposits (Figs. 11.2 and 11.3).

The Danian and Upper Cretaceous sequences have a maximum thickness of about 2000 m in south-west Skåne (Fig. 11.4). They are underlain by Lower Cretaceous, Jurassic, Triassic, and Cambro-Silurian rocks, and the maximum thickness of the sedimentary sequence can exceed 2500 m. Along the north-eastern border of the area, the lower parts of the Upper Cretaceous, mainly sandstone and marlstone, subcrop in a narrow strip. South-west of this strip, the Maastrichtian white chalk occurs directly below Quaternary deposits in a zone 1–3 km wide. The Maastrichtian rocks have a maximum thickness of more than 600 m, with the Lower and Middle Maastrichtian sediments being usually limestones with clayey intercalations (Brotzen 1945). The limestones in the Upper

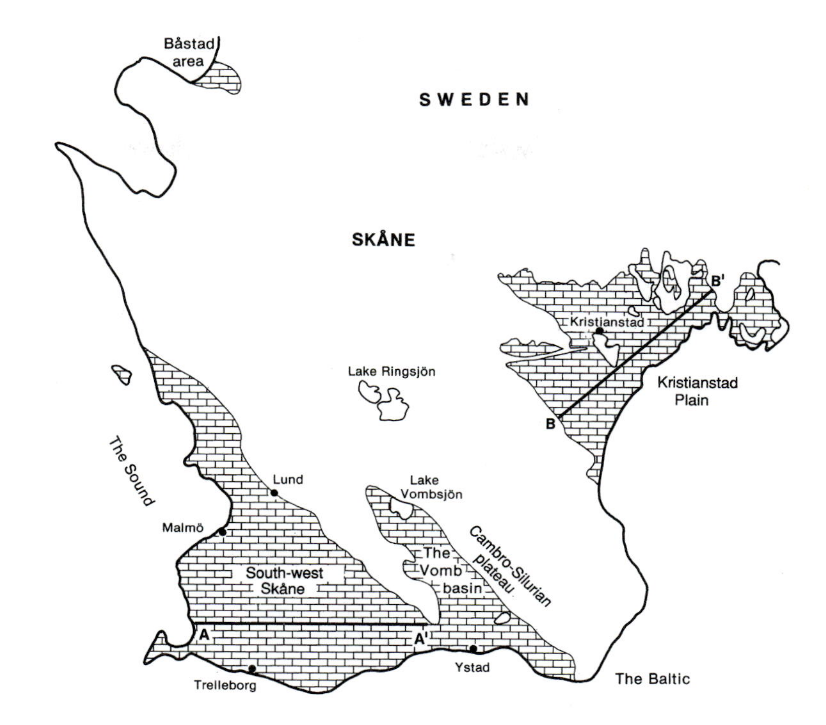

Fig. 11.1. Distribution of the Upper Cretaceous and Tertiary rocks in Sweden.

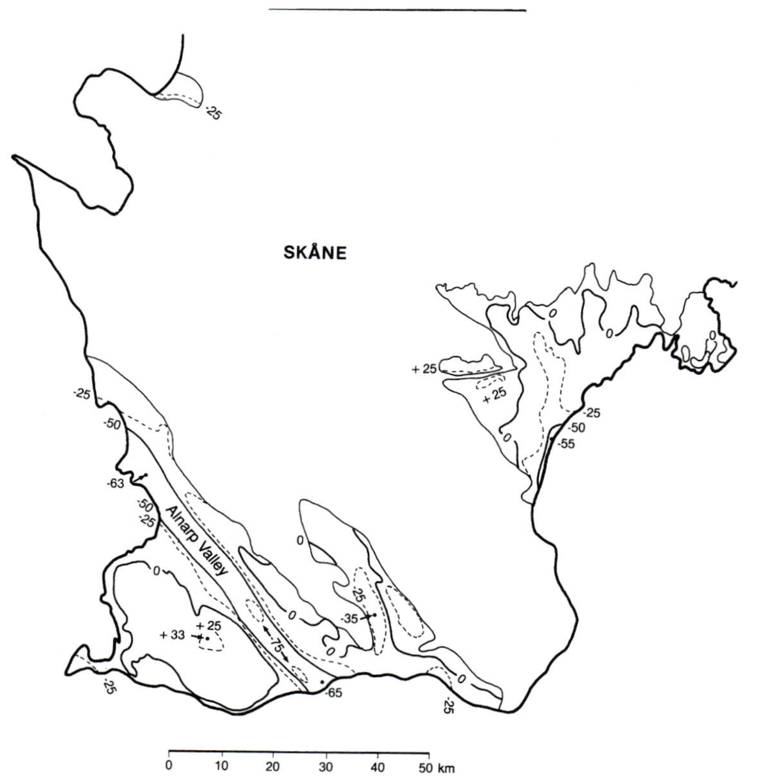

Fig. 11.2. Contours on the top of the Upper Cretaceous and Tertiary (in metres relative to sea level).

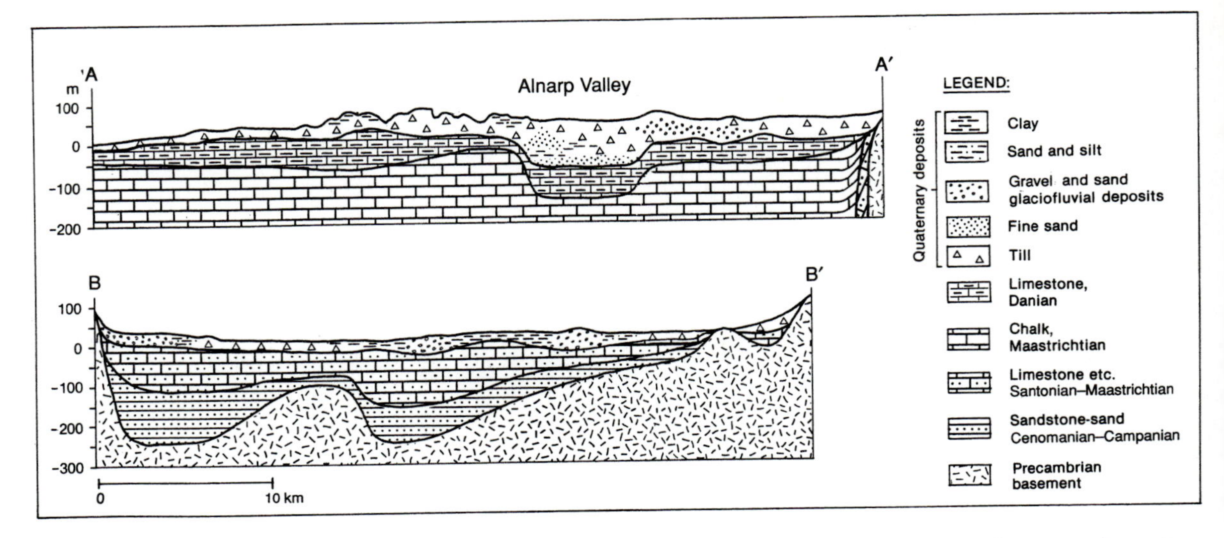

Fig. 11.3. Simplified geological sections of south-west Skåne and the Kristianstad Plain. Lines of sections shown in Fig. 11.1.

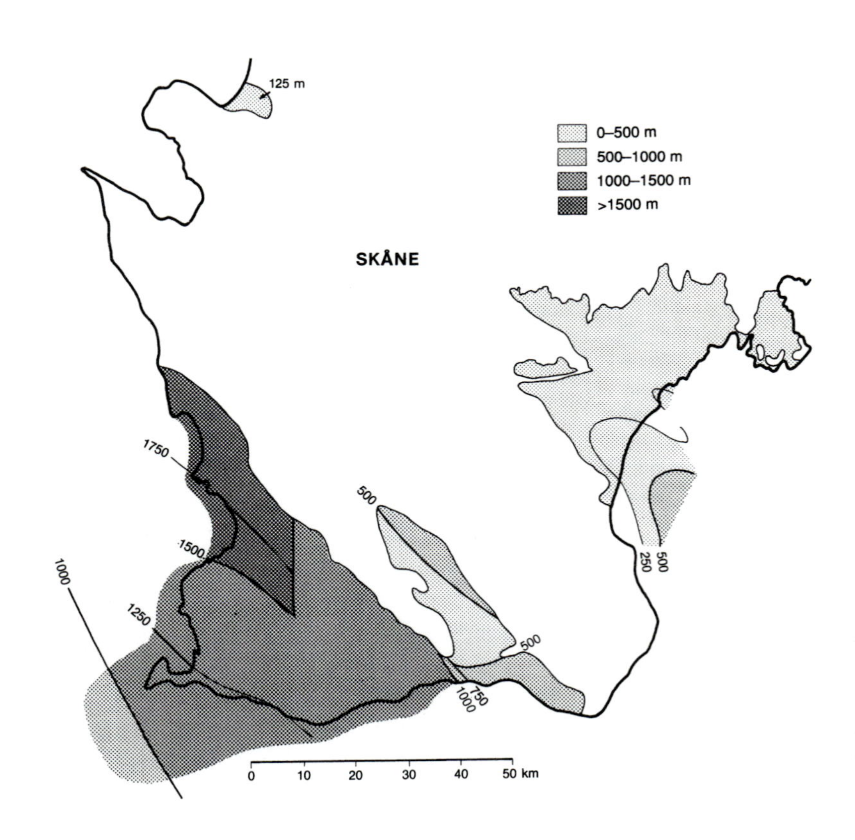

Fig. 11.4. Thickness of the Upper Cretaceous and Tertiary.

Maastrichtian are chalks with many flints towards the top. Over most of the area, the subcrop below the Quaternary is formed by a rather thin (usually 50–150 m) Danian limestone or chalk. Flint is very common in some parts of the limestone and can occur as layers 1–2 m thick.

The Danian and Cretaceous sediments are covered by younger deposits over the entire area. Thin (up to 25 m) marine Lower Tertiary sediments consisting of calcareous sandstones, siltstones and claystones have been preserved in a few small areas. Quaternary deposits provide the main cover with a thickness ranging from a few metres to more than 180 m (Fig. 11.5). The interior of the area is a hummocky moraine region, bordered by coastal moraine plains. The most common deposit is a clayey till or a clay till. From a hydrogeological point of view the so-called Alnarp sediments consisting mainly of sand, that occur at the bottom of the Alnarp Valley, are of special interest. Extensive gravel and sand deposits below the upper till in the south-eastern part of the area are also important aquifers.

The Vomb Basin

The Vomb Basin is a narrow, elongated graben (Bergström *et al.* 1982). The eastern margin is defined by a major fault, which forms the western boundary of the Cambro–Silurian plateau. The Jurassic sediments along this fault are more or less vertical or have even been slightly overthrust. To the west, the basin is bordered by the Romeleåsen Horst. The basin is affected by a large number of faults with different trends which have formed many minor troughs and uplifts as well as producing significant variations in the thickness of the sediments.

The bedrock surface below the Quaternary deposits in the Vomb Basin is mainly of Late Cretaceous age (Campanian–Maastrichtian) with a narrow rim of Late Triassic and Jurassic rocks. The Cretaceous bedrock consists of marlstone with minor limestones and sandstones. The maximum thickness of the Upper Cretaceous sediments is about 700 m (Fig. 11.4).

The bedrock is overlain by Quaternary deposits,

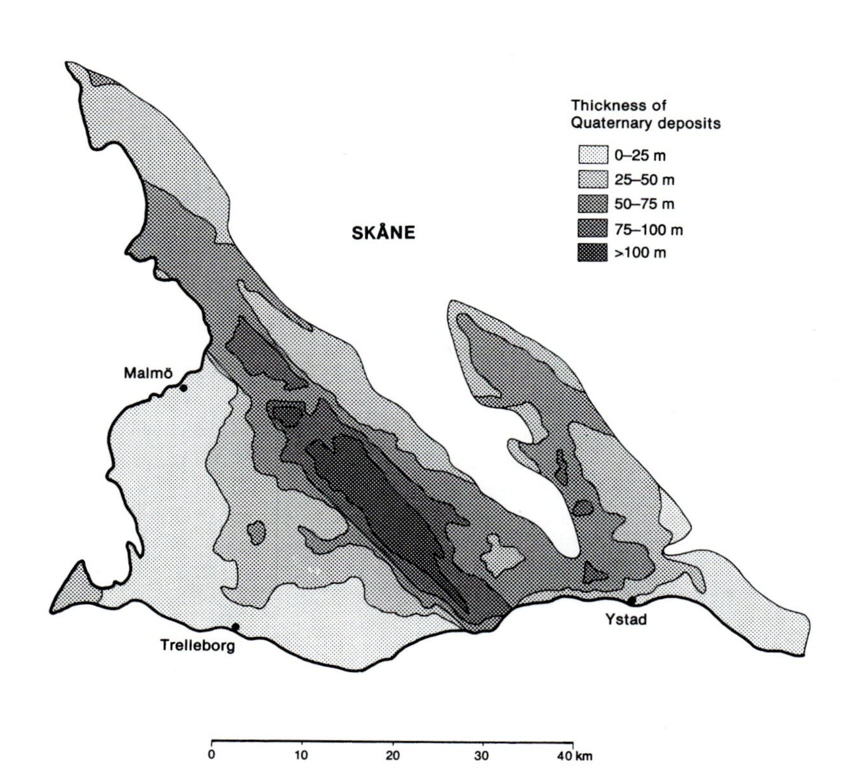

Fig. 11.5. Thickness of Quaternary deposits in south-west Skåne and the Vomb Basin.

usually 25–75 m thick (Fig. 11.5). In the south there is an area of hummocky moraine and to the north extensive deposits of gravel and sand occur.

The Kristianstad Plain

This is a depression in the Precambrian basement filled with Cretaceous and Quaternary deposits. To the south the plain is bounded by tectonic fault blocks, while along the northern boundary the bedrock is deeply eroded, creating a comparatively irregular topography. In the western part of the plain, there is probably an isolated Cretaceous deposit, possibly separated from the rest of the plain by Tertiary river erosion. There are also several small isolated Cretaceous deposits north of the plain.

Most of the sedimentary bedrock on the Kristianstad Plain belongs to the Upper Cretaceous, Cenomanian–Maastrichtian stages, (Kornfält *et al.* 1978). In the lower parts of the sequence an unconsolidated glauconitic sandstone (or sand) is commonly found (Fig. 11.3) and its thickness exceeds 50 m in the two deepest depressions of the basement. The upper parts of the bedrock consist of calcareous sediments, either limestone, sandy limestone or calcareous sandstone. However, only about 1 per cent of the total volume of the sequence is pure limestone, that is, with at least 90 per cent $CaCO_3$ (Nilsson 1966). Flint is rather common in the north-east of the plain. The calcareous sediments are usually well consolidated and between 50 and 120 m thick.

The Quaternary deposits cover the entire area and are usually 20–40 m thick in the centre of the plain. Till is the most common deposit on the plain but there are also several glaciofluvial deposits (eskers), surrounded by glacial clay and beach deposits of sand and silt. Sand dunes along the coast in the east are noticeable features of the landscape.

The Båstad area

The Late Cretaceous sediments in the Båstad area are believed to be of Cenomanian and Campanian age (Wikman and Bergström 1987). They consist mainly of limestone with a maximum thickness of about 125 m. In the extreme south-west of the area, they are covered by thin Quaternary deposits, but usually the thickness of these deposits is about 50 m.

Groundwater levels and flow paths

South-west Skåne

In south-west Skåne the groundwater flow is from inland areas towards the Baltic and the Sound (Fig. 11.6). North-west of the groundwater divide in the Alnarp Valley, the groundwater flows to the north-west, and is referred to as the Alnarp stream, and on the other side of the divide the flow is directed south-eastwards, and referred to as the Skivarp stream. In some areas the groundwater contours clearly indicate upward leakage from the bedrock into the overburden and eventually into surface streams.

As a result of groundwater abstraction, there are some cones of depression, especially around the municipal wells in the north-west, but in most of the area abstraction is rather limited and the contours show no influence of pumping.

The Vomb Basin

An east–west groundwater divide separates the Vomb Basin into two parts with flow directed inland and to the Baltic, respectively. The groundwater abstraction from the bedrock is small and there are no signs of any fall in levels caused by water withdrawal.

The Kristianstad Plain

On the Kristianstad Plain groundwater flow is from the upland areas towards the lowlands and the Baltic (Fig. 11.6). In some low-lying western parts of the plain the groundwater contours clearly show evidence of leakage from the bedrock aquifer into the Quaternary deposits and then into surface streams. In one small area, around the city of Kristianstad, the groundwater level has declined below sea level due to groundwater abstraction for municipal and industrial purposes. Pumping from a limestone quarry in the north-west of the plain has caused a limited decline of water levels. Over the remainder of the plain the groundwater levels are influenced by pumping only in the summer when

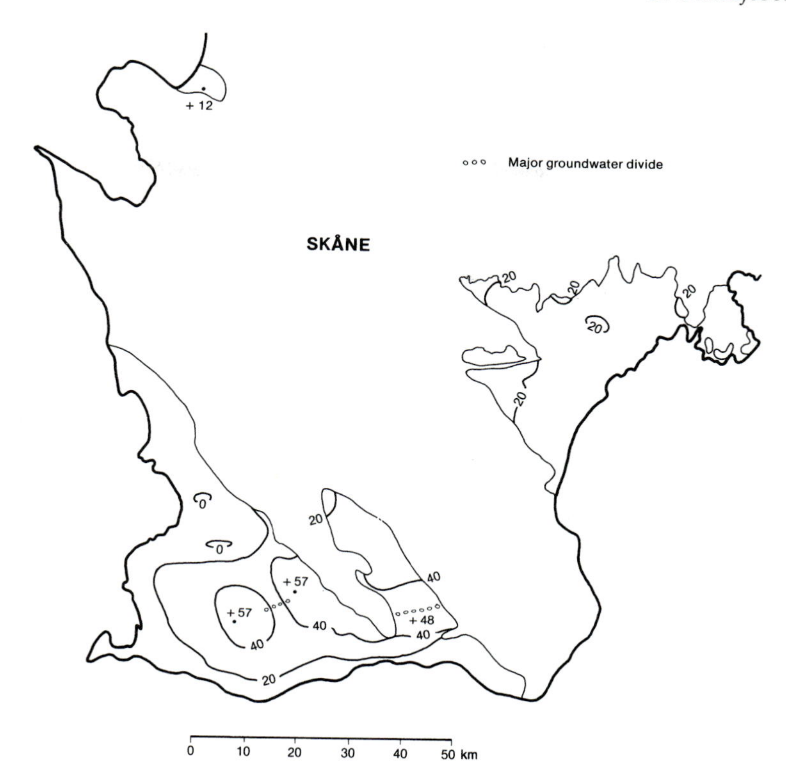

Fig. 11.6. Groundwater levels of the Upper Cretaceous and Tertiary (in metres above sea level).

withdrawals for irrigation cause local cones of depression over short periods of time.

Regional aquifer properties and karstic features

South-west Skåne

The aquifers of the superficial deposits above the Upper Cretaceous and Danian are usually unconfined. Discharge from them takes place to rivers and streams, to the sea and by leakage into underlying confined aquifers. The confined Chalk aquifers are mainly recharged inland and in the north-west of the area. Discharge is in coastal areas and to some extent offshore (Fig. 11.7).

The transmissivity of the Chalk aquifer along the coast, where the water-yielding capacity is at its best, usually ranges from 10^{-3} to $3 \times 10^{-3}\,m^2\,s^{-1}$. In areas where the conditions are less favourable values range between 10^{-5} and $10^{-4}\,m^2\,s^{-1}$. The storage coefficient is generally between 10^{-4} and 5×10^{-4}.

An example of direct recharge of the Chalk aquifer from the surface occurs in south-west Skåne where, about 3 km east of Trelleborg, the water of a small stream disappears into a swallow-hole in the summer.

The Vomb Basin

The confined Cretaceous aquifer in the Vomb Basin is usually covered by till. In the north, superficial gravel and sand deposits, representing extensive aquifers, are underlain by clay and till. Very few pumping tests have been performed in the bedrock aquifer. In the south-east the transmissivity seems to be between 10^{-3} and $5 \times 10^{-3}\,m^2\,s^{-1}$ though values are most likely to be significantly lower in the north of the basin.

The Kristianstad Plain

The Cretaceous aquifer is mainly confined in the plain except in small parts of the recharge areas. In the north-west and south-west of the plain, direct

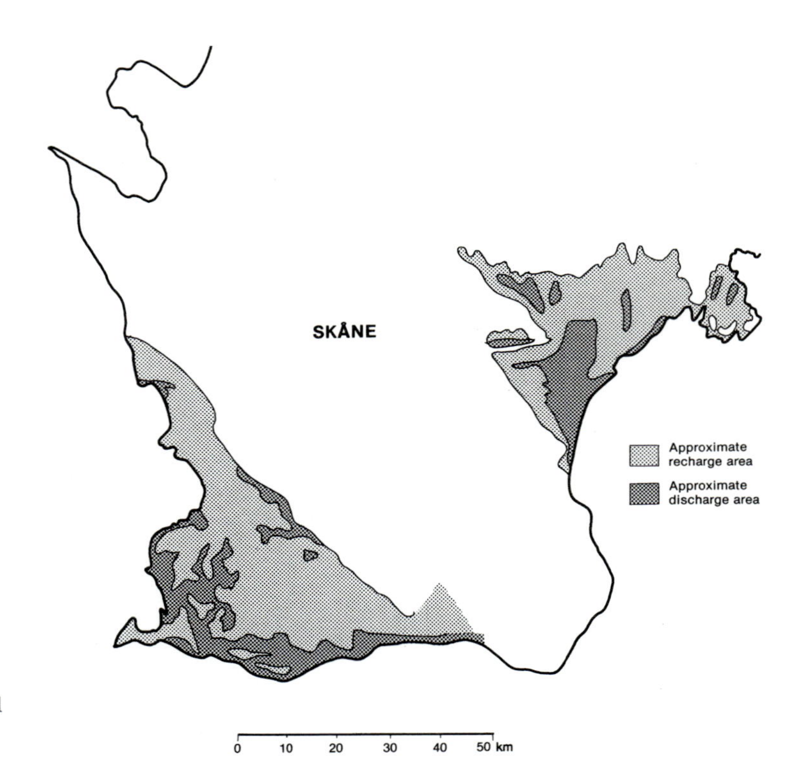

Fig. 11.7. Recharge and discharge areas in south-west Skåne and on the Kristianstad Plain.

recharge of surface water from small rivers and streams occurs. Dolines are known in the north-west as well as in the north-east of the plain. The transmissivity of the limestone aquifer is variable but commonly is in the range of 10^{-3} to $2 \times 10^{-3}\, m^2\, s^{-1}$.

The Båstad area

In the south of the Båstad area, the Cretaceous sequence abruptly rises some 50 m and occurs at the surface at a number of localities. As there is no sign of a fault in this area, the slope probably represents a Tertiary cliff, subsequently rounded by erosion (Wikman and Bergström 1987). Recently, a local underground collapse of the cliff produced a deep hole in the ground surface.

Hydrochemistry

Chemical properties

The groundwater from the Chalk aquifers in Skåne is usually hard or very hard and of the calcium bicarbonate type; it often has a rather high or high iron content. The fluoride content is generally low and below values that give protection against dental caries. The salinity is also usually low except in a few small areas (Fig. 11.8).

South-west Skåne

A chloride content of more than 300 mg l^{-1} is rather common in the upper parts of the limestone in the Alnarp Valley and its surrounding area (Fig. 11.9). This is generally the case in the north-west of the area and is probably always caused by fossil saline water rising from the deeper parts of the bedrock. Groundwater from the bedrock has a rather high natural ammonium content but usually lacks nitrate, which is common in aquifers at shallow depths in the Quaternary.

The Vomb Basin

One well with a rather high chloride content is known from the north of the Vomb Basin (Fig. 11.9). High values are also found in some wells penetrating the Triassic and the Precambrian

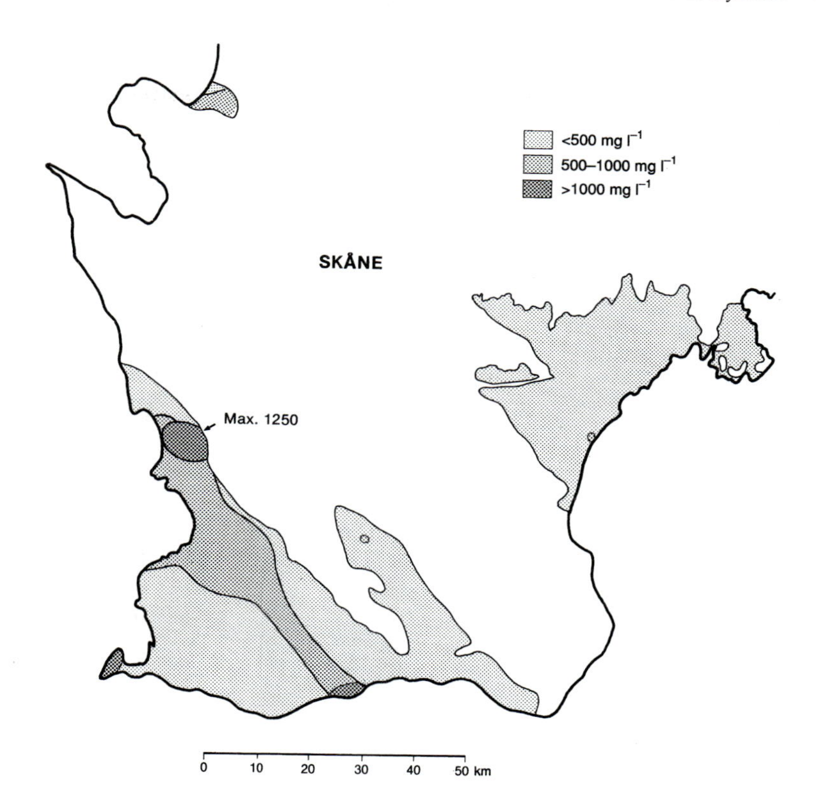

Fig. 11.8. Total ionic content of groundwater in the Upper Cretaceous and Tertiary.

basement in a small area along the north-western boundary of the basin. Locally, high nitrate contents are found in the south-east of the area, where the Quaternary sequence is thin.

The Kristianstad Plain

The groundwater from the glauconitic sandstone is usually less hard and has a lower iron content than the water from the limestone. Low chloride contents are common and there are very few signs of any rising values caused by saltwater intrusion from the Baltic. The only exceptions are two small areas in the north-east and east of the plain, where high chlorides do occur (Fig. 11.9). The nitrate content is high in the upper Quaternary aquifers and is also rising in the Cretaceous aquifers in some of the recharge areas (Gustafsson *et al.* 1988).

The Båstad area

The groundwater quality of the Cretaceous aquifer is virtually unknown, but both fresh and brackish groundwaters do exist near the coast.

Tritium

South-west Skåne and the Vomb Basin

Tritium measurements usually show very low values in the bedrock aquifers, indicating that the water has a long residence time. In areas with high-producing wells and fairly thin Quaternary cover, the tritium content has been found to be somewhat higher (Gustafsson 1972; 1978). The infiltration rate might have been influenced by the high level of abstraction in these areas.

The Kristianstad Plain

Tritium analyses give varying values from different parts of the plain. In the recharge areas near the boundaries of the plain, the values indicate a groundwater age of 5–10 years in the Cretaceous aquifers (Gustafsson *et al.* 1988). On the other hand, in the discharge areas near the coast, the tritium content is very low. A carbon-14 analysis has given an approximate age of more than 4000 years for groundwater from the glauconitic sand-

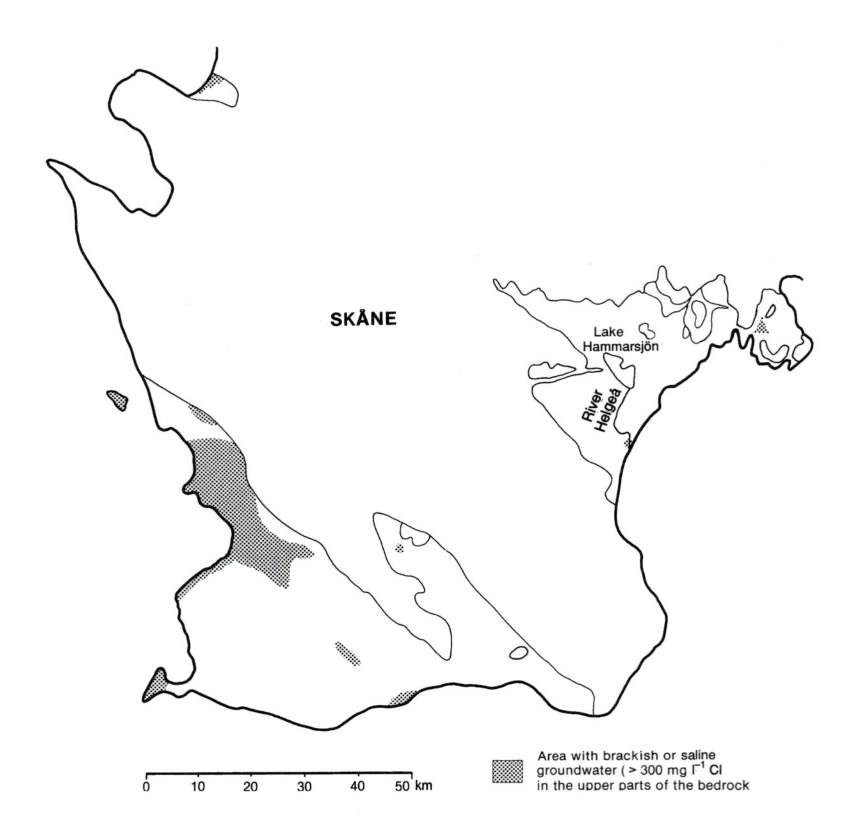

Fig. 11.9. Areas with brackish or saline groundwater in the upper parts of the Upper Cretaceous and Tertiary in Skåne.

stone at a depth of about 200 m. The tritium values indicate a faster rate of recharge to the aquifer around the city of Kristianstad than in surrounding areas due to the high withdrawal of water to supply the community.

Pollution problems

South-west Skåne and the Vomb Basin

Because of the good natural protection afforded by the Quaternary deposits, there are very few pollution problems in the deep-seated Tertiary and Upper Cretaceous aquifers. In exceptional cases high nitrate contents are known in areas with a thin cover, but on the whole the groundwater is unaffected by pollution.

The Kristianstad Plain

The primary groundwater problem on the Kristianstad Plain is nitrate pollution in some of the recharge areas. The pollution, which mainly originates from agricultural practices, requires the introduction of special measures to protect the quality of the groundwater against contamination caused by the use of fertilizers. These steps need to be taken as soon as possible in order to reduce the nitrogen input to the aquifers.

Saltwater intrusion

South-west Skåne

As mentioned previously, saline groundwater is a problem in some parts of south-west Skåne, near the coast as well as inland (Fig. 11.9). Most of the saline groundwater is considered to be fossil water originating in the deeper parts of the sedimentary bedrock where very saline groundwater occurs. Near the coast, the saline water may have a different origin, being the result of seawater intrusion caused by groundwater abstraction.

During the early 1970s groundwater withdrawal from the Alnarp Valley reached its peak. At that time, groundwater levels were lowered below sea level over a rather large area in the north-west of the valley. In order to control possible saltwater intrusion, three observation wells were drilled in 1972 and 1974–5 to a depth of 150 m (Brinck and Leander 1981). The movement of the saltwater front was then observed regularly over 10 years. The measurements showed that very small changes of the saltwater front took place and that no significant saltwater intrusion occurred. During the 1980s groundwater abstraction was reduced and, with rising groundwater levels during the last decade, there is now no risk of saltwater intrusion.

The Kristianstad Plain

In the north-east of the Kristianstad Plain, appreciable groundwater abstraction from a limited coastal area has increased the chloride content of the groundwater (Fig. 11.9) leading to the abandonment of a muncipal well. In the same part of the plain saline, probably fossil, groundwater, is found very locally but generally the chloride content is low. In the east of the plain there are a few saline, rather shallow, wells in the immediate vicinity of the Helgeå river outlet. These probably reflect saltwater intrusion from the Baltic into the Helgeå river during periods of high sea levels. The saltwater can infiltrate the shallow Quaternary aquifers and then flow to greater depths into bedrock aquifers.

Groundwater resources and abstraction

Hydrometeorology

The precipitation over Skåne province ranges from about 850 mm a^{-1} in the north-west to 650 mm a^{-1} in the south-west and east of the area. The annual evaporation is calculated to be 400–550 mm, which gives an annual effective rainfall, for disposal as surface runoff and/or infiltration, of about 300 mm. The area is drained by several small rivers and streams.

Recharge and discharge areas

South-west Skåne

The principal area for recharge to the deep-seated aquifers is in the hummocky upland area (Fig. 11.7) but it also occurs in the north-west where groundwater abstraction is concentrated. The discharge mainly occurs in the plains near the coast, where groundwater leaks from the bedrock into the Quaternary aquifers and eventually into the streams. To some extent discharge probably also occurs offshore.

The Vomb Basin

A similar recharge and discharge pattern to that in south-west Skåne probably occurs in the Vomb Basin, but few details are known.

The Kristianstad Plain

On the Kristianstad Plain recharge occurs into the higher ground and discharge is in the lowlands (Fig. 11.7). As the figure shows, the discharge areas are extensive. Artesian wells are common along the southern part of the coast.

Groundwater resources

South-west Skåne and the Vomb Basin

In Table 11.1 the calculated renewable groundwater resources of the Upper Cretaceous and Danian are shown. The calculations are based on an assumption that an increase in groundwater abstraction from the bedrock aquifers would increase the leakage from the Quaternary in favourable areas (Gustafsson and De Geer 1977). The groundwater levels would decline and recharge would take place over most of the region. An annual recharge of 60–100 mm to the bedrock aquifers would then be probable. The estimates suggest that a considerable increase in abstraction is possible.

The Kristianstad Plain

As for south-west Skåne, maximum abstraction from the aquifers requires an increase in pumping in the discharge areas. Calculations and modelling of the aquifers show that the present abstraction could be increased at least four times, which would mean a yield of possibly 70–80 $\times$ $10^6 m^3 a^{-1}$ (Gustafsson *et al.* 1979; 1988).

Groundwater abstraction

South-west Skåne

The municipal water supply for the city of Malmö and its surrounding area is primarily based on the water-treatment works at Lake Vombsjön in the central part of Skåne, where water is artificially recharged into Quaternary sands. About 75 per cent of the water used in Malmö originates from there. The towns in the north-west of the region are connected with a water conduit from the water treatment works at Lake Ringsjön, the other large lake in Skåne. The water has been brought here since 1987 from Lake Bolmen in the province of Småland, north of Skåne.

The rest of the municipal water supply for the region is dependent on groundwater. Wells drilled in the Alnarp Valley are screened at the bottom of the Quaternary Alnarp sediments. Most groundwater for municipal use is from these sediments. Adjacent to the valley the limestone aquifer is used. In addition to the municipal supply, groundwater is used for local water consumption, industrial purposes and irrigation. The estimated abstraction from the Chalk aquifer is given in Table 11.1.

The Vomb Basin

The area is very important for the water supply of Skåne as the Vomb waterworks is situated here. However, the groundwater from the bedrock is used to only a small extent, as indicated in Table 11.1.

The Kristianstad Plain

The municipal water supply for the entire region is based on groundwater and a great deal of groundwater is also used for industrial purposes. The Kristianstad Plain is the most irrigated area in Sweden and groundwater abstraction for this purpose equals, and in some years even exceeds, the municipal consumption. The present use of groundwater from the Chalk aquifer is given in Table 11.1.

The Båstad area

Very little groundwater is taken from the bedrock aquifer in this area (Table 11.1).

The use of heat pumps

During the early 1980s, heat pumps became widely used in Sweden. In Skåne many factories and houses installed them, generally using groundwater as the source of energy. Usually one production well and one injection well were drilled (Fig. 11.10). The groundwater, with an initial temperature of about 9°C, was cooled by 5–7°C in the heat pumps before reinjection. The good water-yielding properties of the Cretaceous and Danian bedrock in Skåne made this kind of installation very popular. Typical abstraction rates for domestic and industrial purposes are 2000 $m^3 h^{-1}$ (for one-family houses) and 20000 $m^3 h^{-1}$, respectively. In some cases installations were designed to provide air conditioning during the summer. However, problems have arisen in many cases. One reason is the commonly high iron content of the groundwater, which has caused clogging in one or both wells. Another problem is that too many heat pumps have been installed in some densely populated areas. Consequently the temperature of the groundwater has fallen and led to operational difficulties. The possibility of avoiding this problem by reversing the water flow during the summer and adding solar energy from solar panels or air conditioning

Table 11.1 Estimated renewable groundwater resources and current use of groundwater in the Chalk aquifers in Skåne ($\times 10^6 m^3 a^{-1}$)

Area	Municipal use	Other use	Total consumption	Calculated renewable resources
South-west Skåne	5	10	15	60–75
Vomb Basin	0.5	2	2.5	10–15
Kristianstad Plain	8	12	20	70–80
Båstad area	0	0.1	0.1	1
Total	13.5	24	37.5	140–170

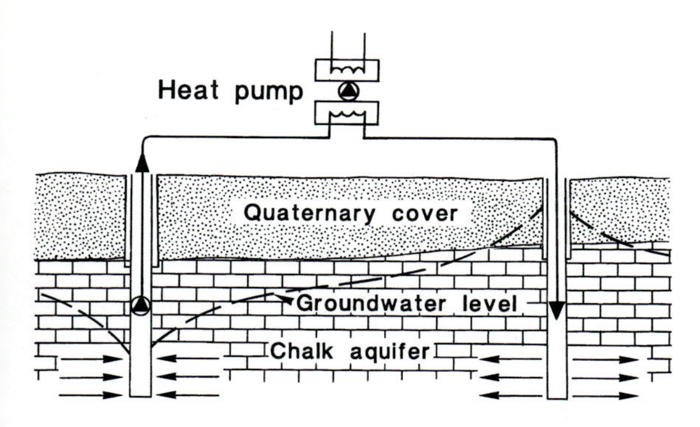

Fig. 11.10. Typical heat-pump installation with reinjection of groundwater.

systems has been discussed. Mainly because of the clogging problems, many heat pumps are now being replaced by other heating systems; nevertheless, the technique is still widely used.

The development of geothermal energy

At Lund, groundwater in the Upper Cretaceous rocks has been developed for geothermal energy. The project started in the early 1980s. It involves abstracting water from an Upper Campanian sandstone, approximately 200 m thick, for use as an energy source. The sandstone is overlain by almost 500 m of Danian and Maastrichtian limestone, chalk, and calcareous claystone. Four production and five injection wells have been drilled to depths of 600–700 m. The water has a temperature of only 20°C but it is cooled to 4°C in two heat pumps with capacities of 20 and 27 MW. The water has a chloride content of about 40 000 mg l^{-1} and the maximum production rate is 1600 $m^3 h^{-1}$.

Typical well yields

South-west Skåne

In this area, well yields are mainly dependent upon the degree of fracturing of the rocks. Fractures are usually most common in the upper 30 m of the sequence, mainly because of the influence of the Pleistocene glaciations. A few metres of the uppermost parts of the bedrock are sometimes soft due to the same cause.

The most favourable conditions for groundwater abstraction are from the Danian bedrock in the south and south-west of the area, where maximum capacities of most wells penetrating the entire aquifer are 10–20 l s^{-1}. A few wells, under exceptionally good conditions, yield 40–50 l s^{-1}. In the area close to the Alnarp Valley the yields are smaller, usually 5–10 l s^{-1}, though in the deeper parts of the Valley, where the bedrock was more protected during the glaciations, well yields are much smaller, usually 2–5 l s^{-1}. Because of the small yields from the bedrock aquifers, all major wells in this area are screened in the basal section of the overlying Alnarp sediments and yields of well-designed wells are 10–40 l s^{-1}.

Wells in the white chalk generally yield 2–5 l s^{-1} except in parts of the northern area. There the conditions are unfavourable and well yields in the Danian limestone and the Maastrichtian chalk are small, usually only 0.5–2 l s^{-1}.

The Vomb Basin

In the Vomb Basin few details are available about well yields. Irrigation wells have been pumped at 10–30 l s^{-1} in the south-east. The aquifer conditions are less favourable in the centre of the region, where the bedrock is finer-grained.

The Kristianstad Plain

On the Kristianstad Plain, well yields are mainly dependent upon the well design. Municipal wells and major irrigation wells are usually screened in the glauconitic sandstone aquifer. These wells yield 50–100 l s^{-1} in areas where the sandstone is thick. Most irrigation wells are drilled through the limestone and a few metres into the sandstone. These wells normally yield 10–20 l s^{-1}. Small wells only penetrate the top 10–30 m of the limestone and they usually yield 1–5 l s^{-1}.

The Båstad area

Very few data are available for the Båstad area, although some wells have been pumped at rates of 5–15 l s^{-1}.

Acknowledgements

Drs Jan De Geer, Mikael Erlström, and Ulf Sivhed, of the Geological Survey of Sweden, have supplied me with additional geological and hydrogeological information and have also critically read the manuscript.

12. The United Kingdom

J.W. Lloyd

Introduction

The Chalk in the UK has been extensively used as an aquifer since the nineteenth century. Prior to that abstraction was generally from springs and hand dug wells in the outcrop areas, developments that in most cases were on a small scale. In the early seventeenth century the City of London was partly supplied by an aqueduct in the Lee Valley, in east London, referred to as the New River. In 1613, this began to carry some 30 000 m^3 of water per day from chalk springs in Hertfordshire (the Chadwell and Amwell springs) to New River Head in Clerkenwell.

Exploitation of the Chalk at depth and below Tertiary deposits commenced in the early nineteenth century from excavated shafts but towards the end of the century yields were supplemented in some areas by driving adits from the wells, as in east London and in the North and South Downs. Eventually development progressed to the use of standard percussion and large diameter caylex drilled wells. By the turn of the century the potential for groundwater abstraction from both confined and unconfined areas of the Chalk, in many parts of eastern and southern England, had been realized, and large numbers of wells were being drilled to depths of up to about 170 m and generally penetrating some 50–100 m of Chalk. Large beam pumps were installed and significant quantities of water abstracted. The early confidence in the potential of the Chalk as a source of water supply was not misplaced and today in the south, south-east and east of the country it represents an important element of the integrated water supply systems that have been developed.

The high permeability but low storage coefficient of the Chalk, while allowing the ready transmission of groundwater, limits the total storage and inhibits the flexible manipulation of the storage. Furthermore, the dominant, secondary permeability character of the aquifer makes it susceptible to relatively rapid contamination. The overall nature of the aquifer, therefore, poses constraints on its development, particularly in a modern society, and careful management is essential for effective and optimum use of the resource.

For many years the development of the Chalk as an aquifer was considered in predominantly engineering terms as a facility to provide water, without due regard to the consequences of abstraction on the aquifer itself and the impact upon associated surface waters. Also the Chalk was seen as 'the aquifer' and the concept of the Chalk as part of a larger hydrogeological system in much of the country was not recognized.

Today the Chalk is reasonably well understood in hydraulic terms, particularly the saturated zone. This understanding has led to sensible management policies developed on a regional basis in many areas. As with most aquifers, a large amount of research is still required to understand the flow mechanisms in the unsaturated Chalk, and in both the unsaturated and saturated zones, the nature of the movement of contaminants is poorly understood.

The Chalk is the most important aquifer in the UK, supplying about 55 per cent of the total volume of groundwater used. It underlies eastern and southern England with the base extending from Flamborough Head in north Humberside to the Dorset coast near Weymouth (Fig. 12.1). It also occurs in Northern Ireland. In the south of England, the Chalk is overlain by Tertiary deposits in the Hampshire and London basins, but has been eroded from the anticlinal structure of the Weald in south-east England (Fig. 12.2). Over much of Hampshire, Dorset, and Wiltshire the outcrop forms a high downland of gently rolling hills. In East Anglia, Lincolnshire, and Humberside the eastern part of the outcrop is largely obscured by glacial deposits. The Chalk's outcrop is characterized by an absence of surface drainage, but the existence of many 'dry valleys' is important in terms of the hydrogeology of the aquifer.

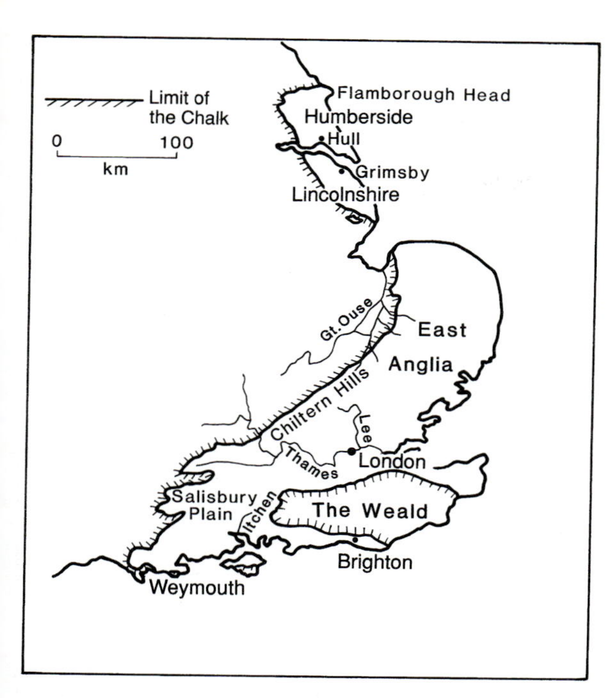

Fig. 12.1. Principal geographical features associated with the Chalk.

In this chapter the main hydrogeological features of the British Chalk are described and the bearing these have on the use of the aquifer for water resources is discussed.

General stratigraphy and structure

A protracted discussion of the geology of the Chalk is outside of the scope of this chapter, but a brief summary is essential in order to understand certain aspects of the hydrogeology. The following résumé is largely based on Hancock (1975*a*), Chadwick (1985), and Rawson (1992).

The Chalk in the UK is entirely Upper Cretaceous in age. During the Late Cretaceous marine transgression, much of England and Wales was submerged from Early Cenomanian times and, at the time of the maximum extent of the Chalk sea in Late Campanian times, the sea possibly covered all the British Isles except the Scottish Highlands and north Wales. A major fall in sea level near the end of the Maastrichtian heralded the Late Cretaceous inversion and led to erosion removing considerable thicknesses of Chalk. Subsequently, Tertiary marine and freshwater sediments were deposited on an eroded Chalk surface, particularly over what are now the London and Hampshire basins. The main folding event that finally created these structural basins was in post-Oligocene times and was related to Alpine tectonic events. As a consequence, Tertiary deposition ceased and an erosional cycle began. The inversion of the Weald Anticline separated the London and Hampshire basins, leading eventually, through erosion, to the formation of the North and South Downs, now major aquifer units of the Chalk.

At the present time the Chalk ranges in thickness between 200 and over 400 m in southern England, although less than 200 m remains in the London Basin because of erosion over the London Brabant massif (or London Platform) at the end of the Cretaceous. In East Anglia, Lincolnshire, and Humberside, thicknesses increase towards the coast from less than 100 m near the western outcrops to over 400 m in Humberside and East Anglia (Fig. 12.3). Maastrichtian chalk is preserved in north-east Norfolk below Tertiary cover and the Campanian is found in East Anglia and the Hampshire Basin; Upper Maastrichtian strata are not found onshore.

The basic structure of the Chalk is comparatively simple. In Humberside, Lincolnshire, and East Anglia it dips gently towards the east but further south the London and Hampshire synclines are the dominant structures (Fig.12.4). Subsidiary east–west flexures are a feature in southern England. In East Anglia, a pronounced trough occurs in the surface of the Chalk. Whether this is of structural or erosional origin is uncertain, but it is a feature that has hydrogeological significance. In Humberside, the Chalk is folded into a shallow NW–SE syncline that plunges towards the south-east.

The Chalk was deposited in two lithological provinces lying respectively to the north and south of East Anglia, which represents an intermediate area over the London Brabant massif where the Chalk is thinner than to the north and south (Table 12.1; Rawson 1992). The Chalk of the northern province is harder than that to the south, possibly because of high palaeotemperatures in Late Cretaceous to Early Tertiary times (Green *et al.* 1993). The Chalk of the northern province was subjected to two phases of

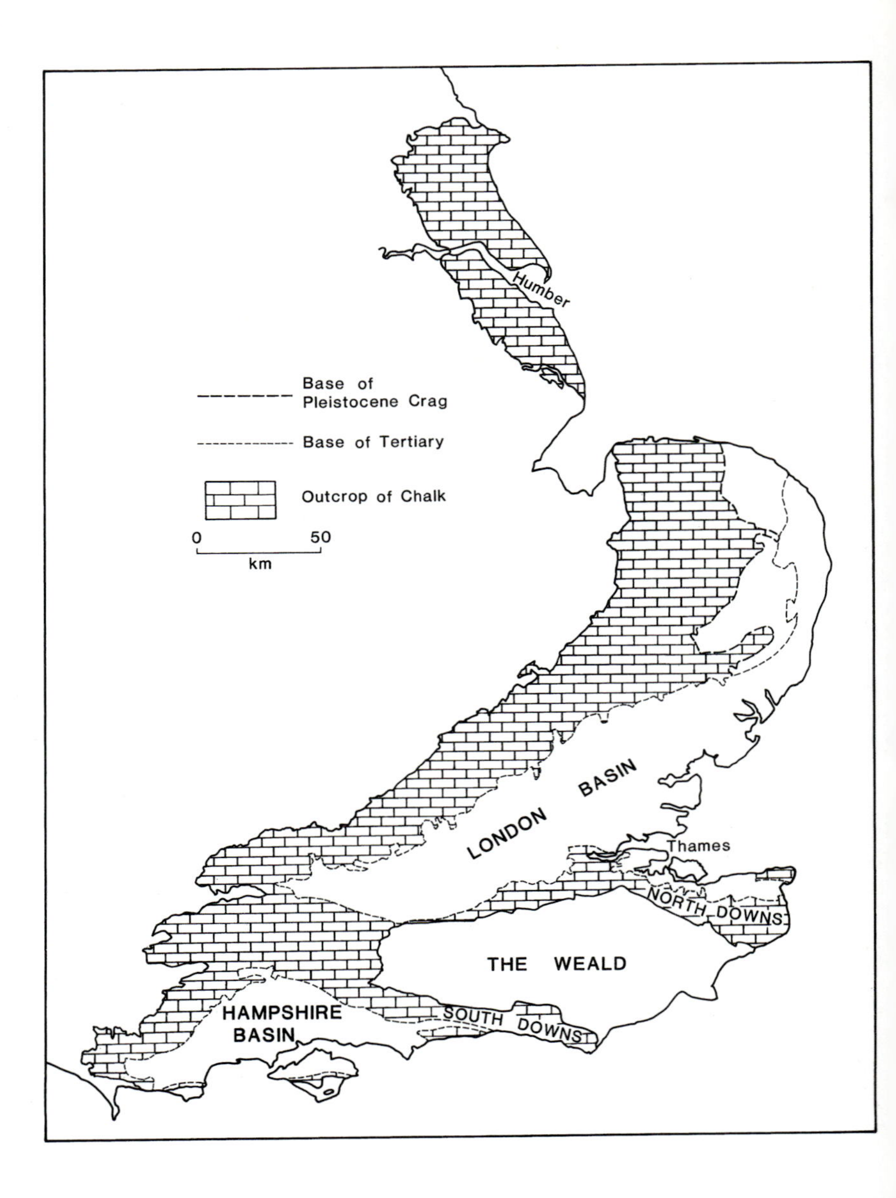

Fig. 12.2 Distribution of the Chalk and overlying formations.

calcite cementation although the first phase only affected nodules (Jeans 1980). Hard grounds are more common in the south where important horizons are the Melbourn and Chalk rocks and, in East Anglia and the Chiltern Hills, the Totternhoe Stone. The Chalk in Northern Ireland is very hard because of thermal diagenesis and in Dorset it has been hardened by tectonic stresses (see Chapter 2).

Geological sequences across England are given in Table 12.2 to illustrate the position of the Chalk in an aquifer system comprising overlying and underlying formations. Because of erosion, or in some instances non-deposition of younger deposits, the Chalk is exposed over large areas and for the same reasons the overlying younger formations are represented by variable thicknesses.

Historically, the Chalk has been divided in the UK into Lower, Middle, and Upper divisions that correspond approximately to the Cenomanian, the Lower, and Middle Turonian, and the Upper Turonian to Maastrichtian, respectively.

The Lower Chalk contains a high proportion of

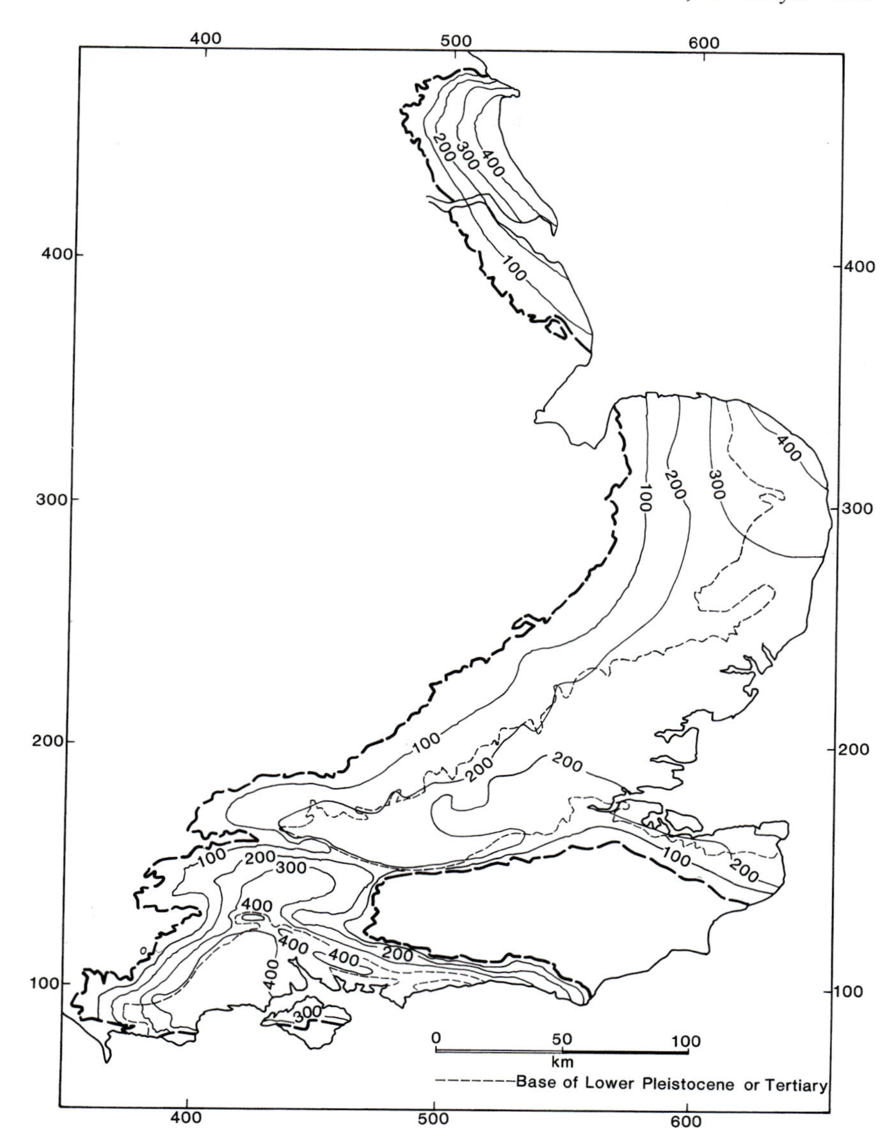

Fig. 12.3 Thickness of the Chalk, in metres (after Whittaker 1985).

terrigenous sediment—as much as 50 per cent at the base, decreasing to 10 per cent at the top. In southern England, the lowest deposit is the Glauconitic Marl which is overlain by marls and chalky limestones of the Chalk Marl and then the marly Grey Chalk; the Totternhoe Stone occurs between the Chalk Marl and Grey Chalk, and the Plenus Marls mark the top of the Lower Chalk. In northern England, interbedded bioclastic chalks and massive and marly chalks rest on the Red Chalk, an Albian limestone.

In Norfolk and areas further south, the base of the Middle Chalk is marked by the Melbourn Rock, a thin, hard, greyish-yellow chalk, which is succeeded by nodular chalk and then massively bedded white chalks. The Middle Chalk contains few flints.

The Upper Chalk is the thickest of the three units. It is a soft white chalk with flint bands and occasional hardgrounds including the Chalk Rock, a thin, hard, nodular limestone which occurs at the base in southern England. Most of the chalk uplands are formed by the Upper Chalk, with the Middle

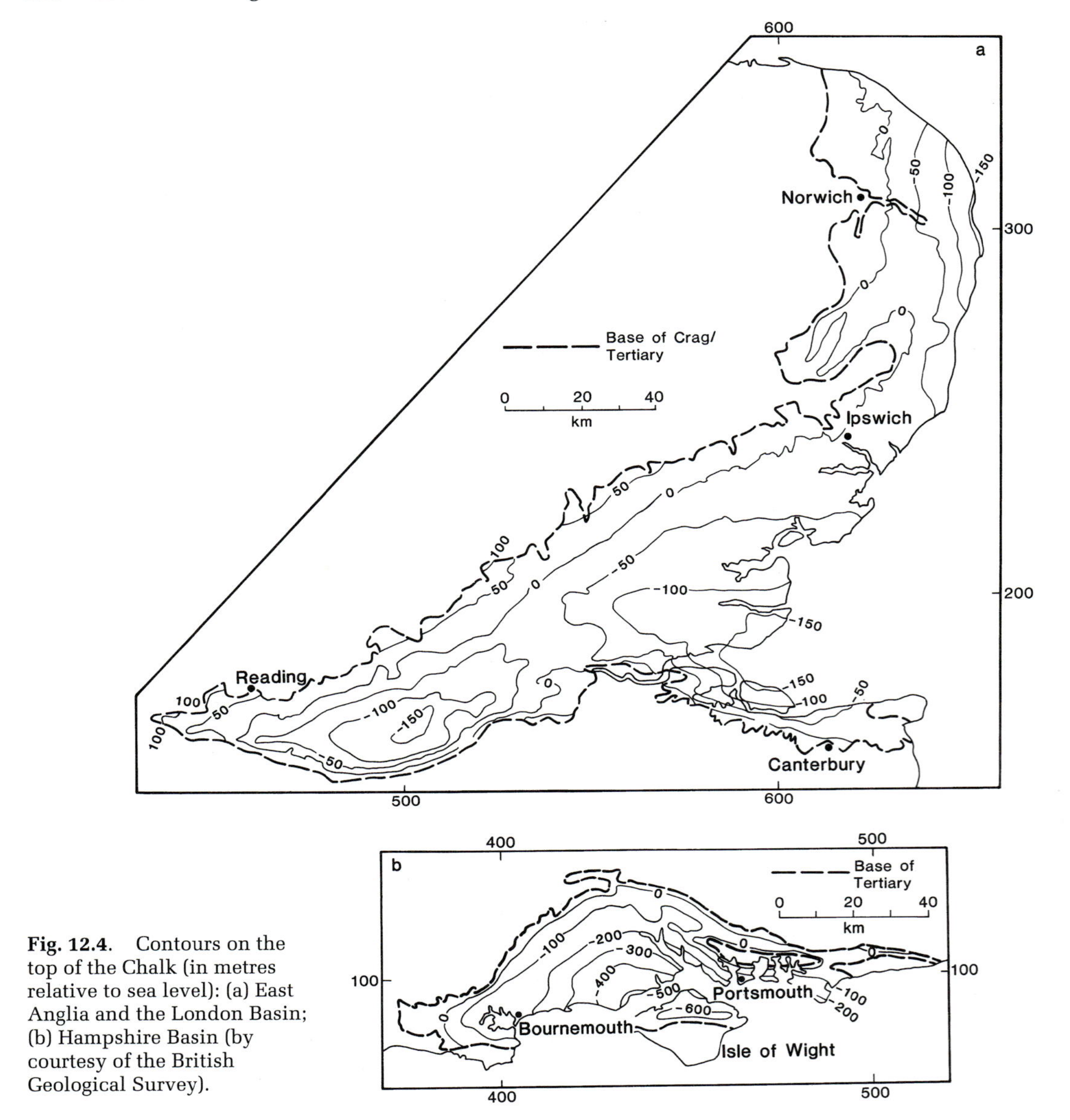

Fig. 12.4. Contours on the top of the Chalk (in metres relative to sea level): (a) East Anglia and the London Basin; (b) Hampshire Basin (by courtesy of the British Geological Survey).

and Lower Chalk occurring on the scarp slopes of the main outcrop or in the bottom of some valleys.

Recently the stratigraphy of the Chalk has been reassessed and a new detailed nomenclature defined (Table 12.1; Wood and Smith 1978; Mortimore 1983; 1986). As shown in Fig. 12.5, the detailed lithostratigraphical units can be recognized on geophysical logs, and thus major discontinuities can be delineated regionally, and these can have hydrogeological significance (Chapter 5; Barker *et al.* 1984; Murray 1986; Mortimore and Pomerol 1987).

The Tertiary deposits resting on the Chalk in the London Basin and the extreme east of East Anglia comprise a series of sands and clays referred to collectively as the Lower London Tertiaries. The lowest formation is the Thanet Formation, an arenaceous deposit that is in hydraulic continuity

Table 12.1 Subdivisions and correlation of the Upper Cretaceous Chalk (after Wood and Smith 1978, and Mortimore 1986; based on Rawson 1992)

NORTHERN PROVINCE LITHOLOGICAL UNIT	STAGE	SOUTHERN PROVINCE LITHOLOGICAL UNIT	
	MAASTRICHTIAN		
Flamborough Chalk Formation 300+ m	CAMPANIAN	Sussex White Chalk Formation	Portsdown Member 30 m
			Culver Member 115m +
Burnham Chalk Formation 150 m	SANTONIAN		Newhaven Member 75 m
	CONIACIAN		Seaford Member 90 m
Welton Chalk Formation 53 m	TURONIAN		Lewes Member 90 m
			Ranscombe Member 85 m
	UPPER CENOMANIAN	Plenus Marls Formation up to 8.5 m	
Ferriby Chalk Formation 28 m	MIDDLE CENOMANIAN	Abbotts Cliff Chalk Formation 22 m	
	LOWER CENOMANIAN	East Wear Bay Formation 58 m	

with the Chalk. In the Hampshire Basin, the basal Tertiary formation is the Reading Formation, which is a clay.

North of the Thames, large areas of the Chalk outcrop are overlain by Pleistocene deposits. In central East Anglia the Lower Pleistocene is represented by the Crag, a predominantly arenaceous sequence, but further west and south-west the sediments resting on the Chalk are a complex series of glacial boulder clays and outwash sands and gravels. A feature in East Anglia is the presence of deep buried channels in the Chalk that are infilled with glacial deposits. These channels, which generally coincide with present day and pre-glacial valleys, are not graded to a base level but have uneven floors representing a series of rock basins.

Table 12.2 Stratigraphical relationship of the Chalk to adjacent formations

General location	Humberside	Lincolnshire and central East Anglia	South-east England	Hampshire Basin
Geological units	Glacial deposits Chalk Red Chalk Gault	Glacial deposits Crag Chalk Red Chalk Carstone Sutterby Marl	London Clay Lower London Tertiaries Chalk Upper Greensand Gault	London Clay Reading Beds Chalk Upper Greensand Gault

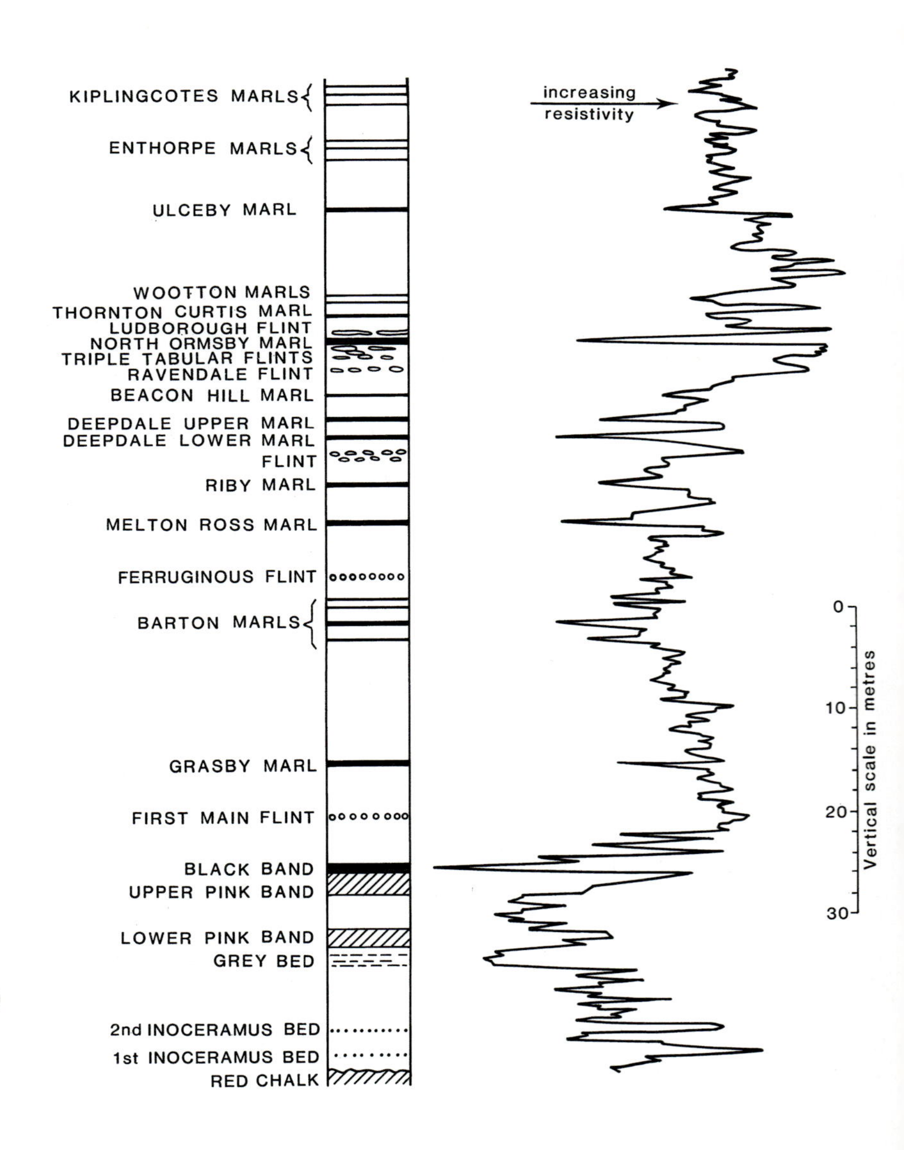

Fig. 12.5. Correlation between detailed Chalk lithology in Lincolnshire (after Wood and Smith 1978) and a typical single-point resistivity log (University of Birmingham 1978).

The deposits infilling the channels vary considerably from sands and gravels to boulder clays (Woodland 1946).

Although the distribution and geometry of the Chalk and associated Tertiary and Cretaceous formations are reasonably well known from drilling, principally for groundwater development, the detailed lithology and thicknesses of the Pleistocene sediments above the Chalk are only poorly known.

Development of hydraulic characteristics

In common with other limestones, the Chalk's ability to transmit and store groundwater is a function of its past history with regard to the flow of waters that were undersaturated with respect to calcium carbonate. It is uncertain how far back in time the dissolution of the Chalk matrix occurred, which created the present-day fissuring and semi-karstic system, but it probably commenced at least in Tertiary times if not Late Cretaceous times.

The recharge–discharge scenario, leading to dissolution, has been due to multiple components related to changing climatic conditions and accompanying variations of base levels. Price (1987) presented a calculation based on the amount of calcite dissolution possible for an average annual infiltration of 100 mm into chalk with a porosity of 40 per cent. He concluded that in the unconfined Chalk aquifer, the present-day fissuring could have developed since the last glacial recession. This is plausible as in the unconfined areas a flow system from recharge mounds to effluent streams or the sea is the classical situation. However, fissuring is also present in the Chalk below the Tertiary interface. In the marginally confined areas the explanation can reasonably be assumed to be the same as in the unconfined area. At greater distances from the outcrop, the flow mechanisms that have created such important secondary permeability have not yet been fully explained and it seems probable that in part it is associated with conditions prior to the last glacial recession (Morel 1980).

In order to appreciate how permeability has development in the eastern part of the country, where the thick drift deposits overlie the Chalk, Hiscock and Lloyd (1992) carried out a sequential palaeohydrogeological modelling study of the Chalk aquifer. They showed that effective permeability increased by about 6–9 per cent between 5000 and 10 000 BP, but that since 5000 BP, with recharge conditions similar to those of today and forest clearance occurring, the permeability has increased between 11 and 67 per cent depending upon depth in the Chalk.

Whatever the origin of the Chalk's permeability, the main development has been in the upper part of the aquifer. Bedding-plane partings particularly have succumbed to dissolution, with the process lessening with depth. In consequence, although the Chalk sequence may be some 300 m thick, the effective aquifer thickness maybe only 30–60 m. Furthermore, as precipitation and recharge were higher in the early part of the last post-glacial recession than today, some of the permeability has developed at levels above present-day water tables, under palaeosaturated conditions.

The fissuring has resulted in high permeability, decreasing with depth through the part of the aquifer affected by dissolution. Specific storage also decreases with depth in parallel with the permeability, but whereas the transmissivity of a section of unconfined fissured Chalk may be very high, the overall specific yield is relatively small (Fig. 12.6). Both these features have an important bearing on the development of water resources, as discussed below. Also, because the control on the development of the hydraulic parameters has not been primarily related to the depositional history of the Chalk, the resulting aquifer does not conform to stratigraphical boundaries and therefore differs from most sandstones, for example. Consequently, it cannot be mapped using conventional geological or geophysical methods. To some extent the difficulty has been overcome by the use of fluid-flow logging techniques developed by Tate *et al.* (1970), later supported by down-hole cameras and packer tests, which can define the depth of effective fissuring.

In addition to the gradational decrease in the hydraulic properties of the Chalk with depth, considerable variation in values also occurs areally (Ineson 1962), as shown in the transmissivity map for the London Basin (Fig. 12.7). Although reasonably uniform direct recharge from precipitation may occur over the outcrop, the concentration of surface waters in valleys permits active indirect recharge and the preferential development of permeability and aquifer storage in such linear features.

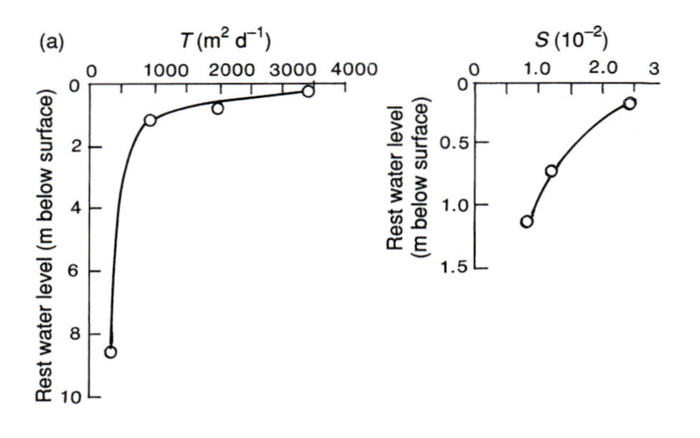

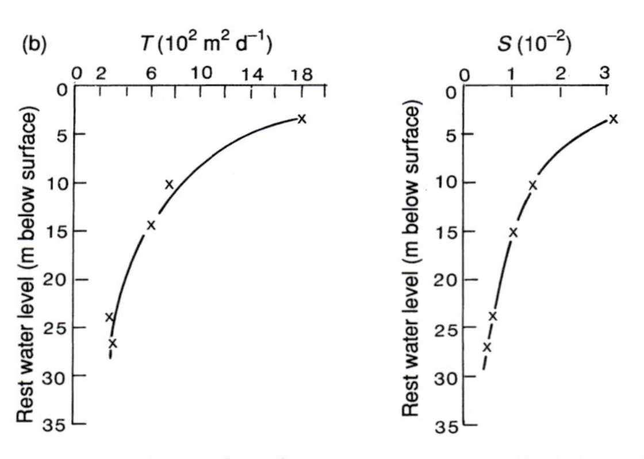

Fig. 12.6. Examples of variations in transmissivity and storage with depth: (a) Hampshire Chalk (after Headworth *et al.* 1982); (b) London Basin Chalk (after Owen 1981).

As a result, markedly differing hydraulic parameters generally occur in unconfined chalk in a section extending from a valley into an interfluve. High transmissivity and relatively high specific storage occur under valleys, but below the interfluves the transmissivity and storage coefficient may be so low that acceptable well yields cannot be obtained. An indication of the marked differences in transmissivity that can occur is shown in Fig. 12.8 for a radial-flow model representation in the Thames Basin.

To conclude this section, values for the aquifer's main hydraulic parameters are summarized. The porosity of the Chalk generally exceeds 20 per cent. Values for the Upper and Middle Chalk range from 40 to 50 per cent, but 20 to 30 per cent is more usual for the Lower Chalk. The harder chalks (hardgrounds) have lower values, generally less than 20 per cent. The Chalk of Humberside and Lincolnshire has a lower porosity than that of southern England and typical values for the White Limestone of Northern Ireland are only about 5 per cent. The variation in porosity reflects the extent of secondary diagenetic cementation. Although the harder rock bands in the Chalk, such as the Totternhoe Stone, and the Melbourn and Chalk rocks have lower porosities than the softer chalks, they are better water-yielding horizons as they are more fissured.

The specific yield of the aquifer is small because of the very fine-grained nature of the matrix which retains most of its saturated water content. Values lie in the range 0.5–5 per cent, although most tend to be around 1–2 per cent and higher values are generally only found in the zone near the water table, over which the groundwater level fluctuates seasonally, and where greater dissolution of the rock has commonly occurred.

A characteristic feature of the Chalk's matrix is its essentially isotropic nature. The permeability is generally about 10^{-8} m s^{-1} but the presence of fissures increases this many times. In perennial and dry valleys, the transmissivity may be 2×10^{-2} to 3.5×10^{-2} m^2 s^{-1}, reducing to 10^{-4} to 2×10^{-4} m^2 s^{-1} below the interfluves and also, as previously discussed, with increasing depth, particularly below about 50–60 m. The high transmissivity range just quoted, over a thickness of 50 m, implies an average hydraulic conductivity of 3 to 6×10^{-4} m s^{-1}. However, it must be appreciated that at an individual site most of the transmissivity may be provided by only one or two fissures and most of the Chalk does have a low hydraulic conductivity.

Groundwater flow systems

In outcrop areas the groundwater head distributions may show classical relationships to topography, but flows are strongly directionalized within the valley section because of the transmissivity distributions described above. The resulting concentration of groundwater flow produces in many areas an intimate relationship between groundwater in the aquifer, discharge from the aquifer as base flow, and surface runoff. Groundwater–surface-water interrelationships are susceptible to variations in recharge

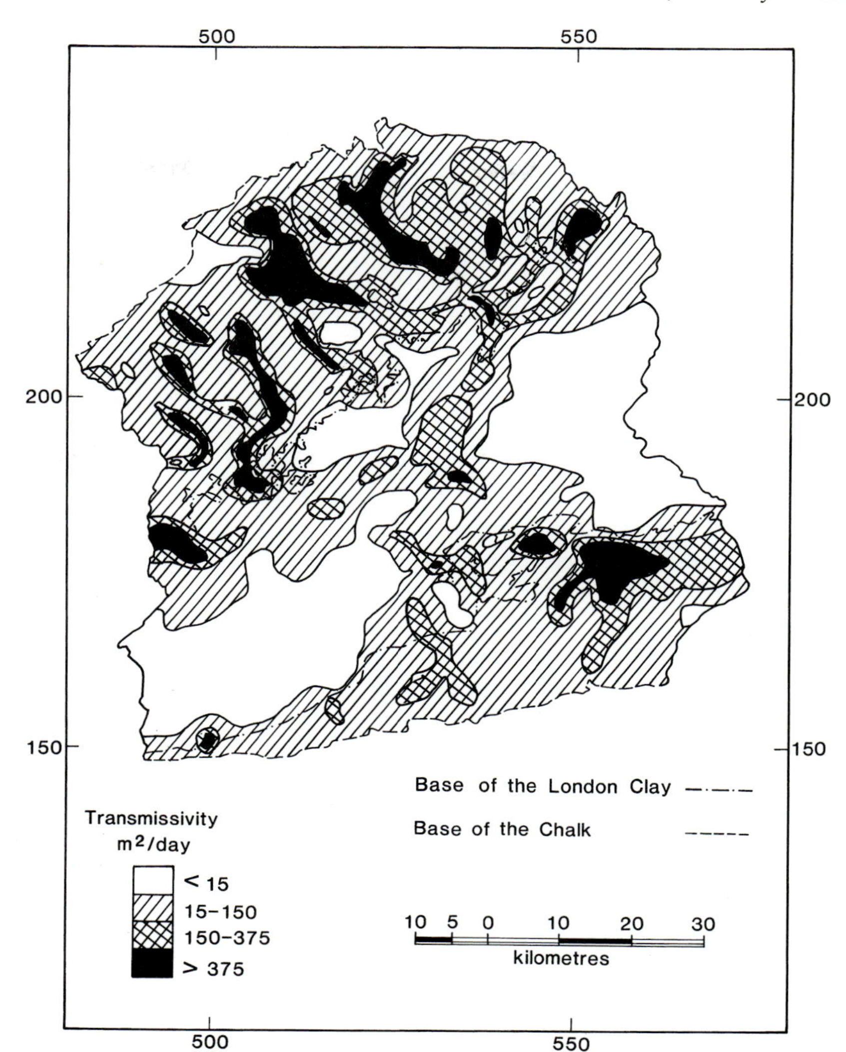

Fig. 12.7. Variation in transmissivity in the London Basin (after Water Resources Board 1972).

and can, of course, be seriously disrupted by groundwater abstraction. The groundwater head distribution for the Chalk aquifer is shown on Fig. 12.9.

Regional flows in the Chalk aquifer can be understood in terms of Darcian flow, but the nature of the permeability distribution permits very variable flow rates within limited sections of the aquifer. This has been illustrated by Atkinson and Smith (1974) who demonstrated, with tracers, groundwater velocities of 2.2 km d^{-1} in the Hampshire Chalk, and concluded that turbulent flow must, on occasions, occur in fissures under natural conditions.

Because the effective aquifer in the Chalk is relatively thin, while the formation as a whole is very thick, it is not normal for the Chalk to have any sensible hydraulic continuity with underlying aquifer units. However, locally where the Chalk is thin, as in parts of Dorset, limited permeability has been developed in the basal Chalk and hydraulic continuity established with the ferruginous sandstones of the Upper Greensand (Avon and Dorset River Authority 1973).

In Lincolnshire, hydraulic continuity exists between the Chalk and the underlying sandstones as a result of deep, glacial inspired erosion. Under

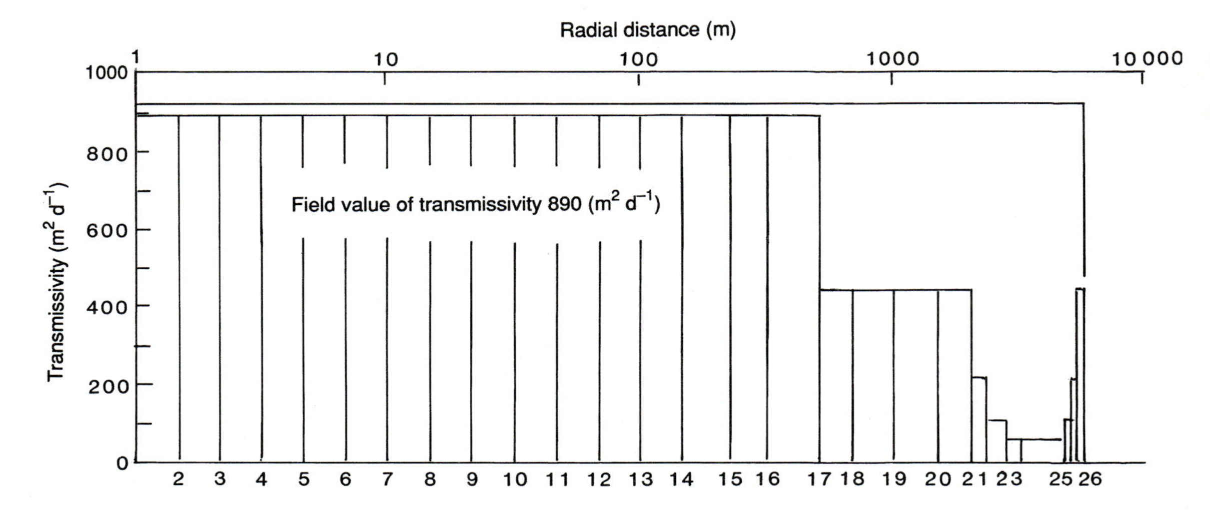

Fig. 12.8. Transmissivity variations used in radial modelling of pumping-test data simulations for the Chalk aquifer (after Connorton and Reed 1978).

natural head conditions, flow occurs from the sandstones into the Chalk aquifer, but these flows are small and have only been recognized from hydrochemical properties (Haines and Lloyd 1985).

Where the Lower London Tertiaries and the London Clay overlie the Chalk in south-east England, the London Clay or the basal clay in the Lower London Tertiaries act as confining units. However, the Lower London Tertiaries consist of a variable sequence of sands, silts and clays, and as such possess significant storage which can represent the principal source of water for wells drawing from the Chalk, demonstrating the importance of overlying units with high storage capacity in determining the viability of the Chalk as an aquifer. For example, in east London over-abstraction lowered water levels close to the base of the Lower London Tertiaries in the 1960s and as a result the yield of wells completed in the Chalk rapidly reduced.

The main area of confined Chalk in southern England lies in the westerly trending synclinal London Basin which has been comprehensively described by the Water Resources Board (1972) and Downing *et al.* (1979). As noted above, the upper part of the Chalk is fissured below the Tertiary in this area and hydrochemical and isotopic information confirm extensive natural groundwater flow along preferred paths below valleys into confined areas. The original natural outlet was in the Lower Thames valley. Hydraulic continuity also exists, through the London Clay, with overlying alluvial deposits by means of a set of geological features termed 'scour hollows', which have several possible origins. Generally they are associated with deep post-Tertiary channels which penetrate the London Clay and are back-filled with alluvial deposits. Others are believed to have evolved because diapiric movements in the Lower London Tertiaries, created by high groundwater heads, have disrupted thin sections of the London Clay (Berry 1979; Hutchinson 1980).

To the north of the London Basin, glacial deposits cover the Chalk over extensive areas, although in the centre of East Anglia the aquifer is directly overlain by Crag deposits of Pleistocene age. The glacial deposits have a variable lithology, consisting of boulder clays, sands and gravels, and they exert a fundamental control on the Chalk as an aquifer. Few detailed hydraulic data sets are available for these deposits, because groundwater is abstracted from the Chalk and usually little attention is paid to them during drilling and well testing. The mechanisms controlling groundwater movement and abstraction in the Chalk overlain by glacial deposits have, therefore, largely been inferred from hydrochemical data (Lloyd *et al.* 1981) and hydraulically confirmed to some extent by groundwater modelling of the Chalk aquifer

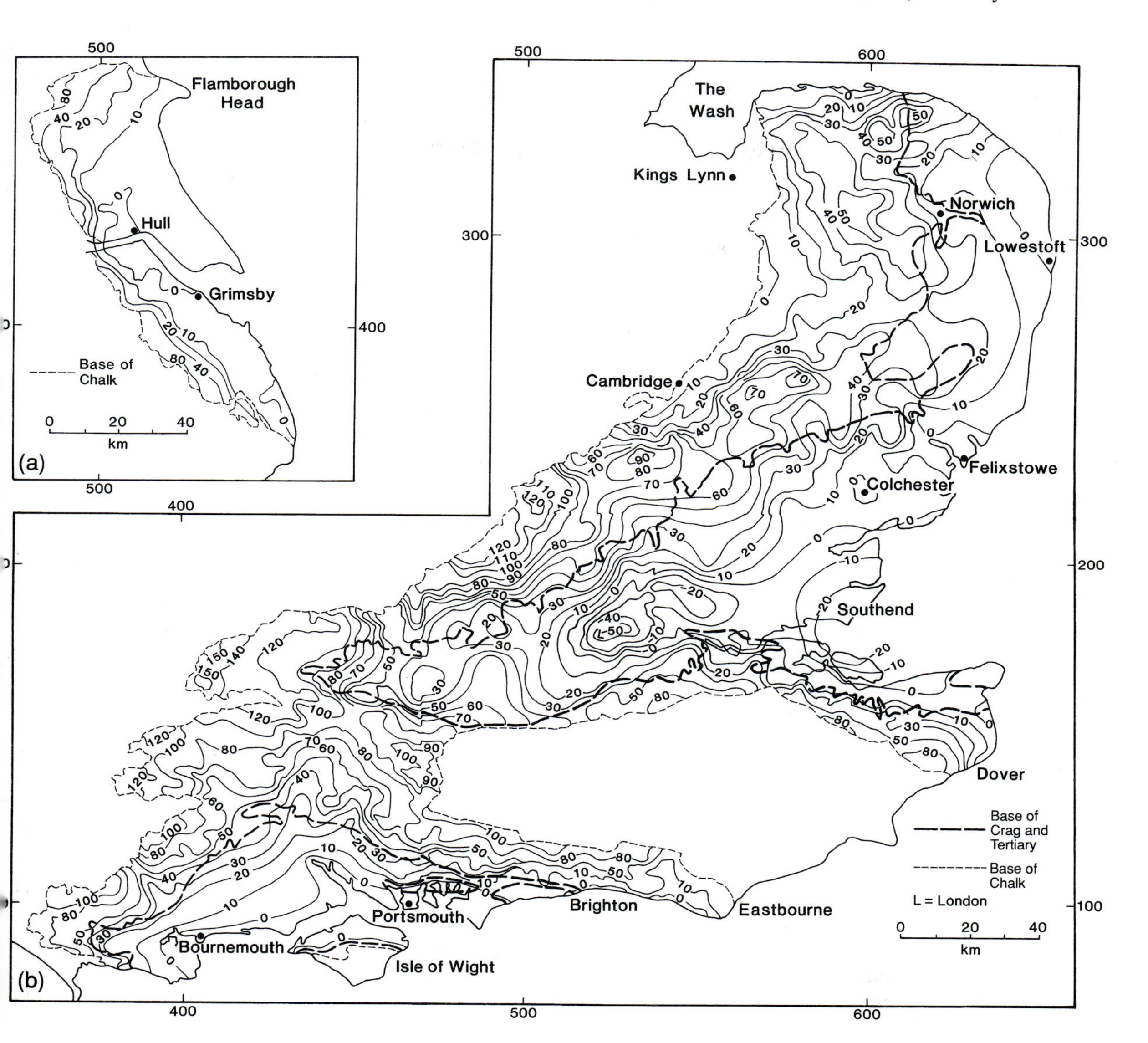

Fig. 12.9. Groundwater levels in the Chalk for December 1991: (a) Humberside and Lincolnshire; (b) South-east England in metres relative to sea level (by the British Geological Survey and by courtesy of the National Rivers Authority).

(Jackson and Rushton 1987; Lloyd and Hiscock 1990).

In Fig. 12.10, the hydrogeological conditions appertaining to the aquifer system in Essex are shown. Although a classical topography-related groundwater head distribution exists within the Chalk, the presence of thick boulder clays on the interfluves precludes any sensible recharge in those areas. The potential for recharge progressively increases towards the valleys so that both groundwater and surface runoff are concentrated in the valleys to produce an enhanced transmissivity in the Chalk along the valleys where the glacial deposits have been eroded or not deposited. Very low transmissivity values are found under the interfluves. The distribution of the hydraulic

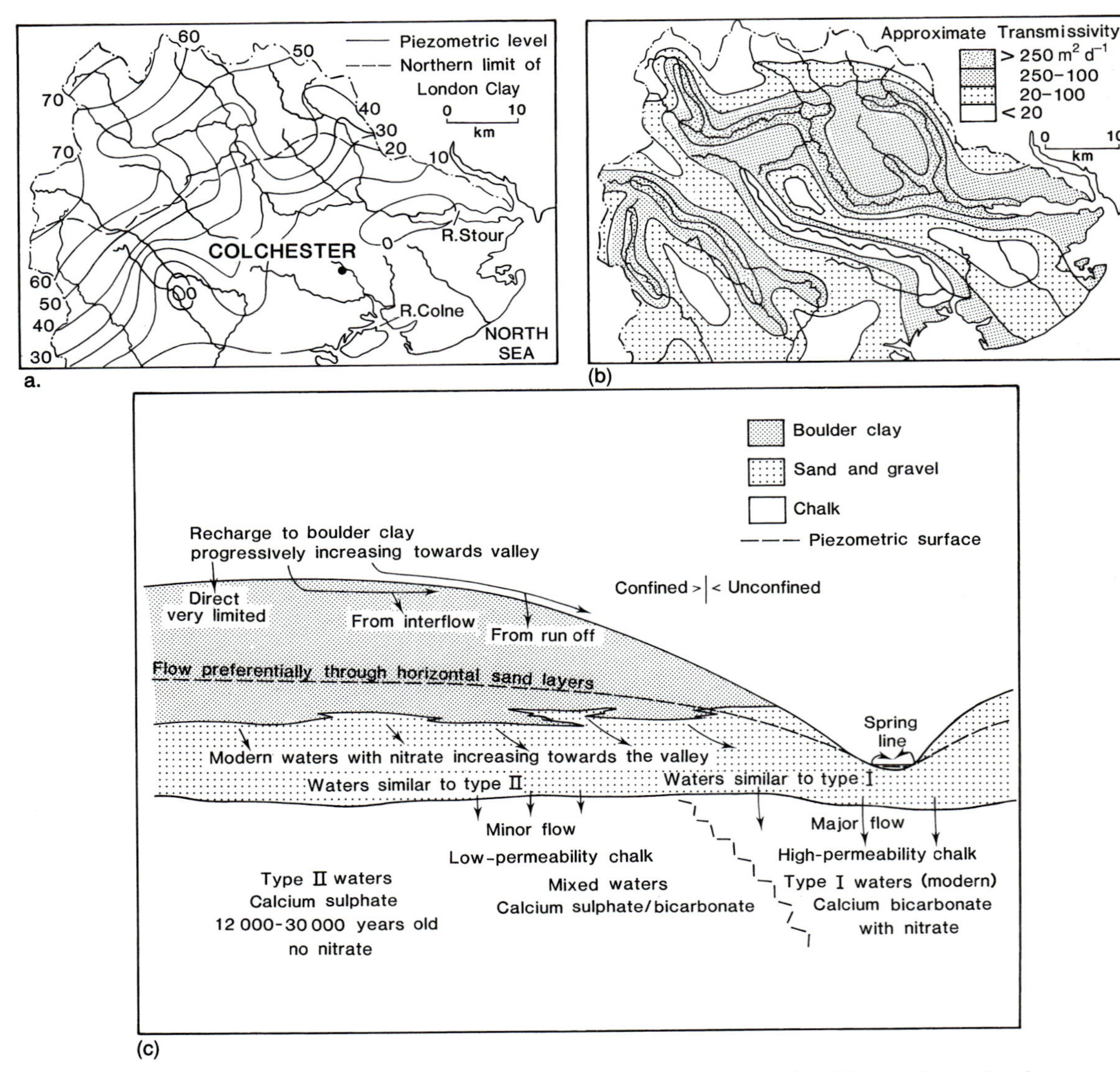

Fig. 12.10. Hydrogeological features of drift covered Chalk in southern East Anglia: (a) groundwater head distribution; (b) transmissivity distribution; (c) conceptual model of recharge and groundwater movement (after Lloyd *et al.* 1981).

parameters is therefore analogous to that described above for the unconfined chalk. But the presence of gravels in the drift has a profound influence upon the availability of groundwater resources in that as groundwater levels decline seasonally, the gravels progressively dewater away from valleys into the interfluves, and their higher specific yield maintains flows to the underlying Chalk, safeguarding well yields and base flows in the river system. The bulk specific yield of the Chalk is typically 1 per cent, but in areas where sands and gravels overlie the aquifer the storage values for the system as a whole are increased to some 5–15 per cent.

Throughout East Anglia and to the north in Lincolnshire and Humberside, boulder clays restrict recharge of the Chalk where they are present towards the coast, but sands and gravels influence flow and recharge mechanisms as in Essex. The glacial deposits can also influence the groundwater chemistry and quality with respect to criteria that

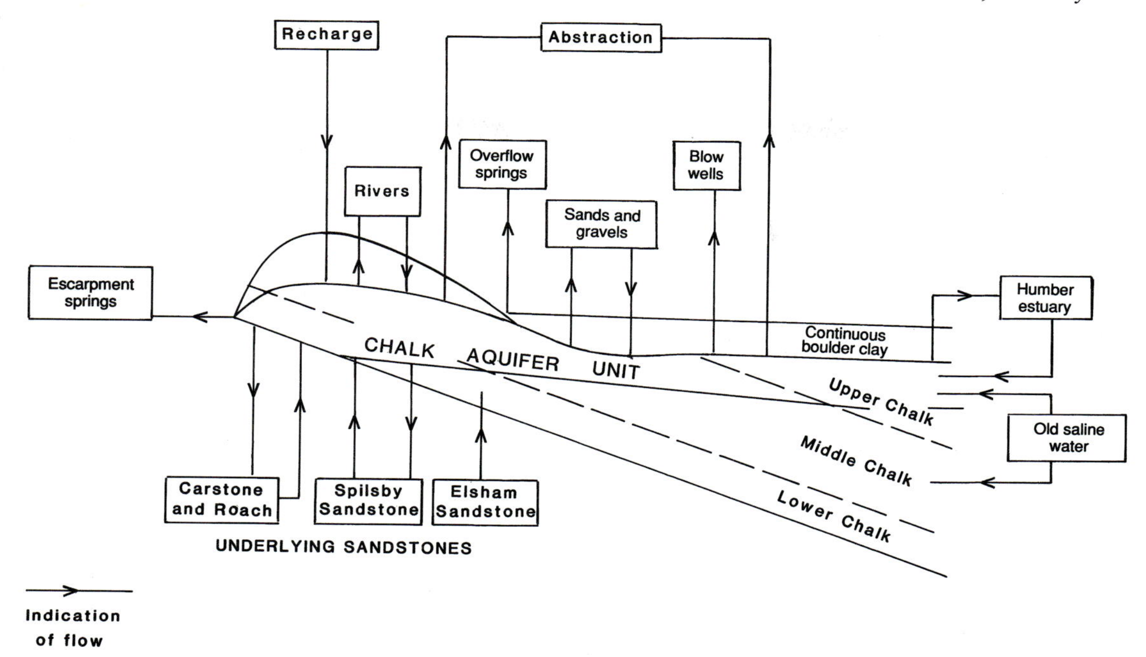

Fig. 12.11. Conceptual model for the hydrogeological conditions in the Lincolnshire Chalk (after University of Birmingham 1978).

determine the use of the water. An example of the hydrogeological framework in which the Chalk aquifer can be set is given in Fig. 12.11 for north Lincolnshire, where the aquifer is an essential source of water for industrial and public use.

The palaeohydrogeological evolution of north Lincolnshire is complex and is now recognized as such in the present-day interpretation of the hydrogeology and in assessing the abstraction potential of the Chalk and associated deposits. The evolution is shown in Fig. 12.12; during the Ipswichian interglacial, some 130 000 years BP, incursion of the proto-North Sea eroded a wave-cut platform into the Chalk and deposited a shingle beach against the ensuing cliff-line. A fresh–saline groundwater interface is believed to have existed below the cliff-line. In the Early to Middle Devensian the sea retreated and it is postulated that limited recharge, under permafrost conditions, only marginally moved the Ipswichian saline groundwaters eastward. Some sand dunes formed against the cliff-line.

In Lincolnshire, the westerly movement of the Late Devensian ice sheet stopped against the Ipswichian cliff-line. Snow-melt deposited some gravel and locally the base of the cliff-line was cut down through the Chalk. With the melting of the ice sheet, boulder clays were deposited and the Ipswichian saline groundwaters entrapped. These Ipswichian waters can be distinguished from modern saline groundwater in that they have distinctly higher iodine signatures (Lloyd *et al.* 1982). At their interface with fresh Chalk groundwater, carbon-14 ages in the latter are up to 20 000 years.

At the present day, groundwater flow from the recharge mound in the unconfined Chalk to the west is controlled by discharge at the unconfined–confined interface between the Chalk and boulder clay, and by localized groundwater discharge from 'blow-wells', or springs, that emerge through 'lenses' of gravel in the boulder clay, down-gradient of this interface. A modern active groundwater flow zone therefore exists through the Chalk to the 'blow-wells'. Ancient saline groundwaters are entrapped to the east beneath the boulder clays, which also severely restrict the intrusion of modern seawater. At the unconfined–confined interface, gravels are present, which are dewatered when the Chalk

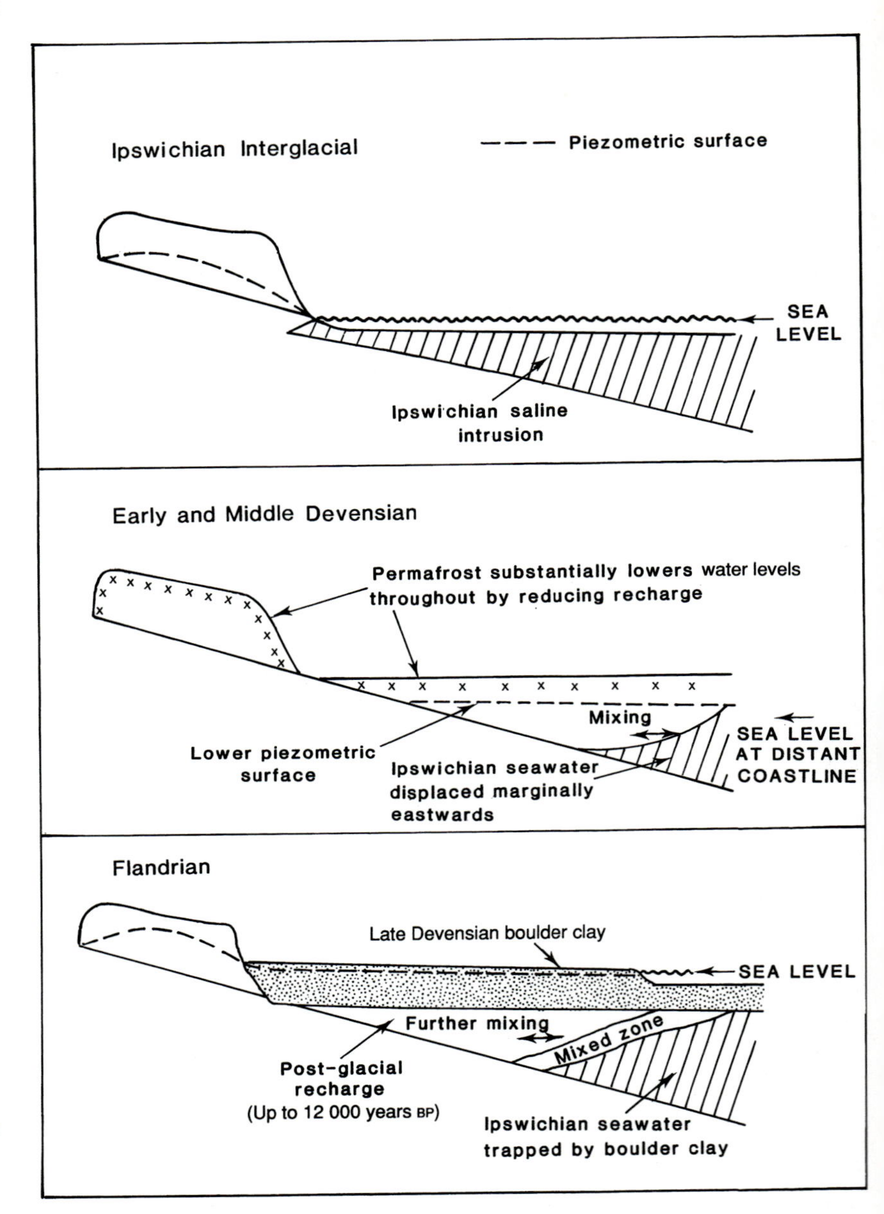

Fig. 12.12. Palaeohydrogeological evolution of the Chalk aquifer system in northern Lincolnshire.

aquifer is developed (Fig. 12.13), and thus maintain well yields in what would otherwise be a low-yielding situation, very similar to that described above for part of Essex (Lloyd 1980).

On a regional scale, groundwater in the Chalk drains from the major groundwater divide along the high downland that extends from Humberside to Dorset (Fig. 12.9). Most of the flow is below the dip slope towards the North Sea or the English Channel. Local drainage systems are to the main river outlets including the Thames and Humber estuaries and the rivers of East Anglia. Maximum groundwater levels are in the south where over 100 m above sea level (asl) is attained, for example, below Salisbury Plain and the South Downs. In East Anglia, the highest levels are only about 50–70 m asl, reflecting the

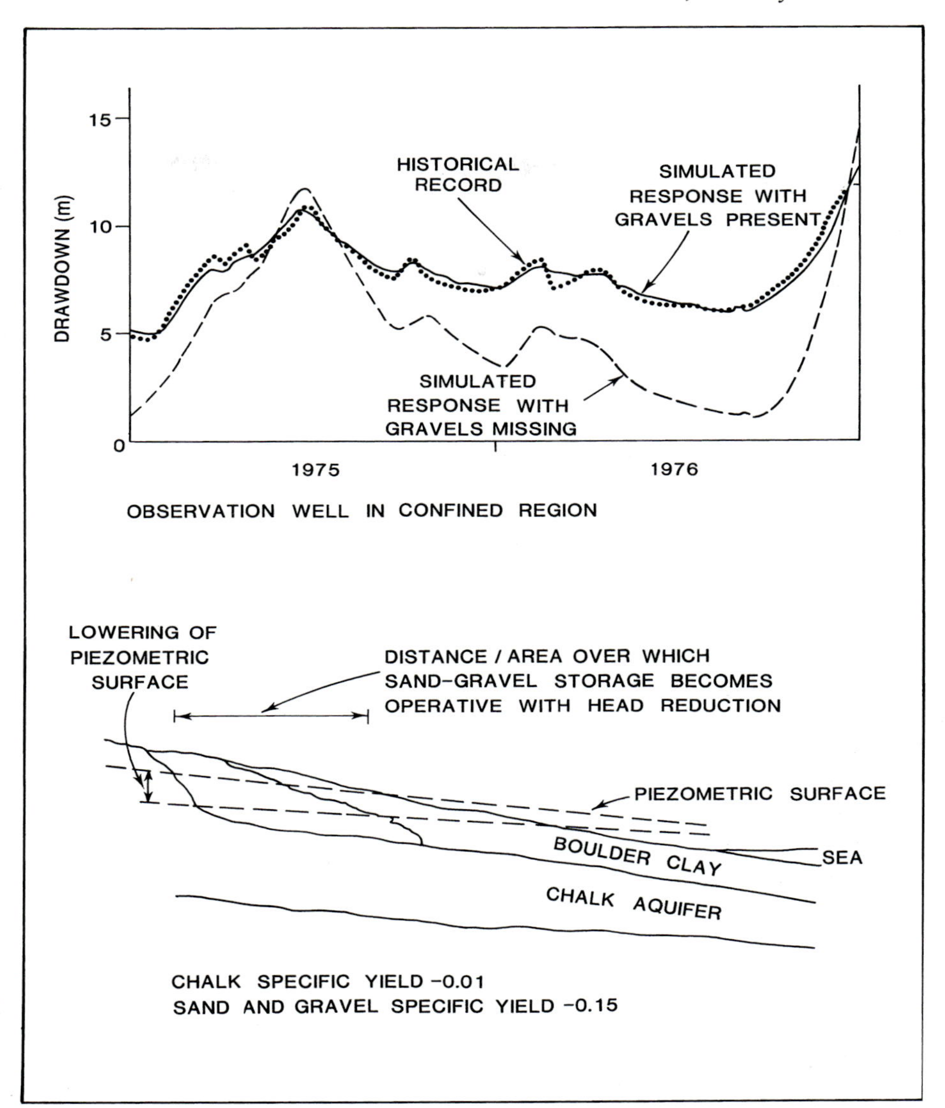

Fig. 12.13. Significance of sand and gravel specific yield in maintaining groundwater heads in the Chalk in northern Lincolnshire.

lower elevations of the ground surface, but further north, along the Lincolnshire and Yorkshire Wolds, the highest points are over 80 m because the ground is higher. Below the higher ground, the water table is commonly 40–60 m deep and in some localities over 100 m.

Around the coast, abstraction of groundwater has lowered levels to below sea level at a number of localities. In the London Basin, below the confining Tertiary cover, the piezometric surface is below sea level (bsl) over an extensive area. The minimum levels are now 50 m bsl in central London, some 30 m higher than recorded in the mid-1960s; the rise is due to a reduction in abstraction.

The hydraulic gradient is determined to a significant degree by the distance between the crest of the Chalk uplands and the coast and hence it is steeper in the North and South Downs than, for example, in East Anglia.

The annual fluctuation of the water table between the maximum levels in the spring and minimum levels in the autumn, can be as much as 30 m below the uplands, but generally below the river interfluves it is much less and 5–6 m would be more typical, grading to very small values in valleys containing perennial streams. In East Anglia, where ground elevations are not particularly high, the maximum fluctuation below the high ground is about 20 m but over much of the region it is less than 3 m.

Groundwater flow through the Chalk is diffuse and generally the discharge from the aquifer is from small springs and seepages into rivers. However, individual large springs do occur, for example the Chadwell Spring in the Lee Valley, that had an average flow of 9000 to 13 000 $m^3 d^{-1}$ under natural conditions before groundwater was developed in the London Basin. Many large springs, flowing at 5000 $m^3 d^{-1}$ or more, are used for public water supplies, as for example at Bedhampton, near Portsmouth, and Marham and Hunstanton in Norfolk.

Hydrochemistry

In general terms, the Chalk can be considered to have a relatively homogeneous lithology. Consequently, in any particular district, the groundwater tends to have a fairly constant composition. Solution reactions control the composition where the aquifer crops out and the water is of the calcium bicarbonate type, usually with variable amounts of sulphate and small amounts of magnesium (Table 12.3). Typical concentrations of calcium and bicarbonate are 100 and 200–300 mg l^{-1} respectively, while sodium, chloride and sulphate are respectively 10, 10–25, and 20–30 mg l^{-1}; the total ionic content (or total dissolved solids, TDS) is typically between 250 and 400 mg l^{-1}. Near the margin of the Tertiary cover, acidic, sulphate-rich runoff from the Tertiary rocks (that are dominated by clays) recharges the Chalk's saturated zone, commonly very rapidly and focused on swallow-holes or major fissures. The nature of this water increases the bicarbonate and sulphate contents of the groundwater near and for some distance down-gradient from the base of the Tertiary outcrop.

With increasing distances from the outcrop of the Chalk, the calcium ions are replaced by sodium as ion exchange becomes a dominant reaction. Calcium concentrations can decline to as little as 5–10 mg l^{-1}, while sodium increases to more than 250 mg l^{-1}. The chloride content increases down-gradient because of an increasing proportion of relict marine-derived water; values of the order of 100–150 mg l^{-1} are found in the centre of the London Basin. The fluoride ion also increases to between 2.5–3.0 mg l^{-1}. Changes in other trace

Table 12.3 Representative chemical analyses of groundwater in the Chalk (all figures in milligrams per litre, except pH)

Situation	TDS	pH	Ca	Mg	Na	K	HCO_3	SO_4	Cl	NO_3
Humberside unconfined	497	7.3	124	4	7	1.1	203	93	20	45
Lincolnshire unconfined	338	7.8	87	17	8	0.3	240	56	28	22
Lincolnshire confined	265	8.0	55	15	20	6.3	259	25	11	4
Norfolk unconfined	648	7.0	114	9	41	6.0	314	87	66	11
Norfolk confined	686	7.4	96	11	47	8.0	404	65	55	0
Kent unconfined	587	7.4	125	4	21	3.2	367	21	31	15
Kent confined	765	7.2	95	38	65	11.1	367	55	134	0
London Basin unconfined	397	7.2	93	2	9	1.1	241	12	12	27
London Basin confined	786	7.7	35	20	173	9.9	313	51	184	22
Sussex unconfined	415	7.3	93	3.2	27	2.0	206	34	41	8.8

elements and further details of the sequence of chemical changes along flow paths below the Tertiary cover are discussed in Chapter 3. The composition of the water in the confined aquifer of the London Basin has been influenced by the leakage of sulphate-rich water from the overlying Tertiary deposits as Chalk groundwater levels have declined.

Below the Tertiary deposits in east Suffolk and Norfolk, groundwater in the Chalk is saline with chloride values in excess of 1000 mg l^{-1} in a zone about 6 km wide. Saline waters also extend into the synclinal tract north-east of Stowmarket. These waters are fossil waters and not mixes with modern seawater.

Post-war changes in agricultural practices, involving in particular the much greater use of nitrogenous fertilizers, have had a profound effect on the composition of groundwater in the Chalk at outcrop in the form of marked increases in the nitrate content (Foster *et al.* 1986; Parker *et al.* 1991). The average loss of nitrogen from arable soils by leaching is 40–80 kg ha^{-1} a^{-1}. The impact on the Chalk is particularly severe where it is overlain by thin permeable soils which are common over much of the outcrop. In the drier central parts of the outcrop, the lower infiltration rates result in higher concentrations of solutes in the unsaturated zone, and where the zone is thin the nitrate content in groundwater already exceeds 50 mg l^{-1}, with peak values of 200 mg l^{-1}. Some wells have been closed down or the water has been blended with other sources. The quality of the groundwater in the Chalk tends to be stratified, with the highest concentrations occurring immediately below the water table. Because of the accumulation of nitrate in the unsaturated zone and its slow movement down towards the water table, nitrate concentrations in the Chalk where it crops out will continue to increase slowly for many years, and will eventually exceed 50 mg l^{-1} in most arable farming areas and 100 mg l^{-1} in areas with low effective rainfall. The western parts of East Anglia and Lincolnshire, where the infiltration is less than 200 mm a^{-1}, are particularly vulnerable. So far there is little evidence of high nitrate concentrations in the confined aquifer where nitrate reduction is probably taking place, at least to some extent.

Evidence is accumulating that in many areas now affected by increasing nitrate concentrations, pesticide levels are becoming or will become a problem for water engineers. During a detailed study of the distribution of pesticides in Chalk groundwater in the Granta catchment in Cambridgeshire, a widespread presence was detected of atrazine and simazine in low concentrations (generally up to about 0.1 μg l^{-1}). In addition, two uron herbicides, isoproturon and chlortoluron, were found in water from boreholes sited close to streams which flow seasonally due to runoff from upland boulder-clay areas. These streams recharge the Chalk and this may be the means of entry of the herbicides into the Chalk (Gomme *et al.* 1992). Locally the Chalk is also vulnerable to contamination from leaking tanks and pipelines containing oils, petroleum products and organic chemicals, from surface runoff from paved areas, and from landfills that rely on the 'dilute and disperse' disposal philosophy.

Where the Chalk aquifer is confined by boulder clay, natural groundwater flows are only moderate and chemically reducing conditions are typically present. In such areas nitrate concentrations are usually low, unlike in many unconfined areas, but dissolved iron species (Song Lin-Hua and Atkinson 1985) reach concentrations that frequently necessitate treatment.

An indication of the hydrochemical variations resulting from the influence of glacial deposits on Chalk groundwaters is given in Fig. 12.10 for Essex. Such influences, however, are found universally in the UK where glacial deposits are an important overlying unit. An example from north Norfolk (Hiscock 1984; Lloyd and Hiscock 1990) shows that, in addition to exerting general hydrochemical controls, the composition of the individual types of glacial deposit can impose distinct hydrochemical signatures on the Chalk groundwaters.

Figure 12.14a shows the geology of the glacial deposits for the area of north Norfolk in the vicinity of Norwich. The area is completely underlain by the Chalk which dips gently to the east-north-east. Tertiary deposits in the east unconformably overlie the Chalk. Pleistocene deposits are extensive and include the ferruginous Norwich Crag Series, east of Norwich, and widespread glacial deposits.

Two types of till are present, the North Sea Drift and the Lowestoft Till. The North Sea Drift is composed of a high proportion of sand, together with intervening tills of brown loamy clay. The Lowestoft Till contains less sand and a greater

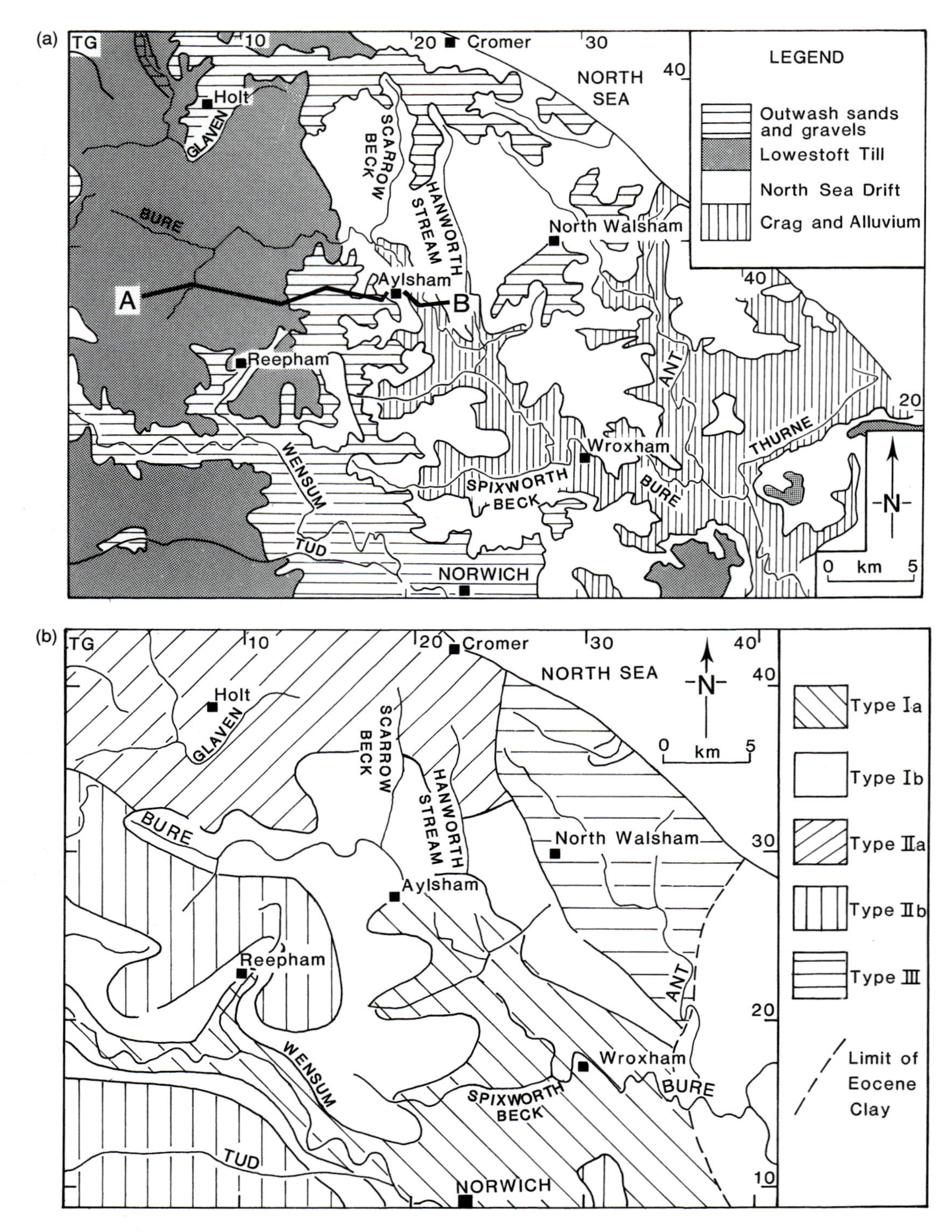

Fig. 12.14. Geology and groundwater chemistry in northern Norfolk: (a) Pleistocene and other superficial deposits; (b) hydrochemical types.

proportion of clay and carbonate debris. The tills are difficult to distinguish mineralogically; both contain illite, kaolinite and smectite.

Extensive groundwater sampling in the area has revealed how the two types of till are important in determining the underlying Chalk water chemistry. The distribution of the chemical water types is shown in Fig. 12.14b and is summarized as follows:

Type Ia: Modern water with high concentrations of calcium and bicarbonate (80–150 mg l^{-1} and 250–350 mg l^{-1}, respectively). High concentrations of nitrate, sulphate and chloride are encountered.

Type Ib: Similar in composition to Type Ia, except for lower calcium and bicarbonate concentrations (60–100 mg l^{-1} and 150–300 mg l^{-1}, respectively). Sulphate and chloride levels remain high, although nitrate is reduced to zero for an Eh value of 120–200 mV.

Type IIa: Beneath regions covered by the low calcite North Sea Drift. The water type has low concentrations of calcium and bicarbonate, with mean values of 85 and 210 mg l^{-1}, respectively. The dissolved iron content averages 0.4 mg l^{-1} for an Eh in the range 90–170 mV. This water is chemically reducing with low sulphate (<30 mg l^{-1}) and zero nitrate concentrations.

Type IIb: Beneath the thick accumulation of Lowestoft Till in the west of the catchment. Hydrochemically distinct from Type IIa in terms of calcium and bicarbonate concentrations (between 80–130 mg l^{-1} and 270–400 mg l^{-1}, respectively). The water is chemically reducing and contains high dissolved iron (0.6 mg l^{-1}) and low sulphate (<20 mg l^{-1}) concentrations.

Type III: This water exists beneath the North Sea Drift in the east. The water chemistry is similar to Type IIa, but contains higher concentrations of sodium (60 mg l^{-1}), magnesium (15 mg l^{-1}), sulphate (75 mg l^{-1}), and chloride (85 mg l^{-1}). The increase in ionic concentrations cannot be explained by a change in the lithology of the glacial deposits, but relates to the past history of seawater intrusion during the early Pleistocene.

In regions with a thick boulder clay less groundwater is developed, and in areas covered by sand and gravel oxidizing conditions prevail. These conditions affect iron speciation in the Chalk groundwater such that Type Ia waters fall within the amorphous iron hydroxide Eh-pH field, while samples from beneath cover of tills approach the boundary between ferric and ferrous iron fields. Under appropriate redox conditions, iron can enter solution and is responsible for the high iron concentrations encountered. The iron is thought to be contributed by the oxidation of disseminated pyrite contained in the glacial deposits.

Water-well construction

The majority of wells drilled into the Chalk have been constructed using percussion methods, methods which can still be practiced very effectively, particularly for small-production wells. Two features favour percussion drilling: first, the fissure permeability of the upper part of the aquifer, particularly in unconfined areas above the water table, poses no problems, but it leads to lost circulation in conventional mud flush drilling; second, the presence of tabular and nodular flints, common in the upper and to some extent middle divisions of the Chalk, is not particularly detrimental to percussion bits, but can cause considerable damage to tricone bits. For major modern production wells, foam-on-air drilling techniques are now favoured, as the method readily overcomes the lost circulation difficulties and is infinitely more rapid than percussion drilling. Appropriate drilling bits, however, have to be specified for sections that include many flint bands. Air-assisted reverse-circulation drilling is also proving effective.

The conventional approach to well design in the Chalk is for casing to be inserted in the top part of the borehole with open-hole completion for the producing section of the aquifer. The normal procedure is to case out the unsaturated zone and/or other sequences overlying the Chalk, with the casing base located at a limited nominal depth into the saturated zone. In confined sections in East Anglia, the upper part of the Chalk, below the glacial drift, is commonly disturbed and fragmented by permafrost, and in a saturated state forms a mobile 'putty-chalk'. This requires stabilizing and thus has to be cased out. If the Chalk is unstable because of fissuring at the water table, or because it is just below the base of the Tertiary or below Pleistocene Crag deposits, casing may have to be inserted into the high-yielding, upper fissured part of the aquifer and this reduces well yields and well efficiencies. Where large seasonal groundwater-

Table 12.4 Groundwater abstractions ($\times 10^3 m^3 d^{-1}$) from the Chalk compared with other abstractions by the Regional Water Authorities in 1985 (modified from Halcrow 1988).

	Anglian	Southern	South West	Thames	Wessex	Yorkshire
Total water abstraction	1946	1353	655	4160	894	1898
Total groundwater abstraction	986	996	114	1802	408	348
Abstraction from the Chalk aquifer	690	857	23	1622	265	59
Chalk abstraction as % total groundwater abstraction	70	86	20	90	65	17

level variations occur in unconfined areas, and the casing is set near or below the lowest seasonal level, this effect can become pronounced at times of high groundwater levels.

Chalk wells are commonly treated with acid after drilling has been completed to remove chalk that has been smeared on the sides of the wells by the drilling process and which tends to block fissures (Stow and Renner 1965). The method is particularly effective, leading to high-yielding, efficient wells. The effect of acid treatment varies considerably but yields are commonly doubled. As a guide, 6 tonnes of 31 per cent by weight hydrochloric acid is required to treat 40 m of chalk in a well 600 mm in diameter (Cruse, 1986).

Groundwater resources and abstraction

The volumes of groundwater abstracted from the Chalk in 1985 are given in Table 12.4. In national terms the aquifer provides some 55 per cent of the total volume of groundwater used in the UK (Halcrow 1988). The average annual infiltration to the Chalk amounts to $4.631 \times 10^9 m^3$, of which $1.477 \times 10^9 m^3$ (32 per cent) were abstracted in 1985 (Table 12.5). In the Thames Basin the abstraction exceeded 60 per cent of the infiltration although, of course, the abstraction figure also includes water derived by infiltration from rivers. Under drought conditions, when the annual infiltration may be reduced to less than 50 per cent, and possibly as low

Table 12.5 Average annual infiltration to the Chalk and abstraction in 1985 ($\times 10^6 m^3$)

	Infiltration*	Abstraction**	Abstraction as % of infiltration
Anglian	953	252	26
Southern	1231	313	25
South-West	202	8	4
Thames	976	592	61
Wessex	947	97	10
Yorkshire	322	22	7
Total	4631	1477	32

* From Monkhouse and Richards (1982)
** From Halcrow (1988)

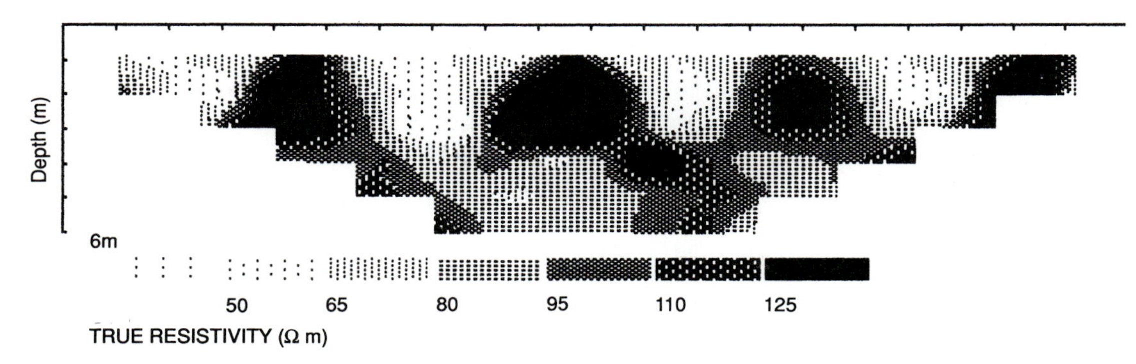

Fig. 12.15. Variations in the geological character of shallow unsaturated Chalk as determined by surface resistivity measurements. Such variations must have a very significant influence on recharge.

as 30 per cent, of the average value, abstraction represents a very significant proportion in the Anglian, Southern and Thames water authority areas.

The majority of abstractions is from individual sources rather than wellfields and the water is used directly for supply, with only limited treatment necessary as a public health safeguard, normally only chlorination, although now denitrification and activated carbon are becoming necessary for some sources. Management policies are, therefore, related to simplistic safe-yield assessments for individual sites which may appear unimaginative but are to a large extent dictated by the hydrogeological character of the Chalk, that is, a relatively thin effective aquifer with a high permeability and a low specific yield.

Two main difficulties exist in assessing the resources of the Chalk and their subsequent management: first, assessing the infiltration (or recharge); and second, the seasonal response of the aquifer to abstraction. These are irrespective of the complexities introduced by flow from overlying beds in the total groundwater system, saline intrusion, and so on.

Considerable attention has been paid to the mechanisms by which infiltration enters and moves through the unsaturated zone of the Chalk, mechanisms which are still not clearly understood. Emphasis has been placed upon studies using lysimeters (Kitchen and Shearer 1982) and vertical profiles of tritium and moisture contents (Smith *et al.* 1970; Wellings 1984), which, while instructive on a site-specific basis, can take no account of the ubiquitous marked inhomogeneity found in shallow weathered chalk on a regional scale (Fig. 12.15). Simple, evenly distributed vertical infiltration cannot occur universally and localized concentrations of recharge through the unsaturated zone must be a feature.

Groundwater resources of the Chalk are determined on a regional basis by estimating the infiltration from rainfall measurements and the calculation of evaporation using the method developed by Penman (1948; 1950) and extended by Grindley (1967). The actual evaporation is calculated by taking into account the type of soil and vegetation and using the concept of a root constant (commonly taken as 75–200 mm), which represents the water that can be freely drawn on by a crop before it begins to wilt and transpiration is curtailed. Studies of chalk catchments in Hampshire, in southern England (Headworth 1970), have shown that for thin chalk soils a root constant of 25–50 mm is more appropriate. The Penman–Grindley method for deriving a soil-moisture balance has been widely applied to assess regional groundwater resources of the Chalk (Headworth 1970; Water Resources Board 1972; University of Birmingham 1978). As there is virtually no direct runoff from the Chalk, the difference between the mean annual rainfall and the actual evaporation gives the recharge to the aquifer. Such data can be refined empirically against groundwater hydrographs in a numerical groundwater model (Spink and Rushton 1979; Rushton and Ward 1979). Attempts have also been made to re-

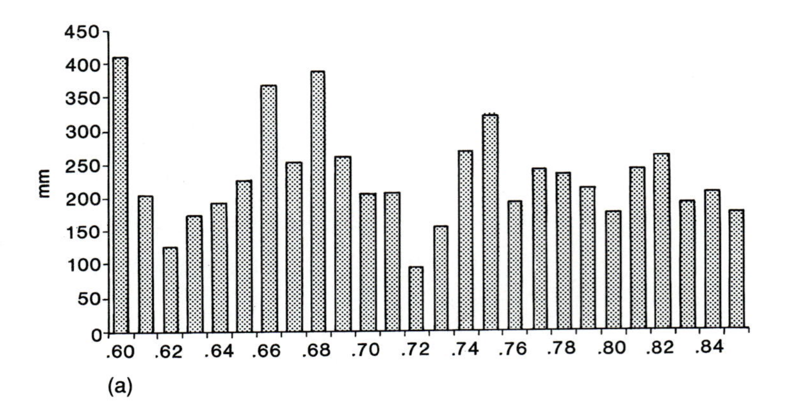

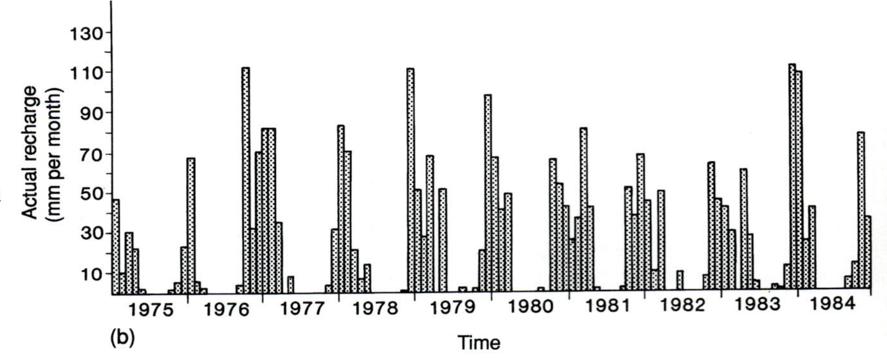

Fig. 12.16. Recharge values for the Chalk aquifer. (a) Annual totals for the Chalk in north Kent (after Southern Water 1989*b*). (b) Monthly totals for the Chalk in northern Humberside (after University of Birmingham 1985).

late infiltration directly to rainfall in south-east England and one formula, derived by Lapworth (1948) is:

$$\text{Infiltration} = 0.9\ \text{Rainfall} - 343\ \text{mm}$$

As would be expected, recharge to the Chalk varies considerably, as shown in Fig. 12.16, which gives annual recharge values for north Kent and the monthly variations found in north Humberside. Values over the Chalk's outcrop range from over 350 mm a^{-1} on Salisbury Plain to less than 125 mm a^{-1} in the extreme north-west of East Anglia (Rodda *et al.* 1976).

From the above discussion of the development and distribution of hydraulic parameters in the Chalk, and from Figs. 12.6 and 12.8, it is clear that the yield of a well is very dependent upon groundwater levels. When high levels occur, well yields and specific capacities are often extemely good, but they can decline rapidly as drawdowns in wells increase in combination with the seasonal decline of groundwater levels. Furthermore, failure to re-establish the groundwater levels in the subsequent recharge period can result in continued drawdown during the following seasonal recession with a consequent accentuated decline in well yields and specific capacities, as has occurred in many areas during the recent extended dry period. The nature of the Chalk requires the interpretation of pumping-tests during periods of both high and low groundwater levels, but in most cases source management is based empirically upon long-term abstraction data and experience.

Groundwater is now developed in the Chalk by wells up to 1 m in diameter and up to 120–150 m deep. Such wells can yield 15000 $m^3 d^{-1}$, and 5000 $m^3 d^{-1}$ is not uncommon from sites in valleys. Away from the valleys, yields are much lower and only exceptionally is more than 500–1000 $m^3 d^{-1}$ obtained from boreholes 250–300 mm in diameter. Boreholes 100 mm in diameter generally produce about 100 $m^3 d^{-1}$ (1 l s^{-1}). Yields from the Chalk, where it is overlain by Tertiary deposits, are less than where the aquifer crops out. Near the margin of the Tertiary some 1000 $m^3 d^{-1}$ may be obtained but this declines to 500 $m^3 d^{-1}$ or less as the thickness of cover increases. In Northern Ireland, where the Chalk is almost entirely overlain by Tertiary basalts, recharge is by leakage from fissures in the basalts. Yields are lower than in England and average 1100 $m^3 d^{-1}$ (Monkhouse and Richards 1982).

Yields from Chalk wells can vary widely, even in restricted areas, because of the heterogeneous nature of the aquifer, linked to the fissure density. As previously mentioned, supplies from large-diameter wells were in the past supplemented by driving adits often for considerable distances in different directions from different levels in a well. Individual wells were connected by adits, which were driven to intersect the principal fissure or joint directions. Adits were typically 1.2 m wide and 1.8 m high. In the South Downs, near Brighton, there are 13.6 km, in east London 18 km, and on the Isle of Thanet, in north-east Kent, 14 km of adits.

Use of groundwater storage

The steady increase in the abstraction of groundwater from the Chalk gradually reduced the natural spring flow from the aquifer and the flow of the perennial rivers fed by springs. This was a particular problem where groundwater was abstracted in the upper parts of catchments and piped directly to demand centres in the lower parts or into adjacent catchments. It was accentuated in the summer and autumn and during prolonged dry spells when discharge from the aquifer was naturally very low. The problem came to a head in the 1950s and led to the suggestion that groundwater storage should be developed during the summer in areas sufficiently far from the river to ensure that groundwater discharge to the river was not significantly affected at the time. It was proposed that the water could be used, not only for direct supply, but also to augment low river flows (Ineson and Downing 1964).

In 1965 proposals were published by the Thames Conservancy and Binnie and Partners to regulate the flows of the upper reaches of the Thames and Great Ouse rivers by discharging into them groundwater from the Chalk and abstracting it for water supply in the lower reaches. This led to field-scale pilot studies to investigate the feasibility of the proposal (Thames Conservancy 1971; Great Ouse River Authority 1972). The success of this type of scheme depends upon achieving a high net gain in river flow. With time, pumping groundwater reduces the natural discharge to the river and once the cone of depression reaches the river, water can recirculate into the aquifer because of leakage through the river bed; both factors reduce the net gain. The delay between the beginning of pumping and the effect on the river is a function of the transmissivity and the storage coefficient (that is, the diffusivity of the aquifer), the distance of the well from the river and the geometry of the aquifer (Oakes and Wilkinson 1972). Using a numerical model, these authors related the effects of abstraction to the aquifer response time, T/SL^2 (where T is transmissivity, S is the storage coefficient and L is the distance from the river to the catchment boundary parallel to the river). The Chalk does not have ideal properties for developing storage in the manner described. The high transmissivity, which tends to be restricted to the top of the aquifer in linear zones along valleys, together with the low storage coefficient, means that the aquifer responds rapidly to abstraction. Ideally wells need to be at least 5 km from a river and this is often difficult to achieve in the small catchments so characteristic of the Chalk's outcrop. Nevertheless, field experiments indicated that the average net gain during an abstraction season is 60–70 per cent of the total abstraction (Downing *et al.* 1981). Since the early 1970s, many schemes have been developed in the Chalk to augment river flows either to provide water supplies in the lower reaches of a catchment or to protect the river environment at times of natural low flows (Owen *et al.* 1991). The schemes were designed to be used during dry years, on average about one year in ten.

The scheme in the upper part of the Thames Basin is in operation and involves 33 wells, mainly in the unconfined aquifer, that can pump $115 \times 10^3 m^3 d^{-1}$ into the upper tributaries of the river to give a net gain of about $70 \times 10^3 m^3 d^{-1}$. The scheme was operated in 1976 and 1989. The first stage of the Great Ouse scheme has 52 wells that provide a gross yield of $122 \times 10^3 m^3 d^{-1}$. This scheme is also in use.

The management problems in developing the

Chalk are clearly illustrated in the Itchen catchment of southern England. Two tributaries—the Candover and the Alre—have been investigated. During test pumping of six wells in the Candover catchment, a culminating net gain of 80 per cent was achieved after six months (Southern Water Authority 1979*b*). The catchment is unusual because of the presence of a very permeable zone 6 m thick, immediately below the water table, with a specific yield of 5–7 per cent (Headworth *et al.* 1982). But the Chalk is fissured very uniformly and significant storage was developed in the aquifer in all directions around the pumping wells. Investigation of the Alre catchment revealed a different situation. Although the Chalk is again very permeable, with transmissivities as high as 0.14, and even 0.35 $m^2 s^{-1}$, the specific yield is generally only about 0.5 per cent. The fissure system allows rapid groundwater flow but the fissures are not uniformly developed. Because of the high permeability and low storage, the cones of depression are shallow and extensive and hence the river is more rapidly affected by pumping. Consequently,

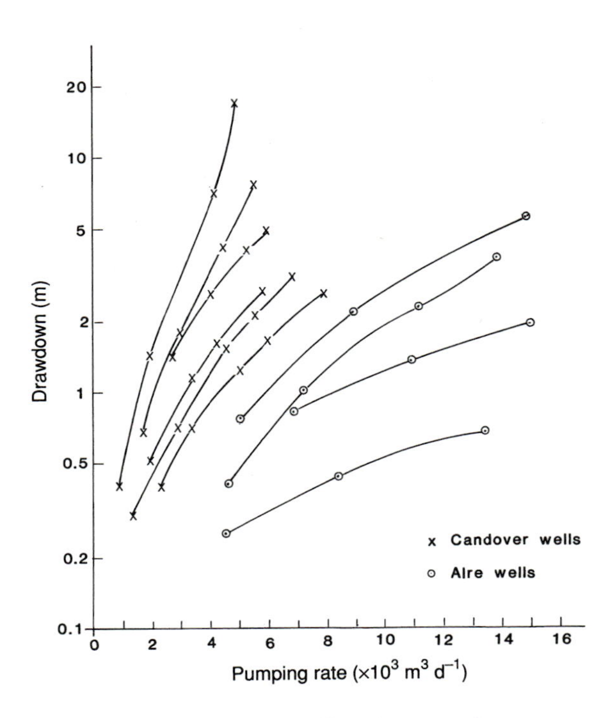

Fig. 12.17. Variations in well performance from two adjacent Chalk valleys in the Itchen catchment of Hampshire (after Giles and Lowings 1990).

the net gain is much lower than in the Candover catchment (Giles and Lowings 1990) and is only about 30 per cent. The Candover scheme is in operation and has been used several times during dry seasons in the late 1980s. The Alre scheme will become operational in the near future although, because of its more immediate impact on river flows, pumping will be intensive but for short periods. The yield–drawdown data for the production wells in the two valleys, shown in Fig. 12.17, illustrate the high permeability in the Alre catchment which, in conjunction with the low specific yield, gives the high aquifer response time.

River augmentation has also been examined for the Chalk in Humberside, which has been historically exploited principally for Kingston upon Hull's public water supply. The city has suffered water shortages at various times since the 1940s and forecasts indicate that new supplies will be necessary by the year 2000 if frequent drought restrictions are to be avoided (Barker *et al.* 1983). Overpumping in the Hull area has caused saline intrusion, but the northern part of the Chalk has been largely unexploited and has therefore been considered in terms of augmentation (Foster and Milton 1976).

Although the Chalk in Humberside has somewhat different properties from that in the south of England, hydraulically it behaves in the same manner. On the northern dip-slope outcrop, perennial streams occur but the spring heads migrate some 2–4 km downstream seasonally. The aquifer has been modelled (University of Birmingham 1985) using transmissivity variations as a function of head and including layering of storage values. An indication of the transmissivity for two locations in the aquifer is shown on Fig. 12.18. Considerable transmissivity variations are required in the model to represent well and spring hydrographs; generally a doubling of values from summer to winter is necessary. Near springs, long-term flow paths have locally increased the permeability up to 2500 $m\ d^{-1}$ (0.03 $m\ s^{-1}$). In model representations, the vertical variation of storage close to the springs does not appear to be a sensitive parameter. These results support the interpretation of the flow in the Chalk from pumping tests (Foster and Milton 1974*b*), with the permeability variation being closely linked to fissures enlarged by dissolution near the water table.

Because of the nature of the Chalk, careful

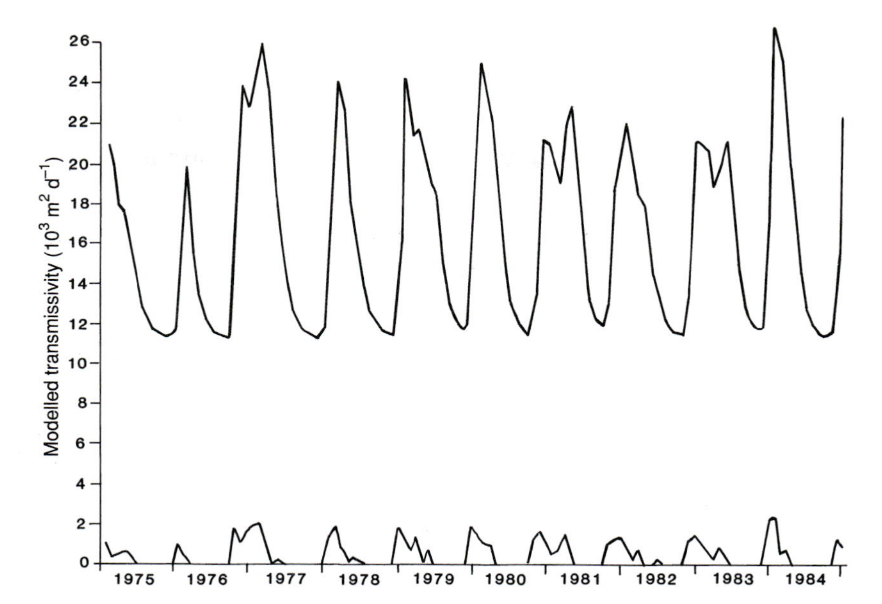

Fig. 12.18. Variations in transmissivity with seasonal groundwater head fluctuations determined through modelling of the Chalk in north Humberside (after University of Birmingham 1985).

management is necessary to manipulate the storage to advantage. Intensive, prolonged pumping in what is essentially a thin, highly permeable aquifer, will dewater the shallow, principal fissures, and further pumping from the underlying aquifer, with a lower permeability and storage coefficient, will accentuate the drawdown. For optimum results in these circumstances, it is necessary to restrict the drawdown around wells, to maintain the transmissivity, by pumping at lower rates but from more wells (Morel 1980).

Over-abstraction

Over-abstraction of parts of the Chalk aquifer was recognized as an important problem in the 1940s, resulting from decades of uncontrolled groundwater development. In parts of Sussex, Kent, Lincolnshire, and Humberside, over-abstraction caused the local intrusion of saline water which has subsequently been controlled by appropriate management methods.

The Chalk of the South Downs has been an important source of water for the major towns of the south coast since the early part of the nineteenth century. The outcrop is divided into 'blocks' by north–south rivers and the rising demand for public water supply in the Brighton 'block', together with the fear that this would increase the salinity of the groundwater, led, in 1957, to a very effective new management policy. Groundwater which discharges into the sea in winter was intercepted by abstracting from wells near the coast (referred to as 'leakage wells'); inland wells (or 'storage wells') were rested at this time, allowing water levels to recover. In the summer, as regional groundwater levels declined seasonally and the natural outflow from the aquifer was thereby reduced, abstraction was largely switched from coastal to inland wells to prevent saline intrusion. At times of low water levels the intrusion in some coastal wells is closely linked to the tidal cycle. High tides provide the head for saline water to move rapidly inland through fissures in the Chalk (Monkhouse and Fleet 1975). Abstraction from coastal wells was, therefore, carefully controlled according to the value of the chloride ion over a tidal cycle and in this manner saline intrusion was limited.

In the 1950s, abstraction from the Chalk near Brighton was $20 \times 10^6 m^3 a^{-1}$, equally distributed between inland and coastal wells. The new management policy actually put a greater emphasis on abstraction from coastal wells to reduce the natural discharge from the aquifer. By 1984 abstraction had risen to $31 \times 10^6 m^3 a^{-1}$, with 65 per cent from coastal wells. Despite the increase in abstraction, average groundwater levels rose by

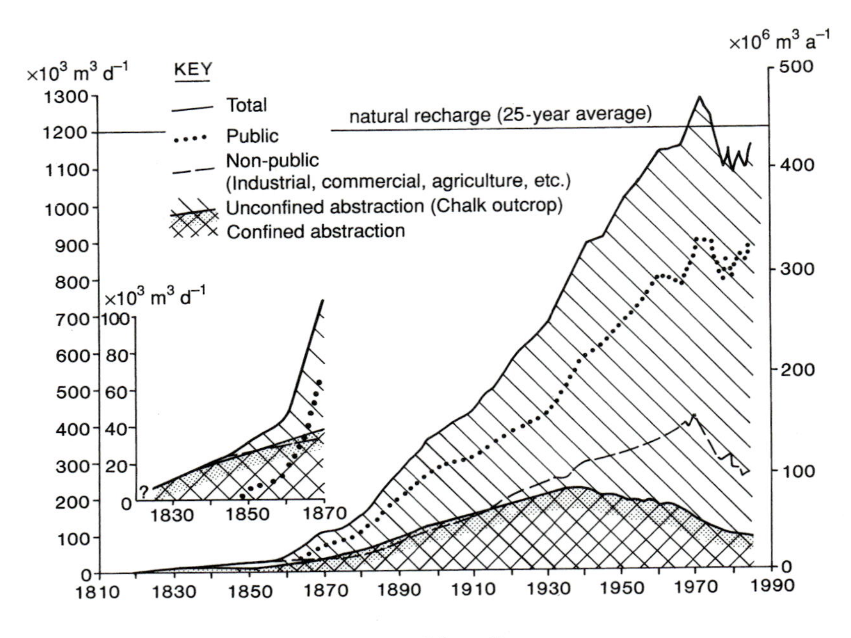

Fig. 12.19. Groundwater abstraction from the Chalk aquifer in the London Basin, 1820–1985 (from CIRIA Special Publication 69).

1.6 m. Currently, the actual abstraction from the aquifer amounts to 70 per cent of the infiltration in a drought year (Downing and Headworth 1990) which is a high level of development, and reflects the effectiveness of the management of the groundwater storage (Headworth and Fox 1986).

In South Humberside, intrusion has occurred along the Humber estuary and near Grimsby. Comprehensive modelling of the Chalk has led to well-authenticated conclusions that the abstraction sites should be moved further from the coast to reduce the potential for saline groundwater movement. However, the location of new sources would also require modification of the mains distribution system with substantial and unacceptable capital costs so that instead, detailed management of individual sources has had to be implemented. Infrastructure and costs have, therefore, taken precedence over recommendations based on hydrogeological considerations. Now the region is supplied with groundwater from the Chalk in conjunction with water from the River Trent, according to the availability of water in the Chalk.

The most extreme example of overabstraction in the Chalk has occurred in the confined area of the London Basin. Abstraction from 1820 to 1985 is shown in Fig. 12.19; a peak of $1300 \times 10^3 \, m^3 d^{-1}$ was reached in the late 1960s, exceeding the overall natural recharge to the confined part of the basin. Despite the high advantageous specific yield of the sands at the base of the Lower London Tertiaries, extensive dewatering of the aquifer occurred, as shown in Fig. 12.20. There were two main areas in which water levels declined, one in central and west London and the other around Romford; the maximum decline in these two areas was about 95 m (in central London) and 61 m, respectively. To the east of London, along the River Thames, large volumes of uncontrolled abstraction occurred both to the north and south of the river because of pumping to dewater Chalk quarries. Abstraction on both banks of the river caused the intrusion of saline riverwater into the Chalk.

The high abstraction rate from the confined aquifer also led to a reduction in the flow of rivers draining the Chalk's outcrop as well as slight subsidence of the ground surface because of the consolidation of the London Clay as the groundwater pressure was lowered in the Chalk. The source of the groundwater that was abstracted from the confined aquifer was principally a reduction in the groundwater storage in the Chalk and

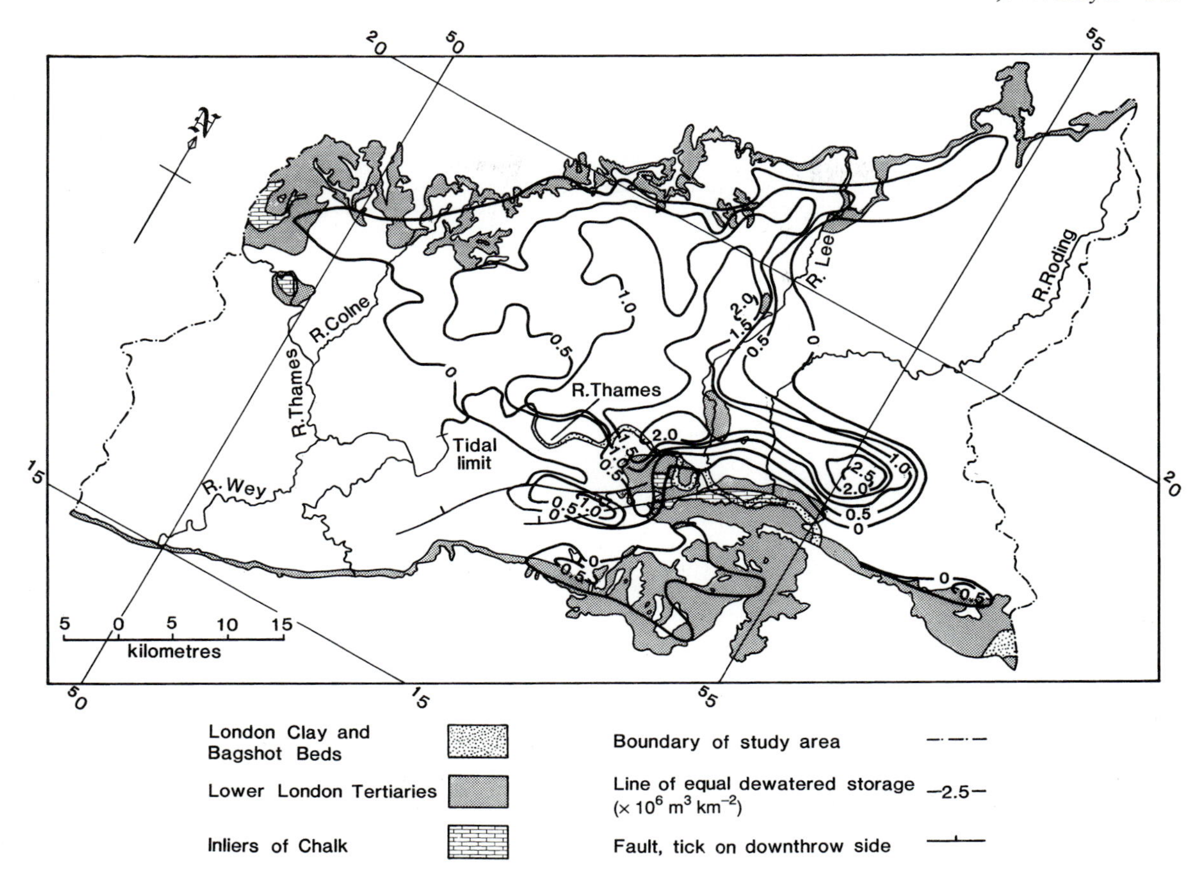

Fig. 12.20. Dewatered storage in the Chalk and Lower London Tertiary sands in 1973 (after Water Resources Board 1974).

Tertiary sands and an increase in the flow through the confined aquifer as a result of the increased hydraulic gradient. The latter provided about 80 per cent of the total, while the change in storage contributed 18 per cent. Between 1800 and 1965 the Chalk and Tertiary sands provided an average of 90 000 $m^3 d^{-1}$ towards the water supply of London which was a great economic advantage, especially in the early part of the period when London was expanding and water was a key factor in this development (Water Resources Board 1972).

The groundwater levels for 1965 are shown on a section through the basin in Fig. 12.21. However, because of the reduction in the use of water by industry, and a reduction in demand as a consequence of the more efficient use of water from the 1970s, groundwater levels have started to recover, as indicated by the levels in 1985 in Fig. 12.21. Rising groundwater levels in the Chalk and overlying strata are perceived as a problem (Simpson *et al.* 1989), principally because of possible structural effects on buildings with deep foundations. The rate of the rise of groundwater levels has been between 0.3 and 2 $m\ a^{-1}$.

High abstraction rates from wells in valleys on the Chalk outcrop around the London Basin have had a significant effect on stream flows. A number of tributaries of the River Thames have been severely depleted. The perennial heads have moved downstream by several kilometres, the rivers dry up in drought years, and summer flows in general have been drastically reduced. The river channels are often no longer able to support fish and the river environment has suffered. The general appearance

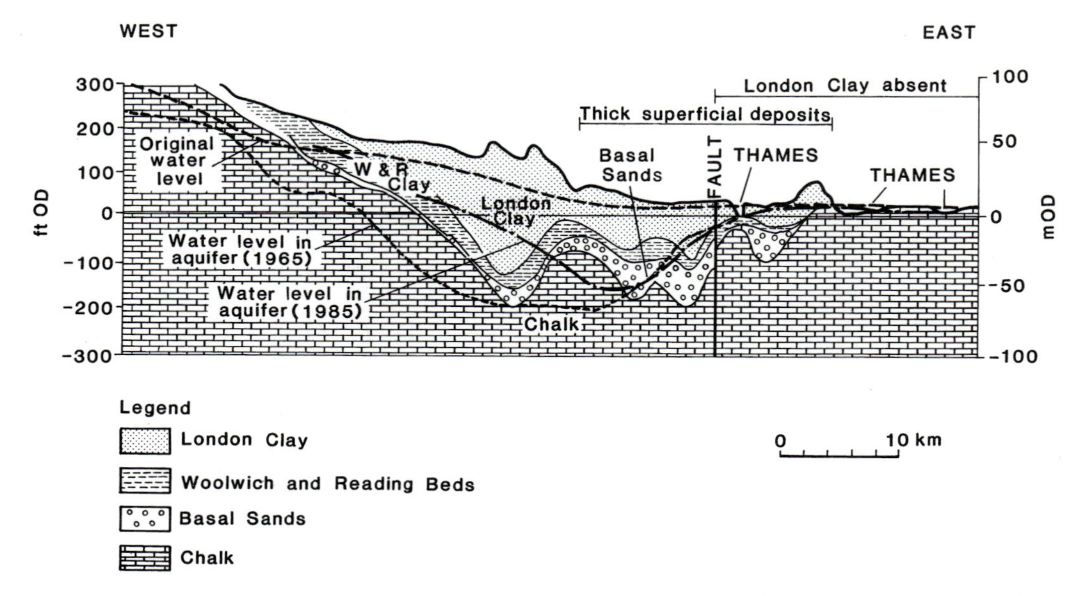

Fig. 12.21. Geological section through the London Basin showing groundwater heads in 1965 and 1985 (from CIRIA Special Publication 69).

of the rivers has deteriorated, trout fishing has almost disappeared, and watercress farming has ceased.

The feasibility of re-establishing adequate perennial flows has been studied (Halcrow 1987) with a view to recreating environmental conditions in the rivers and providing acceptable habitats for aquatic flora and fauna, and associated river-bank species. A number of tributaries have been studied and various options considered. Replacement of existing groundwater supplies by developing alternative sources, recirculating river flows, artificial recharge using winter river flows or sewage effluent, and supplementing river flows by sewage effluent, were all examined and found to be very expensive, and, in the cases of recirculation and the use of sewage effluent, to have potential water-quality problems. The most promising means of increasing river flows was by augmenting the flows with abstraction from the Chalk from new, appropriately placed wells and seasonal redistribution of abstraction. In several of the rivers, leakage through the river bed is a major contributory factor to low flows and this could be alleviated by lining the channel using environmentally acceptable materials such as clays or compacted chalk.

An example of existing conditions with remedial flows calculated for various options is given in Fig. 12.22 for the River Misbourne where abstraction of about 52 per cent of natural recharge has radically affected the river. A target flow profile that is ecologically acceptable is shown, which is substantially below the natural non-derogated flows. The actual flow demonstrates both accretion down the river profile, and leakage losses. Of the various schemes examined, abstraction at a distance from the river would appear to be the most attractive purely in terms of flow. However, it is not substantially more beneficial than local augmentation, which is less than half the cost. The latter has, therefore, been recommended although neither of these options meets the target requirement for the entire length of the river. The results are interesting in that the environmental impact caused by abstraction from the aquifer cannot be adequately alleviated if a viable water supply policy based on the use of the Chalk is maintained.

Artificial recharge

While the redistribution of abstraction and manipulation of groundwater-related flows may be environmentally rewarding, such operations do not

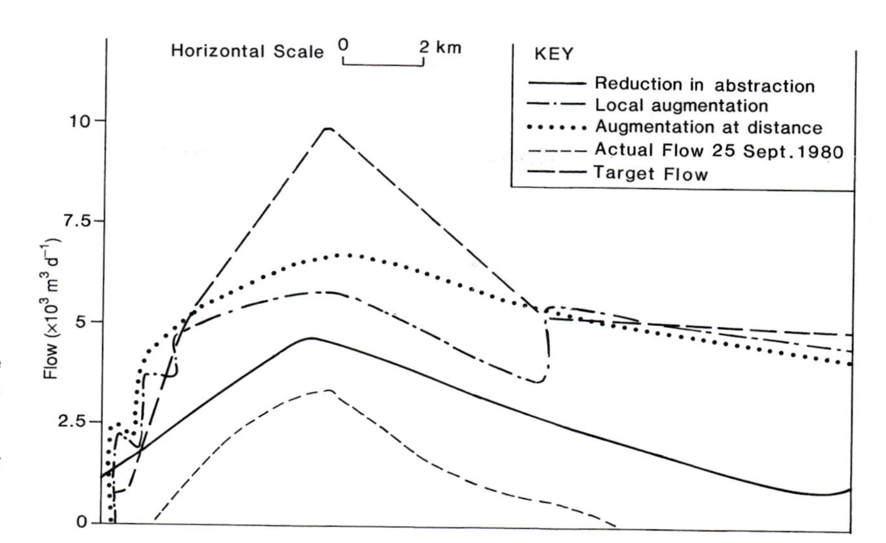

Fig. 12.22. River-flow augmentation proposal for the River Misbourne chalk stream in order partly to overcome derogation resulting from groundwater abstraction (after Halcrow 1987).

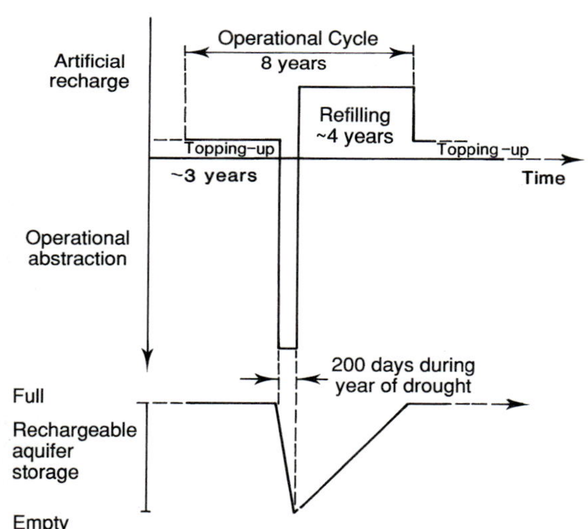

Fig. 12.23. Schematic representation of the operational cycle for the projected artificial recharge schemes in the London Basin (after Connorton 1990).

add fundamentally to the total water resource. To supplement resources, investigations have been carried out to assess the feasibility of artificial recharge of the Chalk and Lower London Tertiaries. These studies have been in the London Basin because of the volume of dewatered Chalk and Tertiary sands that has developed. Connorton (1988) identified a deficit in the area of $138 \times 10^3 m^3 d^{-1}$, representing 7 per cent of the total requirement for the 5.5 million customers, and forecast that this deficit could rise to some $300 \times 10^3 m^3 d^{-1}$ by the year 2011.

A prototype artificial recharge scheme to the north-east of London, in the Lee Valley, has been operated to examine the feasibility of replenishing storage with subsequent abstraction at times of drought. The scheme provides an estimated yield of $80 \times 10^3 m^3 d^{-1}$ over 200 days when fully operational. It is anticipated that artificial recharge schemes in north-east and south London will provide some 250–300 × $10^3 m^3 d^{-1}$ abstracted over 200 days in dry years (Owen *et al.* 1991).

The source of the water for artificial recharge will be the River Thames, treated to potable standards; water will be injected through wells during off-peak demand periods in non-drought years. Under operational drought conditions, the recharge wells will be used to abstract groundwater which will be treated on site and pumped directly into supply. A schematic representation of the operational cycle is shown in Fig. 12.23.

13. The Chalk as a hydrocarbon reservoir

M. D'Heur

Introduction

Below the North Sea, in the central part of the Tertiary basin, the Chalk is a major hydrocarbon reservoir. Oil and gas were first discovered in 1966 in chalks of Cretaceous and Palaeocene ages in Well A-lx drilled in the Danish sector but the find was not considered to be commercial. Further north, in the southern part of the Norwegian sector, Well 2/11-1, drilled in 1969 on the flank of the Valhall structure, penetrated 17 m of hydrocarbon-bearing chalk which flowed at 900 barrels per day. Again, at that time, this was not regarded as a commercial discovery and the Valhall Field was actually not developed until 1976.

In the late 1960s exploration drilling began in Norwegian Block 2/4 on the south-western flank of a large dome called Ekofisk; Palaeocene sandstones were the target. The first well was abandoned before reaching the Palaeocene and replaced by Well 2/4-1A. In this well the Palaeocene sandstones were not encountered but, at a depth of 2865 m below the sea floor, the well penetrated very porous chalk containing hydrocarbons over an interval of 183 m. Flow tests revealed a sustainable yield of more than 10 000 barrels per day. This discovery, at the end of December 1969, was the first of the giant oilfields in the North Sea. Since then, ten fields have been discovered in the Chalk of the Norwegian sector (often referred to as the oilfields of the Greater Ekofisk area) and ten smaller fields in the Danish sector, with at least one in the UK sector (Fig. 13.1). The North Sea chalk fields contain recoverable reserves of over 700×10^6 standard cubic metres (scm) of oil and 430×10^9 scm of gas, with over 80 per cent in the southern part of the Norwegian sector in Blocks 2/4 and 2/7 (Table 13.1). Ekofisk remains by far the largest of these fields, and it is in fact the biggest chalk field in the world. Recoverable reserves are more than 300×10^6 scm of oil and 150×10^9 scm of gas. Two-thirds are in the Danian Ekofisk Formation and the remainder in the Maastrichtian Tor formation. Gas also occurs in the Chalk of Friesland, in the north of the Netherlands, in the Harlingen Field, and in Germany (Boigk 1981).

Geological setting

The rifting phase of the North Sea Basin began in late Triassic times and continued into the mid-Cretaceous. This was succeeded by the post-rifting phase of Tertiary to Recent times (Olsen 1987). The rifting phase created the 'Central Graben' which underlies the Late Cretaceous–Cenozoic 'Central Trough'. The chalk fields occur in the southern part of the Central Trough. The occurrence of oil in the North Sea is due to a series of geological events over a long period of time. During Late Jurassic times a thick sequence of marine black shales was

Table 13.1 Approximate original recoverable reserves of hydrocarbons of the Chalk fields in the North Sea

	Oil ($\times 10^6$ scm)	Gas ($\times 10^9$ scm)
Albuskjell	10	20
Edda	5	2
Ekofisk	350	150
Eldfisk	85	51
Hod	7	1
Tommeliten	8	20
Tor	25	15
Valhall	70	20
West Ekofisk	16	28
Total Norway	576	307
Total Denmark	150	120
Total United Kingdom	10	5

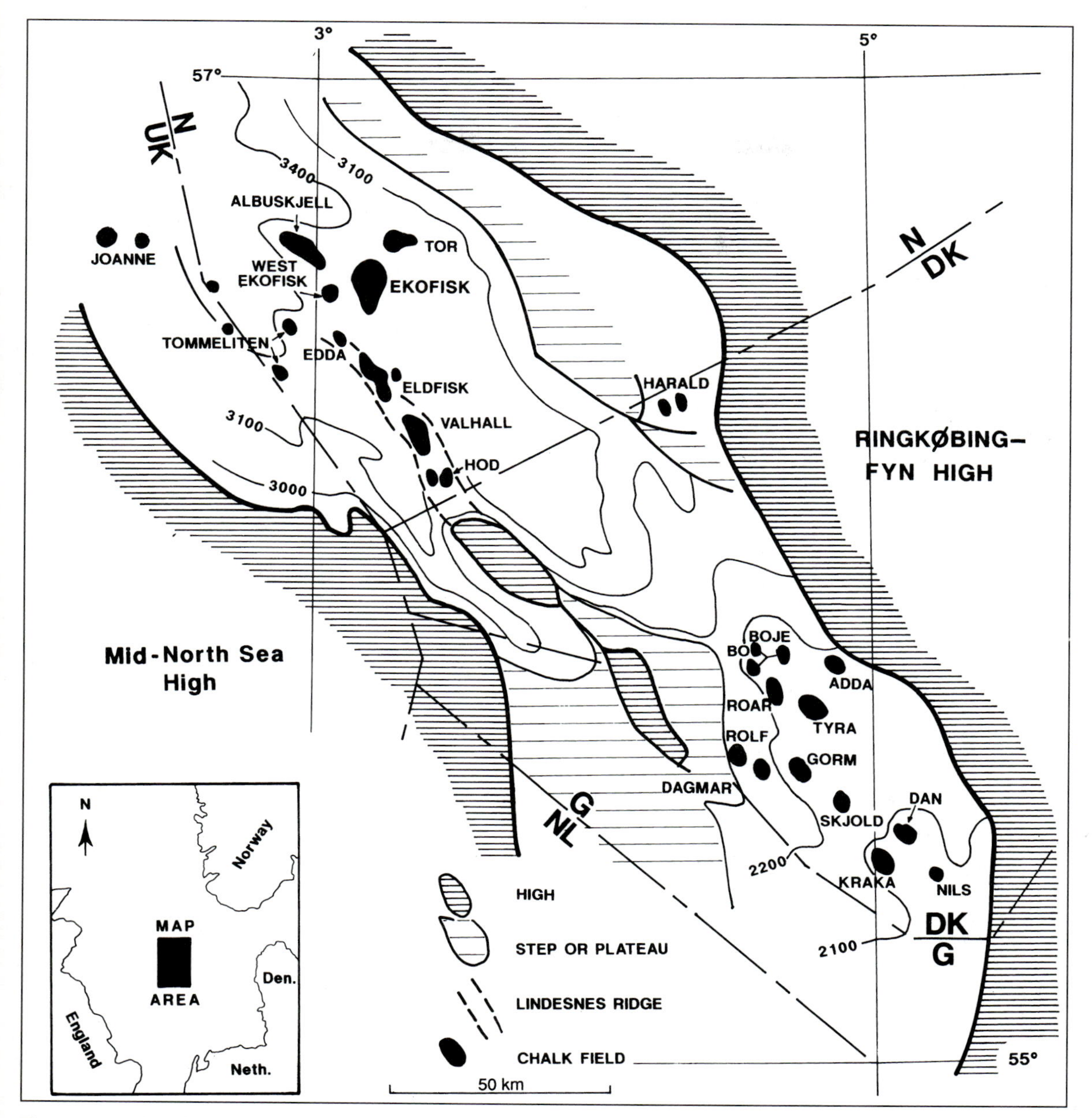

Fig. 13.1. Oil and gasfields in the Chalk of the North Sea. Figures indicate depth to the top of the Chalk in metres. The boundaries of the national sectors are shown.

deposited in the Central Graben; they were to become the source rocks for the hydrocarbons in the Chalk. Rapid subsidence occurred during Cretaceous times and as much as 1000 m of chalk were deposited in the deepest part of the Central Trough. During the Cretaceous, local inversions formed elongated structural uplifts in the Trough. These have had a very significant bearing on the thickness of the Chalk and the distribution of the facies that form the chalk reservoirs.

Tectonic activity in Maastrichtian and Danian times, particularly along the faulted margins of the graben and on the adjacent platform, mobilized unstable carbonate deposits causing slumps, slides, debris flows and turbidites. The material was redeposited in the Central Trough as allochthonous chalks and these are now the best chalk reservoirs.

The Chalk fields have a characteristic domal configuration, caused in some fields (for example, Ekofisk, Tor, Albuskjell and Dan) by halokinesis due to the movement of Permian salt deposits, but in others (Valhall, Eldfisk, Tyra, and Roar) to transpressional inversion features associated with transcurrent faulting (Sorensen *et al.* 1986). These structures provided favourable traps for migrating hydrocarbons. Finally, as subsidence of the basin continued, the accumulation of thick Tertiary clays provided the necessary seal for the oil reservoirs in the Chalk.

The top of the Chalk generally lies at a depth of more than 3000 m in the Norwegian sector while in the Danish sector it rises to about 2000 m in the south of the sector. It is underlain by the Lower Cretaceous resting on thick Jurassic sediments.

The Chalk of the Central Trough is divided into five formations (Fig. 13.2). The main reservoirs are in the Ekofisk and Tor formations of Danian and Maastrichtian ages, respectively, although the Hod Formation is also an important reservoir, especially in the Valhall area. More than 90 per cent of the hydrocarbon reserves in the Chalk are in the Ekofisk and Tor formations of the Maastrichtian and Danian chalks (Fig. 13.2). The Danian and Maastrichtian are generally less argillaceous than chalks of

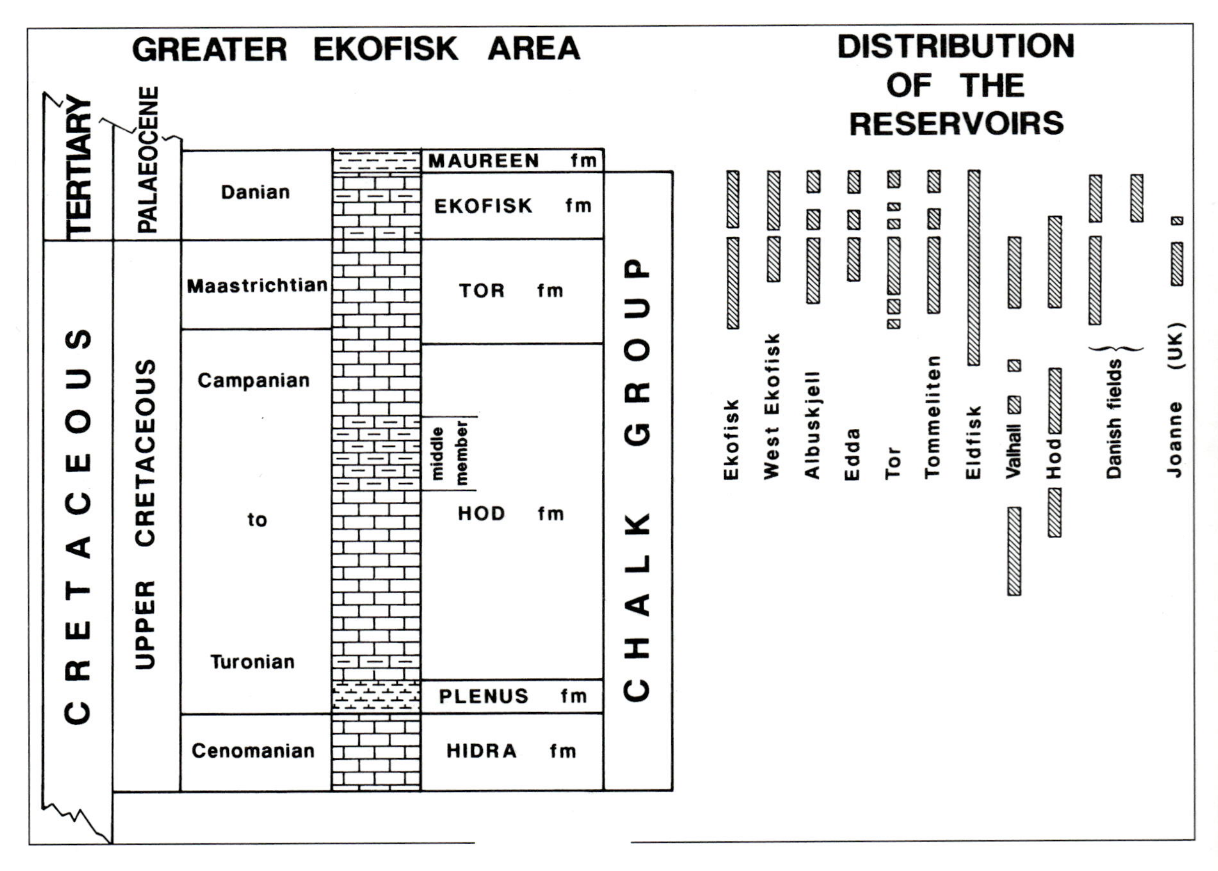

Fig. 13.2. Stratigraphical distribution of hydrocarbons in the Chalk of the North Sea.

Cenomanian to Campanian ages. For a particular porosity the presence of clay has an adverse effect on permeability and it also tends to make the rock less favourable for 'spot welding' which preserves porosity. Chalks with the highest permeability have an insoluble content of less than 15 per cent. Pre-Maastrichtian chalks, with few exceptions, also contain a very small proportion of allochthonous material.

Shaley pelagic chalks of the Middle Hod Formation generally do not have favourable reservoir characteristics, while the clean pelagic or slightly allochthonous chalks of the Hod Formation have been affected by chemical compaction. They consequently have a high entry pressure and are invaded by hydrocarbons only when more favourable reservoirs do not exist or already contain hydrocarbons.

The chalk of the Tor Formation is generally very pure and contains a high proportion of allochthonous material. The Ekofisk Formation has higher clay and silica contents and this tends to reduce its quality as a reservoir. For a given porosity, clean chalks of the Tor Formation have higher permeabilities than those of the Ekofisk Formation because the average grain size of the chalk in the Tor is 2 μm whereas it is 1 μm in the Ekofisk Formation.

There are several sedimentological zones in the Danian which are very relevant from the point of view of developing hydrocarbons. The Ekofisk Tight Zone, some 3–15 m thick, at the base of the Ekofisk Formation, is composed of pelagic chalks, periodites and mud cloud deposits interbedded with distal turbidites and hardgrounds. It is not a reservoir rock and tends to act as a permeability barrier between reservoirs in the Danian and Maastrichtian, although its continuity is broken by faults and it is now recognized that fluid transfer across the zone is possible.

In Early Danian times the so-called 'reworked Maastrichtian' unit was deposited as a slump complex, made up, as the name implies, of a very high content of original Maastrichtian sediments. It has a porosity–permeability relationship more akin to the Tor Formation than to the rest of the Ekofisk Formation.

At the top of the Ekofisk Formation, about 15 m of pelagic chalk and mud cloud deposits, with thin turbidites containing clay and silica towards the top, eventually pass into typical periodites. Despite a high porosity, the interval tends to have a very low permeability.

On the basis of the nature of the chalk, linked to its origin and sedimentation type, the Ekofisk and Tor formations have been divided into eight members each of which can be distinguished by uniform petrophysical characteristics (Fig. 13.3). The distribution of the main chalk facies can also be recognized in the Greater Ekofisk area on an areal basis (Fig. 13.4). In Area A, adjacent to the fault complex marking the eastern boundary of the Central Trough, most of the reservoir rock consists of mass-flow deposits displaying evidence of extensive slumping and sliding. Local unconformities or non-sequences occur, mainly in the Ekofisk Formation. In Area B, the sea floor was virtually flat and stable during Danian and Maastrichtian times and the current was much reduced. Thick, continuous debris flows were deposited. Massive turbidites and slump structures are also found but are of less importance than in Area A. The lowest member of the Ekofisk Formation locally consists of the 'reworked Maastrichtian unit', referred to above, which contains about 25 per cent of the reserves of the Ekofisk and West Ekofisk fields. Area C extends along the Lindesnes Ridge, an inverted half-horst, reactivated during the Campanian, where allochthonous deposits tend to be thin. To the west of the Lindesnes Ridge, in Area D, the chalk is mainly developed as pelagic or periodite facies with thin interbeds of debris flow and distal turbidites.

Chalk as a petroleum reservoir

The Chalk below the North Sea would not be expected to be a petroleum reservoir. That it is so is due to a series of unusual geological circumstances. Under normal conditions the porosity of the Chalk at a depth of 3000 m would be reduced, by diagenesis and compaction, to some 10 ± 5 per cent, values that would imply a permeability too low for production. In actual fact the chalk reservoirs of the North Sea have porosities that often exceed 30 per cent, in places attaining 50 per cent. These high values are found in redeposited or allochthonous chalks that were originally pelagic chalks. Debris flows are generally the best reservoirs. Most of the Danian and Maastrichtian chalks are allochthonous.

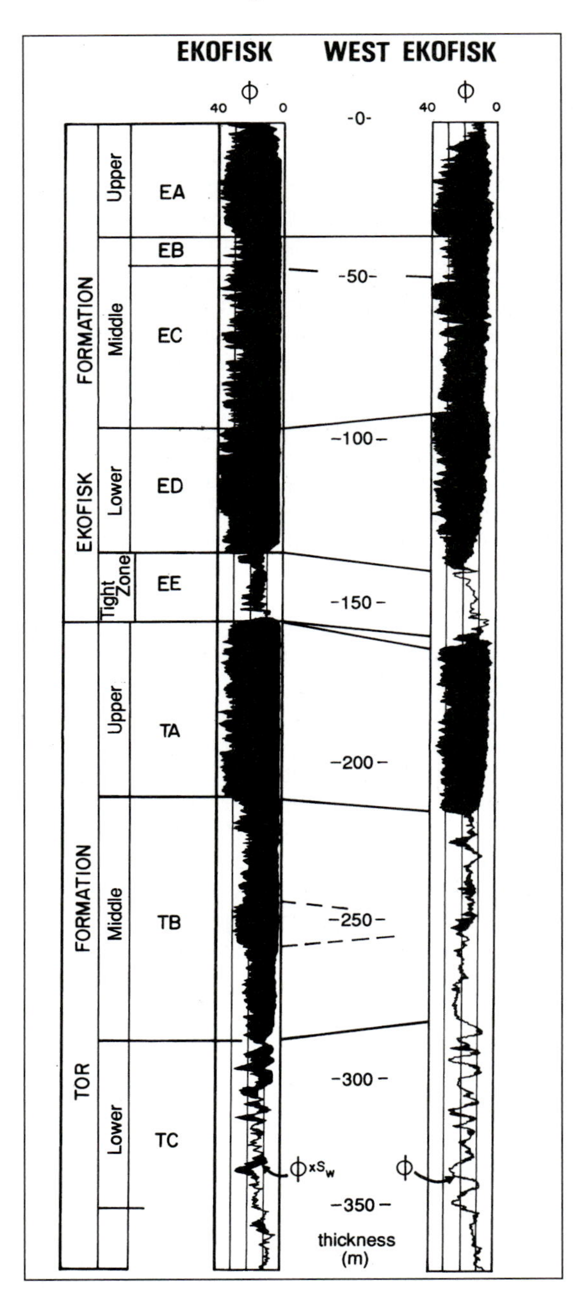

Fig. 13.3. Division of the Ekofisk and Tor formations into eight members based on petrophysical characteristics. The logs on the figure indicate porosity and the black shading the fraction of the porosity invaded by oil (after D'Heur 1986).

In contrast, pelagic chalk deposits do not constitute good reservoirs. In the Ekofisk area 75 per cent of the chalk sequence consists of allochthonous chalk, while in the Danish sector the figure is 50 per cent (Nygaard *et al.* 1983).

The fact that redeposited chalk is the key factor in the creation of oil reservoirs was first suggested by Perch-Nielsen *et al.* (1979) who identified a thick sequence of 'Maastrichtian Chalk' *above* the well-known Danian Ekofisk Tight Zone in the Ekofisk well 2/4-A8. This represented redeposition in early Danian times of a resuspended Maastrichtian pelagic, chalk-ooze. Subsequently, Kennedy (1980) showed that the Tor Formation in the Tor field consists almost entirely of debris flows, and similar redeposited chalks have been identified in other areas, for example in the Valhall-Hod (Hardman and Kennedy 1980) and Albuskjell fields (Watts *et al.* 1980) and in the Danish sector (Nygaard *et al.* 1983). Debris flows occur over the entire Central Trough and individual flows can be correlated between fields (D'Heur 1984). Local tectonic activity, erosion and redeposition have led in the Danish sector to very local depocentres and basins of redeposited chalk, contrasting with the basin-wide deposition more characteristic of the Norwegian sector (Megson 1992).

Preservation of porosity

The disturbance, movement and subsequent rapid deposition of an original pelagic ooze have been the significant factors in the preservation of high porosities in the allochthonous chalks. Nygaard *et al.* (1983) maintained that early lithification is a prerequisite for good reservoir quality in redeposited chalk, but the extent of the disintegration of the matrix is the main factor. Disintegration of chalk as it is redeposited forms intraclasts (or aggregates) with a wide range of sizes and this leads to high porosity and matrix permeability. The chalk aggregates transported over long distances have disintegrated to a much greater extent and the resulting rock has a poor reservoir quality because of denser packing. Hence the proximal redeposited facies provides the best reservoirs. The resuspension and transport of the chalk not only sorts the particle size but also removes clay particles

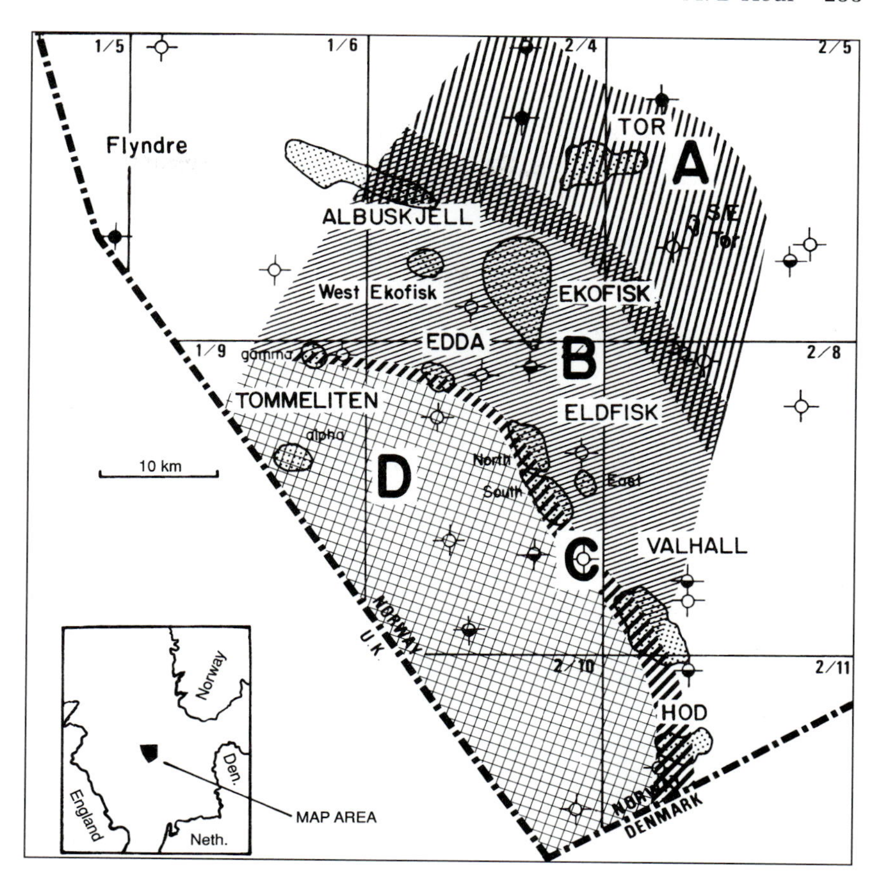

Fig. 13.4. Distribution of the principal chalk facies. The characteristics of areas A, B, C and D are given in the text (after D'Heur 1986).

which would otherwise tend to reduce the spot welding of carbonate particles. In the best reservoirs the total insoluble fraction is not greater than 15 per cent and even in the poorest it is never greater than 35 per cent (Hardman and Eynon 1978). Hardman (1982) found that redeposited chalk tends not to contain aragonite, which is a source for early cement, and also less silica may be available for cementation.

There are at least three reasons for variations in porosity within a field: lateral variation in the chalk facies itself, differential mechanical compaction; and differential diagenesis linked to the timing of the migration of hydrocarbons into the Chalk. The marked differences in porosity of wells on the flanks and between fields is due more to differential diagenesis than to lateral variations in sedimentary facies.

The abnormal porosity appears to have been preserved in the allochthonous chalks for several reasons. The rapid deposition creates a more open framework and a low clay content (Fig. 13.5). Burrowing organisms have not had an opportunity to reduce the porosity by encouraging dewatering. At contact points between grains very effective 'spot welding', representing the first stage of chemical compaction, has created a rigid framework (Mapstone 1975) which resists compaction. However, the development of overpressures and the early migration of hydrocarbons are also important factors. The high overpressures are believed to be primarily caused by rapid burial during the Tertiary, and by loading of the Tertiary sediments leading to compaction. Pore pressures attain about 50 MPa at depths of 3000 m, of which some 16 MPa represents overpressure. The fluid pressure is thus supporting a

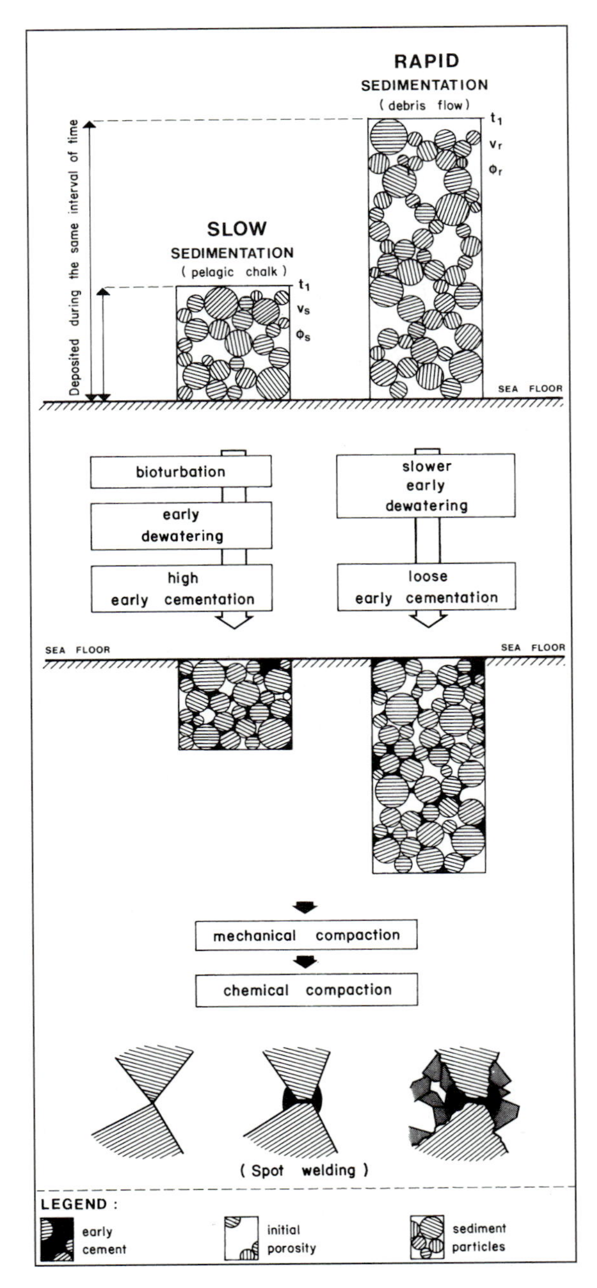

Fig. 13.5. Effect of sedimentation rate on original porosity and a schematic illustration of chemical compaction (after D'Heur 1984).

large part of the lithostatic load of 60 MPa. These high pressures prevent mechanical compaction. The porosity of the chalk reservoirs varies in a complex manner, implying that the pore-fluid pressure is also variable. The excess pressures are localized in the reservoir structures, and this must have been so from an early stage for the preservation of the high porosities required an early and effective seal (Watts 1983a; Jones 1990).

Oil migration into the pores, which reduced or arrested chemical compaction, is considered to have occurred from Eocene to Miocene times. Migration into the Chalk was controlled by capillary effects, and large pore-entry pressures were necessary. It would not have occurred uniformly but rather preferentially into horizons with the largest pores. High pressures would have been necessary in the hydrocarbons before entry into the Chalk was possible. The invasion of hydrocarbons displaced brines downwards and into the limbs of the structures where pressure solution occurred, further reducing the porosity to the present low values of about 5 per cent (Sorensen *et al.*1986; Jones 1990)

Overpressures and migration of oil were each equally important influences in the preservation of porosity. Overpressure was certainly critical until hydrocarbons invaded the pores. But overpressure alone would not have been sufficient to preserve porosity in the thick homogeneous redeposited chalk sequences of the Ekofisk and Tor formations. This is illustrated by the fact that high porosities in these rocks were retained after the invasion of hydrocarbons, though values decreased significantly in the surrounding brine-saturated chalk, where it is now rarely greater than 25 per cent. However, overpressures can be maintained, and the presence of hydrocarbons is not necessary to preserve porosity in thin reservoirs that are interbedded with tight pelagic chalk (D'Heur 1984).

The chalk of the Ekofisk and Tor formations is represented by alternations of porous and less porous beds, with the individual layers traceable laterally from field to field (Fig. 13.6). This reflects the manner in which the individual porous layers formed and the fact that they have responded in a similar manner to cementation. The porosity probably reflects the initial primary porosity distribution of the redeposited chalk. The original unstable pelagic sediments were resuspended and transported before they had been significantly cemented. In the

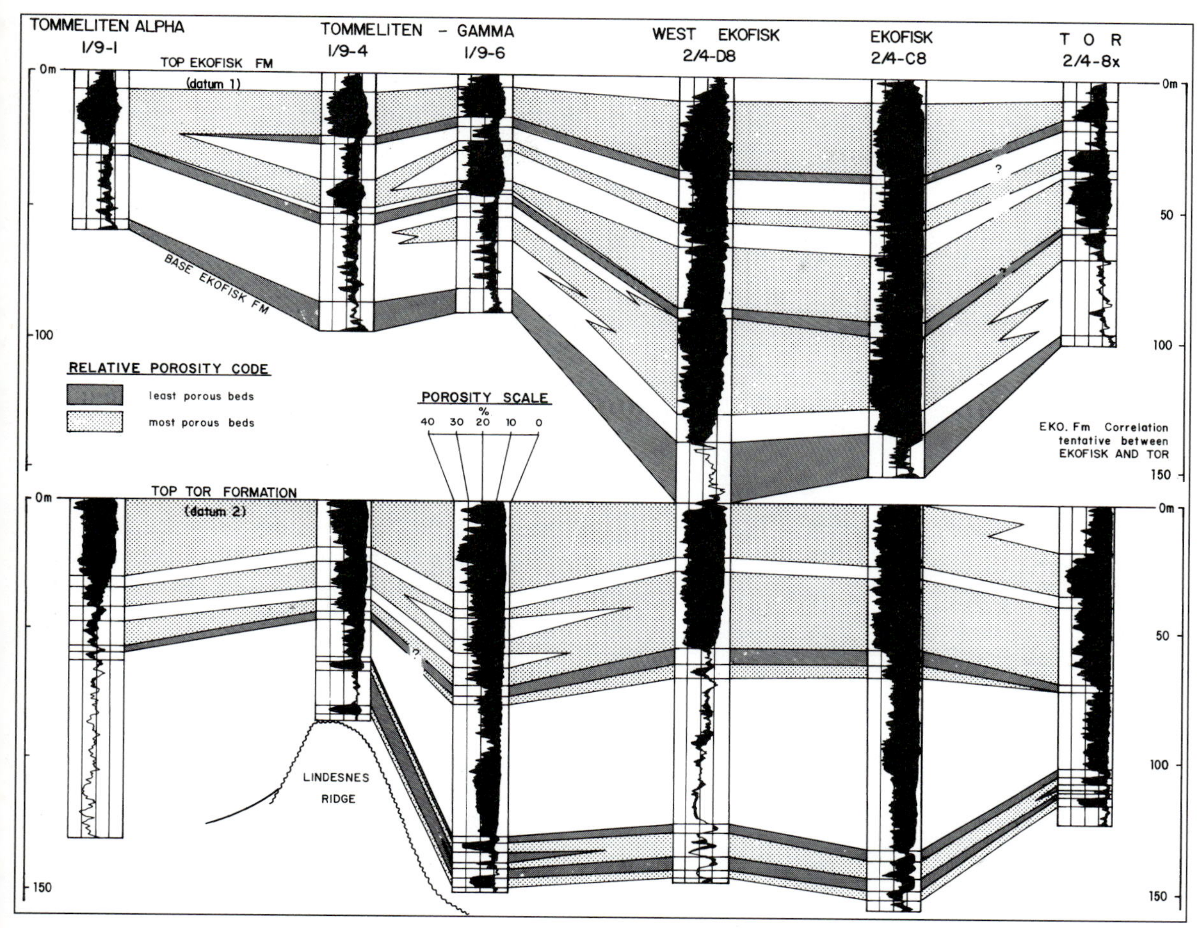

Fig. 13.6. Correlation of the Ekofisk and Tor formations between fields and illustrating the lateral and vertical variation of porosity. The logs indicate porosity and the black shading the fraction invaded by oil (after D'Heur 1986).

redeposited chalk the particle sizes are larger, and the particles less cemented. The layered nature is emphasized as the diagenetic process has been arrested by the migration of oil into the more porous zones. The early porosity differences are masked by late diagenesis in chalks that do not contain hydrocarbons.

There is a general and progressive reduction of porosity with depth, both vertically and laterally, from the crestal part of a structure to the periphery (Fig. 13.7). This seems to be related to the timing of the hydrocarbon migration. Hydrocarbons accumulated first in the higher parts of a structure, thereby inhibiting further chemical compaction. As the infilling of a structure continued, diagenetic reduction of porosity was progressively stopped at deeper and deeper levels. The Upper Tor Formation retains significant porosity, even when it does not contain oil, but generally the porosity is higher in the hydrocarbon zone than in the water-filled zone. It appears that the present variations in thickness of the formations across several fields are mainly caused by differential compaction. The porosity decreases down the flanks of the structures and the variation in thickness represents the degree to which the porosity has been preserved by hydrocarbon migration (D'Heur 1984; Fig. 13.7).

In the Ekofisk and West Ekofisk fields the porosity of the Ekofisk Formation attains more than 40 per cent in the oil zone but it is generally less

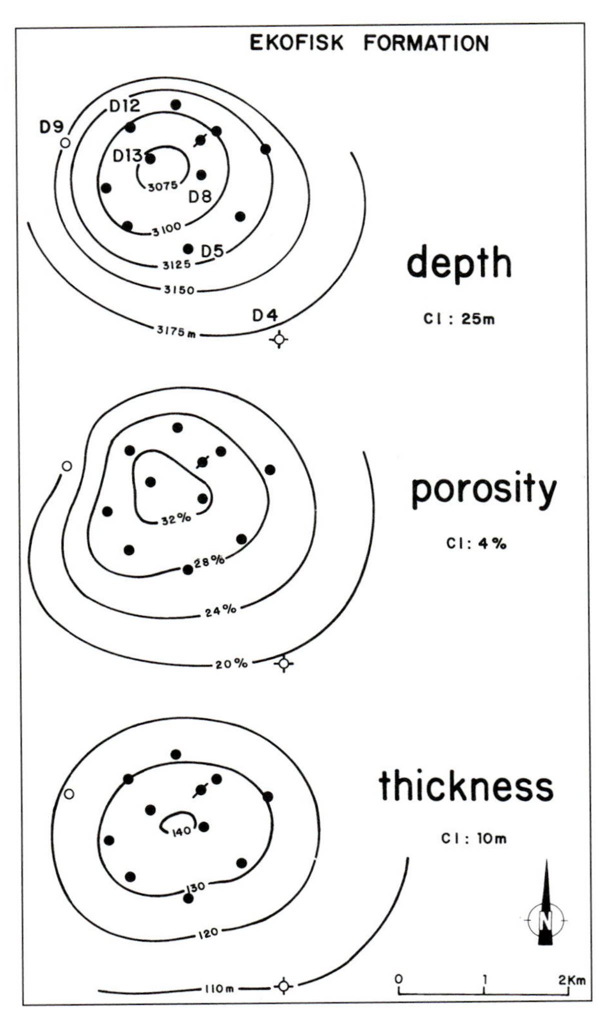

Fig. 13.7. West Ekofisk Field, illustrating the similarity between structure, porosity, and isopachytes for the Ekofisk Formation (after D'Heur 1986).

than 22 per cent where hydrocarbons are not present. Locally, the main section of the Upper Tor Member has values as high as 52 per cent in the hydrocarbon zone and 20–25 per cent where the chalk is water-filled. The Middle Tor Member has been affected by the chemical compaction in the water zone and the porosity rarely exceeds 18 per cent but attains 38 per cent in the oil zone. In contrast, values in the pelagic chalks of the Lower Tor Member never reach 15 per cent even in the oil zone.

Origin of permeability

The Chalk can be a hydrocarbon-producing reservoir because it is intersected by an extensive fracture system which collects hydrocarbons from the matrix and conveys them to wells. The permeability of the matrix typically ranges from less than 1 to 5 mD, although it does attain 10 mD in the most favourable rock types which have porosities of more than 40 per cent.

Tectonic movements occurred in the Central Trough throughout the Upper Cretaceous and Tertiary periods. The Chalk was faulted and fractured by reactivation of the major boundary faults of the graben and faults within the graben. At this time the regional stresses were of a compressional nature, but inversion, subsidence and halokinesis were all associated with these movements. As a consequence, the Chalk was fractured on both micro and macro scales, thereby creating pathways for fluid flow.

The origin of micro-fractures in the Albuskjell Field have been discussed by Watts (1983a,b) in some detail. This field is a large halokinetically induced anticline in the Ekofisk complex (Fig. 13.1). Two types of micro-fractures can be recognized in cores of the Chalk. The first are sub-vertical to vertical, straight to sinuous, narrow fractures (generally less than 1 mm and often less than 0.5 mm wide) commonly arranged in conjugate sets. They are usually healed by calcite, although kaolinite and barytes also occur. They resemble shear fractures that formed relatively early, prior to advanced pressure solution. The second type is generally, but not exclusively, associated with stylolites. They are microfractures with narrow apertures, largely unhealed, that extend vertically away from the tips of stylolite teeth. Most are less than 0.1 mm wide but, where partially filled with diagenetic minerals, they may be up to 2 mm wide. The fractures terminate at stylolite surfaces, and they appear to postdate the formation of the stylolites. The fractures are invariably vertical, and they are believed to be tension fractures. Shear fractures occur throughout the Danian and Maastrichtian sequences but the tension fractures are more common in the harder, more brittle chalk of the

Maastrichtian. The development of fractures is very dependent on rock properties; the more argillaceous, softer chalks generally display a poor development.

The shear fractures are believed to be associated with either early Tertiary movements along the major north-west to south-east faults of the Central Graben system or halokinetic reactivation of these major faults. The tension fractures possibly formed at a burial depth of 1370 m (in Early Miocene times) as a consequence of a combination of overpressures and doming caused by halokinesis. They formed after hydrocarbon emplacement. Model studies suggest that tension fractures may be both radially and concentrically oriented with respect to the dome structures and concentrated in crestal areas. Calculations suggest tension fractures may be open to depths of about 3500 m and that the optimum fracture spacing in brittle chalks in crestal areas is 45–65 mm. Despite the small size and extent of the fractures and the fact that interconnection between them is generally poor, they probably contribute to field permeability.

Brewster *et al.* (1986) recognized three main types of fractures in the Ekofisk and Tor formations of the Ekofisk Field. Two are the healed and stylolite-associated fractures identified by Watts, while the third are tectonic fractures which are planar features forming well-defined parallel and conjugate sets. They occur in zones, and one set is dominant. Essentially they are small-scale normal faults, the displacements across the fractures being very small, only up to a few centimetres. The fracture spacing is quite dense (the average is 10–15 cm) and when they are open they provide the interconnected networks necessary for large-scale fluid movement. As Watts recognized, the tension fractures associated with stylolites are of limited extent. They probably form zones of enhanced horizontal permeability, but do not represent a pervasive network throughout the reservoir.

More recently, Duval *et al.* (1990) have studied the Ekofisk Field and questioned the view that it is a salt-dome-related structure producing oil through radial fractures that reduce in intensity down the flanks of the structure. Their view is that the field results from Cretaceous inversion and transpressional folding and associated faulting along the major pre-Zechstein faults. Late Eocene or Miocene halokinesis may have occurred but the fracture pattern has developed by reactivation of basement faults. The major fault trend is NNE–SSW, with a secondary NW–SE alignment; the major trend extends over the entire field (Duval *et al.* 1990, Figs. 9 and 10). The secondary system is restricted to the crestal, and north-western parts of the field. It postdates the main trend and may have developed in response to doming associated with diapirism. The major faults appear to be open, at least in the crestal part of the field. Thus, these fracture systems coincide with Caledonian and Hercynian fault lines reactivated during a compressional phase of the Laramide and Alpine orogenies. Post-Alpine halokinesis may have accentuated the displacements.

The natural fracture system created by the various tectonic processes has enhanced the matrix permeability by a factor of up to 50 or even 100. The relationship between porosity and matrix permeability for the Ekofisk Field (Fig. 13.8) indicates that the maximum permeability that can be expected from the matrix only is about 8 mD, with typical values of no more than 1 mD. Although fractures can be recognized in cores, laboratory tests (Fig. 13.9), and on well logs, the principal reliable method of detecting those that contribute significantly to flow is from well tests (Brown 1987), which reveal effective permeabilities of up to between 150 and 200 mD. It is quite likely that in most wells the majority of the flow is derived from one or two main fractures connected to a ramifying system of smaller fractures.

The reservoir in the Machar Field in the UK sector comprises fractured chalk and Palaeocene sands above a steeply dipping salt structure. The fine-grained pelagic chalk matrix has a primary porosity of between 12 and 35 per cent and a permeability generally less than 1 mD; the secondary fracture porosity is probably less than 1 per cent. Production tests of fractured intervals indicated a permeability in excess of 1000 mD. The average spacing of open fractures measured in cores is 75 mm, and about a third of the fractures are open (Foster 1993).

In the southern part of the Danish group of fields, halokinesis has been a factor in the formation of the structural traps but in the north structural factors only have determined the forms of the fields. The permeability varies from 1 to 1000 mD, and the highest permeabilities are found in the very fractured fields developed over salt-domes. The Dan Field is over a salt-swell; the porosities are typically

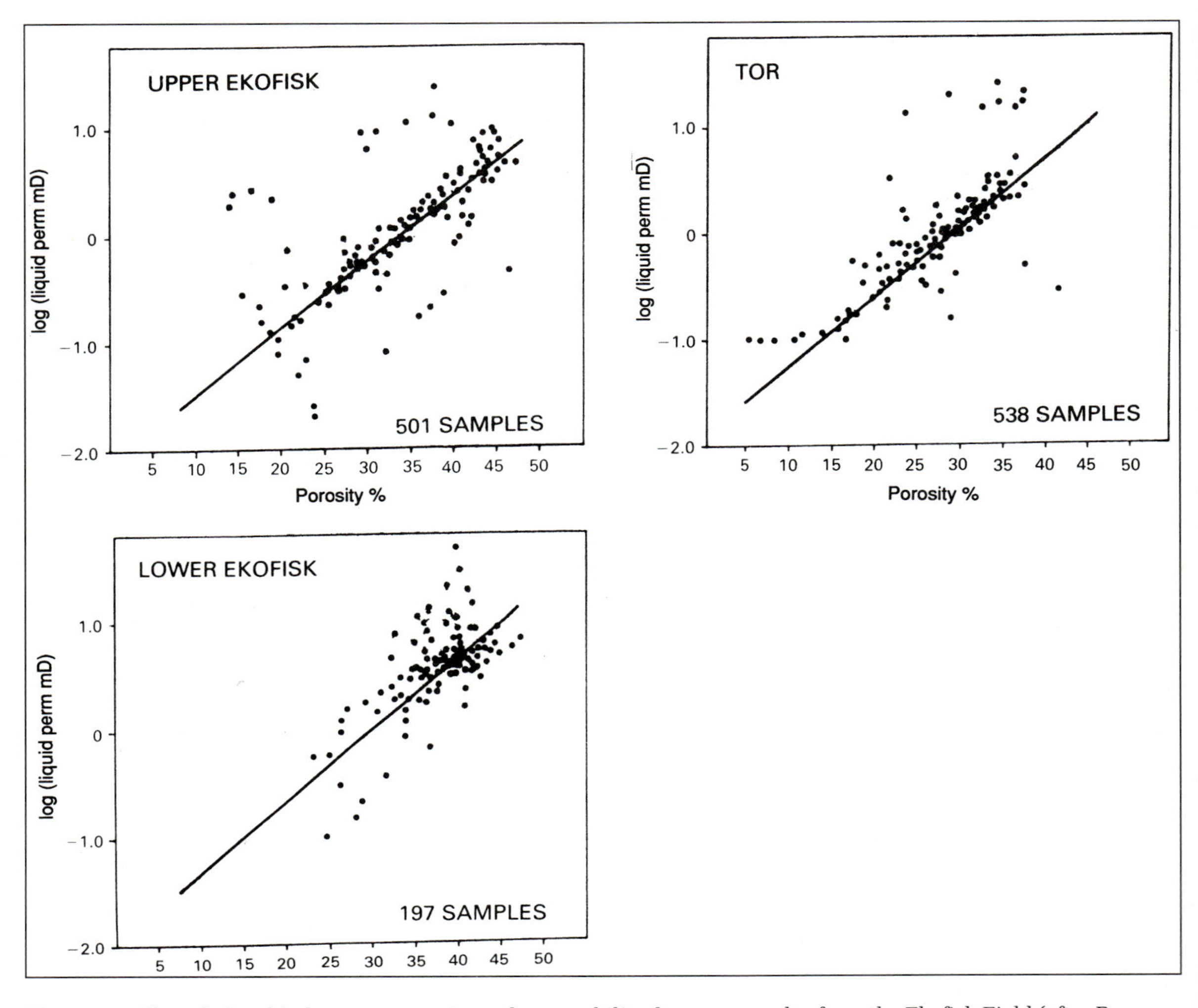

Fig. 13.8. The relationship between porosity and permeability for core samples from the Ekofisk Field (after Brown 1987).

22–34 per cent, but the average permeability is only 1.75 mD. Because of the lack of natural fracturing the field is currently produced by horizontal hydraulically fractured wells (Megson 1992).

Reservoir compaction and its consequences

The reduction in pore-fluid pressure, as a result of the production of hydrocarbons leads to the collapse of the largest pores in the most porous chalk intervals and a consequent reduction in thickness. The problem is compounded by the fact that the Chalk is compactable, it is very thick, and the development of hydrocarbons extends over large areas (Berget *et al.* 1990). Compaction and subsidence in the chalk fields and its consequences have been reviewed by Jones (1990).

The chalk reservoirs are very overpressured but there is no support by pressure from external renewable sources of overpressured fluids because the reservoirs are sealed by Tertiary clays above and low-porosity chalks below and on the flanks of

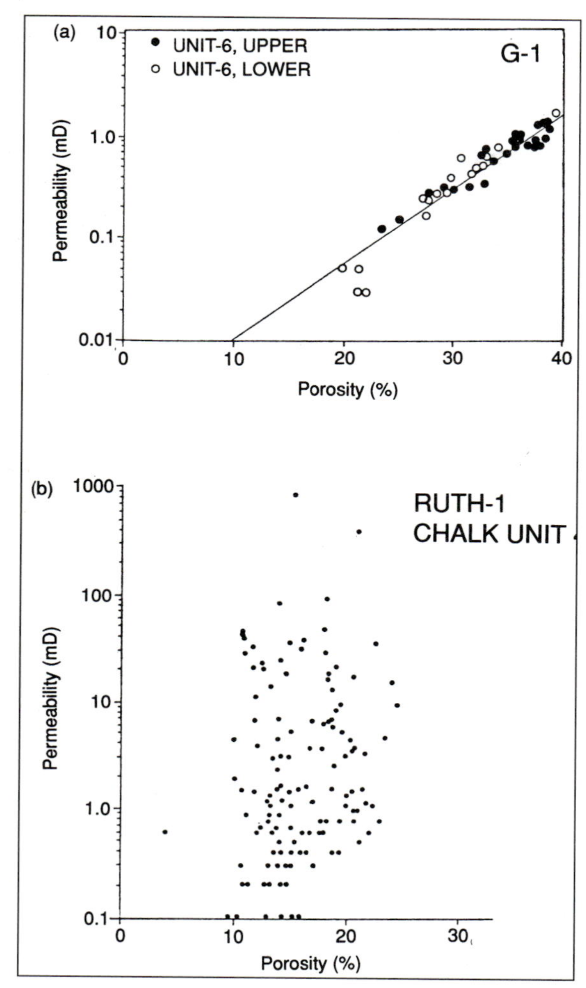

Fig. 13.9. (a) Permeability versus porosity measured on cores from Well G-1 for Chalk Unit 6, upper and lower parts (equivalent to Sequence A, see Fig.1.4, in the Danish sector. The logarithmic relationship indicates that the permeability relates to the chalk's matrix. (b) Permeability versus porosity measured on cores from Well Ruth-1 for Chalk Unit 4 (approximately equivalent to Sequences D and E/F, see Fig. 1.4), in the Danish sector. The deviation from a logarithmic relationship indicates microfracturing in the cores. The cluster of points in the lower right of the figure gives the relationship between porosity and matrix permeability. These do not exceed 25 per cent and 1.5 mD, respectively (after Nygaard *et al.* 1983).

structures. Where the porosity of the Chalk exceeds 35 per cent, as it does over significant parts of Ekofisk and surrounding fields, the material is weakly bonded and it behaves as a weak elastoplastic rock, in fact responding in a similar manner to that of a sensitive-bonded clay. An increase in the effective stress in the reservoir, as a result of fluid production, progressively induces elastoplastic strains in chalks of decreasing original (pre-production) porosity. Elastic bonding in the Chalk (which, together with excess pore-fluid pressure, resisted the loading of the overburden) is destroyed. The compaction process comprises an elastic, reversible component and a plastic, irreversible component; the latter is commonly referred to as *pore collapse* (Sulak *et al.* 1991).

Reservoir compaction can damage the reservoir rock and cause problems with the collapse of well casings and subsidence of the sea floor. The sea bed below the Ekofisk complex had subsided by some 3 m by the end of 1984 as a result of compaction of the chalk reservoir of about 4.5 m (Sulak, 1991). It has been predicted that vertical displacement of the sea floor at Ekofisk could exceed 6 m if reservoir pressures were allowed to continue to decline. However, reinjection of natural gas and injection of seawater, and possibly nitrogen (Berget *et al.* 1990), should maintain reservoir pressures at present levels so that subsidence will not be too critical; maximum subsidence rates have been about 0.3–0.5 m a^{-1}. Subsidence to this extent created potential hazards with regard to the safety and stability of the production platforms. This has been overcome by imaginative, but expensive, engineering works that raised the height of the platforms above the sea floor by an additional 6 m (Aam 1988).

Well casings can be deformed by compaction. The wells are subject to vertical compression which produces shortening in the case of a vertical well and oblique shortening in the case of a deviated well. Directly above the reservoir crest the strains are extensional (Jones *et al.* 1990), subjecting wells to extensional rather than compressional forces, while above the flanks of structures deviated wells may be subject to shear displacements. However, it must be appreciated that many wells do not fail as a result of compaction.

Not uncommonly, wells produce significant amounts of solid material. This appears to follow rapid reduction in pressure in a well that had been 'shut in'. The chalk adjacent to the casing perforations fails—it 'liquefies'—and flows into the well. The rapid decrease in well-fluid pressure leads to shear stresses caused by the undrained nature of the chalk. On the other hand, a slow reduction in well-fluid pressure leads to drained loading; larger shear stresses are needed for shear failure under drained than under undrained conditions. Thus, it appears that the production of excessive material in some wells is due to a liquefaction type of failure caused by the behaviour of the chalk in the presence of relatively large shear stresses at low mean effective stresses (Leddra and Jones 1990). The problem has been overcome by completing wells with a gravel pack and by controlling pressure drawdown in wells that have a history of such problems (Jones 1990).

Model studies showed that reservoirs with large surface areas, that are not particularly elongate and have a distribution of high-porosity chalks (for example, the Ekofisk Field), are prone to subsidence, but smaller reservoirs (for example, West Ekofisk) and more elongate structures (for example, Valhall) are less susceptible (Jones *et al.* 1990).

Compaction reduces the porosity and may be expected to have a consequential similar effect on permeability. However, data from Ekofisk have not shown this to be so. Experimental results (Leddra *et al.* 1990) indicate that generally the change in matrix permeability is small even for large changes in porosity. The explanation may be that pore-throat sizes do not change significantly during compaction and so the reduction in porosity is independent of the resistance to fluid flow. It would appear that compaction does not have a major impact on the fracture system. Possibly part of the fracture system is favourably orientated and is dilated during compaction (Jones 1990).

Chemical nature of the formation waters

Very little information has been published about the composition of the formation waters in the Chalk that are associated with the hydrocarbons, but Egeberg and Aagaard (1989) have given some analyses and discussed the origin and evolution of the waters.

The total salinity ranges from about 50 to 100 g l^{-1} and, in general, increases with depth. The waters are all Na-Cl brines, the only other ion present in any significant concentration being calcium. Apart from sulphate, which has a very low concentration, all the ions can be directly correlated with chloride. Egeberg and Aagaard considered that the linear Na/Cl relationship indicates that the waters are the result of mixing a brine, with a salinity of over 200 g l^{-1} and a dilute water. The brine is believed to be a bromide-enriched water formed by the evaporation of seawater beyond the point of halite precipitation, rather than by the dissolution of halite. The dilute end member could be either seawater or meteoric water, but data for stable isotopes imply that it is actually a meteoric water.

Before entering the Chalk the brine evolved from a Na-Mg-Cl-SO_4 to a Na-Ca-Cl water. Overall there was a reduction in K, Mg, SO_4, but an increase in Ca and Sr. The reactions modifying the brine are believed to be associated with dolomitization, precipitation of gypsum, dissolution of celestite, and the formation of K feldspar. The source of the brine is considered to be connate water from Permian evaporites which migrated from the deeper parts of the Central Graben in response to compaction of the sediment column and possibly diffusion. The meteoric end member is believed to have entered the Chalk in Late Palaeocene times during the formation of the regional unconformity associated with the Laramide tectonic event, or by flow from surrounding clastic rocks that were recharged with meteoric water during Late Jurassic or Late Tertiary uplifts (Egeberg and Saigal 1991). However, it does seem possible that the brines may have had a rather more complex evolution than a simple mix of two end members. Other components, such as saline Jurassic and Lower Cretaceous formation waters that migrated upwards into the Chalk as a result of compaction, Tertiary waters that percolated downwards at the beginning of the Tertiary transgression, as well as the long-distance migration of waters from other formations, through permeable zones and fractures, may also have been involved. Nevertheless, the dominant role of a Permian connate water and its dilution as it migrated into the Chalk is convincing.

It has been proposed by some workers (for example Taylor and Lapré 1987; Jensenius 1987) that healed fractures in the Chalk and the cementation of the matrix were the result of hot

CO_2-rich fluids ascending from underlying Permian evaporites. However, Egeberg and Saigal (1991) have argued that the calcitic cements in the fractures and the matrix have precipitated slowly from a solution of nearly constant composition and are the result of diagenesis in a closed or semi-closed system that is independent of any ascending hot-water flux. Their view is partly based on the fact that analyses of formation waters show that pore fluids initially had significantly more negative $\delta^{18}O$ values than previously assumed because they included a meteoric water component. They support the view of Scholle (1977) that the reduction in porosity during burial is caused by local pressure solution and recrystallization. The cementation of the shear fractures probably began soon after they were formed at a depth of about 1 km and continued to the burial depth at the present time of about 3 km.

Source of hydrocarbons and their development

The Upper Jurassic shales are recognized as the source rocks for the chalk fields. They are bituminous, black, marine shales with a very high organic content and a total thickness believed to be in excess of 1000 m. Significant oil generation began in the deeper zones of the Central Trough as early as Palaeocene times, when temperatures exceeded 80°C. Below the shallow structures of the southernmost area of the Norwegian sector, such as Valhall, peak oil generation is occurring at the present time. Commonly, the top of the Upper Jurassic is at depths of more than 5000 m and the source rocks are mature enough to produce the rich gas condensates of the Tommeliten, Albuskjell, and West Ekofisk fields (D'Heur 1986).

Several hundred metres of Lower Cretaceous shales, which would normally be regarded as impermeable, separate the Upper Jurassic source rocks from the Chalk reservoirs. The fluids in the source rocks would have expanded as the temperature increased with burial and the resulting rise of pressure may have been sufficient to fracture the overlying shales creating vertical migration paths into the Chalk. The process would no doubt have been facilitated by faults (Campbell and Ormaasen 1987) and by diaparic salt movement. Thus hydrocarbons probably migrated upwards from the Upper Jurassic through open fractures into the most porous allochthonous units of the Ekofisk and Tor formations. Where these favourable facies were missing, or poorly developed, oil entered poorer-quality reservoir rocks such as the Hod Formation. Lateral secondary migration must also have occurred over long distances, from the deep central part of the Central Trough towards its margins, through the permeable matrix of widespread massive debris flows. During early secondary migration, the permeability barriers on the flanks and in the lower part of the structures, that exist now, had not been formed by diagenesis. For example, oil migrated into the Maastrichtian reservoir of the Valhall Field, preserving the extremely high porosity, even though oil generation below the structure itself started at a much later date. Oil has accumulated in the Chalk because of an absence of reservoir rocks in the Upper Jurassic and Lower Cretaceous, and also because the Palaeocene sands are not present over the chalk fields.

Fields in the Danish sector, for example Kraka, Dan, and Tyra, have a sloping oil–water contact because of a complex interplay of structural tilting after hydrocarbon emplacement and lateral diagenesis, and hydrodynamics. As a result of regional structural tilting to the west and north in Tertiary times, the Danish fields lie considerably updip of the Norwegian fields. This has contributed to the development of a hydrodynamic gradient in the Chalk (Fig. 13.10). Pressure decreases to the south-east, and there is a very slow movement of water in this direction leading to the possibility of oil occurring in hydrodynamic traps downdip of structural crests (Megson 1992).

A peculiarity of the chalk fields is that the oil–water transition zone is commonly much thicker on the flanks of the structures than in the central areas. This is partly due to the reduction in porosity on the flanks of the structures but, in addition, the water saturation often increases significantly towards the periphery for the same porosity and at the same height above the free-water level. This is attributed to a modification of the pore morphology and throat size, possibly due to variations of the stress regime during the development of the domal structures. The rock was under tension on the crests, opening pores or creating microbrecciation, while compression was a feature in synclines, reducing pore-throat size and flattening pores. In the West

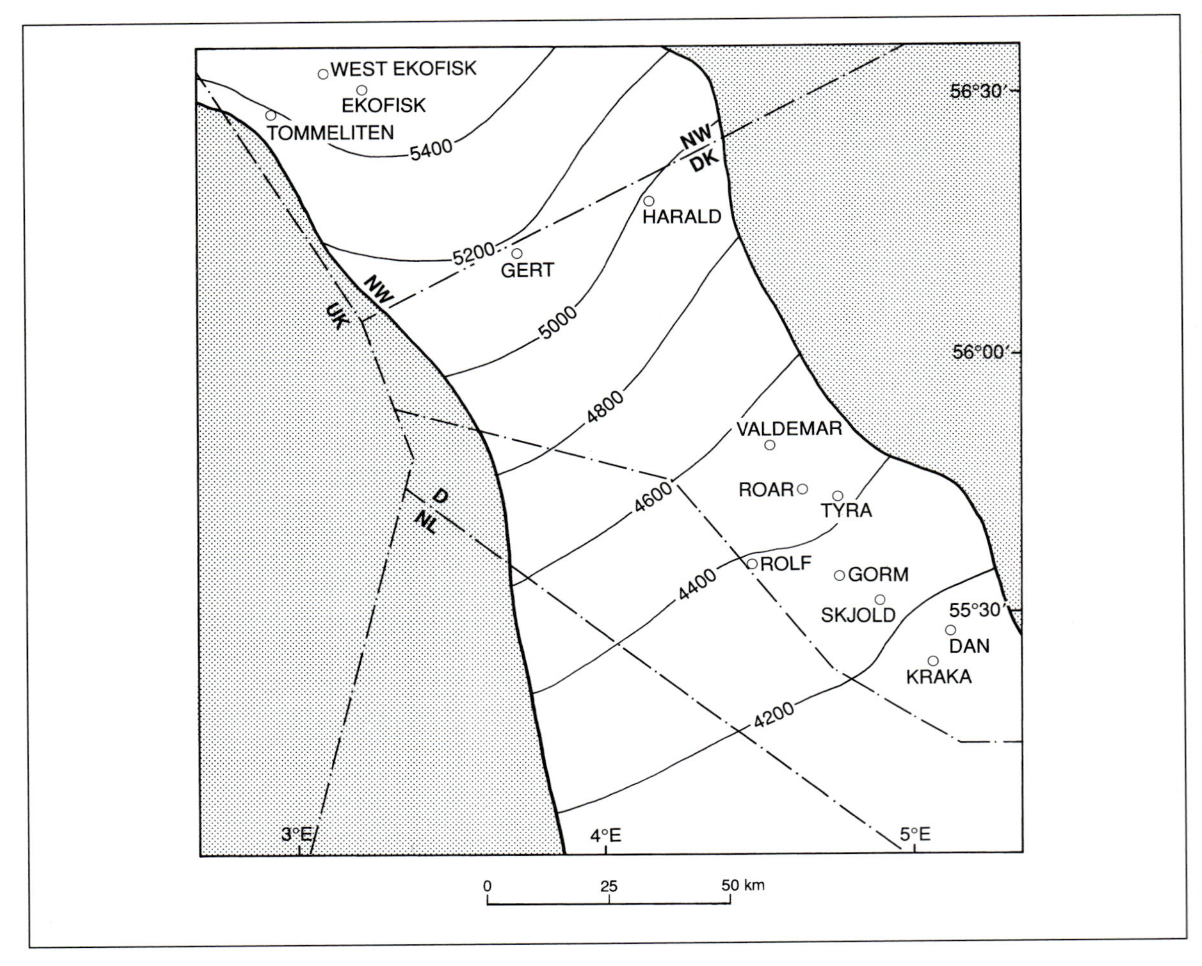

Fig. 13.10. Regional variation of water pressure in the Chalk in the Central Trough, in pounds per square inch measured at a depth of about 2100 m; to convert to freshwater head in metres multiply by 0.7. Field names indicate a data point (after L. N. Jørgensen in Megson 1992).

Ekofisk Field, the base of the oil-pay zone rises over 60 m in less than 2 km, from the axis of the structure to the boundary of the field, as the oil–water transition zone thickens because of large variations in capillary pressure. Generally, water below the oil zone does not flow, and hence water pressure in the water zone does not contribute very much to the recovery of hydrocarbons.

For convenience, porosity, rather than permeability, is used to define a reservoir rock; this is because porosity data are more readily obtained. But in making decisions about the thickness of pay zones several other factors are also taken into account. The type of chalk, its origin and position in the sequence and on the structure are obviously critical factors. For example, the uppermost member of the Ekofisk Formation is generally tight when its porosity is less than 22 per cent, while, at mid-structure level, the porosity cut-off for underlying Ekofisk Formation members may vary from 15 to 18 per cent. Under similar conditions the cut-off for the Tor Formation is between 10 and 15 per cent but at these values the Hod Formation would be tight. The height above the free-water level, reflecting pressure distribution in the reservoir, distance from the structural axis, and fracture distribution, are also very relevent. For example, in the West Ekofisk Field the middle of the Ekofisk Formation does not produce on the flank of the structure above the free-water level even though the porosity exceeds 22 per cent. As porosity decreases towards the flanks and because the chalk comprises alternations of more and less porous material, the porosity cut-off distribution can be quite complex. Although individual reservoirs appear to be distinct, it is now recognized that varying degrees of capillary continuity probably exist (Berget *et al.* 1990)

The oil column in chalk fields can be 300–400 m thick, which emphasizes the efficiency of the seal provided by Tertiary deposits. Production rates can attain 15 000–20 000 barrels per day and such rates imply the presence of very fractured rock with a permeability in excess of 200 mD, although, as already pointed out, the matrix permeability may be only about 1 mD. The original pore-fluid pressure in the chalk fields was about 48 MPa; currently it is about 25 MPa. Maximum advantage has been taken of the large overpressures to lift the hydrocarbons to the surface. The fluids (either oils with a high concentration of dissolved gas or gas condensates) are highly compressible and in a very compressed state initially. Only a limited reduction in fluid pressure is necessary to produce large volumes of oil and/or gas. From the point of view of production, the consequent compaction of the reservoir provided an additional increment to pore-fluid pressures but, of course, the detrimental effects of a reduction in fluid pressure are overriding and steps are now taken to maintain pressures by injecting either gas or seawater. Reinjection of natural gas has an economic penalty as the gas is a saleable commodity. Therefore injection of seawater, or possibly nitrogen, to displace oil and natural gas, is a more attractive alternative.

Under primary depletion conditions only the fracture network produces initially, but as the volume in the fractures is rapidly depleted and as the fluid pressure in them reduces, the matrix yields fluid to the fractures. The pressure drop in the matrix is controlled by the permeability of the fracture system and its ability to allow fluid to flow to wells. Ultimately the production rate is controlled by the rate of flow from the matrix. The principal mechanism for expelling the hydrocarbons is the expansion of the pore fluids following the relief of pressure by the production process.

Above the bubble-point pressure, only oil and water expand but below the bubble point a free gas phase provides further energy. As the pressure is lowered, gas continues to be evolved from the oil and eventually it forms a continuous phase through the matrix. Instead of expanding to expel oil into the fracture system, it begins to be produced itself. The path of least resistance is through the continuous gas phase rather than through the oil, and as a result the gas traps oil in the matrix (Brown 1987).

In a fractured reservoir a potential gradient must be created so that water migrates into the matrix from fractures to displace hydrocarbons. The matrix blocks in the chalk are not large and the flow of water into the matrix from the fractures is by capillary imbibition (Fig. 13.11). A stage is reached when water will no longer imbibe into the rock without the application of an external force (Brown 1987, Fig. 15).

Studies of water injection to supplement production in the Ekofisk Field have been described by Brown (1987). The imbibition characteristics of the chalk of the Ekofisk and Tor formations are affected by the presence of an organo-silicate film,

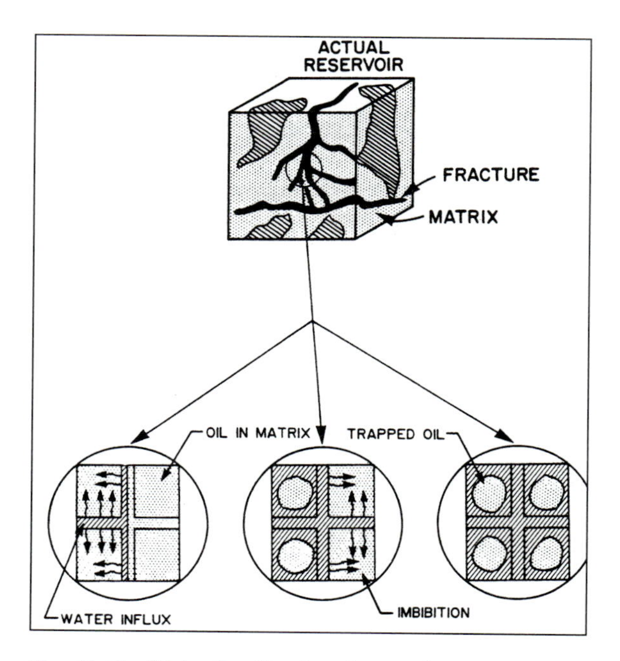

Fig. 13.11 Water flooding in a fractured reservoir (after Brown 1987).

20–30 nm thick, which may have migrated into the reservoir with the oil. A secondary coating, some 0.5 μm thick, which is possibly a hydrocarbon residue, has also been identified on the surface of particles of the Ekofisk Formation. These surface coatings can be removed by strong solvents, thereby improving the imbibition properties.

In April 1981 a pilot project began injecting treated seawater into the Tor Formation of the Ekofisk Field. Water was injected into one well, at rates of up to 5700 $m^3 d^{-1}$, until July 1984; three nearby wells were used for observation. A total of $3.6 \times 10^6 m^3$ were injected with a terminal well-head injection pressure of 25.8 MN m^{-2}. Water breakthrough occurred in two of the observation wells, 400 and 183 m from the injection well, after 13 and 27 months, respectively; the third observation well, at a distance of 150 m, was not affected. These results confirmed the anisotropy of the fracture orientation. The Tor Formation imbibed about 50 per cent of its water pore volume (Brewster *et al.* 1986).

The pilot studies revealed that water could be injected into the Chalk at commercial rates without rapid breakthrough to producing wells (Brown 1987). The field studies gave more optimistic results than the laboratory studies (Berget *et al.* 1990; Sulak *et al.* 1990). Water flooding has now been applied to the Tor and Lower Ekofisk formations of the Ekofisk Field and may be extended to the Upper Ekofisk Formation in due course (Sulak 1991).

Primary recovery of oil from the Chalk is typically only 10–20 per cent. Therefore, enhanced recovery is an essential part of effective management of chalk fields. Many techniques will be, and are being, used but seawater flooding and injection of gas, possibly including nitrogen, are likely to be most important. In tight chalk, as found in the Albuskjell Field, stimulation of the chalk to open existing incipient fractures or create new ones may be necessary (Berget *et al.* 1990).

Away from the recognized fields, the Chalk does contain hydrocarbons in the Central Trough but the production potential is low because it is not fissured. The porosity is less than 20 per cent and the permeability very low, commonly less than 0.1 mD because of the rarity of favourable allochthonous facies. Such chalks are generally not overpressured. The difference between the areas of high and low productivity may be the different diagenetic histories that are controlled in turn by the structural setting. In low-productivity areas there have not been any postdepositional structural movements to give favourable conditions for production. Development of hydrocarbons in such areas will require enhancement of permeability by hydrofracturing the matrix, opening weak zones along stylolites and cleaning fracture surfaces with acid (Sorensen *et al.* 1986).

The chalk forming the Harlingen gasfield in Friesland, 70 km west of Groningen, has an average permeability of only 1.5 mD. The age of the reservoir is Early Maastrichtian and Campanian. Some intervals are fractured and these appear to be in the most competent beds. The porosity has been enhanced along the fracture zones by dissolution. Pore space developed by this means varies in size from 30 to 300 μm and is concentrated near fractures. In contrast, the pore size of chalk comprising intact coccoliths is between 3 and 10 μm, and 1–2 μm for chalk consisting of compacted coccolith fragments. In the Harlingen Field the porosity is between 25 and 40 per cent, the permeability ranges from 0.7 to 8.0 mD and the reservoir is overpressured at 14.3 MPa (compared with a hydrostatic pressure of 10 MPa). The fractured intervals do not improve the effective permeability. The gas in the field is probably derived from underlying Carboniferous rocks and migrated into the Chalk in Oligocene times. Test wells have yielded 60 000 $m^3 d^{-1}$ (van den Bosch 1983).

The effective management of the oil- and gasfields in the Chalk depends on a realistic understanding of the properties of the Chalk, how they have developed and how they control and influence the migration of multiphase fluids in the rock. The challenge is to use the information to abstract a high proportion of the hydrocarbon resources.

14. An aquifer at risk

R. A. Downing, M. Price, and G. P. Jones

A glance at an atlas of North-west Europe reveals the strong negative correlation between annual precipitation and population density. This is particularly evident in areas such as the Netherlands, south-east England, and the Paris Basin. The high demand for water in these densely populated and often highly developed areas is met largely from groundwater and frequently from the Chalk, which currently provides water supplies totalling almost $8.5 \times 10^6 m^3 d^{-1}$.

The Chalk is a considerable economic asset to North-west Europe in other ways. It provides some of the richest farmland of the region, and it forms a major hydrocarbon reservoir below the North Sea. The Chalk contains recoverable reserves of over $700 \times 10^6 m^3$ of oil and $430 \times 10^9 m^3$ of gas. In the first 20 years of operation of the Ekofisk Field alone nearly $142 \times 10^6 m^3$ of oil and $80 \times 10^9 m^3$ of gas were produced.

But it is as a freshwater reservoir, both now and in the long term, that its importance is indisputable, and this is not always recognized. England and France abstract 80 per cent (that is, almost $7 \times 10^6 m^3 d^{-1}$) of the total volume of water abstracted from the Chalk. In Denmark, 60 per cent of the recharge to the Chalk is abstracted, in England it is 33 per cent, and in Sweden 25 per cent. In many parts of North-west Europe the resources of the Chalk are heavily committed for public and industrial water supplies.

The permeable portion of the Chalk forms a relatively thin aquifer that displays significant lateral heterogeneity; the highly permeable areas follow the valley systems. The high fissure permeability makes the aquifer very vulnerable to contamination. The high porosity of the matrix encourages the diffusion of solutes, including contaminants, from fissures into the matrix, and the fine-grained nature of the matrix encourages the retention of those solutes. Once contaminants have entered the matrix, centuries or even millennia will be required for them to diffuse out again naturally, and artificial methods such as pumping have little effect.

Many examples are known of contamination of the Chalk, but the principles and the long-term consequences are illustrated well by the contamination caused by drainage water from coalmines in east Kent, England (Headworth *et al.* 1980). This is one of the worst incidents of groundwater contamination of the Chalk on record. The aquifer overlies a deep concealed coal basin. The first mineshafts were sunk at the beginning of the century and saline drainage-water from the mines was disposed of either into streams or into lagoons excavated in the chalk. As the depth of the mine workings increased, the chloride content of the drainage water increased to $6000 mg l^{-1}$. Between 1906 and 1974, when disposal on to the Chalk finally ceased, some $187 \times 10^6 m^3$ of water containing 318 000 t of chloride had been discharged, of which over 80 per cent remained in the Chalk matrix in 1974. The top 40–50 m of the unsaturated and saturated zones of the aquifer is contaminated but at greater depths the contamination is localized around occasional fissures. The aquifer is polluted over an area of $27 km^2$, thereby sterilizing a groundwater resource equivalent to $20\,000 m^3 d^{-1}$. It will take some 30 years for the chloride value to decline from over $1500 mg l^{-1}$ to $200 mg l^{-1}$, and possibly 150 years to decline to $100 mg l^{-1}$, a value that would still be too high for potable water supply.

This is a sobering case history, for chloride is a very conservative and non-reactive ion; the retardation and storage of many contaminants, such as hydrocarbons and organic solvents, would be much greater. The interpretation of tritium and carbon-14 analyses of groundwater, although fraught with difficulty in such a complex dual-porosity carbonate aquifer as the Chalk, does nevertheless indicate that the mean age of the water in the outcrop zone is at least several decades, and in the confined aquifer is typically thousands of years, thereby emphasizing the very long residence time for solutes and contaminants in the aquifer.

The example just quoted of pollution of the Chalk is but one of many similar cases across North-west Europe. This reinforces the overwhelming need to protect such a vital aquifer from contamination, a need that is not widely appreciated outside the relatively small circle of scientists and engineers intimately concerned with providing water supplies from the Chalk.

Aquifer protection policies are pursued in many countries. They involve interaction with the planning authorities and aim, by persuasion and legislation, to prevent pollution occurring. They provide guidance in terms of the necessary constraints that must be adhered to if groundwater pollution is to be avoided. A weakness is that legal action against an offender who has caused groundwater pollution can only be invoked when there is a case to answer. The long retention time of pollutants in the Chalk may mean that the evidence may not come to light for many years after the incident took place, by which time it may be impossible to identify the culprit.

The Chalk is at risk from contamination by a host of activities such as agricultural practices, landfills, leakage from tanks and pipelines, and the careless disposal of wastes and effluents. The nitrate content of groundwater in the outcrop zone in many areas is now over $50\,mg\,l^{-1}$, and in areas of low infiltration is likely eventually to exceed $100\,mg\,l^{-1}$. More than 2 million people in England and France have been supplied with groundwater containing more than $50\,mg\,l^{-1}$ over periods of many years.

Towns and villages commonly originated on the Chalk as valley settlements utilizing the topographically controlled communication lines and groundwater supplies from springs, streams, and shallow wells. Historically, there was an understandable tendency to locate water-supply sources adjacent to demand centres, and thus a series of local urban contamination centres was inadvertently created. Septic tanks, cesspits, drains and soakaways, not forgetting leaking piped sewerage systems, all increased the potential for pollution incidents in such communities, and this was heightened if utilities or industry were concentrated in the areas. It is no exaggeration to equate sites of past gasworks with sources of phenolic pollution of groundwater; the frequency of hydrocarbon accumulations at water-supply sources in the Chalk is greater than may be imagined. The World War II legacy of organic contamination from degreasing agents in the Chalk of rural Hampshire may be minor compared to that in industrial areas such as the English Midlands, but the nature of the Chalk ensures that the remediation problem is more exacting.

In the 1970s, the design of landfills for the disposal of domestic and industrial wastes emphasized the value of filters of natural materials, such as sand and gravel, to act as effective barriers for bacterial and organic contaminants, and it was assumed that chemical contaminants would be diluted by the large store of water in aquifers. Although both processes do have beneficial effects, the 'dilute and disperse' policy is now discredited. It is questionable whether landfills should be allowed on the Chalk or its permeable cover deposits at all but, if pragmatism wins the day, then they should certainly be containment sites with the leachates collected for careful treatment and disposal.

Over the Chalk outcrop many examples of point sources of pollution by hydrocarbons are known and many more must occur but are not as yet recorded. Generally it can be assumed that urban areas are centres of groundwater pollution. Roads and motorways, railways, and airports (civil and military), along with petrol stations and storage depots, provide a ready-made communication network over the Chalk that has created the means by which contamination can occur. Leaks from underground storage tanks used for storing chemicals, petroleum and oils pose a significant threat to groundwater quality in the Chalk. When 4500 l of aviation fuel leaked into the Chalk from a ruptured pipe in eastern England, it cost over £0.5 million to clean up the aquifer and provide alternative supplies. Groundwater in the Chalk below Luton and Dunstable, north-west of London, is contaminated by chlorinated solvents from industrial processes (Longstaff *et al.* 1992). Low concentrations ($3–10\,\mu g l^{-1}$) are widespread in the aquifer but locally maximum concentrations exceed $500\,\mu g l^{-1}$. Water from public-supply wells in the urban area contains high concentrations and requires treatment by the air-stripping technique, which reduces the concentration by about 95 per cent, before distribution. Centres of chemical pollution occur in the Lower Seine Valley and near Lille in northern France (Chapter 7), and in Denmark, in Copenhagen, pollution caused by chlorinated solvents is a problem (Chapter 10). Greater care in handling chemicals and disposing of wastes on aquifers such as the Chalk is clearly necessary. Chemical storage tanks should have double-wall construction with a mechanism to prevent overfilling. Pipelines should be laid to high specifications.

Advantage is taken of the high infiltration capacity of the Chalk to dispose of drainage from roads and motorways into soakaways in the Chalk. Although on major roads interceptors are incorporated to retain liquids such as oil and petrol, which are less dense than and generally insoluble in water, the possibility remains that water-soluble pollutants or liquids denser than water could pass through interceptors and enter a soakaway. Such substances include phenols, pesticides and chlorinated organic liquids which are not acceptable in drinking water at concentrations greater than about one part per thousand million. Recently the risk to water supplies from road drainage has been investigated at a site north of London using tracers (Price *et al.* 1992). This once again revealed the complexity of flow in the Chalk. Part of the flow was predominantly through relatively large fissures at velocities of about $2\,km\,d^{-1}$, rapid enough to inhibit the retarding effects of diffusion. However, part of the flow path appeared to be through smaller fissures, and diffusion had a significant retarding and diluting effect, but with the risk of pollution extending over a very much longer period. Soakaways may be a desirable way of disposing of road drainage as they provide recharge to the aquifer, but their use needs careful consideration of the risks from pollution. If they are to be permitted, increased storage capacity should be provided between the road drainage system and the soakaways to prevent pollutants entering the ground.

Groundwater protection is concerned with taking actions and making decisions that will prevent pollution. The value of the freshwater resource stored in the Chalk is so great that stringent measures are necessary to protect it from contamination. Activities that are likely to cause pollution should be prohibited. This need is emphasized by the high cost of attempting to rehabilitate the aquifer, a process which may not be very successful in an aquifer with properties such as those of the Chalk. Failure to prevent the deterioration of groundwater quality can lead to the loss of a water source with the inconvenience and cost of replacing it, or a possible health hazard caused by the degradation in quality. The monetary value of a lost groundwater source can be given in terms of replacement cost, but the monetary value of a decline in health or the quality of life from a degradation of water quality is more difficult to give. Nevertheless, groundwater protection policies are primarily intended to provide a social benefit through the maintenance of water quality. The problem is that the benefits to a population of using a good-quality groundwater are intangible; value judgements are required. Some rights are perhaps inviolate and should not be subjected to cost–benefit analysis that simply redistributes limited funds. If this argument is accepted, the protection of groundwater in a major aquifer such as the Chalk must fall in that category. The cost of remedial action, or water treatment, after the aquifer is polluted inevitably exceeds the cost of preventing the pollution in the first place.

The true economic cost is concerned with establishing the real cost of a policy for conserving and protecting groundwater. Policy decisions involve trade-offs between two desirable benefits (for example, increased agricultural production and prevention of pollution by nitrate or pesticides; or reduced industrial cost and reduced pollution from organic solvents). To prevent pollution of the groundwater in the Chalk it may be necessary to impose limitations on land use, reduce crop yields, extend piped sewerage systems, reduce the production of wastes by increasing sorting and recycling, pre-treat wastes before disposal, and severely restrict the disposal of many types of waste. All of these measures will involve tangible costs or disadvantages in the forms of loss of production, reduced productivity, increased costs of production, and taxation. The principle that 'the polluter pays' is desirable in theory, but in practice the costs of meeting the payment will be passed on either in the form of increased prices or increased taxes, or in social costs such as unemployment if producers in other parts of the world can avoid paying to meet environmental standards and so compete unfairly.

The benefits of the protective measures will be less tangible but extremely valuable. In addition to the preservation of the water quality in the Chalk and the associated benefits of good health and peace of mind in an increasingly stressful world, they include more effective use of resources, improved quality of life and leisure, and safer working environments. Unfortunately, the accountants and economists who increasingly rule our lives seem less able to take account of 'hidden' benefits than they do of costs, whether real or hidden.

Safe drinking water, which was available from the Chalk among other sources, has been very cheap. Food, energy, and industrial goods were, until recently, relatively expensive. The history of the nuclear power industry has shown that the promise of cheap, clean power was illusory; when the full costs of disposing of nuclear wastes and of decommissioning nuclear reactors are taken into account, nuclear power—like other forms of power—is expensive. In the same way, the increased efficiency of production of food and manufactured items has reduced their costs, but threatens the continuing supply of cheap, safe water from aquifers such as the Chalk. If money is to be saved in the one area, some at least needs to be invested in the other, to preserve a water resource whose value can only increase in real terms in the future, as populations and water consumption increase and the stability of the global climate appears increasingly at risk.

The last few years have seen the groundwater resources of the Chalk under stress because of the drought that has affected extensive areas of North-west Europe. In the chalk areas of England and France, the seasonal cycle of groundwater-level fluctuations has departed significantly from the normal pattern. The relatively dry winters of the four years from 1988 to 1992 led to very low groundwater levels (Figs. 14.1 and 7.9). In some years the winter infiltration was less than a third of the mean. Between March 1988 and March 1992, parts of southern and eastern England received less than 80 per cent of the average rainfall. The high rates of evaporation and in particular the high soil moisture deficits in the autumn of each year reduced the recharge of the Chalk by the limited winter rainfall (Marsh and Monkhouse, 1993). During the entire period, accumulated recharge to the aquifer was less than 50 per cent of the average over Lincolnshire, East Anglia, and parts of the North Downs and Humberside (Fig. 14.2).

The low water levels caused concern for water-supply undertakings, but with few exceptions the deep wells used for public supply were not seriously affected. Much more at risk were shallow wells and boreholes used for agricultural, horticultural, and isolated domestic supplies, and springs and seepages feeding streams and wetlands and providing baseflow to rivers. The environmental and ecological effects of the drought in England have been particularly emphasized by local pressure groups and the media, which have often unfairly blamed water companies for 'drying up' the aquifer, sometimes seeming to ignore the fact that water companies do not make use of the water themselves but only supply it to consumers.

The very nature of the unconfined Chalk aquifer—its shallow highly permeable zones and rather low specific yield—makes it liable to yield reduced supplies when recharge is markedly reduced. The greater drawdowns of water levels reduce the transmissivity in the vicinity of wells. This tendency could be ameliorated by drilling more wells over a wider area, with each pumping at a lower rate. A bigger worry for water suppliers, however, is the amount of water remaining in storage at drought-water levels (Price 1990). The uncertainty is compounded by the lack of information on the way aquifer properties in the Chalk—particularly the specific yield—vary with depth. It is likely that there is still a significant amount of usable storage—probably equivalent to at least a normal winter's recharge—remaining when the water table is at the minimum recorded level, but this water can be abstracted only by causing further reduction in baseflows to streams and discharges to wetlands. There may soon be a need for society to make a conscious decision on which it values most: the maintenance of wetland ecology or the provision of cheap supplies of high-quality drinking water. In some respects the Chalk is almost too good an aquifer: the aquifer below river valleys is so permeable that abstraction from it quickly interferes with stream flows. However, the storage beneath the interfluves is not easy to exploit because of the way permeability is reduced in these areas.

Since the beginning of the 1970s, droughts have become more frequent and the 1988–92 drought was the most severe for over 100 years. They have often been caused by virtually stationary anti-cyclonic weather systems over western Europe that divert the more variable weather patterns, normally

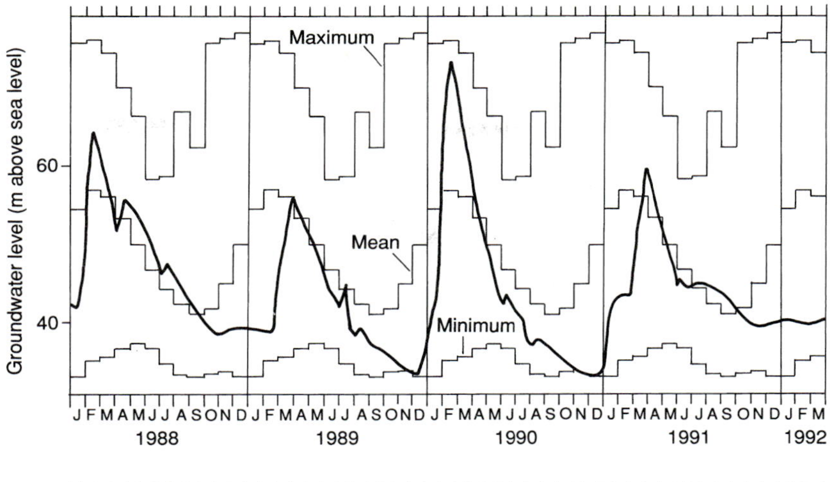

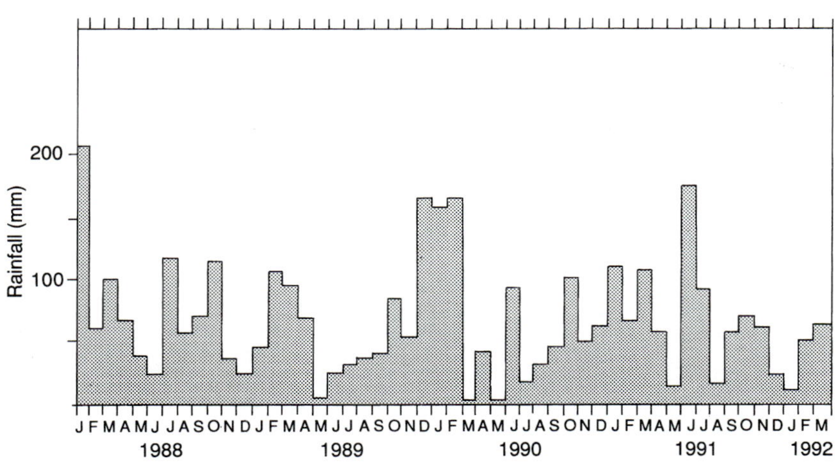

Fig. 14.1. Groundwater level and rainfall in the South Downs, England, between 1988 and early 1992, a period that included four winters with unusually low rainfall over extended intervals. The groundwater level is for the Chilgrove House Well, near Chichester. Water levels in this well have been measured since 1836, probably the longest continuous record for a well in the world. The maximum and minimum monthly values and the mean monthly value for the period 1836–1989 are shown. (Reproduced from data provided by the British Geological Survey and the National Rivers Authority.)

experienced, further north. They have coincided with a period when world temperatures have risen and concern about possible global warming has increased.

The publication in 1987 of the report of the World Commission on Environment and Development, chaired by the Prime Minister of Norway, Mrs Brundtland, (and which became known as the 'Brundtland Report'), at last produced some reaction to the dangers of global warming and its consequent effect on the climate and environment. Forecasting the impact of changes due to rising temperatures remains imprecise, but the consensus opinion based on climate models is that the global average surface temperature will rise by about 1°C by 2030 and by 3°C by 2100, assuming that present emissions of greenhouse gases continue (Houghton *et al.* 1990). The question is what effect this will have on groundwater resources in the Chalk. One view is that, although the winters will be wetter, the summers will be drier and more prolonged; droughts will be more common. Winter recharge may increase, but soil-moisture deficits in the autumn could be higher and groundwater-level recessions probably more prolonged. In this situation dry winters would be of much greater significance and conditions similar to 1988–92 may become more frequent. A reduction in infiltration is also likely to increase the risk of contamination in the Chalk, as has already been experienced.

If drought conditions do become more frequent, there will be increased pressure to incorporate the

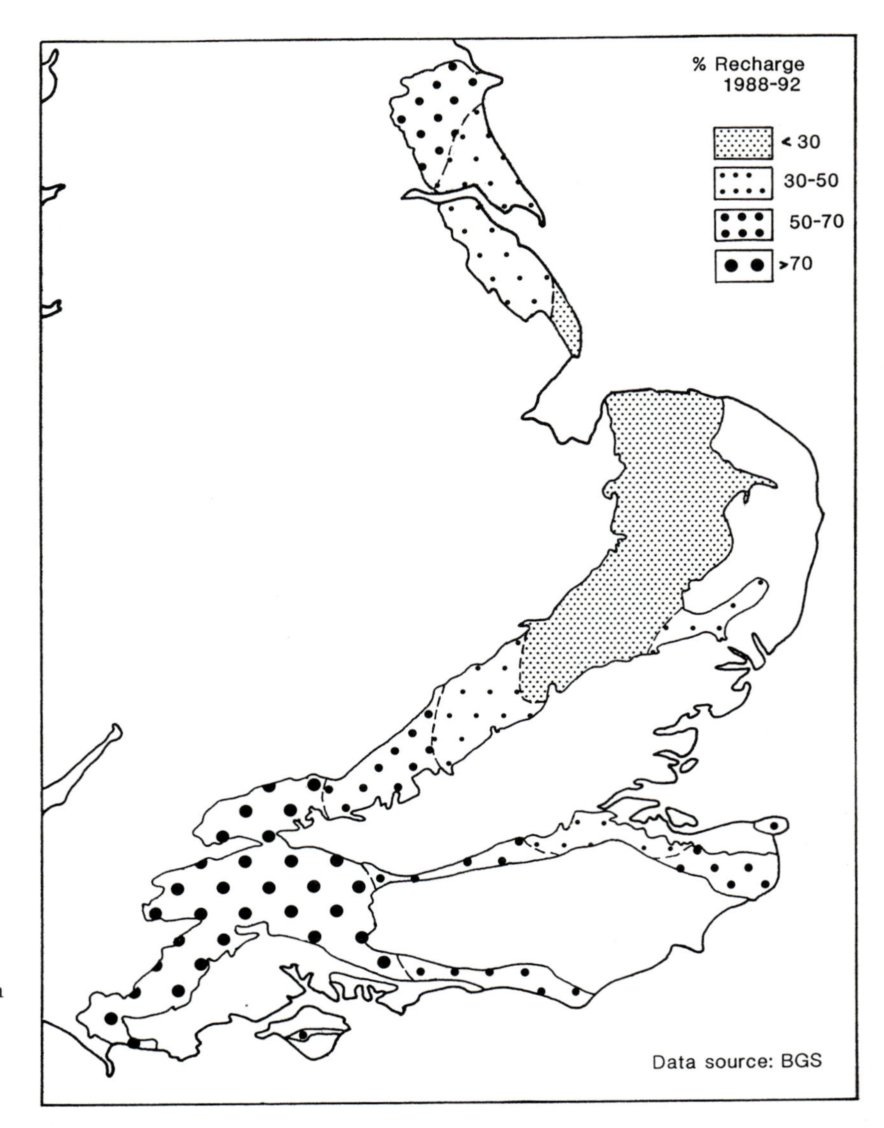

Fig. 14.2. Estimated recharge to the Chalk between 1988 and early 1992 as a percentage of the long-term average (after Marsh and Monkhouse, 1993).

use of groundwater storage in the Chalk within more complex water-resource management systems. Such a relatively thin, highly fissured aquifer, with a linear distribution coinciding with river valleys, requires careful management to ensure optimum results. Development will need to be in conjunction with surface-water resources. The final stage of aquifer development, artificial recharge, is being promoted in England and France. Such schemes will become more common to take advantage of the Chalk's storage capacity which, although much less than that of a sandstone aquifer, is nevertheless very large in comparison with the volume of water stored in surface reservoirs.

Managing the aquifer efficiently, as part of complex water-resource systems, requires detailed monitoring of storage and quality. Reliable data about the baseline chemical quality are necessary to assess changing trends. Carefully designed field studies should provide realistic estimates of aquifer properties including dispersion. The scale and extent of fissure systems in the unsaturated zone need to be evaluated to assess more reliably the future risk from contamination. Historical studies of the behaviour of the aquifer should provide information on both local and regional scales. Contamination problems commonly require detailed local studies, whereas water resources problems are often investigated at the regional scale.

The groundwater resources of the Chalk are at risk from conflicts of interest such as the need to increase yields of agricultural crops with fertilizers, or to dispose of wastes. But they are particularly at risk from those who do not understand the significance of the long residence times of water in the aquifer, extending to centuries, if not millennia. A greater understanding of the importance of the Chalk, its complex flow mechanism, and its vulnerability to pollution, can maintain this resource as a good-quality reliable water supply, and more imaginative management of the resource will ensure its continued availability for future generations.

References

Aam, K. (1988). Ekofisk subsidence: the problem, the solution, and the future. In *Proc. Int. Conf. on behaviour of offshore structures*, Vol. 2, 19–40, Trondheim.

Académie des Sciences (1991). *Pollution des nappes d'eau souterraine en France*, Report no. 29. Académie des Sciences, Paris.

Adams, B. and Foster, S. S. D. (1992). Land-surface zoning for groundwater protection. *J. Inst. Water Environ. Managmnt*, **2**, (6), 312–19.

Adelseck, C. G., Jr. and Berger, W. H. (1975). On the dissolution of planktonic foraminifera and associated microfossils during settling and on the sea floor. In *Dissolution of deep-sea carbonates* (ed. W. V. Sliter, A. W. H. Bé, and W.H. Berger), pp. 70–81. Special Publication No. 13. Cushman Foundation for Foraminiferal Research, Washington, DC.

Agence de Bassin Seine-Normandie (1988). *La qualité physico-chimique des eaux souterraines du Bassin Seine-Normandie*. Paris.

Albers, H. J. and Felder, W. M. (1979). Litho-, Biostratigraphie und Paläökologie der Oberkreide und des Alttertiärs (Präobersanton-Dan/Paläozän) von Aachen-Südlimburg (Niederlande, Deutschland, Belgien). In *Aspekte der Kreide Europas* (ed. J. Wiedmann), pp. 47–84. International Union of Geological Sciences, Series A, No. 6.

Albinet, M. (1967). *Carte hydrogéologique du Bassin de Paris. 1/500 000*. BRGM., Orléans.

Alexandersson, E. T. (1978). Destructive diagenesis of carbonate sediments in the eastern Skagerrak, North Sea. *Geology*, **6**, 324–7.

Allen, D. J. and Price, M. (1990). Hydraulic conductivity of the chalk at shallow depths in north-west Norfolk, UK. In *Chalk*, pp. 577–82. Thomas Telford, London.

Ameen, M. S. (1990). Macrofaulting in the Purbeck–Isle of Wight monocline. *Proc. Geol. Assoc.*, **101**, 31–46.

Ameen, M. S. and Cosgrove, J. W. (1990). A kinematic analysis of meso-fractures from Studland Bay, Dorset, *Proc. Geol. Assoc.*, **101**, 303–14.

Andersen, G. (1990). Grundvandsmoniteringsnet i Danmark. *NPo-forskning*, No. B18. Miljøstyrelsen, Copenhagen.

Andersen, G. (1992). Kortslutning gennem lange boringsfiltre. Undersøgelse ved Ringsted vedrørende vandindvinding. *Datadokumentation*, No. 13, Geol. Surv. Denmark.

Andersen, K. H., and Ødum, H. (1923). En Salt-Flora ved Slagsmose ved Rislev. *Bot. Tidsskr.* **38**, 57–68. Dansk Botanisk Forening, Copenhagen.

Andersen, L. J. (1973) *Cyclogramme technique for geological mapping of borehole data*. Series III, no. 41, Geol. Surv. Denmark.

Andersen, L. J. and Gosk, E. (1987) Applicability of vulnerability maps. In *Vulnerability of Soil & Groundwater to Pollutants*, Proc. RIvM Intl. Conf., 321–32. Noorwijk, Netherlands.

Andersen, L. J., Kallesøe, R., and Kærgaard, H. (1975). Grundvandets Potentialeforhold. In *Norvand, Grundvandsundersøgelse ved Esrom Sø*. Norvand Udvalget, Copenhagen.

Andersen, S. (1930). Nye fund af halofyter i Storebæltsmorådets Indland. *Bot. Tidsskr.*, **41**, 100–12. Dansk Botanisk Forening Copenhagen.

Anon, (1973). La Craie. *Bull. de Liason des Laboratoires des Ponts et Chaussées*, special vol.

Arnoult, P. (1981). Gestion quantitative et qualitative des eaux souterraines en zone agricole. Application à la nappe de la craie de I'Artois et du Cambrésis. Thèse 3ème cycle, Univ. Lille I.

Aspinwall, R. (1974a). Phenolic pollution in the Chalk at Beckton, Essex. In *Groundwater Pollution in Europe*, pp. 299–302. Water Info Centre, Port Washington, NY.

Aspinwall, R. (1974b). Naphtha leakage at Purfleet and spillage of leaded fuel at South Stifford, Essex. *In Groundwater Pollution in Europe*, pp. 338–43. Water Info Centre, Port Washington, NY.

Atkinson, T. C. and Smith, D. I. (1974). Rapid groundwater flow in fissures in the Chalk: an example from South Hampshire. *Q. J. Eng. Geol.*, **7**, 197–205.

Aubry, M. -P. (1976). L'induration de la craie. *Travaux du Laboratoire de Micropaléontologie, Paris*, **7**, 49–67.

Avon and Dorset River Authority, (1973). *Upper Wylye investigation; a report on the effects of abstraction from a Chalk/Upper Greensand aquifer*. Poole, Dorset.

Bækgaard, A. and Knudsen, J. (1979). Grundvandspotentiale og Transmissivitet. In *Hydrogeologisk Kortlægning af Vestsjællands Amtskommune, Fase 3* (ed. K. Binzer, A. Bækgaard, J. Knudsen, H. Kristiansen, and A. Villumsen). Geol. Surv. Denmark.

Bahat, D. (1990). Genetic classification of joints in chalk and their corresponding fracture characteristics. In *Chalk*, pp. 79–86. Thomas Telford, London.

Baker, P. A., Gieskes, J. M., and Elderfield, H. (1982). Diagenesis of carbonates in deep-sea sediments—evidence from Sr/Ca ratios and interstitial dissolved Sr^{2+} data. *J. Sedimentary Petrology*, **52**, 71–82.

Ballif, J. L. (1978). Porosité de la craie, appréciation de la taille et de la répartition des pores. *Ann. Agron.*, **29**, 123–31.

Barenblatt, G. I., Zheltov, I. P., and Kochina, I. N. (1960). Basic concepts in the theory of homogeneous liquids in fissured rocks. *J. App. Math. Mech.* (English translation), **24**, 1286–303.

Barker, I., Chada, D. S., and Courchee, R. (1983). *Groundwater resources of the Chalk of East Yorkshire, Kilham pumping test.* Yorkshire Water Authority, Leeds.

Barker, J. A. (1982). Laplace transform solutions for solute transport in fissured aquifers. *Adv. Water Resour.*, **5**, 98–104.

Barker, J. A. (1985a). Block-geometry functions characterizing transport in densely fissured media. *J. Hydrol.*, **77**, 263–79.

Barker, J. A. (1985b). Modelling the effects of matrix diffusion on transport in densely fissured media. *Memoirs of the 18th Congress of the International Association of Hydrogeologists*, **1**, 250–69.

Barker, J. A. and Black, J. H. (1983). Slug tests in fissured aquifers. *Water Resour.* Res., **19**, 1558–64.

Barker, J. A. and Foster, S. S. D. (1981). A diffusion exchange model for solute movement through fissured rock. *Q. J. Eng. Geol.*, **14**, 17–24.

Barker, R. D., Lloyd, J. W., and Peach, D. W. (1984). The use of resistivity and gamma logging in lithostratigraphical studies of the Chalk in Lincolnshire and South Humberside. *Q. J. Eng. Geol.*, **17**, 71–80.

Barrois, C. (1877). Recherches sur le terrain crétacé supérieur de l'Angleterre et de l'Irlande. *Mémoires de la Société Géologique du Nord*, 1, 1–232.

Barton, M. E. (1990). Stability and recession of the chalk cliffs at Compton Down, Isle of Wight. In *Chalk*, pp. 541–44. Thomas Telford, London.

Bath, A. H. and Edmunds, W. M. (1981). Identification of connate water in interstitial solution of Chalk sediment. *Geochim. Cosmochim. Acta*, **45**, 1449–61.

Bath, A. H., Downing, R. A., and Barker, J. A. (1985). *The age of groundwaters in Chalk and Pleistocene sands of north-east Suffolk*, Report no. WD/ST/85/1. British Geological Survey, Keyworth.

Baxter, K. M., Edworthy, K. J., Beard, M. J., and Montgomery, H. A. C. (1981). Effects of discharging sewage to the Chalk. *Sci. Total Environ.*, **21**, 77–83.

Beall, A. O. and Fischer, A. G. (1969). Sedimentology. *Initial Reports of the Deep Sea Drilling Project*, **1**, 521–93.

Beard, M. J. and Giles, D. M. (1990). Effects of discharging sewage effluents to the chalk aquifer in Hampshire. In *Chalk*, pp. 597–604. Thomas Telford, London.

Beckelynck, J. (1981) Traitement régionalisé des paramètres contribuant à la gestion des nappes. Application à la modélisation de la nappe de la craie dans le bassin de l'Aa et de la moyenne Lys. Thèse 3ème cycle, Univ. Lille 1.

Berget, O. P., Dretvik, Ø., Tonstad, K., and Mathiesen, E. (1990). Production of oil and gas from chalk reservoirs in the Norwegian part of the North Sea. In *Chalk*, pp. 649–54. Thomas Telford, London.

Berkaloff, E. (1970). Interprétation des pompages d'essais. Cas des nappes captives avec une strate conductrice d'eau privilegiée. *Bull. BRGM*, 2nd series, 3, (1), 33–53.

Bergström, J. (1985). Zur tektonischen Entwicklung Schonens (Süd-Schweden). *Zeitschrift für Angewandte Geologie*, **31**, 277–80.

Bergström, J., Holland, B., Larsson, K., Norling, E., and Sivhed, U. (1982). *Guide to excursions in Scania.* Ser. Ca No 54. Geol. Surv. Sweden.

Bernard, D. (1979). Contribution à l'étude hydrogéochimique de la nappe de la craie dans le Nord de la France. Etat et acquisition du chimisme de l'eau. Thèse 3ème cycle, Univ. Lille 1.

Berry, G. F. (1979). Late Quaternary scour hollows and related features in central London. *Q. J. Eng. Geol.*, **9**, 255–63.

Berthelsen, O. (1987). *Geologi i Aalborgområdet, Råstoffer —Fundering—Vandindvinding.* Geol. Surv. Denmark.

Bertrand, L. and Gringarten, A. C. (1978). *Détermination des caractéristiques hydrauliques des aquifères fissurés par pompages d'essai en régime transitoire: application aux nappes de la craie,* Report no. 78 SGN 669 GEG. BRGM, Orléans.

Bevan, T. G. and Hancock, P. L. (1986). A Late Cenozoic regional mesofracture system in southern England and northern France. *J. Geol. Soc.*, **143,** 355–62.

Bibby, R. (1981). Mass transport studies of solutes in dual-porosity media. *Water Resour. Res.*, **17**, 1075–81.

Birch, G. P. (1990). Engineering geomorphological mapping for cliff stability. In *Chalk*, pp. 545–50. Thomas Telford, London.

Black, J. H. and Kipp, Jr, K. L. (1983). Movement of tracers through dual-porosity media—experiments and modelling in the Cretaceous Chalk. *J. Hydrol.*, **62**, 287–312.

Black, M. (1972). British Lower Cretaceous Coccoliths. I. Gault Clay. Part 1. *Palaeontographical Society Monographs.*

Black, M. (1980). On Chalk, Globigerina ooze and aragonite mud. In *Andros Island, Chalk and Oceanic Oozes* (ed. C. V. Jeans and P. F. Rawson), pp. 54–85. Yorkshire Geological Society, Occasional Publication No. 5

Blakey, N. C. and Towler, P. A. (1988). The effect of unsaturated–saturated zone properties upon the hydrogeochemical and microbiological processes involved in the migration and attenuation of landfill leachate components. *Water Sci. Tech.* **20**, 119–28.

Bodelle, J. and Margat, J. (1980). L'eau souterraine en France. In *Les objectifs scientifiques de demain.* Masson, Paris.

Bodenhausen, J. W. A. and Ott, W. F. (1981). Habitat of the Rijswijk oil province, onshore, The Netherlands. In *Petroleum geology of the continental shelf of north-west Europe* (ed. L. V. Illing and G. D. Hobson), pp. 301–9, Heyden, London.

Bogli, A. (1964). Mischungkorrosion: ein Beitrag zum Verkarstungsproblem. *Erdkunde*, 18, 83–92.

Boigk, H. (1968). Gedanken zur Entwicklung des Niedersächsischen Tektogens. *Geologisches Jahrbuch*, 85, 861–900.

Boigk, H. (1981). *Erdöl und Erdgas In der Bundesepublik Deutschland. Erdölprovinzen, Felder, Förderung,*

Vorräte, Lagerstättentechnik. Ferdinand Enke Verlag, Stuttgart.

Bonnemaison, M. (1989). *Indices de diagenèse liés nannofossiles calcaires dans le Crétacé des Pyrenées.* Documents du Bureau de Recherches Géologique et Minières, no. 170. BRGM. Orléans.

Bracq, P., Hanich, L., Delay, F., and Crampon, N. (1992). Mise en évidence par traçage d'une relation rapide entre la surface et les eaux souterraines, liée à des phénomènes de dissolution dans la craie du Boulonnais (Nord de la France). *Bull. Soc. Géol. France*, **163**, 195–203.

Brewster, J., Dangerfield, J., and Farrell, H. (1986). The geology and geophysics of the Ekofisk Field water flood. *Marine and Petroleum Geol.*, **3**, 139–69.

Brinck, S. and Leander, B. (1981). The Alnarp Stream. Compilation of Reports 1970–1980 (in Swedish). Co-operation Committee for the Alnarp Stream, Malmö.

Bromley, R. G. (1965). Studies in the lithology and conditions of sedimentation of the Chalk Rock and comparable horizons. Ph.D. thesis, University of London.

Bromley, R. G. (1967). Some observations on burrows of thalassinidean crustacea in chalk hardgrounds. *Q. J. Geol. Soc. London,* **123**, 157–82.

Bromley, R. G. (1975). Trace fossils at omission surfaces. In *The study of trace fossils* (ed. R. W. Frey), pp. 399–428. Springer-Verlag, New York.

Bromley, R. G. (1980). Enhancement of visibility of structures in marly chalk: modification of the Bushinsky oil technique. *Bull. Geol. Soc. Denmark*, **29**, 111–18.

Bromley, R. G. and Gale, A. S. (1982). The lithostratigraphy of the English Chalk Rock. *Cretaceous Res.* **3**, 273–306.

Brotzen, F. (1945). *The geological results of the drillings at Höllviken. Part I: The Cretaceous* (in Swedish with English summary). Ser. C No 465. Geol. Survey of Sweden.

Brown, D. A. (1987). The flow of water and displacement of hydrocarbons in fractured chalk reservoirs. In *Fluid flow in sedimentary basins and aquifers* (ed. J. C. Goff and B. P. J. Williams), pp. 201–18. Special Publication 34, Geological Society, London.

Bruun-Peterson, J. (1975). Upper Cretaceous shelf limestone from Ignaberga, Scania (Sweden), and its diagenesis. *Int. Cong. Sedimentology, No. 9 (Nice)*, **7**, 33–8.

Burki, P. M., Glasser, L. S. D., and Smith, D. N. (1982). Surface coatings on ancient coccoliths. *Nature* (London), **297**, 145–7.

Calembert, L. and Monjoie, A. (1975*a*) Observations hydrogéologiques dans la vallée de la Haine. *Compte rendu Académique des Sciences,* series D, **280**, 2637–9.

Calembert, L. and Monjoie A. (1975b) Modification du régime hydraulique des eaux souterraines dans la vallée de la Haine, pp. 369–80 *Colloque Comité Belge de Géologie de l'Ingénieur*, Brussels.

Campbell, C. J. and Ormaasen, E. (1987). The discovery of oil and gas in Norway: an historical synopsis. In *Geology of the Norwegian oil and gas fields* (ed. A. M. Spencer, E. Holter, C. J. Campbell, S. H. Hanslien, P. H. H. Nelson, E. Nysaether, and E. G. Ormaasen), pp. 1–37. Graham and Trotman, London.

Caous, J. Y. and Comon, D. (1987). *Atlas hydrogéologique de l'Oise*. BRGM, Orléans.

Caous, J. Y. and Roux, J. C. (1981). Ressources en eau souterraine de la Picardie (France). *Bull B.R.G.M.* Sect III, (1), 19–52.

Caous, J. Y., Caudron, M. and Dumont, D. (1978a). Pollution industrielle de la nappe de la Craie par le chrome hexavalent dans le Vimeu (Somme). In *Hydrogéologie de la Craie du Bassin de Paris*, pp. 89–94. BRGM, Orléans.

Caous, J. Y., Caudron, M. and Dumont, D. (1978b). Pollution accidentelle par hydrocarbures de la nappe de la Craie en Picardie. In *Hydrogéologie de la Craie du Bassin de Paris*, pp. 85–8. BRGM, Orléans.

Caous, J. Y., Caudron, M. and Mercier, E. (1983). *Atlas hydrogéologique de l'Aisne, et carte hydrogéologique du département de l'Aisne*. BRGM, Orléans.

Carter, D. J. and Hart, M. B. (1977). Aspects of mid Cretaceous stratigraphical micropalaeontology. *Bull. Brit. Museum, Natural History (Geology),* **29**, 1–135.

Carter, P. G. and Mallard, D. J. (1974). A study of the strength, compressibility, and density trends within the Chalk of south east England. *Q. J. Eng. Geol.,* **7**, 43–55.

Carter, R. M. (1972). Aadaptations of British Chalk Bivalvia. *J. Paleontology*, **46**, 325–40.

Castany, G. and Mégnien, C. (1974). *Les bassins de la Seine et des cours d'eau normands. Fasc. 4 eaux souterraines et atlas (cartes hydrogéologiques à 1/500000).* Agence Financière de Bassin Seine-Normandie, Paris.

Catt, J. A. and Penny, L. F. (1966). The Pleistocene deposits of Holderness, East Yorkshire. Proc. *Yorks. Geol. Soc.*, **35**, 375–420.

Caudron, M. and Cremille, L. (1985). Impact des bassins d'infiltration d'eaux industrielles sur la nappe de la Craie à Estrées-Mons (Somme). *Hydrogéologie*, **1**, 65–73.

Caulier, P. (1974). Etude des faciès de la craie et de leurs caractéristiques hydrauliques de la région du Nord. Thèse 3ème cycle, Univ. Lille I.

Cěch, S., Klein, V., Kříž, J., and Valečka, J. (1980). Revision of the Upper Cretaceous stratigraphy of the Bohemian Cretaceous Basin. *Věstník Ústředního ústavu geologického*, **55**, 277–96.

Chadwick, R. A. (1985). Cretaceous sedimentation and subsidence. In: *Atlas of onshore sedimentary basins in England and Wales* (ed. A. Whittaker), pp. 57–60. Blackie, London.

Chadwick, R. A. (1986). Extension tectonics in the Wessex Basin, southern England. *J. Geol. Soc.*, **143**, 465–88.

Chemin, J. and Holé, J. P. (1980). *Atlas hydrogéologique de l'Eure*. BRGM, Orléans.

Chemin, J. and Holé, J. P. (1981). *Atlas hydrogéologique de la Seine-Maritime*. BRGM, Orléans.

Chilton, P. J. and Foster, S. S. D. (1991). Control of groundwater nitrate pollution in Britain by land-use changes. In *Nitrate Contamination: Exposure, Consequences and Control*, Proc. NATO Conf., pp. 333–47. Lincoln, NB.

Chilton, P. J., Lawrence, A. R., and Barker, J. A. (1990). Chlorinated solvents in chalk aquifers: some preliminary observations on behaviour and transport. In Chalk, pp. 605–10. Thomas Telford, London.

Clark, L., Blakey, N. C., Foster, S. S. D., and West, J. M. (1991). Microbiology of aquifers. In *Applied Groundwater Hydrology—A British Perspective* (ed. R. A. Downing and W. B. Wilkinson), pp. 164–76. Clarendon Press, Oxford.

Clarke, R. H. (1977). Earthworks in soft chalk: performance and prediction. The *Highway Engineer,* **24**, (3), 18–21.

Clausen, C. K., Nygaard, E., Andersen, C., Møller, C., and Stouge, S. (1990). Porosity variations within the Chalk Group, Danish North Sea. *Geonytt*, **17**, 36–37 (abstract).

Clayton, C. J. (1986). The chemical environment of flint formation in Upper Cretaceous chalks. In *The scientific study of flint and chert* (ed. G. de G. Sieveking and M. B. Hart), pp. 43–54. Cambridge University Press.

Clayton, C. R. I. (1977). Chalk in Earthworks—Performance and Prediction. *The Highway Engineer,* **24**, (2) 14–20.

Colbeaux, J. P. and Mania, J. (1976). Relations entre la fracturation et l'écoulement des eaux superficielles et souterraines en pays crayeux au Cran d'Escalles—Application à l'Artois. Deuxième colloque d'hydrologie en pays calcaire. *Ann. Scient. de l'Université de Besançon.* Fascicule 25, 3rd series, 1976, 179–94.

Colbeaux, J. P., Beugnies, A., Dupuis, C., Robaszynski, F., and Somme, J. (1977). Tectonique de blocs dans le sud de la Belgique et le nord de la France. *Annales de la Société géologique du Nord, Lille*, **97**, 191–222.

Colloque (1988). *Karst et Quaternaire de la Basse-Seine.* Actes du Museum de Rouen.

Colloque AIDEC (1984). *Sur la pollution par les nitrates, quels responsables? Causes et préventions.* Assoc. Int. Des Entretiens Ecologiques, nos 21 and 22.

Colloque Régional de Rouen (1978). *Hydrogéologie de la craie du Bassin de Paris.* BRGM, Orléans.

Connorton, B. J. (1988). Artificial recharge in the London Basin. 17th Int. Water Supply Congress. *Technical Papers*, 5513–17 Rio de Janeiro, Brazil.

Connorton, B. J. (1990). Groundwater resources: a Thames water utilities perspective. In Water: *availability, quality and cost.* IBC Tech. Serv. Ltd., London.

Connorton, B. J. and Reed, R. N. (1978). A numerical model for the prediction of long term well yield in an unconfined chalk aquifer. *Q. J. Eng. Geol.*, 127–38.

Cooper, J. D., Gardner, C. M. K., and MacKensie, N. (1990). Soil water controls on recharge to aquifers. *J. Soil Sci.*, **41**, 613–30.

Cooper, M. A. and Williams, G. D. (1989). *Inversion tectonics.* Special Publication Geological Society, London, 44.

Cope, J. C. W., Ingham, J. K., and Rawson, P. F. (ed.) (1992). Atlas of palaeogeography and lithofacies. *Memoir Geological Society of London*, 13.

Cottez, S. and Dassonville, G. (1965). *Carte de la surface piézométrique de la nappe de la craie dans la region du Nord (1:200000)*, Report No DSGR 65A49. BRGM Orléans.

Coulon, M. (1986). Contribution de la sismique-réfraction à la connaissance de la zone d'altération de la craie et de ses formations superficielles. *Bull. d'Information des Géologues du Bassin de Paris*, **23**, 29–36.

Coulon, M. and Frizon de Lamotte, D. (1988). Les craies éclatées du secteur d'Omey (Marne, France): le résultat d'une bréchification par fracturation hydraulique en contexte extensif. *Bull. Soc. Géol. Fr.*, **8**, 177–85.

Crampon, N. (1983). *Qualité de la nappe de la craie dans les zones rurales de l'Artois et du Cambrésis. Essais de typologie de la pollution azotée.* Report to the Chambre Régionale de l'Agriculture du Nord et du Pas-de-Calais.

Crampon, N., Levassor, A., Colbeaux, J. P., Porel, G., Chesneau, A., and Guyot-Sionnest, D. (1990). Tunnel sous la Manche ou tunnel dans les eaux souterraines? L'aspect hydrogéologique des travaux dans le domaine continental français. *Ann. Soc. Géol. Nord*, **109**, 141–9.

Crawley, J. D. and Pollard, C. (1992). Ground treatment to improve tunnel progress on the Channel Tunnel marine drives. *Ground Engineering*, **25**, 27–35.

Croll, B. T. (1986). The effects of the agricultural use of herbicides on fresh waters. In *Effects of Land Use on Fresh Waters*, Proc. WRC Conf., pp. 201–9. Medmenham, Buckinghamshire.

Croll, B. T. (1991). Pesticides in surface waters and groundwaters. *J. Inst. Water Environ. Mangmnt.*, **5**, 389–95.

Cruse, P. K. (1986). Drilling and construction methods. In *Groundwater: occurrence, development and protection.* (ed. T. W. Brandon), pp. 437–84. Inst. Wat. Engrs. Sci., London.

Curry, D. (1982). Differential preservation of foraminiferids in the English Upper Cretaceous—consequential observations. In *Aspects of micropalaeontology* (ed. F. T. Banner and A. R. Lord), pp. 240–61. Allen & Unwin, London.

Curry, D. (1992). Tertiary. In *Geology of England and Wales* (ed. P. M. D. Duff and A. J. Smith), pp. 389–411. Geological Society of London.

Cuttell, J. C., Lloyd, J. W., and Ivanovich, M. (1986). A study of uranium and thorium series isotopes in Chalk groundwaters of Lincolnshire, UK. *J. Hydrol.*, **86**, 343–65.

d'Arcy, D. (1969). Contribution à l'etude hydrogéologique du Bassin de l'Authie. Thèse 3ème cycle, Univ. Paris.

Darling, W. G. (1985). *Methane in Chalk groundwater of Central London.* Report no. WD/ST/85/3. British Geological Survey, Keyworth.

Darling, W. G. and Bath, A. H. (1988). A stable isotope study of recharge processes in the English Chalk. *J. Hydrol.*, **101**, 31–46.

Dassargues, A. (1991). Paramétrisation et simulation des réservoirs souterrains. Doctoral thesis, Faculté des Sciences Appliquées, Université de Liège, pp. 161–85.

Dassargues, A., Radu, J. P., and Charlier, R. (1988). Finite element modelling of a large water table aquifer in transient conditions. *Adv. Water Resour.*, **11**, 58–66.

Deegan, C. E., and Scull, B. J. (1977). *A standard lithostratigraphic nomenclature for the central and northern North Sea*, Report 77/25. Inst. Geol. Sci., London.

de la Querière, P. (1972). Esquisse hydrogéologique du Noyonnais et du Soissonais. Relations entre les nappes perchées de l'Eocène, la nappe de la craie, sous les plateaux et dans les vallées de l'Oise et de l'Aisne. Thèse 3ème cycle. Univ. Paris.

Department of the Environment (1978). *Hazardous wastes in landfill sites.* DoE Policy Review Committee Final Report. HMSO, London. 169pp.

Department of the Environment (1986). *Nitrate in water.* DoE Pollution Paper 26.

Derijcke, F. (1978). *Carte piézométrique de la nappe des craies du Bassin de Mons, colloque régional, Hydrogéologie de la craie du bassin de Paris*, Vol. 1, pp. 211–19. BRGM., Orléans.

Destombes, J. P. and Shephard-Thorn, E. R. (1971). *Geological results of the Channel Tunnel site investigation 1964–65*, Report, 71/11. Institute of Geological Sciences, London.

Dewey, J. F. (1982). Plate tectonics and the evolution of the British Isles. *J. Geol. Soc., London*, **139**, 371–412.

de Wit, R. G. (1988). *De invloed van discontinuiteiten op de grond water stroming in de Kalk steen formaties van Zuid Limburg.* Memoirs of the Centre for Engineering Geology in the Netherlands, 64.

D'Heur, M. (1984). Porosity and hydrocarbon distribution in the North Sea Chalk reservoirs. *Marine and Petroleum Geology*, **1**, 211–38.

D'Heur, M. (1986). The Norwegian chalk fields. In *Habitat of hydrocarbons on the Norwegian continental shelf*, pp. 77–89. Graham and Trotman, London.

D'Heur, M. (1990). Chalk and petroleum in North-west Europe. In *Chalk*, pp. 631–40. Thomas Telford, London.

Dines, H. G. and Edmunds, F. H. (1933). *The geology of the country around Reigate and Dorking*, Mem. Geol. Surv. HMSO, London.

Dolfuss, G. F. (1890). Hypsometrical map of the surface of the Chalk in the Paris Basin. *Bull. Services de la Carte Géologique de France,* **2**(14).

Dolfuss, G. F. (1910). On the classification of the beds in the Paris Basin. *Proc. Geol. Assoc.*, **21**, 101–18.

Downing, R. A. and Headworth, H. G. (1990). Hydrogeology of the Chalk in the UK: the evolution of our understanding. In *Chalk*, pp. 555–70. Thomas Telford, London.

Downing, R. A. and Penn, I. E. (1992). *Groundwater flow during the development of the Wessex Basin and its bearing on hydrocarbon and mineral resources.* Research report SD/91/1 British Geological Survey. Keyworth.

Downing, R. A., Smith, D. B., and Warren, S. C. (1978). Seasonal variations of tritium and other constituents in groundwater in the Chalk near Brighton, England. *J. Inst. Wat. Engrs. Sci.*, **32**, 123–36.

Downing, R. A., Pearson, F. J. and Smith, D. B. (1979). The flow mechanism in the Chalk based on radio-isotope analyses of groundwater in the London Basin. *J. Hydrol.*, **40**, 67–83.

Downing, R. A., Ashford, P. L., Headworth, H. G., Owen, M., and Skinner, A. C. (1981). The use of groundwater for river angmentation. In *A survey of British hydrogeology*, pp. 153–72. The Royal Society, London.

Dreybrodt, W (1990). The role of dissolution kinetics in the development of karst aquifers in limestone: a model simulation of karst evolution. *J. Geology*, **98**, 639–55.

Droz, B. (1985). Influence de la structure et de la nature des terrains du valenciennois sur la qualité de la nappe de la craie (Nord de la France). Apport du krigeage à l'hydrochimie régionale; gestion qualitative des eaux souterraines. Thèse 3ème cycle, Univ. Lille I.

Drummond, P. V. O. (1970). The mid-Dorset swell. Evidence of Albian–Cenomanian movements in Wessex. *Proc. Geol. Assoc.*, **81**, 679–714.

Duermael, G, Rampon, G, Morfaux, P, Picot, G., and Laurentiaux, D. (1966). *Hydrogéologie de la région Champagne-Ardenne. Vertus—Châlons-sur-Marne—Vitry-le-François—Féré-Champenoise.* Cartes 1/100000. BRGM, Orléans.

Duval, C., Doucet, P., Vidal, J., and Brewster, J. (1990). Faults and fractures in Ekofisk Field. In *Chalk*, pp. 655–68. Thomas Telford, London.

Edmunds, W. M. (1986). Groundwater chemistry. In *Groundwater: occurrence, development and protection* (ed. T. W. Brandon), pp. 49–107. Institution Wat. Engrs and Scientists, London.

Edmunds, W. M., Lovelock, P. E. R., and Gray, D. A. (1973). Interstitial water chemistry and aquifer properties in the Upper and Middle Chalk of Berkshire, England. *J. Hydrol.*, **19**, 21–31.

Edmunds, W. M., Cook, J. M., and Miles, D. L. (1984). A comparative study of sequential redox processes in three British aquifers. In *Hydrochemical balances of freshwater systems* (ed. E. Eriksson), pp. 55–70. Publication No. 150, Int. Assoc. Hyrol. Sciences, IAHS Press, Wallingford.

Edmunds, W. M., Cook, J. M., Darling, W. G., Kinniburgh, D. G., Miles, D. L., Bath, A. H., et al. (1987). Baseline geochemical conditions in the Chalk aquifer, Berkshire, UK: a basis for groundwater quality management. *Appl. Geochem.*, **2**, 251–74.

Edmunds, W. M., Darling, W. G., Kinniburgh, D. G., Dever, L., and Vachier, P. (1993). *Chalk groundwater in England and France: hydrogeochemistry and water quality.* Research Rpt SD/92/2, British Geological Survey, Keyworth.

Edworthy, K. J., Wilkinson, W. B., and Young, C. P. (1978). The effects of the disposal of effluents and sewage sludge on groundwater quality in the Chalk of the United Kingdom. *Prog. Water Tech.*, **10**, 479–92.

Egeberg, P. K. and Aagaard, P. (1989). Origin and evolution of formation waters from oil fields on the Norwegian shelf. *Appl. Geochem.*, **4**, 131–42.

Egeberg, P. K. and Saigal, G. C. (1991). North Sea chalk diagenesis: cementation of chalks and healing of fractures. *Chemical Geology*, **92**, 339–54.

Ehrmann, W. U. (1986). Zum Sedimenteintrag in das zentrale nordwesteuropäische Oberkreidemeer. *Geologisches Jahrbuch, A*, **97**, 3–139.

Ekdale, A. A. and Bromley, R. G. (1983). Trace fossils and ichnofabric in the Kjølby Gaard Marl, uppermost Cretaceous, Denmark. *Bull. Geol. Soc. Denmark*, **31**, 107–19.

Ekdale, A. A. and Bromley, R. G. (1988). Diagenetic microlamination in chalk. *J. Sediment. Petrol.*, **58**, 857–61.

Eller, M. G. (1981). The Red Chalk of eastern England: a Cretaceous analogue of Rosso Ammonitico. In *Rosso Ammonitico Symposium Proceedings* (ed. A. Farinacci and S. Elmi), pp. 207–31. Edizioni Tecnoscienza, Rome.

Elliot, T. (1990). Geochemical indicators of groundwater ageing. Unpublished Ph. D. thesis, Univ. of Bath.

E N P C (1989). Le Tunnel sous La Manche. Géologie et géotechnique. *Actes des journées d'études organisées à l'Ecole. Nat. des Ponts et Chaussées.* 31 May–1 June. Paris.

Evans, C. J. (1988). The stress-field in the United Kingdom. *Invest. Geotherm. Potent. UK.* British Geological Survey, Keyworth.

Felder, W M (1975). Lithostratigrafie van het Boven-Krijt en het Dano-Montien in Zuid-Limburg en het aangrenzende gebied. In *Toelichting bij geologische overzichtskaarten van Nederland* (ed. W. H. Zagwijn and C. J. van Staalduinen), pp. 63–72. Neth. Geol. Survey, Haarlem.

Felder, W. M. (1983). De kalkstenen uit het Boven-Krijt en Dano-Montien in het Maasdal bij Maastricht. Unpublished report OP 5730-1, Neth. Geol. Survey.

Flavin, R. J. and Joseph, J. B. (1983). The hydrogeology of the Lee Valley and some effects of artifical recharge. *Q. J. Eng. Geol.*, **16**, 65–82.

Fookes P. G. and Denness, B. (1969). Observational studies on fissure patterns in Cretaceous sediments of southeast England. *Geotechnique*, **19**, 453–77.

Fookes, P. G. and Parrish, D. G. (1969). Observations on small-scale structural discontinuities in the London Clay and their relationship to regional geology. *Q. J. Eng. Geol.*, **1**, 217–40.

Foster, P. T. (1993). The evolution of a fractured chalk reservoir—Machar Oilfield, U.K. North Sea. In *Petroleum geology of Northwest Europe: Proceedings of the 4th Conference.* (ed J. R. Parker). Geological Society, London.

Foster, S. S. D. (1975). The Chalk groundwater tritium anomaly—a possible explanation. *J. Hydrol.*, 25, 159–65.

Foster, S. S. D. (1976). The vulnerability of British groundwater resources to pollution by agricultural leachates. *Tech. Bull.* **32**, 68–91. Min. Agric. Fish and Food, London.

Foster, S. S. D. (1987). Fundamental concepts in aquifer vulnerability, pollution risk and protection strategy. In *Vulnerability of Soil & Groundwater to Pollutants*, Proc. RIvM Int. Conf., pp. 69–86. Noordwijk, Netherlands

Foster, S. S. D. (1989). Diffuse pollution of groundwater by agriculture—lessons learnt and future prospects. In *Recent Advances in Groundwater Hydrology*, Proc. AIH Int. Cong., pp. 185–94. Tampa, FL.

Foster, S. S. D. and Bath, A. H. (1983). The distribution of agricultural soil leachates in the unsaturated zone of the British Chalk. *Environ. Geol.*, **5**, 53–9.

Foster, S. S. D. and Crease, R. I. (1974). Nitrate pollution of Chalk groundwater in East Yorkshire—a hydrogeological appraisal. *J. Inst. Wat. Eng.*, **25**, 178–94.

Foster, S. S. D. and Hirata, R. C. A. (1989). *Groundwater pollution risk assessment—a methodology using available data.* WHO–PAHO–CEPIS Publication. CEPIS, Lima, Peru.

Foster, S. S. D. and Milton, V. A. (1974). The permeability and storage of an unconfined chalk aquifer. *Hydrol. Sci. Bull.*, **19**, 485–500.

Foster, S. S. D. and Milton, V. A. (1976). *Hydrological basis for large-scale development of groundwater storage capacity in the East Yorkshire Chalk.* Inst. Geol. Sci., London.

Foster, S. S. D. and Smith-Carington, A. K. (1980). The interpretation of tritium in the Chalk unsaturated zone. *J. Hydrol.*, **46**, 343–64.

Foster, S. S. D. and Young, C. P. (1979). Conséquences de l'utilisation agricole des sols sur la qualité de l'eau souterraine et notamment sur sa teneur en nitrate. In *Hydrogéologie Britamique: progrés recents*, 245–56. Unesco, Paris.

Foster, S. S. D. and Young, C. P. (1980). Groundwater contamination due to agricultural land-use practices in the United Kingdom. *IHP Studies and Reports in Hydrogeology*, **30**, 268–82. Unesco, Paris.

Foster, S. S. D. and Cripps, A. C., and Smith-Carington, A. K. (1982). Nitrate leaching to groundwater. *Phil. Trans. R. Soc. London*, **196**, 477–89.

Foster, S. S. D., Geake, A. K., Lawrence, A. R., and Parker, J. M. (1985a) Diffuse groundwater pollution: lessons of the British experience. In *Memoirs 18th IAH Congress*, **3**, 168–77.

Foster, S. S. D., Kelly, D. P., and James, R. (1985b). The evidence for zones of biodenitrification in British aquifers. In *Planetary Ecology,* pp. 356–69. Van Nostrand, NY.

Foster, S. S. D., Bridge, L. R., Geake, A. K., Lawrence, A. and Parker, J. M. (1986). *The groundwater nitrate problem.* Hydrogeol. Rpt. 86/2. British Geological Survey, Keyworth.

Foster, S. S. D., Chilton, P. J., and Stuart, M. E. (1991). Mechanisms of groundwater pollution by pesticides. *J. Inst. Water Environ. Managmnt*, **5**, 186–93.

Frederiksen, P., Andersen, L. J., and Kelstrup, N. (1991). Groundwater in Denmark. In *Ground Water in Western and Central Europe*, Natural Resources Water Series, No. 27. United Nations, New York.

Freeze, R. A. and Cherry, J. A. (1979) *Groundwater.* Prentice Hall, Englewood Cliffs, NJ.

Froelich, P. N., Klinkhammer, G. P., Bender, M. L., Luedtke, N. A., Heath, G. R., Cullen, D., Dauphin, P., Hammond, D., Hartman, B., and Maynard, V. (1979). Early oxidation of organic matter in pelagic sediments of the eastern equatorial Atlantic: suboxic diagenesis. *Geochimica et Cosmochimica Acta*, **43**, 1075–90.

Gale, A. S. (1980). Penecontemporaneous folding, sedimentation and erosion in Campanian Chalk near Portsmouth, England. *Sedimentology,* **27**, 137–51.

Gale, A. S. (1990). A Milankovitch scale for Cenomanian time. *Terra Nova*, **1**, 420–6.

Gale, A. S. and Smith, A. B. (1982). The palaeobiology of the Cretaceous irregular echinoids *Infulaster* and *Hagenowia*. *Palaeontology*, **25**, 11–42.

Gardner, C. M. K., Cooper, J. D., Wellings, S. R., Bell, J. P., Hodnett, M. G. Boyle, S. A., and Howard, M. J. (1990). Hydrology of the unsaturated zone of the Chalk of south east England. In *Chalk*, pp. 611–8. Thomas Telford, London.

Garrison, R. E. (1981). Diagenesis of oceanic carbonate sediments: a review of the DSDP perspective. In *The Deep Sea Drilling Project: a decade of progress* (ed. J. E. Warme, R. G. Douglas and E. L. Winterer), pp. 181–207. Society of Economic Paleontologists and Mineralogists, Special Publication, No. 32.

Garrison, R. E. and Kennedy, W. J. (1977). Origin of solution seams and flaser structure in Upper Cretaceous chalks of southern England. *Sedimentary Geology*, **19**, 107–37.

Gerasimov, P. A., Meegacheeva, E. E., Naidin, D. P., and Sterlin, B. P. (1962). *Jurassic* and *Cretaceous sediments of the Russian Platform,* Ocherki Regional'noi Geolgii SSSR, Moskva, No. 5 (in Russian).

Geake A. K. and Foster S. S. D. (1989). Sequential isotope and solute profiling of the unsaturated zone of the British Chalk. *Hydrol. Sci. J.* **34**, 79–95.

Giles, D. M. and Lowings, V. A. (1990). Variation in the character of the chalk aquifer in east Hampshire. In *Chalk*, pp. 619–26. Thomas Telford, London.

Glasbergen, P (1987). Oorsprong, kwaliteit en winningsmogelijk-heden van grondwater in de Krijtafzettingen te Oploo. Unpublished report 728619001. RIVM, Bilthoven.

Glasser, L. S. D. and Smith, R. (1986). Siliceous coatings on fossil coccoliths—how did they arise? In *The scientific study of flint and chert* (ed. G. de G. Sieveking and M. B. Hart), pp. 105–9, Cambridge University Press.

Glennie, K. W. and Boegner, P. L. E. (1981). Sole Pit inversion tectonics. In *Petroleum geology of the continental shelf of north-west Europe* (ed. L. V. Illing and G. D. Hobson), pp. 110–20. Heyden, London.

Gosselet, M. J. (1904). Les assises crétaciques et tertiaires dans les fosses et sondages du Nord de la France. Fascicule 1, région de Douai. *Imprimerie Nationale Paris.*

Godfriaux, Y., and Rorive, A., (1987). L'aquifère des craies du bassin de Mons, In *Les eaux souterraines en Wallonie: Bilans et perspectives*, Colloque ES087, pp. 66–77.

Gomme, J., Shurvell, S., Hennings, S. M., and Clark, L. (1992). Hydrology of pesticides in a chalk catchment: groundwaters. *J. Inst. Wat. Environ. Man.*, **6**, 172–8.

Gosk, E., Bull, N., and Andersen, L. J. (1981). *Hydrogeological Programme, Mors Salt Dome, Hydrogeological Main Report.* Geol. Surv., Denmark.

Gosk, E., Nygaard, E., and Lundsgaard, A. (1990). *Status for grundvand og drikkevand i Danmark, Vandmiljøplanens Overvågningsprogram.* Int. Rep. No. 45, Geol. Surv. Denmark

Gray, D. A. (1958). Electrical resistivity marker bands in the Lower and Middle Chalk of the London Basin. *Bull. Geol. Surv. G. B.*, no. 15, 85–95.

Gray, D. A. (1965). The stratigraphical significance of electrical resistivity marker bands in the Cretaceous strata of the Leatherhead (Fetcham Mill) Borehole, Surrey. *Bull. Geol. Surv. G. B.*, no. 23, 65–115.

Gray, E. M. and Morgan-Jones, M. (1980). A comparative study of nitrate levels at three adjacent groundwater sources in a Chalk catchment area west of London. *Ground Water*, **18**, 159–67.

Great Ouse River Authority (1972). *Great Ouse Groundwater Pilot Scheme, Final Report.* Cambridge.

Green, J. C. (1986). Biomineralization in the algal class Prymnesiophyceae. In *Biomineralization in Lower Plants and Animals* (ed. B. S. C. Leadbeater and R. Riding), pp. 173–88. Systematics Association, Special Volume no. 30.

Green, P. F., Duddy, I. R., and Bray, R. (1993). Elevated palaeotemperatures in the Late Cretaceous to Early Tertiary throughout the UK region identified by AFTA: implications for hydrocarbon generation. In *Petroleum geology of North-west Europe: Proceedings of the 4th Conference.* (ed J. R. Parker). Geological Society, London.

Greene, L. A. and Walker, P. (1970). Nitrate pollution of Chalk water. *Proc. Soc. Water Treat. Exam.*, **19**, 169–82.

Grindley, J. (1967). The estimation of soil moisture deficits. *Meteorological Magazine*, **76**, 97–108.

Grisak, G. E. and Pickens, J. F. (1980). Solute transport through fractured media. *Water Resour. Res.*, **16**, 719–30.

Gulinck, M. (1966). Atlas de Belgique, Hydrogéologie, Plates 16a and 16b, 68 pp, Atlas National, Comité National de Géographie, L'Institut Géographique Militaire, Brussels.

Gustafsson, O. (1972). Description to the hydrogeological map Trelleborg NV and Malmö SV (in Swedish with English summary). Ser. Ag No 4. Geol. Survey of Sweden.

Gustafsson, O. (1978). Description to the hydrogeological map Trelleborg NO and Malmö SO (in Swedish with English summary). Ser. Ag No 6. Geol. Survey of Sweden.

Gustafsson, O. and De Geer, J. (1977). The major groundwater resources of Skåne (in Swedish). Reports and bulletins No 8. Geol. Survey of Sweden.

Gustafsson, O., Andersson, J. -E., and De Geer, J. (1979). Compilation of hydrogeological data from the Kristianstad Plain (in Swedish). Reports and bulletins No 12. Geol. Survey of Sweden.

Gustafsson, O., Jonasson, S. A., Magnusson, E., and Andersson, C. (1988). Groundwater investigations on the Kristianstad Plain 1976–1987 (in Swedish). Reports and bulletins No 52. Geol. Survey of Sweden.

Haines, T. S. and Lloyd, J. W. (1985). Controls on silica in groundwater environments in the United Kingdom. *J. Hydrol.*, **81**, 277–95.

Håkansson, E., Bromley, R., and Perch-Nielsen, K. (1974). Maastrichtian chalk of north-west Europe—a pelagic shelf sediment. In *Pelagic Sediments: on Land and under the Sea* (ed. K. J. Hsü and H. C. Jenkyns), pp. 211–23. Special Publications of the International Association of Sedimentologists, No. 1.

Halcrow (Sir W. Halcrow and Partners)(1987). *Studies of an elevation of low river flows resulting from groundwater abstraction.* Report to Thames Water, Reading.

Halcrow (Sir W. Halcrow and Partners)(1988). *Assessment of groundwater quality in England and Wales.* U.K. Dept. of Environment, Research Contract Dept. PECD7/7/227.

Hallam, A. (1984). Pre–Quaternary sea-level changes. *Ann. Rev. Earth Planet. Sci.*, **12**, 205–43.

Hancock, J. M. (1961). The Cretaceous system in Northern Ireland. *Q. J. Geol. Soc. London*, **117**, 11–36.

Hancock, J. M. (1975*a*). The sequence of facies in the Upper Cretaceous of northern Europe compared with that in the western interior. In *The Cretaceous System in the Western Interior of North America* (ed. W. G. E. Caldwell), pp. 83–118. Special Papers of the Geological Association of Canada, No. 13.

Hancock, J. M. (1975*b*). The petrology of the Chalk. *Proc. Geol. Assoc.*, **86**, 499–535.

Hancock, J. M. (1990). Cretaceous. In *Introduction to the petroleum geology of the North Sea*, (ed. K. W. Glennie), pp. 255–72. Blackwell Scientific, Oxford.

Hancock, J. M. and Kauffman, E. G. (1979). The great transgressions of the Late Cretaceous. *J. Geol. Soc. London*, **136**, 175–86.

Hancock, J. M. and Kennedy, W. J. (1967). Photographs of hard and soft chalks taken with a scanning electron microscope. *Proc. Geol. Soc. London*, **1643**, 249–52.

Hancock, J. M. and Scholle, P. A. (1975). Chalk of the North Sea. In *Petroleum and the Continental Shelf of north-west Europe. I Geology* (ed. A. W. Woodland), pp. 413–27. Applied Science Publishers, Barking.

Hardman, R. F. P. (1982). Chalk reservoirs of the North Sea. *Bull. Geol. Soc. Denmark*, **30**, 119–37.

Hardman, R. F. P. and Eynon, G. (1978). Valhall Field—a structural/stratigraphic trap. In *Mesozoic northern North Sea symposium.* Norwegian Petroleum Society, Geilo.

Hardman, R. F. P. and Kennedy, W. J. (1980). Chalk reservoirs of the Hod fields, Norway. In *The sedimentation of the North Sea reservoir rocks.* Norwegian Petroleum Society (NPF), Geilo.

Hardy, A. C. (1956). *The Open Sea. Its natural history: the world of plankton.* Collins, London.

Harman, D. G. S. and Vadgama, N. (1980). Site investigation at Portobello Pumping Station Telscombe, Brighton. Terresearch Limited Report No. S. 30/717. For Southern Water Authority West Sussex Water and Drainage Division.

Haswell, C. K. (1969). Thames Cable Tunnel. *Proc. Inst. Civ. Eng.*, **44**, 323–40.

Haswell, C. K. (1975). Problems of tunnelling in Chalk. *Tunnels and Tunnelling*, **7**, 40–3.

Hattin, D. E. (1975a). Petrology and origin of fecal pellets in Upper Cretaceous strata of Kansas and Saskatchewan. J. *Sedimentary Petrology*, **45**, 686–96.

Hattin, D. E. (1975b). Stratigraphy and depositional environment of Greenhorn Limestone (Upper Cretaceous) of Kansas. *Kansas Geol. Surv., Bull.*, **209**, 1–128.

Hattin, D. E. (1981). Petrology of the Smoky Hill Chalk Member, Niobrara Chalk (Upper Cretaceous), in the type area, western Kansas. *Bull. Am. Assoc. Petrol. Geol.*, **65**, 831–49.

Hattin, D. E. (1982). Stratigraphy and depositional environment of Smoky Hill Chalk Member, Niobrara Chalk (Upper Cretaceous) of the type area, western Kansas. *Kansas Geol. Surv., Bull.*, **225**, 1–108.

Headworth, H. G. (1970). The selection of root constants for the calculation of actual evaporation and infiltration for Chalk catchments. *J. Inst. Wat. Eng.*, **24**, 431–46.

Headworth, H. G. (1972). The analysis of natural groundwater level fluctuations in the Chalk of Hampshire. *J. Inst. Water Eng.*, **26**, 107–24.

Headworth, H. G. (1978). Hydrogeological characteristics of artesian boreholes in the Chalk of Hampshire. *Q. J. Eng. Geol.*, **11**, 139–44.

Headworth, H. G. and Fox, G. B. (1986). The South Downs Chalk aquifer; its development and management. *J. Inst. Wat. Eng. Sci.*, **40**, 345–62.

Headworth, H. G., Puri, S., and Rampling, B. H. 1980. Contamination of a Chalk aquifer by mine drainage at Tilmanstone, East Kent, UK. *Q. J. Eng. Geol.*, **13**, 105–17.

Headworth, H. G., Keating, T., and Packman, M. J. (1982). Evidence for a shallow, highly permeable zone in the Chalk of Hampshire. J. *Hydrol.*, **55**, 93–112.

Heathcote, J. A. (1981). Hydrochemical aspects of the Gipping Chalk salinity investigation. Ph. D. thesis, University of Birmingham.

Heederik, J. P. *et al.* (1989). Geothermische reserves Centrale Slenk, Nederland. Unpublished report OS 89–18, Groundwater-survey/TNO, Delft.

Heinberg, C. (1979). Evolutionary ecology of nine sympatric species of the pelecypod *Limopsis* in Cretaceous chalk. *Lethaia*, **12**, 325–40.

Henry, F. D. C. (1986). *The design and construction of engineering foundations.* Chapman and Hall, London.

Hester, S. W. (1980). *A revision of the geological map of the Moor Park area of Hertfordshire.* Institute of Geological Sciences Open File Report, 1980/1.

Heybroek, P. (1974). Explanation to tectonic maps of the Netherlands. *Geologie en Mijnbouw*, **53**, 43–50.

Heybroek, P. (1975). On the structure of the Dutch part of the Central North Sea Graben. In *Petroleum and the continental shelf of northwest Europe. 1. Geology* (ed. A. W. Woodland), pp. 339–51. Applied Science Publishers, Barking.

Higginbottom, I. E. (1966). The Engineering Geology of Chalk. In *Chalk in Earthworks and Foundations*, Proc. Symp. Inst., Civ. Engrs, London, pp. 1–13.

Hill, D., (1984). Diffusion coefficients of nitrate, chloride, sulphate and water in cracked and uncracked Chalk. *J. Soil Sci.*, **351**, 27–33.

Hiscock, K. M. (1984). Groundwater chemistry study in the vicinity of Norwich. Unpublished Ph.D. Thesis, University of Birmingham, U.K.

Hiscock, K. M. and Lloyd, J. W. (1992). Palaeohydrogeological reconstructions of the north Lincolnshire Chalk, UK for the last 140,000 years *J. Hydrol.*, **133**, 313–42.

Honjo, S. (1975). Dissolution of suspended coccoliths in the deep-sea water column and sedimentation of coccolith ooze. In *Dissolution of deep-sea carbonates* (ed. W. V. Sliter, A. W. H. Bé, and W. H. Berger), pp. 114–28. Special Publication, No. 13. Cushman Foundation for Foraminiferal Research, Washington, DC.

Houghton, J. T., Jenkins, G. I., and Ephraums, J. J. (ed.) (1990). *Climate Change. The IPCC Scientific Assessment.* Cambridge University Press.

Howard, K. W. F. and Lloyd, J. W. (1983). Major ion characterisation of coastal saline groundwaters. *Ground Water,* **21**, 429–37.

Hume, W. F. (1894). The genesis of the Chalk. *Proc. Geol. Assoc.*, **13**, 211–46.

Hunter-Blair, A. (1978). Oil pollution of a Chalk aquifer: a case history. In *Intl Symp on Groundwater Pollution by Oil Hydrocarbons*, pp. 19–38. Stavebni Geologie, Prague

Hunter-Blair, A. (1980). Groundwater pollution by oil products. J. *Inst. Water Engrs Sci.*, **34**, 557–69.

Hure, A. (1933). Monographie des craies turonienne et sénonienne de l'Yonne et tectonique du Sénonais. *Bulletin de la Société des Sciences Historiques et Naturelles de l'Yonne* (1931), 1–85.

Hutchinson, J. N. (1969). A reconsideration of the coastal landslides at Folkestone Warren, Kent. *Geotechnique*, **19**, 6–38.

Hutchinson, J. N. (1979). Various forms of cliff instability arising from coast erosion in south-east England. *Fjellspregningsteknikk ergmekanikk/Geoteknikk*, **19**.1–19.32.

Hutchinson, J. N. (1980). Possible late Quaternary pingo remnants in central London. *Nature*. **284**, 253–5.

Ineson, J. (1959). Yield depression curves of discharging wells, with particular reference to Chalk wells, and their relationship to variations in transmissibility. *J. Inst. Wat. Engrs.*, **13**, 119–63.

Ineson, J. (1962). A hydrogeological study of the permeability of the Chalk. *J. Instn. Wat. Engrs,* **16,** 449–63

Ineson, J. and Downing, R. A. (1963). Changes in the chemistry of groundwaters in the Chalk passing beneath argillaceous strata. *Bull. Geol. Surv. Great Britain*, no. 20, 176–92.

Ineson, J. and Downing, R. A. (1964). The groundwater components of river discharge and its relationship to hydrogeology. *J. Inst. Wat. Engrs.*, **18**, 519–41.

Institute of Hydrology/British Geological Survey. (1991). *Hydrological data United Kingdom: 1990 Yearbook.* Wallingford, Oxfordshire.

Jackson, D. and Rushton, K. R. (1987). Assessment of recharge components for a Chalk aquifer unit. *J. Hydrol.*, **92**, 1–15.

Jacobsen, J., and Kelstrup, N. (1981). *Suså Hydrogeologi: Belysning af sammenhængen mellem grundvandsjemi og geologi.* Rep. Suså-H, no. 8, English abstract. Dansk Komite for Hydrologi.

Jacobsen, R. (1991). Hydraulik og stoftransport i en opsprækket kalkbjergart. *Lossepladsprojektet, Rep. H9*, English summary. Miljøstyrelsen, Copenhagen.

Jannasch, H. W. and Wirsen, C. O. (1977). Microbial life in the deep sea. *Scientific American*, **236**, 42–52.

Japsen, P., and Langtofte, C. (1991). *Geological map of Denmark 1:400,000, 'Base Chalk' and the Chalk Group.* Map Ser., no. 29. Geol. Surv. Denmark.

Jarvis, I. (1980). The initiation of phosphatic chalk sedimentation—the Senonian (Cretaceous) of the Anglo-Paris Basin. In *Marine Phosphorites—Geochemistry, Occurrence Genesis* (ed. Y. K. Bentor), pp. 167–92. Special Publications of the Society of Economic Paleontologists and Mineralogists, no. 29.

Jeans, C. V. (1968). The origin of the montmorillonite of the European Chalk with special reference to the Lower Chalk of England. *Clay Minerals*, **7**, 311–29.

Jeans, C. V. (1973). The Market Weighton structure: tectonics, sedimentation and diagenesis during the Cretaceous. *Proc. Yorks. Geol. Soc.*, **39**, 409–44.

Jeans, C. V. (1980). Early submarine lithification in the Red Chalk and Lower Chalk of eastern England: a bacterial control model and its implications. *Proc. Yorks. Geol. Soc.*, **43**, 81–157.

Jeans, C. V., Merriman, R. J., Mitchell, J. G., and Bland, D. J. (1982). Volcanic clays in the Cretaceous of southern England and Northern Ireland. *Clay Minerals*, **17**, 105–56.

Jefferies, R. P. S. (1963). The stratigraphy of the *Actinocamax plenus* Subzone (Turonian) in the Anglo-Paris Basin. *Proc. Geol. Assoc.*, **74**, 1–33.

Jenkyns, H. C. (1974). Origin of red nodular limestones (Ammonitico Rosso, Knollenkalke) in the Mediterranean Jurassic: a diagenetic model. In *Pelagic Sediments: on land and under the sea* (ed. K. J. Hsü and H. C. Jenkyns), pp. 249–71. International Association of Sedimentologists, Special Publications, No. 1.

Jensenius, J. (1987). High-temperature diagenesis in shallow chalk reservoir, Skjold oilfield, Danish North Sea: evidence from fluid inclusions and oxygen isotopes. *Bull. Amer. Assoc. Petrol. Geol.*, **71**, 1378–86.

Jenyon, M. K. (1987). Seismic expression of real and apparent buried topography. *J. Petrol. Geol.*, **10**, 41–58.

Jones, D. K. C. (1981). *Southeast and Southern England.* London, Methuen.

Jones, D. K. C., Griffiths, J. S., and Lee, E. M. (1988). Geomorphological development of the terrain in the vicinity of the Channel Tunnel Terminal and Portal. *Notes to accompany the Engineering Group of the Geological Society Field Visit, 19th November, 1988.*

Jones, D. L. and I. M. Morrison, I. M. (1990). Driven piles in Chalk. In *Chalk*, pp. 359–64. Thomas Telford, London.

Jones, M. E. (1990). Hydrocarbon production from the North Sea Chalk: geotechnical considerations. In *Chalk*, pp. 641–7. Thomas Telford, London.

Jones, M. E., Bedford, J., and Clayton, C. (1984). On deformation mechanisms in the Chalk. *J. Geol. Soc. London*, **141**, 675–83.

Jones, M. E., Leddra, J., and Potts, D. (1990). Ground motions due to hydrocarbon production from the chalk. In *Chalk*, pp. 675–81. Thomas Telford, London.

Jørgensen, N. O. (1975). Mg/Sr distribution and diagenesis of Maastrichtian white chalk and Danian bryozoan limestone from Jylland, Denmark. *Bull. Geol. Soc. Denmark*, **24**, 299–325.

Jørgensen, N. O. (1986). Geochemistry, diagenesis and nannofacies of Chalk in the North Sea Central Graben. *Sed. Geol.*, **48**, 267–94.

Jørgensen, N. O. (1987). Oxygen and carbon isotope compositions of Upper Cretaceous chalk from the Danish subbasin and the North Sea Central Graben. *Sedimentology*, **34**, 559–70.

Joseph, J. B. and Clark, L. (1982). Pig slurry applied as a fertilizer: the effects on groundwater quality in the Yorkshire Chalk. *Proc. North Eng. Soil Discuss. Grp*, **18**, 47–59.

Juignet, P. and Kennedy, W. J. (1974). Structures sédimentaires et mode d'accumulation de la craie du Turonien Supérieur et du Sénonien du Pays de Caux. *Bulletin du Bureau de Recherches Géologigues et Miniéres*, 2nd series, Section 4, 19–47.

Kelstrup, N., Bækgaard, A., and Andersen, L. J. (1982). Grundvandsressourcer i Danmark. In *Groundwater resources of the European Community, Synthetic Report* (Coord. J. J. Fried). European Community, Brussels.

Kemper, E. (1979). Die Unterkkreide Nordwestdeutschlands. Ein Überblick. In *Aspekte der Kreide Europas* (ed. J. Wiedmann), pp. 1–9. International Union of Geological Sciences, Series A, No. 6.

Kennedy, W. J. (1969). The correlation of the Lower Chalk of south-east England. *Proc. Geol. Assoc.*, **80**, 459–560.

Kennedy, W. J. (1980). Aspects of chalk sedimentation in the southern Norwegian offshore. In The *sedimentation of the North Sea reservoir rocks*. Norwegian Petroleum Society (NPF), Geilo.

Kennedy, W. J. (1987). Late Cretaceous and early Palaeocene Chalk Group sedimentation in the Greater Ekofisk area, North Sea Central Graben. *Bulletin des Centres de Recherches Exploration-Production Elf-Aquitaine, Pau*, **11**, 91–126.

Kennedy, W. J. and Garrison, R. E. (1975). Morphology and genesis of nodular chalks and hardgrounds in the Upper Cretaceous of southern England. *Sedimentology*, **22**, 311–86.

Kennedy, W. J. and Juignet, P. (1974). Carbonate banks and slump beds in the Upper Cretaceous (Upper Turonian–Santonian) of Haute Normandie, France. *Sedimentology*, **21**, 1–42.

Kimpe, W. F. M. (1960). Le chimisme des eaux de la Craie du Limbourg néerlandais. *Ann. Soc. Géol. du Nord*, **80**, 285–95.

Kimpe, W. F. M. (1963). Geochimie des eaux dans le houiller du Limbourg (Pays Bas). *Verh. Kon. Ned. Geol. Mijnbouwk. Genootschap*, Geol. Ser., 21–2, 25–45.

Kirkaldy, J. F. (1950). Solution of the Chalk in the Mimms Valley, Herts. *Proc. Geol. Assoc.*, **61**, 219–24.

Kitchen, R. and Shearer, T. R. (1982). Construction and operation of a large undisturbed lysimeter to measure recharge to the chalk aquifer, England. *J. Hydrol.*, **58**, 267–77.

Klein, V. and Soukup, J. (1966). The Bohemian Cretaceous basin. In *Regional Geology of Czechoslovakia. I: The Bohemian Massif* (ed. J. Svoboda *et al.*), pp. 487–512. Ústrední Ústav Geologický, Prague.

Koestler, A. G. and Ehrmann, W. U. (1987). Fault-induced microtextural changes in Upper Cretaceous chalks, northern Germany. In *Petroleum Geology of North West Europe*, Vol.2, (ed. J. Brooks and K. W. Glennie), pp. 1214–15. Graham and Trotman, London.

Kornfält, K. -A., Bergstrüm, J., Carserud, L., Henkel, H., and Sundquist, B. (1978). Description to the map of solid rocks and the aeromagnetic map Kristianstad SO (in Swedish with English summary). Ser. Af No 121. Geol Survey of Sweden.

Krumbein, W. C. and Sloss, L. L. (1956). *Stratigraphy and sedimentation*. Freeman, San Francisco.

Lake, L. M. (1990). Underground excavations in Chalk. In *Chalk*, pp. 461–8. Thomas Telford, London.

Lal, D. and Lerman, A. (1975). Size spectra of biogenic particles in ocean water and sediments. *J. Geophys. Res.*, **80**, 423–30.

Lallemand-Barrès, A. and Landreau, A. (1986). *Carte à 1/1500000 des teneurs en nitrates des nappes phréatiques de la France. Etat des connaissances (2nd end). Report no. 87 .SGN. 237 EAU*. B R G M, Orléans.

Landreau, A. and Lemoine, B. (1977). Carte de la qualité des eaux souterraines de la France (à 1/100000) et notice explicative. *Rpt. 77SGN 606 HYD*. BRGM, Orléans.

Lapparent, A. F. de (1943). Révision de la Feuille de Reims au 1/80,000. *Bulletin de la Carte Géologique de France*, **44**, 19–38.

Lapworth, C. F. (1948). Percolation in the Chalk. *J. Instn. Wat. Engrs.*, **2**, 97–108.

Lasne, H. (1890). Sur les terrains phosphatés des environs de Doullens. Etages Sénonien et terrains superposés. Bull. Soc. Géol. France, 3è série, t XVIII, 441–490.

Lasne, E. and Lepiller, M. (1989). Carte piézométrique de l'aquifére de la craie dans l'Est du département du Loiret (1:100000). Université d'Orléans.

Lawrence, A. R. and Foster, S. S. D. (1987). The pollution threat from agricultural pesticides and industrial solvents: a comparative review in relation to British aquifers. *Hydrogeology Research Report* 87/2. British Geological Survey, Keyworth.

Lawrence, A. R. and Foster, S. S. D. (1991). The legacy of aquifer pollution by industrial chemicals—technical appraisal and policy implications. *Quart J Eng Geol* **24**, 231–9.

Lawrence, A. R., Foster, S. S. D., and Izzard, P. W. (1983). Nitrate pollution of Chalk groundwater in East Yorkshire—a decade on. *J. Inst. Water Eng. Sci.*, **37**, 410–9.

Leary, P.N. and Wray, D. S. (1989). The foraminiferal assemblages across three middle Turonian marl bands and a note on their genesis. *J. Micropalaeontology,* **8**, 143–8.

Leddra, M. J. and Jones, M. E. (1990). Steady-state flow during undrained loading of chalk. In *Chalk*, pp. 245–52. Thomas Telford, London.

Leddra, M. J., Jones, M. E., Pederstad, K., and Lønøy, A. (1990). Influence of increased effective stress on the permeability of chalks under hydrocarbon reservoir conditions. In *Chalk*, pp. 253–60. Thomas Telford, London.

Legrand, R. (1951). Carte géologique et hypsométrique du socle paléozoïque de la Belgique, complétée par les allures générales du Crétacé. *Bull. Soc. Belge de Géologie*, **59**, 318–41.

Lepiller, M. (1975). Le système karstique de Villequier (Seine-Maritime). Etude hydrogéologique, hydrochimique et sédimentologique d'une circulation souterraine typique du Crétacé Supérieur normand. *Bull. Trim. Soc. Géol. Normandie. amis Mus. Havre,* **62**, 51–85.

Lepiller, M. (1990). Réflexions sur l'hydrogéologie karstique de la craie du bassin Parisien à partir de quelques exemples régionaux (Normandie, Gatinais). Unpublished.

Lepiller, M. and Lasne, E. (1990). Contribution à l'étude de la productivité des forages à l'amélioration de l'exploitation des ressources en eau de la craie. Exemple du Gâtinais, de La-Puisaye et du Berry (Sud du Bassin Parisien- France). Unpublished.

Leriche, M. (1926). Les 'rideaux' du Cambrésis et du Vermandois. *Bull. Service. Carte. Géol. France*, **31**, (166).

Lerman, A., Lal, D., and Dacey, M. F. (1974). Stokes' settling and chemical reactivity of suspended particles in natural waters. In *Suspended solids in water* (ed. R. J. Gibbs), pp. 17–47. Plenum Press, New York.

Letsch, W. J. and Sissingh, W. (1983). Tertiary stratigraphy of the Netherlands. *Geologie en Mijnbouw*, **62**, 305–18.

Lieberkind, K. (1982). Lithostratigraphic subdivision of the Chalk Group onshore Denmark with correlation to the North Sea and Germany. Int. Assoc. Sedimentologists, *3rd. Eur. Mtg.* Copenhagen, Abstr., pp. 35–7.

Lieberkind, K., Bang, I., Mikkelsen, N., and Nygaard, E. (1982). Late Cretaceous and Danian limestone. In *Geology of the Danish Central Graben* (ed. O. Michelsen). Ser. B, No. 8, Geol. Surv. Denmark.

Lippmann, F. (1960). Versuche zur Aufklärung der Bildungsbedingungen von Kalzit und Aragonit. *Fortschritte der Mineralogie, Stuttgart*, **38**, 156–61.

Lippmann, F. (1973). *Sedimentary Carbonate Minerals.* Springer-Verlag, Berlin.

Lloyd, J. W. (1980). The influence of Pleistocene deposits on the hydrogeology of major British aquifers. *J. Inst. Wat. Eng. and Sci.*, **33**, 346–56.

Lloyd, J. W. and Hiscock, K. M. (1990). Importance of drift deposits in influencing chalk hydrogeology. In *Chalk*, pp. 583–90. Thomas Telford, London.

Lloyd, J. W., Harker, D. and Baxendale, R. A. (1981). Recharge mechanisms and groundwater flow in the chalk and drift deposits of southern East Anglia. *Q. J. Eng. Geol.*, **14**, 87–96.

Lloyd, J. W., Howard, K. W. F., Pacey, N., and Tellam, J. H. (1982). The value of iodine as a parameter in the chemical characterization of groundwaters. *J. Hydrol.*, **57**, 247–65.

Long, J. C. S. and Billaux, D. M. (1987). From field data to fracture network modelling: An example incorporating spatial structure. *Water Resour. Res.*, **23**, 1201–16.

Longstaff, S. L., Aldous, P. J., Clark, L., Flavin, R. J. and Partington, J. (1992). Contamination of the Chalk aquifer by chlorinated solvents: a case study of the Luton and Dunstable area. *J. Inst. Water Environ. Managmnt*, **6**, 541–50.

Lowenstam, H. A. (1986). Mineralization processes in monerans and protocists. In *Biomineralization in Lower Plants and Animals* (ed. B. S. C. Leadbeater and R. Riding), pp. 1–17. Systematics Association, Special Volume, no. 30.

Malecha, A. (1966). The basins of southern Bohemia. In *Regional Geology of Czechoslovakia. I: The Bohemian Massif* (ed. J. Svoboda *et al.*), pp. 581–95. Ústrední Ústav Geologický, Prague.

Maloszewski, P. and Zuber, A. (1985). On the theory of tracer experiments in fissured rocks with a porous matrix. *J. Hydrol.*, **79**, 333–58.

Mania, J. (1978). *Gestion des systèmes aquifères. Application au Nord de la France.* Mémoire no. 15. Soc. Géol. Nord, France.

Mapstone, N. B. (1975). Diagenetic history of a North Sea chalk, *Sedimentology,* **22**, 601–13.

Margat, J. (1986). *Abrégé sur les eaux souterraines de la France.* Report no. 86.SGN 623 Eau. BRGM, Orléans.

Markussen, L. M., Møller, H. -M. F., Villumsen, B., Mortensen, J. K., and Selchau, T. (1991): Frederiksberg Kommune, Sikring af drik-kevandsressourcen. *Bull. No. 27*, Rambøll and Hannemann A/S, Copenhagen. Engl. summary.

Marlière, R. (1967). Texte explicative de la feuille Mons-Givry. *Carte géologique de la Belgique à l'échelle 1/25,000*, no. 151.

Marriotti, A., Landreau, A., and Simon, B. (1988). Isotope biogeochemistry and natural denitrification process in groundwater; application to the Chalk aquifer of northern France. *Geochim. Cosmochim. Acta*, **52**, 1869–78.

Marsh, T. J. and Monkhouse, R. A. (1993). Drought in the United Kingdom, 1988–92. *Weather*, **48**, 15–22.

Mary, G. (1988). Piézométrie de la nappe de la craie dans le département de la Sarthe (1:100 000). Université du Maine.

Mary, G. (1990). Hydrogéologie des formations crayeuses du Maine. Unpublished.

Matter, A., Douglas, R. G., and Perch-Nielsen, K. (1975). Fossil preservation, geochemistry, and diagenesis of pelagic carbonates from Shatsky Rise, northwest Pacific. *Initial Reports of the Deep Sea Drilling Project*, **32**, 891–921.

Mégnien, C. (1959). Le karst et la nappe de la craie Turonienne et Sénonienne du Bassin de la Vanne (Yonne). *Bull. Soc. Géol de France*, **7**, 456–60.

Mégnien, C. ed. (1970). *Atlas des nappes aquifères de la Région Parisienne*. BRGM, Orléans.

Mégnien, C. (1979). *Hydrogéologie du centre du Bassin de Paris, contribution à l'étude de quelques aquifères principaux*, Mémoire, no. 98, BRGM, Orléans.

Mégnien, C. (1980). *Synthèse géologique du Bassin de Paris*. Memoire nos. 101–3. BRGM, Orléans.

Megson, J.B. (1987). The evolution of the Rockall Trough and implications for the Faeroe–Shetland Trough. In *Petroleum geology of north-west Europe* (ed. J. Brooks and K. W. Glennie), pp. 653–65. Graham and Trotman, London.

Megson, J. B. (1992). The North Sea Chalk play: examples from the Danish Central Graben. In *Exploration Britain: geological insights for the next decade* (ed R. F. P. Hardman), pp 247– 82. Special Publication No 67, Geol. Soc., London.

Michelsen, O. and Anderson, C. (1983). Mesozoic structural and sedimentary development of the Danish Central Graben. In *Petroleum Geology of the Southeastern North Sea and the adjacent onshore areas* (ed. J. P. H. Kaasschieter and T. J. A. Reijers), pp. 93–102. Geologie en Mijnbouw, Vol. 62.

Middlemiss, F. A. (1967). Analysis of structure in a region of gentle en-echelon folding. *N. Jb. Geol. Paläont. Abh.*, **129**, 137–56.

Middlemiss, F. A. (1983). Instability of Chalk cliffs between the South Foreland and Kingsdown, Kent, in relation to geological structure. *Proc. Geol. Assoc.*, **94**, 115–22.

Mimran, Y. (1975). Fabric deformation induced in Cretaceous chalks by tectonic stresses. *Tectonophysics,* **26**, 309–16.

Mimran, Y. (1978). The induration of Upper Cretaceous Yorkshire and Irish chalks. *Sed. Geol.* **20**, 141–64.

Mimran, Y. (1985). Tectonically controlled freshwater carbonate cementation in chalk. In *Carbonate Cements* (ed. N. Schneidermann and P. M. Harris), pp. 371–9. Special Publications of the Society of Economic Paleontologists and Mineralogists, no. 36.

Monciardini, C. (1989). Profil ECORS nord de la France: corrélations biostratigraphiques entre quarante-six sondages sismiques intra-crétacés et implications structurales. *Géologie de la France*, **4**, 39–47.

Monjoie, A. (1967). Observations nouvelles sur la nappe aquifère de la craie en Hesbaye (Belgique), *Mémoires de l'Association Internationale des Hydrogéologues* (I.A.H.), Istanbul.

Montgomery, H. A. C., Beard, M. J. and Baxter, K. M. (1984). Effects of the recharge of sewage effluents upon the quality of Chalk groundwater. *J. Water Poll. Control,* **83**, 349–66.

Monkhouse, R. A. and Fleet, M (1975). A geophysical investigation of saline water in the Chalk of the south coast of England. *Q. J. Eng. Geol.,* **8**, 291–302.

Monkhouse, R. A. and Richards, H. J. (1982). *Groundwater resources of the United Kingdom.* Commission of the European Communities. Th. Schafer GmbH, Hanover.

Morel, E. H. (1980). A numerical model of the Chalk aquifer in the Upper Thames Basin. *Tech. Note 35*, Central Water Planning Unit, Reading.

Morfaux, P. (1976). *Esquisse hydrogéologique de la Champagne crayeuse (1:320 000).* BRGM, Orléans.

Morgan-Jones, M. (1977). Mineralogy of the non-carbonate material from the Chalk of Berkshire and Oxfordshire, England. *Clay Minerals*, **12**, 331–44.

Mortimore, R. N. (1979). The relationship of stratigraphy and tectonofacies to the physical properties of the White Chalk of Sussex. Unpublished Ph.D., Brighton Polytechnic, 5 vols.

Mortimore, R. N. (1983). The stratigraphy and sedimentation of the Turonian–Campanian in the southern province of England. *Zitteliana*, **10**, 27–41.

Mortimore, R. N. (1986). Stratigraphy of the Upper Cretaceous White Chalk of Sussex. *Proc. Geol. Assoc.*, **97**, 97–139.

Mortimore, R. N. (1987). Upper Cretaceous Chalk in the North and South Downs, England: a correlation. *Proc. Geol. Assoc.*, **98**, 77–86.

Mortimore, R. N. (1990). Chalk or chalk? In *Chalk,* pp. 15–45. Thomas Telford, London.

Mortimore, R. N. and Fielding, P. M. (1990). The relationship between texture, density and strength of chalk. In *Chalk*, pp. 109–32. Thomas Telford, London.

Mortimore, R. N. and Jones, D. L. (1989). SPT's in chalk: a re-think. In *Conference on Penetration Testing in the United Kingdom.* Thomas Telford, London.

Mortimore, R. N. and Pomerol, B. (1987). Correlation of the Upper Cretaceous White Chalk (Turonian to Campanian) in the Anglo-Paris Basin. *Proc. Geol. Assoc.*, **98**, 97–143.

Mortimore, R. N. and Pomerol, B. (1991). Upper Cretaceous tectonic disruptions in a placid Chalk sequence in the Anglo- Paris Basin. *J. Geol. Soc. London*, **148**, 391–404.

Mortimore, R. N. and Wood, C. J. (1986). The distribution of flint in the English Chalk, with particular reference to the 'Brandon Flint Series' and the high Turonian flint maximum, In *The scientific study of flint and chert*, Proceedings of the fourth International Flint Symposium, Brighton Polytechnic, 1983 (ed. G. de G. Sieveking and M. B. Hart), pp. 7–20. Cambridge University Press.

Mortimore, R. N., Pomerol, B., and Foord, R.J. (1990a). Engineering stratigraphy and palaeogeography for the Chalk of the Anglo-Paris Basin. In *Chalk*, pp. 47–62. Thomas Telford, London.

Mortimore, R. N., Roberts, L. D., and Jones, D. L, (1990b). Logging of chalk for engineering purposes. In *Chalk*, pp. 133–52. Thomas Telford, London.

Mott, Hay and Anderson (1987). Channel Tunnel Development Study (XVII), Geotechnical investigations for tunnels. Internal Report Transmanche-Link.

Mudge, D. C. and Rashid, B. (1987). The geology of the Faeroe Basin area. In *Petroleum geology of north-west Europe* (ed. J. Brooks and K. W. Glennie), pp. 751–63. Graham and Trotman, London.

Muir Wood, A. M. and Casté, G. (1970). In situ testing for the Channel Tunnel. In *In situ investigation in soils and rocks*. British Geotechnical Society, London.

Muller, E. (1987). Modelling groundwater pollution at a site near Cambridge. research report no 7. NERC Water Research Unit, University of Newcastle upon Tyne.

Murray, K. H. (1986). Correlation of electrical resistivity marker bands in the Cenomanian and Turonian Chalk from the London Basin to east Yorkshire. *British Geological Survey Report,* Vol. 17, No. 8. HMSO. London.

Mustchin, C. J. (1974). Brighton's water supply from the Chalk, 1834 to 1956—a history and description of the heading systems. Unpublished report, Brighton Corporation Water Department.

NAM and RGD (1980). Stratigraphic nomenclature of the Netherlands. *Verh. Kon, Ned. Geol. Mijnbouwk. Genootschap*, **32**.

Neal, C. Kinniburgh, D. G., and Whitehead, P. G. (1991). Shallow groundwater systems. In *Applied Groundwater Hydrology* (ed. R. A. Downing and W. B. Wilkinson), pp. 77–95. Clarendon Press, Oxford.

Neilson, J. E. (1990). Experimental investigation of controls on cementation in carbonates. *J. Geol. Soc. London*, **147**, 949–58.

Němejc, F. and Kvaček, Z. (1975). *Senonian plant macrofossils from the region of Zliv and Hluboká (near České Budějovice) in south Bohemia.* Universita Karlova, Prague.

Neretnieks, I. (1981). Age dating of groundwater in fissured rock: Influence of water volume in micropores. *Water Resour. Res.*, **17**, 421–2.

Neugebauer, J. (1974). Some aspects of cementation in chalk. In *Pelagic Sediments: on land and under the sea* (ed. K. J. Hsü and H. C. Jenkyns), pp. 149–76. International Association of Sedimentologists, Special Publications, no. 1.

Neugebauer, J. (1975). Foraminiferen-diagenese in der Schreibkreide. *Neues Jahrbuch für Geologie und Paläontologie, Stuttgart, Abhandlungen,* **150**, 182–206.

Nielsen, L. H. and Japsen, p. (1992). *Deep Wells in Denmark 1935–1990, Lithostratigraphic subdivision.* Ser. A, no. 31. Geol. Surv. Denmark.

Nielsen, O. B., Sorensen, S., Thiede, J., and Skarbo, O. (1986). Cenozoic differential subsidence of the North Sea. *Bull. Amer. Assoc. Petrol. Geol.*, **70**, 276–98.

Nilsson, K. (1966). Geological data from the Kristianstad Plain, southern Sweden. Ser. C Nc 605. Geol. Survey of Sweden.

Nutbrown, D. A., Downing, R. A., and Monkhouse, R. A., (1975). The use of a digital model in the management of the Chalk aquifer of the South Downs, England. *J. Hydrol.*, **27**, 127–42.

Nygaard, E., (ed.), (1991). *Grundvand, Overvågning og Problemer. Ser. D, no. 8*, Appended Data Lists, Engl. Abstr. Geol. Surv. Denmark.

Nygaard, E. and Frykman, P. (1981). Alloktone aflejringer i Maastrichtien kalken på Mors. *Meddelelser fra Dansk Geologisk Forening,* Årsskrift for 1980, pp. 57–60 (with English abstract).

Nygaard, E., Lieberkind, K., and Frykman, P. (1983). Sedimentology and reservoir parameters in the Chalk Group in the Danish Central Graben. In *Petroleum Geology of the Southeastern North Sea and the adjacent onshore areas* (ed. J. P. H. Kaasschieter and T. J. A. Reijers), pp. 177–90. Geologie en Mijnbouw, Special Edition, 62.

Nygaard, E., Andersen, C., Møller, C., Clausen, C. K., and Stouge, S. (1990). Integrated multidisciplinary stratigraphy of the Chalk Group: an example from the Danish Central Trough. In *Chalk*, pp. 195–202. Thomas Telford, London.

Oakes, D. B. (1977). The movement of water and solutes through the unsaturated zone of the Chalk of the United Kingdom. In *Theoretical and applied hydrology*, Proc. 3rd Int. Hydrol. Symp., pp. 447–59. Colorado State University, Fort Collins.

Oakes, D. B. (1982). Nitrate pollution of groundwater resources — mechanisms and modelling. In *Non-point Pollution of Municipal Water Supply Sources: Issues of Application and Control*, Collaboration Proceedings Series CP-82-S4, pp. 207–30. IIASA, Vienna.

Oakes, D. B. (1990). The impact of agricultural practices on groundwater nitrate concentrations. In *World Water* '89, pp. 45–9. Thomas Telford, London.

Oakes, D. B. and Wilkinson, W. B. (1972). *Modelling of groundwater and surface water systems. Theoretical relationships between groundwater abstraction and streamflow.* Water Resources Board, Reading.

Oakes, D. B., Young, C. P,. and Foster, S. S. D., (1981). The effects of farming practices on groundwater quality in the United Kingdom. *The Science of the Total Environment*, **21**, 17–30.

Ødum, H. and Christensen, W. (1936). *Danske grundvandstyper og deres geologiske Øptræden. Rk. III, no. 26,* Geol. Surv. Denmark.

Olsen, J. C. (1983). The structural outline of the Horn Graben. In *Petroleum Geology of the Southeastern North Sea and the adjacent onshore areas* (ed. J. P. H. Kaasschieter and T. J. A. Reijers), pp. 47–50. Geologie en Mijnbouw, Special Edition, 62.

Olsen, J. C. (1987). Tectonic evolution of the North Sea region. In *Petroleum geology of north-west Europe* (ed. J. Brooks and K. W. Glennie), pp 403–17. Graham and Trotman, London.

Owen, M. (1981). The Thames groundwater scheme. In *Case-Studies in Groundwater Resources Evaluation.* (ed. J. W. Lloyd), pp. 186–202. Clarendon Press, Oxford.

Owen, M. and Robinson, V. K. (1978). Characteristics and

yield of fissured chalk. In *Thames Groundwater Scheme*. Institution of Civil Engineers, London.

Owen, M., Headworth, H. G., and Morgan-Jones, M. (1991). Groundwater in basin management. In *Applied Groundwater Hydrology* (ed. R. A. Downing and W. B. Wilkinson), pp. 16–34. Clarendon Press, Oxford.

Pacey, N. R. (1984). Bentonites in the Chalk of central eastern England and their relation to the opening of the northeast Atlantic. *Earth Planet. Sci. Lett.*, **67**, 48–60.

Pacey, N. R. (1989). Organic matter in Cretaceous Chalk from eastern England. *Chem. Geol.*, **75**, 191–208.

Panetier, J. M. (1966). Etude de la surface piézométrique de la nappe de la craie dans le Sénonais et le Gâtinais. CREGR-BRGM Doc. no. DS 66. A 113.

Parke, M. (1961). Some remarks concerning the class Chrysophyceae. *British Phycological Bulletin*, **2**, 47–55.

Parker, J. M. and James, R. (1985). Autochthonous bacteria in the Chalk and their influence on groundwater quality in East Anglia. *J. App. Bact. Symp. Series* **5**, 15–25.

Parker, J. M. Booth, S. K., and Foster, S. S. D. (1987). Penetration of nitrate from agricultural soils into the groundwater of the Norfolk Chalk. *Proc. Inst. Civ. Eng.*, II, **83**, 15–32.

Parker, J. M., Young, C. P., and Chilton, P. J. (1991). Rural and agricultural pollution of groundwater. In *Applied Groundwater Hydrology* (ed. R. A. Downing and W. B. Wilkinson), pp. 149–63. Clarendon Press, Oxford.

Parra, M., Delmont, P., Ferrage, A., Latouche, C., Pons, J. C., and Puechmaille, C. (1985). Origin and evolution of smectites in Recent marine sediments of the NE Atlantic. *Clay Minerals*, **20**, 335–46.

Parsley, A. J. (1990). North Sea hydrocarbon plays. In *Introduction to the petroleum geology of the North Sea* (ed. K. W. Glennie), pp. 362–88. Blackwell, Oxford.

Patsoules, M. G. and Cripps, J. C. (1990). Survey of macro- and micro-fracturing in Yorkshire chalk. In *Chalk*, pp. 87–93. Thomas Telford, London.

Penman, H. L. (1948). Natural evaporation from open water, bare soil and grass. *Proc. R. Soc.*, **193**, 120–46.

Penman, H. L. (1950). The water balance of the Stour catchment. *J. Inst. Wat. Eng.*, **4**, 457–69.

Perch-Nielsen, K., Ulleberg, K., and Evensen, J. A. (1979). Comments on the terminal Cretaceous event, a geologic problem with an oceanic solution. In *Proc. Cretaceous-Tertiary Boundary Symposium*, Vol. 2, pp. 16–32. Univ. Copenhagen.

Person, J. (1978). Protection d'eaux souterraines provenant de nappes de la Craie. In *Hydrogéologie de la Craie du Bassin de Paris* **1**, pp. 477–85. BRGM, Orléans.

Petersen, K. S. (1989). Landet der rejste sig. *Forskning og Samfund*, **15**, no. 4, 24–6. Forskningssekretariatet, Copenhagen.

Petersen, K. S., (1991). *Syndfloden. DGU information*, October, no. 3, p. 5. Geol. Surv. Denmark.

Pitman, J. I. (1978a). Carbonate chemistry of groundwater from chalk, Givendale, East Yorkshire. *Geochim. Cosmochim. Acta*, **42**, 1885–97.

Pitman, J. I. (1978b). Chemistry and mineralogy of some Lower and Middle Chalks from Givendale, East Yorkshire. *Clay Minerals*, **13**, 93–100.

Pozaryski, W. (1962). *Atlas Geologiczny Polski Zagadnienia Stratygraficzno-Facjalne*. Instytut Geologiczny, Warsaw.

Pozaryski, W. and Brochwicz-Lewiński, W. (1978). On the Polish Trough. *Geologie en Mijnbouw*, **57**, 545–58.

Pomerol, B. (1984). *Géochimie des craies du bassin de Paris*. Mémoires des Sciences de la Terre, No. 84–21. Université Pierre et Marie Curie, Paris.

Pomerol, C. (1975). *Stratigraphie et paléogéographie. Ere Mésozoïque*. Doin, Paris.

Pomerol, C. (1989). *The Wines and Winelands of France, Geological Journeys*. Robertson McCarta Ltd, London.

Price, M. (1985). *Introducing groundwater*. George Allen & Unwin, London.

Price, M. (1987). Fluid flow in the Chalk of England. In *Fluid flow in sedimentary basins and aquifers* (ed. J. C. Goff and B. P. J. Williams), pp. 141–56. Spec. Pubn 34, Geol. Soc. London.

Price, M. (1990). Hydrogeology. In *Chalk*, pp. 553–4. Thomas Telford, London.

Price, M., Bird, M. J., and Foster, S. S. D. (1976). Chalk pore-size measurements and their significance. *Water Services*, **80**, 596–600.

Price, M., Robertson, A. S., and Foster, S. S. D. (1977). Chalk permeability—a study of vertical variation using water injection tests and borehole logging. *Water Services*, **81**, 603–10.

Price, M., Morris, B. L., and Robertson, A. S. (1982). A study of intergranular and fissure permeability in Chalk and Permian aquifers using double-packer injection testing. *J. Hydrol.*, **54**, 401–23.

Price, M., Atkinson, T. C., Barker, J. A., Wheeler, D., and Monkhouse, R. A. (1992). A tracer study of the danger posed to a chalk aquifer by contaminated highway run-off. *Proc. Inst. Civ. Eng. Wat. Marit. and Energy*, **96**, 9–18.

Purrer, W., King, J. R. J., Crighton, G., Myers, A., and Wallis, J. (1990). Field Engineering under the sea: excavating the UK crossover chamber. *Tunnels & Tunnelling*, **24**, 15–18.

Quaghebeur, D. and Wulf, E. de (1978). Polynuclear aromatic hydrocarbons in the main Belgian aquifers. *Sci. Total Environ.*, **10**, 231–7.

Quine, M. and Bosence, D. (1992). Stratal geometries, facies and sea-floor erosion in Upper Cretaceous chalk, Normandy, France. *Sedimentology*, **38**, 1113–52.

Rasplus, L. and Alcaydé, G. (1991). Hydrogéologie de la craie en Touraine et ses abords. Unpublished.

Rawson, P. F. (1992). The Cretaceous. In *Geology of England and Wales* (ed. P. M. D. Duff and A. J. Smith), pp. 355–88. Geological Society, London.

Reeve, T., Pearman, H., Dillis, R., Dinn, R., and Jennings, S. (1980). Beachy Head Cave Systems Survey. *Records*, **9**.

Reeves, M. J. (1979). Recharge and pollution of the English Chalk: some possible mechanisms. *Eng. Geol.* **14**, 231–40.

Reid, R. E. H. (1971). The Cretaceous rocks of north-eastern Ireland. *The Irish Naturalists' Journal*, **17**, 105–29.

Reinfelder, J. R. and Fisher, N. S. (1991). The assimilation of elements ingested by marine copepods. *Science*, **251**, 794–6.

Rhoades, R. and Sinacori, M. N. (1941). Pattern of groundwater flow and solution. *J. Geol.*, **49**, 785–94.

Rhys, G. H., Lott, G. K., and Calver, M. A. (1982). *The Winterborne Kingston borehole, Dorset, England.* Institute of Geological Sciences Report 81/3. HMSO, London.

Ricour, J. (1985). La craie, ses caractéristiques physiques, minéralogiques, chimiques, ses usages. *Note interne*, BRGM–SGRNPC, Lezennes.

Ridd, M. F. (1981). Petroleum geology west of the Shetlands. In *Petroleum geology of the continental shelf of north-west Europe* (ed. L. V. Illing and G. D. Hobson), pp. 414–25. Heyden, London.

Ridehalgh, H. (1958). Shoreham Harbour development. *Proc. Inst. Civ. Eng.*, **11**, 285–96.

Robbie, J. A. (1950). The Chalk Rock at Winterborne Abbas, Dorset. *Geological Magazine*, **87**, 209–13.

Roberts, T. O. L. and Preene, M. (1990). Case studies of construction dewatering in chalk. In *Chalk* pp. 571–6., Thomas Telford, London.

Robertson, J. C. (1974). Nitrate pollution in North Lindsey Chalk boreholes. In *Groundwater Pollution in Europe*, pp. 290–1. Water Info Center, Port Washington, NY.

Robinson, N. D. (1986). Fining-upward microrhythms with basal scours in the Chalk of Kent and Surrey, England, and their stratigraphic importance. *Newsletters on Stratigraphy*, **17**, 21–8.

Rodda, J. C., Downing, R. A., and Law, F. M. (1976). *Systematic hydrology*, Newnes-Butterworths, London.

Rodet, J. (1978). Characteristiques du karst crayeux en Haute-Normandie. In *Hydrogeologie de la Craie du Bassin de Paris*, **1**, 513–22. BRGM, Orléans.

Rorive, A. (1987) Le modèle mathématique de l'aquifère des craies du bassin de Mons, In *Les eaux souterraines en Wallonie: Bilans et perspectives*, Colloque ESO87 pp. 132–40.

Roth, P. H. and Bowdler, J. L. (1981). Middle Cretaceous calcareous nannoplankton biogeography and oceanography of the Atlantic Ocean. In *The Deep Sea Drilling Project: A decade of progress* (ed. J. E. Warme, R. G. Douglas and E. L. Winterer), pp. 517–46. Society of Economic Paleontologists and Mineralogists, Special Publications, no. 32.

Roux, J. C. (1963). Contribution à l'étude hydrogéologique du Bassin de la Somme. Thèse 3ème cycle, Univ. Paris, Lab. géo. phys. géol. dyn.

Roux, J. C. (1977). Pollution de la nappe de la Craie sous les zones industrielles de la vallée de la Seine dans l'agglomération rouennaise. In *Proc. Colloque Effet de l'urbanisation et de l'industrialisation sur le régime hydrologique et sur la qualité de l'eau*, pp. 408–20. AIMS Publication no. 123.

Roux, J. C. (1991). Evolution des pollutions d'eau souterraines en France. *J. Inst. Water Environ. Managmnt Symp. Series*, **11**, 1–38.

Roux, J. C. and Tirat, M. (1968). Carte de la surface piézométrique de la nappe de la craie en Picardie (1:200 000). B.R.G.M., Orléans.

Roux, J. C., Caous, J. Y., and Common, D. (1978a). *Atlas hydrogéologique de la Somme. Notice synthèse et index.* Cartes hydrogéologiques du département de la Somme, 1980. BRGM Orléans.

Roux, J. C; Artis, H., and Tremembert, J. (1978b). *Atlas hydrogéologique départemental de la Seine-Maritime. Synthèse générale des données sur les nappes au 1/100000.* BRGM, Orléans.

Rushton, K. R. and Ward, C. J. (1979). The estimation of groundwater recharge. *J. Hydrol.*, **41**, 345–61.

Samson, A., Retkowsky, M., Ardin, M., and Le Reund, D. A. (1986). *L'utilisation des eaux souterraines en Ile-de-France.* Agence Financière de Bassin Seine-Normandie, Direction Régionale de l'Industrie et de la Recherche d'Ile-de-France. Paris.

Schlanger, S. O. and Douglas, R. G. (1974). The pelagic ooze–chalk–limestone transition and its implications for marine stratigraphy. In *Pelagic Sediments: on land and under the sea* (ed. K. J. Hsü and H. C. Jenkyns), pp. 117–48. Special Publications of the International Association of Sedimentologists, no. 1.

Schlanger, S. O., Jenkyns, H. C., and Premoli-Silva, I. (1981). Volcanism and vertical tectonics in the Pacific Basin related to global Cretaceous transgressions. *Earth Planet. Sci. Lett.*, **52**, 435–49.

Schmid, F. (1963). Die Kreide von Lüneburg und die Aufschürfung des Alb-profiles im Kreidebruch am Zeltberg. *Zeitschrift der Deutschen Geologischen Gesellschaft*, **114**, 419–22.

Schmid, F. (1982). Das erweiterte Unter-/Ober-Maastricht-Grenzprofil von Hemmoor, Niederelbe (NW-Deutschland). *Geologisches Jahrbuch, A*, **61**, 7–12.

Scholle, P. A. (1974). Diagenesis of Upper Cretaceous chalks from England, Northern Ireland, and the North Sea. In *Pelagic Sediments: on land and under the sea* (ed. K. J. Hsü and H. C. Jenkyns), pp. 177–210. Special Publications of the International Association of Sedimentologists, no. 1.

Scholle, P. A. (1977). Chalk diagenesis and its relation to petroleum exploration: oil from chalks, a modern miracle? *Bull. Am. Assoc. Petrol. Geol*, **61**, 982–1009.

Scholle, P. A., Arthur, M. A., and Ekdale, A. A. (1983). Pelagic environments. In *Carbonate depositional environments* (ed. P. A. Scholle, D. G. Bedout, and C. H. Moore), pp. 619–91. Memoirs of the American Association of Petroleum Geologists, no. 33.

Scholle, P. A. and Kennedy, W. J. (1974). Isotopic and petrophysical data on hardgrounds from Upper Cretaceous chalks from western Europe. *Abstracts with Programs of the Geological Society of America*, **6**, 943.

Schönfeld, J. and Grube, F. (1990). Chloride distribution pattern and fracturing in the white chalk of Lägerdorf/Holstein (N. W. Germany). In *Chalk*, pp. 591–6. Thomas Telford, London.

Schwille, F., (1988). *Dense chlorinated solvents in porous and fractured media* (trans. J. Pankow). Lewis Publishers, Chelsea, MI.

Seibertz, E. and Vortisch, W. (1979). Zur Stratigraphie, Petrologie und Genese einer Benthonit-lage aus dem oberen Mittel-Turon (Oberkreide) des südöstlichen Münsterlandes. *Geologisiche Rundschau*, **68**, 649–79.

Seki, H. and Taka, N. (1965). Microbiological studies on the decomposition of chitin in the marine environment. 1. Occurrence of chitinoblastic bacteria in the neritic region. *Journal of the Oceanographical Society of Japan*, **19**, 101–8.

Sellwood, B. W. (1979). The Wealden rivers and chalky seas of Cretaceous Britain. In A *dynamic stratigraphy of the British Isles* (ed. R. Anderton, P. H. Bridges, M. R. Leeder, and B. W. Selwood), pp. 227–44. George Allen & Unwin, London.

Shumenko, S. I. (1980). Comparative electronmicroscopic study of the typomorphism of zeolites and accompanying minerals of sedimentary and volcanic-sedimentary deposits. In *Scientific bases and utilization of the typomorphism of minerals*, Proceedings of the 11th general meeting of IMA, Novosibirsk, 4–10 September 1978. Academya Nauk SSSR, Moscow (in Russian).

Simpson, B., Blower, T., Craig, R. N., and Wilkinson, W. B. (1989). *The engineering implications of rising groundwater levels in the deep aquifer beneath London.* Special Publication 69, Construction Industry Research and Information Association, London.

Skilton, H. E., Perkins, M. A., Wheeler, D. J. and Lloyd, B. J. (1985). The migration and survival of viruses through groundwater systems. *J. App. Bact*, **59**, 13.

Skovbro, B. (1983). Depositional conditions during chalk sedimentation in the Ekofisk area, Norwegian North Sea. In *Petroleum Geology of the Southeastern North Sea and the adjacent onshore areas* (ed. J. P. H. Kaasschieter and T. J. A. Reijers), pp. 169–75. Geologie en Mijnbouw, Special Edition, 62.

Smayda, T. J. (1971). Normal and accelerated sinking of phytoplankton in the sea. *Marine Geology*, **11**, 105–22.

Smith, D. B. and Richards, H. J. (1972). Selected environmental studies using radioactive tracers. In *Peaceful Uses of Atomic Energy* 14, 467–80. Int. Atomic Energy Agency, Vienna.

Smith, D. B., Wearn, P. L., Richards, H. J., and Rowe, P. C. (1970). Water movement in the unsaturated zone of high and low permeability strata using natural tritium. In *Isotope Hydrology* pp. 73–87. Int. Atomic Energy Agency, Vienna.

Smith, D. B., Downing, R. A., Monkhouse, R. A., Otlet, R. L., and Pearson, F. J. (1976). The age of groundwater in the Chalk of the London Basin. *Water Resour. Res.*, **12**, 392–404.

Smith, W. E. (1957). The Cenomanian Limestone of the Beer district, south Devon. *Proc. Geol. Assoc.*, **68**, 115–35.

Song Lin-Hua and Atkinson, T. C. (1985). Dissolved iron in Chalk groundwaters from Norfolk, England. *Q. J. Eng. Geol.*, **18**, 261–74.

Sorensen, S., Jones, M., Hardman. R. P. F., Leutz, W. K., and Schwarz, P. H. (1986). Reservoir characteristics of high and low productivity chalks from the central North Sea. In *Habitat of hydrocarbons on the Norwegian continental shelf* (ed. A. M. Spencer, E. Holter, C. J. Campbell, S. H. Hanslien, P. H. H. Nelson, E. Nysaether, and E. G. Ormaasen), pp. 91–110. Graham and Trotman, London.

Sorgenfrei, T. (1945). Træk af Alnarp Dalens geologiske Opbygning. *Medd. Dansk Geol. Foren.*, **10**, 617–30. Copenhagen.

Sorgenfrei, T., and Buch, A. (1964). *Deep Tests in Denmark 1935–1959.*, Ser. III, 36, Geol. Surv. Denmark.

Southern Water, (1989). *North Kent groundwater development scheme.* Mid-Kent Water Co., Chatham.

Southern Water Authority (1979*a*). *South Downs Investigation*, Third Progress Report. Worthing, Sussex.

Southern Water Authority (1979b). The *Candover Pilot Scheme, Final Report.* Worthing.

Southern Water Authority (1985). *Aquifer protection policy.* Worthing.

Spink, A. E. F. and Rushton, K. R. (1979). The use of aquifer models in the assessment of groundwater recharge. *Proc. XVIII Cong.* **5**, 169–76. IAHR, Cagliari.

Springer, N. (1989). Conventional Core Analysis for DOPAS, Vell Stenlille-5, Core no. 1., Confidential report no. 5. Geol. Surv. Denmark.

Stenestad, E. (1972). Troek af det danske bassins udvikling i Øvre Kridt. *Årsskrift Danske Geologiske Forening*, 1971, 63–9.

Stow, A. H. and Renner, L. (1965). Acidizing boreholes. *J. Inst. Water. Eng.*, **19**, 557–72.

Sulak, R. M. (1991). Ekofisk Field: the first 20 years. *J. Petrol. Technology*, **43**, 1265–71.

Sulak, R. M., Nossa, G. R., and Thompson, D. A. (1990). Ekofisk Field enhanced recovery. In *North Sea oil and gas reservoirs*, II, 281–95. Graham and Trotman, London.

Sulak, R. M., Thomas, L. K., and Boade, R. R. (1991). 3-D reservoir simulation of Ekofisk compaction drive. *J. Petrol. Technology*, **43**, 1272–8.

Surlyk, F. (1971). Skrivekridtklinterne på Møn. *Geologi på Øerne*, Vary Ekskursionsfører No. 2, 4–24. Varv, Copenhagen.

Surlyk, F. (1972). Morphological adaptations and population structures of the Danish Chalk brachiopods (Maastrichtian, Upper Cretaceous). *Kongelige Danske Vedenskabernes Selskab Biologiske Skrifter*, **19**, 1–57.

Surlyk, F. and Christensen, W. K. (1974). Epifaunal zonation on an Upper Cretaceous rocky coast. *Geology*, **2**, 529–34.

Takahashi, T. (1975). Carbonate chemistry of sea water and the calcite compensation depth in the oceans. In *Dissolution of deep-sea carbonates* (ed. W. V. Sliter, A. W. H. Bé, and W. H. Berger), pp. 11–26. Special Publication, no. 13. Cushman Foundation for Foraminiferal Research, Washington, DC.

Tate, T. K., Robertson, A. S., and Gray, D. A. (1970). The

hydrogeological investigation of fissure-flow by borehole logging techniques. *Q. J. Eng. Geol.*, **2**, 195–215.

Taylor, S. R. and Lapré, J. F. (1987). North Sea chalk diagenesis: its effect on reservoir location and properties. In *Petroleum Geology of North West Europe* (ed. J. Brooks and K. W. Glennie), pp. 483–95. Graham and Trotman, London.

Ter-Borch, N. (1991). *Geological map of Denmark, 1:500,000, Structural map of the Top Chalk Group.* Map Ser. No. 7. Geol. Surv. Denmark.

Thames Conservancy (1971). *Report on the Lambourne Valley pilot scheme.* 1967–1969. Thames Conservancy, Reading.

Thierstein, H. R. and Okada, H. (1979). The Cretaceous/Tertiary boundary event in the North Atlantic. *Initial Reports of the Deep Sea Drilling Project*, **43**, 601–16.

Thomsen, R. (1987). Vandressourcerne og klimasvingninger, Foreløbig analyse af nedbøren 1860–1985. Miljøprojekt nr. 89, *Miljøstyrelsen,* Copenhagen.

Thomson, D. H. (1938). A 100 years record of rainfall and water levels in the Chalk at Chilgrove, West Sussex. *Trans. Instit. Wat. Engrs*, **43**, 154–81.

Thorling, L., Hundahl, M., Boutrup, S., and Søndergaard, V. (1990). Statusrapport 1990, Vandmiljø—Overvågning af Grundvandet. *Teknisk Rapport*, Miljøkontoret, Århus Amtskommune.

Tinkler, R. (1976). Discussion on problems of tunnelling in Chalk by Haswell, 1975. *Tunnels & Tunnelling*, **8**, 61–5.

Towler, P. (1982). The geochemistry of nitrogen species in groundwaters. Ph.D. thesis, University of Bath.

Towler, P. A., Blakey, N. C., Irving, T. E., Clark, L., Maris, P. J., Baxter, K. M. and Macdonald, R. M. (1985). A study of the bacteria of the Chalk aquifer and the effect of landfill contamination at a site in eastern England. In *Memoirs 18th IAH Congress* III, pp. 84–97.

Tröger, K.-A. (1975). Fazies—Biofazies am Beispiel der sächsischen und subherzynen Kreide. *Zeitschrift für Geologisiche Wissenschaften, Berlin*, **3**, 1265–77.

Tröger, K.-A. and Christensen, W. K. (1991). *Upper Cretaceous (Cenomanian–Santonian) Inoceramid bivalve faunas from the island of Bornholm, Denmark.* Ser. A, no. 28, Geol. Surv. Denmark.

Trotter, J. G., Thompson, D. M., and Paterson, T. J. M. (1985). First mined hydrocarbon storage in Great Britain. *Tunnelling '85*, Paper 17, Institution Mining and Metallurgy, Brighton.

Tsang, Y. W. and Tsang, C. F., (1987). Channel model of flow through fractured media. *Water Resour. Res.*, **23**, 467–79.

Tucker, M. E. (1974). Sedimentology of Palaeozoic pelagic limestones: the Devonian Griotte (southern France) and Cephalopodenkalk (Germany). In *Pelagic Sediments: on land and under the sea* (ed. K. J. Hsü and H. C. Jenkyns), pp. 71–92. International Association of Sedimentologists, Special Publications, no. 1.

University of Birmingham. (1978). *South Humberbank salinity research project.* Final Report to Anglian Water Authority Dept. Geol. Sci. and Civ. Eng.

University of Birmingham. (1985). *Yorkshire Chalk groundwater model study.* Dept. of Geol. Sci., Birmingham.

Vachier, P., Devere, L., and Fontes, J. C. (1987). Mouvement de l'eau dans les zones non saturées et alimentation de la nappe de la craie de Champagne (France): approches isotopique et chimique. In *International symposium on the uses of isotope techniques in water resources development.* IAEA, Vienna.

Valeton, I. (1960). Vulkanische Tuffiteinlagerung in der nordwestdeutschen Oberkreide. *Mitteilungen aus dem Geologischen Staatsinstitut in Hamburg*, **29**, 26–41.

van den Bosch, W. J. (1983). The Harlingen Field, the only gas field in the Upper Cretaceous chalk of the Netherlands. *Geologie en Mijnbouw*, **62**, 145–56.

Varley, P. M. (1990). Machine excavation of chalk rock at the first South Killingholme gas cavern, South Humberside. In *Chalk*, pp. 485–92. Thomas Telford, London.

Villumsen, A. and Binzer, K. (1979). Grundvandskemisk basisdatakort. In *Hydrogeologisk Kortlægning af Vestsjællands Amtskommune*, Fase 3 (ed. K. Binzer, A. Bækgaard, J. Knudsen, H. Kristiansen, and A. Villumsen). Geol. Surv. Denmark.

Voigt, E. (1959). Die ökologische Bedeutung der Hartgründe ('Hardgrounds') in der oberen Kreide. *Paläontologische Zeitschrift*, **33**, 129–47.

Voigt, E. (1962). Frühdiagenetische Deformation der Turonen Plänerkalke bei Halle/Westf. *Mitteilungen aus dem Geologisichen Staatsinstitut in Hamburg*, **31**, 146–275.

Voigt, E. (1963). Über Randtröge vor Schollenrändern und ihre Bedeutung im Gebiet der mitteleuropäischen Senke und angrenzender Gebiete. *Zeitschrift der Deutschen Geologischen Gesellschaft*, **114**, 378–418.

Voigt, E. and Häntzschel, W. (1964). Gradierte Schichtung in der Oberkreide Westfalens. *Fortschritte in der Geologie, von Rheinland und Westfalen*, **7**, 495–548.

Wakeling, T. R. M. (1970). A comparison of the results of standard site investigation methods against the results of a detailed geotechnical investigation in Middle Chalk at Mundford, Norfolk. In *In situ investigations in soils and rocks*, pp. 17–22. British Geotechnical Society, London.

Wanless, H. R. (1979). Limestone response to stress: pressure solution and dolomitization. *J. Sed. Petrol.*, **49**, 437–62.

Ward, W. H., Burland, J. B., and Gallois, R. W. (1968). Geotechnical assessment of a site at Mundford, Norfolk, for a large proton accelerator. *Geotechnique*, **18**, 399–431.

Warren, J. E. and Root, P. J. (1963). Behaviour of naturally fractured reservoirs. *J. Soc. Petrol. Eng.*, **228**, 245–55.

Water Resources Board. (1972). *The hydrogeology of the London Basin.* Reading.

Water Resources Board (1974). *Artificial recharge of the London Basin. Economic and engineering desk studies.* Reading.

Watts, N. L. (1983a). Microfractures in chalks of Albuskjell

Field, Norwegian sector, North Sea: possible origin and distribution. *Am. Assoc. Petrol. Geol. Bull.*, **67**, 201–34.

Watts, N. L. (1983b). Fractures in North Sea Chalks: Geological modelling and a case example from the Albuskjell Field. *JAPEC Chalk Seminar 15th December*. Geol. Soc. London.

Watts, N. L., Lapré, J. F., van Schindel-Goester, F. S., and Ford, A. (1980). Upper Cretaceous and Lower Tertiary chalks of the Albuskjell area, North Sea: deposition in a slope and base-of-slope environment. *Geology*, **8**, 217–21.

Waverijn, C. G. (1990). *De geohydrologische evaluatie van het watervoerende kalksteenpakket in de ENCI groeve*. Memoirs of the Centre for Engineering Geology in the Netherlands, 82.

Weir, A. H. and Catt, J. A. (1965). The mineralogy of some Upper Chalk samples from the Arundel area, Sussex. *Clay Minerals*, **6**, 97–110.

Wellings, S. R. (1984). Recharge of the Upper Chalk aquifer at a site in Hampshire, England, 1. Water balance and unsaturated flow. *J. Hydrol.*, **69**, 259–73.

Wellings, S. R. and Bell, J. P. (1982). Physical controls of water movement in the unsaturated zone. *Q. J. Eng. Geol.*, **15**, 235–41.

Wellings, S. R. and Cooper, J. D. (1983). The variability of recharge of the English Chalk aquifer. *Agric. Water Mangmn*t, **6**, 243–53.

Westbroek, P., de Jong, E. W., van der Wal, P., Borman, A. H., de Vrind, J. P. M., Kok, D., de Bruijn, W. E., and Parker, S. B. (1984). Mechanism of calcification in the marine alga *Emiliana huxleyi. Philosophical Transactions of the Royal Society, London, B*, **304**, 435–44.

Whitaker, W. and Thresh, J. C. (1916). *The water supply of Essex from underground sources*. Mem. Geol. Surv. G. B.

Whitehurst, J. (1786). *An inquiry into the original state and formation of the Earth: deduced from facts and the laws of Nature* (2nd edn). London.

Whitelaw, K. and Edwards, R. A. (1980). Carbohydrates in the unsaturated zone of the Chalk of England. *Chem. Geol.*, **29**, 281–91.

Whitelaw, K. and Rees, J. K. (1980). Nitrate-reducing and ammonium oxidising bacteria in the vadose zone of the Chalk aquifer of England. *Geomicrobiol. J.*, **2**, 179–87.

Whittaker, A. (ed.) (1985). *Atlas of onshore sedimentary basins in England and Wales*. Blackie, London.

Wikman, H. and Bergström, J. (1987). Description to the map of solid rocks Halmstad SV (in Swedish with English summary). Ser. Af No 133. Geol. Survey of Sweden.

Williams, G. M., Young, C. P., and Robinson, H. D. (1991). Landfill disposal of wastes. *In Applied Groundwater Hydrology* (ed. R. A. Downing and W. B. Wilkinson), pp. 114–33. Clarendon Press, Oxford.

Wilson, L. (1973). Variations in mean annual sediment yield as a function of mean annual precipitation. *Am. J. Sc.*, **273**, 335–49.

Winterer, E. L., Ewing J. I. *et al.* (1973). Initial Reports of the Deep Sea Drilling Project, **17**, 5–45. US Govt. Printing Office. Washington DC.

Wood, C. J. and Smith, E. G. (1978). Lithostratigraphical classification of the Chalk in North Yorkshire, Humberside and Lincolnshire. *Proc. Yorks. Geol. Soc.*, **42**, 263–87.

Woodland, A. W. (1946). Water supply from underground sources of Cambridge-Ipswich District. Part X, General discussion. *Wartime Pamphlet No 20*. Geol. Surv. Gt. Britain.

Wooldridge, S. W. and Kirkaldy, J. F. (1937). The geology of the Mimms Valley. *Proc. Geol. Assoc.*, **48**, 307–15.

Wooldridge, S. W. and Linton, D. L. (1955). *Structure, Surface and Drainage in south-east England*. George Phillip & Son, London.

Young, B. R. (1965). X-ray examination of insoluble residues from the Chalk. *Bull. Geol. Surv. Great Britain*, no. 23, 110–14.

Young, C. P., Hall, E. S., and Oakes, D. B. (1976a). *Nitrate in groundwater: studies on the Chalk near Winchester, Hampshire*. Tech. Rept no. TR 31. Water Research Centre, Medmenham.

Young, C. P., Oakes, D. B., and Wilkinson, W. B. (1976b). Prediction of future nitrate concentrations in groundwater. *Ground Water*, **14**, 426–38.

Younger, P. L. (1989). Devensian periglacial influences on the development of spatially variable permeability in the Chalk of southeast England. *Q. J. Eng. Geol.*, **22**, 343–54.

Ziegler, P. A. (1981). Evolution of sedimentary basins in North-west Europe. In *Petroleum geology of the continental shelf of North-west Europe* (ed. L. V. Illing and G. D. Hobson) pp. 3–39. Institute of Petroleum, London.

Ziegler, P. A. (1990). *Geological atlas of western and central Europe*. Shell Internationale Petroleum Maatschappij BV.

Index